Stochastic
H₂/H∞ Control

A Nash Game Approach

Stochastic H_2/H_∞ Control

A Nash Game Approach

Weihai Zhang
Lihua Xie
Bor-Sen Chen

CRC Press
Taylor & Francis Group
Boca Raton London New York

CRC Press is an imprint of the
Taylor & Francis Group, an **informa** business

CRC Press
Taylor & Francis Group
6000 Broken Sound Parkway NW, Suite 300
Boca Raton, FL 33487-2742

First issued in paperback 2020

ISBN 13: 978-0-367-57330-0 (pbk)
ISBN 13: 978-1-4665-7364-2 (hbk)

Version Date: 20170630

Visit the Taylor & Francis Web site at
http://www.taylorandfrancis.com

and the CRC Press Web site at
http://www.crcpress.com

Contents

Preface

H_∞ control is one of the most important robust control approaches, aiming to attenuate exogenous disturbances by a prescribed level. It is well known that the H_∞ control is a worst-case design and is more suitable for dealing with system robustness against disturbances with no prior knowledge other than being energy bounded, parametric uncertainties and unmodeled system dynamics. Generally speaking, for a given level of disturbance attenuation, H_∞ controller is not unique. That is to say, there exists more than one controller that can provide the required level of H_∞ performance. On the other hand, H_2 control has been popular since the 1960s and has been well practiced in engineering to optimize system performance, but it could be sensitive to uncertainties. Mixed H_2/H_∞ control is thus a natural yet desirable control design method that not only restrains the effect of the external disturbance, but also minimizes an H_2 cost functional under the disturbance input. A powerful technique to deal with the mixed H_2/H_∞ control is the so-called Nash game approach initiated in [113]. Compared with other robust H_2/H_∞ control methods, the Nash game-based method can lead to a necessary and sufficient condition for the existence of H_2/H_∞ control. Different from many other methods where the H_2 optimal control is designed assuming zero external disturbance, the H_2 optimization discussed in this book is carried out in the presence of the worst-case disturbance, which could make the Nash game-based H_2/H_∞ control more appealing for real applications.

Before 1998, most researches on the H_∞ control and mixed H_2/H_∞ control were focused on deterministic systems. In 1998, the references [84] and [169] independently developed the H_∞ control for linear Itô-type stochastic differential systems. Soon after [84], the H_∞ control for linear discrete-time stochastic systems with multiplicative noise was discussed in [65]. In order to generalize the classical H_2/H_∞ control to stochastic Itô systems, there was a need to develop a stochastic bounded real lemma (SBRL), which, however, was challenging. In 2004, we succeeded in obtaining an SBRL and solving the stochastic counterpart of [113] for linear Itô systems with state-dependent noise. Since then, a series of related works have appeared for linear discrete-time stochastic systems [207, 208], Markov jump systems [88, 89, 90, 139, 143, 165], and nonlinear H_∞ and mixed H_2/H_∞ control for Itô systems [213, 214, 209, 219].

This book presents our latest results in stochastic H_2/H_∞ control and filtering based on the Nash game approach. To make the book self-contained, Chapter 1 provides some basics of stochastic differential equations, stochastic stability, stochastic observability and detectability. Chapter 2 solves the H_2/H_∞ control for linear Itô systems and establishes a relationship between the H_2/H_∞ control and two-person non-zero sum Nash game. It lays a foundation for the rest of the book. Chapter 3 is on the H_2/H_∞ control for linear discrete time-invariant systems with multiplicative noise, which reveals some essential differences from Chapter 2. Chapter 4 continues the development of Chapter 3 by focusing on linear discrete time-varying systems. Chapter 5 is more general than the previous chapters, and discusses the mixed H_2/H_∞ control for linear Markov jump systems with multiplicative noise. Both continuous-time and discrete-time systems are investigated in this chapter. In order to develop nonlinear stochastic H_∞ control of Itô systems, the equivalence between stochastic dissipativity and the solvability of nonlinear Lure equations is established in Chapter 6. The completion of squares technique is used to present a new type of second-order nonlinear partial differential equation associated with nonlinear stochastic H_∞ control. Chapter 6, to some extent, can be viewed as an extension of deterministic nonlinear H_∞ control. Chapter 7 is concerned with nonlinear H_∞ and H_2/H_∞ filters of Itô systems, which have potential applications in communication and signal processing. In the last chapter, we present some further research topics in stochastic H_2/H_∞ control including the H_2/H_∞ control of the following systems: (i) stochastic Itô systems with random coefficients; (ii) nonlinear discrete-time stochastic systems with multiplicative noise; (iii) singular stochastic Itô systems and singular discrete-time systems with multiplicative noise; and (iv) mean-field stochastic systems. At the end of each chapter, a brief review of related background knowledge and further research topics are presented for the benefit of reader.

The book summarizes a number of research outcomes arising from collaborations with our co-authors including Gang Feng from City University of Hong Kong, Wei Xing Zheng from Western Sydney University, and Huanshui Zhang from Shandong University. In particular, Weihai Zhang sincerely appreciates the collaborations and contributions of his collaborators and students.

Last, but not least, we would like to thank Ruijun He, an editor of CRC Press, who has given us a lot of help during the course of this project. The authors are also grateful to their family members for their long-term understanding and support. Specifically, Weihai Zhang wants to thank his wife Ms. Guizhen Tian for her continuous encouragement in writing this book, otherwise, the book might have been delayed for a long time. Lihua Xie would like to thank his wife Meiyun for her unwavering support and encouragement. The first author's work has been supported by the National Natural Science Foundation of China (No. 61573227), the Research Fund for the Taishan Scholar Project of Shandong Province of China, and the Research Fund of Shandong University of Science and Technology (No.2015TDJH105).

Weihai Zhang,
College of Electrical Engineering and Automation,
Shandong University of Science and Technology,
Qingdao 266590, Shandong Province, China

Lihua Xie,
School of Electrical and Electrical Engineering,
Nanyang Technological University, Singapore

Bor-Sen Chen,
Department of Electrical Engineering,
National Tsing Hua University, Hsinchu 30013, Taiwan

List of Tables

List of Figures

Symbols and Acronyms

$\mathcal{R}^n$:	n-dimensional real Euclidean space.
$\sigma(L)$:	spectral set of an operator or a matrix L.
$\mathcal{R}^{n \times m}$:	set of $n \times m$ real matrices.
$\mathcal{S}_n$:	set of $n \times n$ symmetric matrices.
I_n:	$n \times n$ identity matrix.
$\mathcal{C}^-$:	the open left-hand side of the complex plane.
$\mathcal{C}^{-,0}$:	the closed left-hand side of the complex plane.
$\mathcal{E}$:	mathematical expectation.
$\mathcal{L}^2_{\mathcal{F}_t}(\Omega, X)$:	the family of X-valued $\mathcal{F}_t$-measurable random variables with bounded variances, i.e., for any ξ from the family, $\mathcal{E}\|\xi\|^2 < \infty$.
$\mathcal{L}^\infty_{\mathcal{F}}([0,T], X)$:	space of all X-valued $\{\mathcal{F}_t\}_{t \geq 0}$-adapted uniformly bounded processes.
$\mathcal{L}^2_{\mathcal{F}}([0,T], \mathcal{R}^{n_v})$:	space of nonanticipative stochastic processes $v(t) \in \mathcal{R}^{n_v}$ with respect to an increasing σ-algebra $\{\mathcal{F}_t\}_{t \geq 0}$ satisfying $\mathcal{E} \int_0^T \|v(t)\|^2 \, dt < \infty$.
$\mathcal{L}^2_{\mathcal{F}}(\mathcal{R}^+, \mathcal{R}^{n_v})$:	space of nonanticipative stochastic processes $v(t) \in \mathcal{R}^{n_v}$ with respect to an increasing σ-algebra $\{\mathcal{F}_t\}_{t \geq 0}$ satisfying $\mathcal{E} \int_0^\infty \|v(t)\|^2 \, dt < \infty$.
$\mathcal{C}^{2,1}(U \times [0,T]; X)$:	class of X-valued functions $V(x,t)$ which are twice continuously differentiable with respect to $x \in U$, except possibly at the point $x = 0$, and once continuously differential with respect to $t \in [0,T]$.

$\mathcal{C}^{1,2}([0,T] \times U; X)$:	class of X-valued functions $V(t,x)$ which are once continuously differentiable with respect to $t \in [0,T]$, and twice continuously differential with respect to $x \in U$, except possibly at the point $x = 0$.
$\mathcal{C}^2(U; X)$:	class of X-valued functions $V(x)$ which are twice continuously differentiable with respect to $x \in U$, except possibly at the point $x = 0$.
$P > 0$:	symmetric positive definite matrix $P \in \mathcal{R}^{n \times n}$.
$P \geq 0$:	symmetric positive semidefinite matrix $P \in \mathcal{R}^{n \times n}$.
$P \geq Q$:	$P - Q \geq 0$ for symmetric $P, Q \in \mathcal{R}^{n \times n}$.
$P > Q$:	$P - Q > 0$ for symmetric $P, Q \in \mathcal{R}^{n \times n}$.
Trace(X):	trace of X.
$\|\cdot\|$:	Euclidean vector norm.
$Re(\lambda)$:	real part of a complex number λ.
$Im(\lambda)$:	imaginary part of a complex number λ.
SDE :	stochastic differential equation.
ARE :	algebraic Riccati equation.
$GARE$:	generalized algebraic Riccati equation.
$BSRE$:	backward stochastic Riccati equation.
GLI:	generalized Lyapunov inequality.
GLO:	generalized Lyapunov operator.
$GDRE$:	generalized differential or difference Riccati equation.
LMI:	linear matrix inequality.
LQR:	linear quadratic regulator.
$BSDE$:	backward stochastic differential equation.
$\mathcal{R}^N_{n \times m}$:	set of all N sequences $V = (V_1, \cdots, V_N)$ with $V_i \in \mathcal{R}^{n \times m}$.
$\mathcal{S}^N_n$:	set of all N sequences $V = (V_1, \cdots, V_N)$ with $V_i \in \mathcal{S}_n$.
$\mathcal{S}^{N+}_n$:	set of all N sequences $V = (V_1, \cdots, V_N)$ with $V_i \geq 0$.

$\mathcal{N}_{k_0}^{\infty}$:	$= \{k_0, k_0 + 1, k_0 + 2, \cdots, \}$.
$\overline{N}$:	$= \{1, 2, \cdots, N\}$.
$\mathbb{Z}_{1+}$:	$= \{1, 2, \cdots, \}$.
$\mathcal{N}$:	$= \{0, 1, 2, \cdots, \}$.
N_T:	$= \{0, 1, 2, \cdots, T\}$.
$\text{diag}\{A_1, A_2, \cdots, A_n\}$:	block diagonal matrix with A_j (not necessarily square) on the diagonal.
$Col(x_1, x_2, \cdots, x_n)$:	Stack vectors $x_1, x_2, \cdots, x_n$ to form a high dimensional vector.
X^T:	transpose of matrix X.
X^*:	complex conjugate transpose of matrix X.

1

Mathematical Preliminaries

In this chapter, we present some mathematical preliminaries in stochastic differential equations (SDEs) and introduce basic concepts such as stochastic observability, detectability and stabilization. In Section 1.1, we introduce stochastic Itô-type differential equations, the Itô integral and stochastic stability. All these materials can be found in the recent monographs [130, 145, 196] and classical book [87]. In Section 1.2, we define a generalized Lyapunov operator (GLO) $\mathcal{L}_{F,G}$ that was first introduced in [205] and further developed in [206, 211] and [212]. Based on the spectra of $\mathcal{L}_{F,G}$, a necessary and sufficient condition for the mean square stability of linear stochastic time-invariant systems is presented. The spectrum analysis approach is a powerful technique in the analysis and synthesis of stochastic systems, and has led to many interesting research topics including stochastic spectral assignment, which is parallel to the pole placement in linear system theory. Section 1.3 introduces exact observability, exact detectability and mean square stabilization for linear stochastic systems. Mean square stabilization is not new and can be found in the early reference [177]. Exact observability and detectability are new concepts that were recently reported and studied in [57, 117, 118, 120, 205, 206, 212]. All these concepts can be viewed as extensions of the notions of complete observability, complete detectability and stabilization in linear system theory, and are very useful in studying infinite-time linear quadratic (LQ) optimal stochastic control, stochastic H_2/H_∞ control and Kalman filtering, etc.

1.1 Stochastic Differential Equations

To make this book self-contained, we introduce some basic knowledge about stochastic differential equations which will be used in the subsequent chapters.

1.1.1 Existence and uniqueness of solutions

Let Ω be a nonempty set, $\mathcal{F}$ a σ-field consisting of subsets of Ω, and $\mathcal{P}$ a probability measure, that is, $\mathcal{P}$ is a map from $\mathcal{F}$ to $[0, 1]$. We call the triple $(\Omega, \mathcal{F}, \mathcal{P})$ a probability space. In particular, if for any $\mathcal{P}$-null set $A \in \mathcal{F}$ (i.e., $\mathcal{P}(A) = 0$), all subsets of A belong to $\mathcal{F}$, then $(\Omega, \mathcal{F}, \mathcal{P})$ is said to be a complete probability space. For any $(\Omega, \mathcal{F}, \mathcal{P})$, there is a completion $(\Omega, \bar{\mathcal{F}}, \mathcal{P})$ of $(\Omega, \mathcal{F}, \mathcal{P})$, which is a complete

1

probability space [130]. $\{\mathcal{F}_t\}_{t\geq 0}$ is a family of monotone increasing sub-σ-fields of $\mathcal{F}$, which is called a filtration. The filtration $\{\mathcal{F}_t\}_{t\geq 0}$ is said to be right continuous if $\mathcal{F}_t = \mathcal{F}_{t+} = \cap_{s>t}\mathcal{F}_s$. For a complete probability space, if $\{\mathcal{F}_t\}_{t\geq 0}$ is right continuous and $\mathcal{F}_0$ contains all $\mathcal{P}$-null sets, then we say that such a filtration satisfies the usual condition. The quadruple $(\Omega, \mathcal{F}, \{\mathcal{F}_t\}_{t\geq 0}, \mathcal{P})$ is called a filtered probability space. As in [130, 196], for the purpose of simplicity and without loss of generality, in this book, we always assume that $(\Omega, \mathcal{F}, \mathcal{P})$ is a complete probability space, and $\{\mathcal{F}_t\}_{t\geq 0}$ satisfies usual conditions. Now, we give the following definitions:

DEFINITION 1.1   *For a given filtered probability space $(\Omega, \mathcal{F}, \{\mathcal{F}_t\}_{t\geq 0}, \mathcal{P})$, a d-dimensional Brownian motion is an $\mathcal{R}^d$-valued, t-continuous, and $\mathcal{F}_t$-adapted stochastic process $B(t)$ satisfying the following properties:*

(i) $B(t) - B(s)$ obeys normal distribution with mean zero and covariance $(t-s)I$, i.e., $\mathcal{E}[B(t) - B(s)] = 0$ and $\mathcal{E}[(B(t) - B(s))(B(t) - B(s))^T] = (t-s)I, \forall 0 \leq s \leq t < \infty$.

(ii) $B(\cdot)$ has independent increments, i.e., $B(t_{i+1}) - B(t_i)$ and $B(t_{j+1}) - B(t_j)$ are independent for any $0 \leq t_i < t_{i+1} \leq t_j < t_{j+1} < \infty$.

If, in addition, $\mathcal{P}(B(0) = 0) = 1$, then $B(t)$ is said to be a standard Brownian motion. If $\mathcal{F}_t = \sigma(B(s) : 0 \leq s \leq t)$, $\{\mathcal{F}_t\}_{t\geq 0}$ is the so-called natural filtration.

DEFINITION 1.2   *A stochastic process $X(t)(t \geq 0)$ with $\mathcal{E}|X(t)| < \infty$ is called a martingale with respect to $\{\mathcal{F}_t\}_{t\geq 0}$ if $X(t)$ is $\mathcal{F}_t$-adapted and*

$$\mathcal{E}(X(t)|\mathcal{F}_s) = X(s), \forall 0 \leq s \leq t. \qquad (1.1)$$

In (1.1), if $\mathcal{E}(X(t)|\mathcal{F}_s) \geq X(s)$ (resp. $\leq$), $X(t)$ is called submartingale (resp. supermartingale).

The Brownian motion $B(t)$ and the martingale process $X(t)$ have many nice properties. We refer the reader to the recent books [130, 145, 196]. Below, we introduce a stochastic differential equation (SDE) of Itô type which is expressed as

$$\begin{cases} dx(t) = f(x(t), t)\, dt + g(x(t), t)dB(t), \\ x(0) = x_0 \in \mathcal{L}^2_{\mathcal{F}_0}(\Omega, \mathcal{R}^n), \ t \in [0, T] \end{cases} \qquad (1.2)$$

or equivalently, in an integral form as

$$x(t) = x_0 + \int_0^t f(x(s), s)\, ds + \int_0^t g(x(s), s)\, dB(s) \qquad (1.3)$$

where $f(\cdot, \cdot) : \mathcal{R}^n \times [0, T] \mapsto \mathcal{R}^n$ and $g(\cdot, \cdot) : \mathcal{R}^n \times [0, T] \mapsto \mathcal{R}^{n\times d}$ are measurable functions, which are respectively called a drift term and a diffusion term. The second integral of the right-hand side of (1.3) is the so-called Itô integral whose definition can be found in any standard textbook on SDEs. The first question on (1.2) or (1.3) arises: What is a solution to (1.2)? Further, in what sense is the solution unique?

DEFINITION 1.3   *For a given filtered probability space $(\Omega, \mathcal{F}, \{\mathcal{F}_t\}_{t \geq 0}, \mathcal{P})$, $B(t)$ is a $\mathcal{F}_t$-adapted d-dimensional Brownian motion. An $\mathcal{R}^n$-valued $\mathcal{F}_t$-adapted continuous stochastic process $x(t)(t \in [0,T])$ is called a solution to (1.2) if it satisfies the following conditions:*

(i) $x(0) = x_0 \in \mathcal{L}^2_{\mathcal{F}_0}(\Omega, \mathcal{R}^n)$, a.s..

(ii) (1.3) holds almost surely for every $t \in [0,T]$.

(iii) $\int_0^t \|f(x(s), s)\| \, ds + \int_0^t \|g(x(s), s)\|^2 \, ds < \infty$, a.s., $\forall t \in [0,T]$.

SDE (1.2) is said to have a unique solution if any other solution $\tilde{x}(t)$ is indistinguishable from $x(t)$, i.e.,

$$\mathcal{P}(x(t) = \tilde{x}(t), t \in [0,T]) = 1.$$

REMARK 1.1   A solution $x(t)$ defined in Definition 1.3 is called a strong solution to (1.2) due to that $(\Omega, \mathcal{F}, \{\mathcal{F}_t\}_{t \geq 0}, \mathcal{P}, B(\cdot))$ is given *a priori*. The definition for a weak solution to SDE (1.2) can be found in [145, 196]. A strong solution must be a weak one, but the converse is not true. In this book, we are only concerned with the strong solutions of SDEs. The solution process $x(t)$ of (1.2) is also called an Itô process.   ∎

The SDE (1.2) with a given initial condition $x(0)$ is normally called a forward SDE (FSDE). Otherwise, if the terminal state $x(T)$ is given *a priori*, such an SDE is called a backward SDE (BSDE). There is no difference between forward and backward ordinary differential equations (ODEs) via the time-reversing transformation $t \mapsto T - t$. The requirement for $x(t)$ to be $\mathcal{F}_t$-adapted is an important restriction which makes FSDEs and BSDEs essentially different; see the following example:

Example 1.1
Consider the one-dimensional BSDE

$$\begin{cases} dx(t) = dB(t), \\ x(T) = \xi \in \mathcal{L}^2_{\mathcal{F}_T}(\Omega, \mathcal{R}), \end{cases} \tag{1.4}$$

where $B(t)$ is a standard one-dimensional Brownian motion, $\mathcal{F}_t := \sigma(B(s) : 0 \leq s \leq t)$. BSDE (1.4) does not have an $\mathcal{F}_t$-adapted solution, because $x(0) = \xi - B(T)$ is not $\mathcal{F}_0$-measurable. However, for any $x(0) \in \mathcal{L}^2_{\mathcal{F}_0}(\Omega, \mathcal{R})$, the FSDE

$$\begin{cases} dx(t) = dB(t), \\ x(0) = \xi \in \mathcal{L}^2_{\mathcal{F}_0}(\Omega, \mathcal{R}) \end{cases} \tag{1.5}$$

always admits a unique $\mathcal{F}_t$-adapted solution $x(t) = \xi + B(t)$.   ▯

Parallel to BSDEs, we now consider an example of backward stochastic difference equations corresponding to Example 1.1.

Example 1.2
Consider the one-dimensional backward stochastic difference equation

$$\begin{cases} x_{k+1} = x_k + w_k, \\ x_T = \xi \in \mathcal{L}^2_{\mathcal{F}_T}(\Omega, \mathcal{R}), \ k \in N_T = \{0, 1, \cdots, T\}, \end{cases} \quad (1.6)$$

where $\{w_k\}_{k \in N_T}$ are independent white noises, $\mathcal{E}w_k = 0$, $\mathcal{E}w_k^2 = 1$, $\mathcal{F}_k = \sigma\{w_0, w_1, \cdots, w_k\}$, $\mathcal{F}_{-1} = \{\Omega, \phi\}$. In the theory of stochastic difference equations, we always require that x_k be $\mathcal{F}_{k-1}$-adapted. Obviously, (1.6) does not have any $\mathcal{F}_{k-1}$-adapted solution x_k. However, the following forward stochastic difference equation with the deterministic initial state x_0

$$\begin{cases} x_{k+1} = x_k + w_k, \\ x_0 \in \mathcal{R}, \ k \in N_T = \{0, 1, \cdots, T\} \end{cases} \quad (1.7)$$

always has an $\mathcal{F}_{k-1}$-adapted solution x_k.   ▯

The existence and uniqueness of $\mathcal{F}_k$-adapted solutions of linear and nonlinear BSDEs were first addressed in [22] and [147], respectively. We shall not pay much attention to BSDEs but refer the reader to [196, 140] for the general theory of BSDEs. It is worth noting that the study on backward stochastic difference equations just started in recent years. For example, in [123], a couple of backward stochastic difference equations were introduced to study the maximum principle of a class of discrete-time stochastic control systems with multiplicative noises.

The following theorem on the existence and uniqueness of a solution of an FSDE is standard, and its proof can be found in any textbook on SDEs.

THEOREM 1.1
Consider SDE (1.2). If there are constants $\lambda_1, \lambda_2 > 0$ such that

$$\|f(x,t)\| + \|g(x,t)\| \leq \lambda_1(1 + \|x\|), \forall\, x \in \mathcal{R}^n, t \in [0,T], \quad (1.8)$$

$$\|f(x,t) - f(y,t)\| + \|g(x,t) - g(y,t)\|$$
$$\leq \lambda_2\|x - y\|, \forall\, x, y \in \mathcal{R}^n, t \in [0,T], \quad (1.9)$$

then SDE (1.2) admits a unique strong solution $x(t)$ which belongs to $\mathcal{L}^2_{\mathcal{F}}([0,T], \mathcal{R}^n])$. Moreover, there exists a constant $K_T > 0$ depending only on T such that

$$\mathcal{E} \sup_{0 \leq s \leq T} \|x(s)\|^2 \leq K_T(1 + \mathcal{E}\|x_0\|^2), \quad (1.10)$$

$$\mathcal{E}\|x(t) - x(s)\|^2 \leq K_T(1 + \mathcal{E}\|x_0\|^2)|t - s|. \tag{1.11}$$

REMARK 1.2 (1.8) is called a linear growth condition, while (1.9) implies that $f(x,t)$ and $g(x,t)$ satisfy the Lipschitz condition with respect to x. Generally speaking, the linear growth condition (1.8) guarantees the existence of solution for SDE (1.2), i.e., the solution $x(t)$ does not have a finite time escape. The Lipschitz condition (1.9) is used to ensure that the SDE (1.2) has a unique solution on $[0, T]$. If any one of the conditions (1.8) and (1.9) does not hold, SDE (1.2) may have many or no solution on $[0, T]$; see [130, 145]. ∎

REMARK 1.3 In many numerical examples in the existing literature, conditions (1.8) and (1.9) are often neglected and not carefully checked, leading to meaningless simulation results. ∎

We note that in (1.2), f and g do not depend on the event $\omega \in \Omega$ directly, but they indirectly do via the solution $x(t)$. So, we also call (1.2) an SDE with deterministic coefficients. When f and g explicitly depend on ω, i.e., $f(\cdot, \cdot, \cdot) : \mathcal{R}^n \times [0, T] \times \Omega \mapsto \mathcal{R}^n$ and $g(\cdot, \cdot, \cdot) : \mathcal{R}^n \times [0, T] \times \Omega \mapsto \mathcal{R}^{n \times d}$ are $\mathcal{F}_t$-adapted Borel measurable functions, then we obtain the following SDE with random coefficients:

$$\begin{cases} dx(t) = f(x(t), t, \omega) \, dt + g(x(t), t, \omega) dB(t), \\ x(0) = x_0 \in \mathcal{L}^2_{\mathcal{F}_0}(\Omega, \mathcal{R}), \ t \in [0, T]. \end{cases} \tag{1.12}$$

SDE (1.12) plays an important role in the study of optimal stochastic control problems, especially stochastic dynamic programming theory; see [196]. Similar to Theorem 1.1, we list the following theorem.

THEOREM 1.2
Consider SDE (1.12). If (1.8) and (1.9) hold almost surely, i.e.,

$$\|f(x,t,\omega)\| + \|g(x,t,\omega)\| \leq \lambda_1(1 + \|x\|), a.s. \ \forall \, x \in \mathcal{R}^n, t \in [0, T], \tag{1.13}$$

$$\|f(x,t,\omega) - f(y,t,\omega)\| + \|g(x,t,\omega) - g(y,t,\omega)\|$$
$$\leq \lambda_2\|x - y\|, a.s. \ \forall \, x, y \in \mathcal{R}^n, t \in [0, T], \tag{1.14}$$

then SDE (1.12) admits a unique strong solution $x(t)$ which belongs to $\mathcal{L}^2_{\mathcal{F}}([0, T], \mathcal{R}^n])$. Moreover, (1.10) and (1.11) hold.

Next, the so-called Itô's integral plays an important role in the study of SDEs and has many nice properties which are listed below.

PROPOSITION 1.1
For $f \in \mathcal{L}^2_{\mathcal{F}}([0, T], \mathcal{R}^n)$, we have

(i) $\int_0^t f(s)\,dB(s)$ is a martingale with respect to $\mathcal{F}_t$ for $t \geq 0$.
(ii) $\mathcal{E} \int_0^t f(s)\,dB(s) = 0$, $\forall t \in [0,T]$.
(iii) $\mathcal{E}\| \int_0^t f(s)\,dB(s)\|^2 = \mathcal{E} \int_0^t \|f(s)\|^2\,ds$, $\forall t \in [0,T]$.
(iv) For $\lambda > 0$, we have

$$P\left(\sup_{0 \leq t \leq T} \left\| \int_0^t f(s)\,dB(s) \right\| \geq \lambda \right) \leq \lambda^{-2} \mathcal{E} \int_0^T \|f(s)\|^2\,ds.$$

1.1.2 Itô's formula

For a real function $V(x,t) \in \mathcal{C}^{2,1}(\mathcal{R}^n \times \mathcal{R}^+; \mathcal{R}) : \mathcal{R}^n \times \mathcal{R}^+ \mapsto \mathcal{R}$, $\mathcal{R}^+ := [0,\infty)$, we define

$$V_t = \frac{\partial V}{\partial t}, \; V_x = \left[\begin{array}{ccc} \frac{\partial V}{\partial x_1} & \cdots & \frac{\partial V}{\partial x_n} \end{array} \right]^T, \; V_{xx} = \left(\frac{\partial^2 V}{\partial x_i \partial x_j} \right)_{n \times n} = \left[\begin{array}{ccc} \frac{\partial^2 V}{\partial x_1{}^2} & \cdots & \frac{\partial^2 V}{\partial x_1 \partial x_n} \\ \vdots & \cdots & \vdots \\ \frac{\partial^2 V}{\partial x_n \partial x_1} & \cdots & \frac{\partial^2 V}{\partial x_n{}^2} \end{array} \right].$$

THEOREM 1.3 (Itô's formula)
Let $x(t)$ be an $\mathcal{R}^n$-valued Itô process expressed by SDE (1.2), $V(x,t) \in \mathcal{C}^{2,1}(\mathcal{R}^n \times \mathcal{R}^+; \mathcal{R})$. Then $V(x,t)$ is a one-dimensional Itô process satisfying the following SDE

$$dV(x(t),t) = V_t(x(t),t)\,dt + V_x^T(x(t),t)\,dx(t) + \frac{1}{2} dx^T(t) V_{xx} dx(t) \quad (1.15)$$

where dt and $dB(t)$ obey the rules

$$dt \cdot dt = dt \cdot dB(t) = dB(t) \cdot dt = 0, \; dB(t) \cdot dB^T(t) = I_{d \times d}\,dt.$$

So (1.15) can also be rewritten as

$$\begin{aligned} dV(x(t),t) = &[V_t(x(t),t) + V_x^T(x(t),t)f(x(t),t) \\ &+ \frac{1}{2}\mathrm{Trace}(g^T(x(t),t)V_{xx}(x(t),t)g(x(t),t)]\,dt \\ &+ V_x^T(x(t),t)g(x(t),t)\,dB(t) \end{aligned} \quad (1.16)$$

or equivalently in an integral form as

$$\begin{aligned} V(x(t),t) = V(x_0,0) + &\int_0^t \Big\{ V_t(x(s),s) + V_x^T(x(s),s)f(x(s),s) \\ &+ \frac{1}{2}\mathrm{Trace}[g^T(x(s),s)V_{xx}(x(s),s)g(x(s),s)] \Big\}\,ds \\ &+ \int_0^t V_x^T(x(s),s)g(x(s),s)\,dB(s). \end{aligned} \quad (1.17)$$

In practical applications, we often need to use a more general Itô's formula in which $V(x, t)$ could be not only a real $\mathcal{R}^1$-, but also an $\mathcal{R}^p$- or even an $\mathcal{R}^{p \times q}$-valued map. For example, for $x \in \mathcal{R}^n$, $V(x) = xx^T$ and $V(x) = x \otimes x$ with "$\otimes$" the Kronecker product of two matrices (vectors) are often adopted as Lyapunov function candidates.

It should be noted that in many existing references, $\dot{V}(x(t), t)$ instead of $dV(x(t), t)$ is often adopted in the Itô's formula, however, which is not accurate, because $B(t)$ is continuous but not differentiable almost surely. An Itô-type equation on V can only take either the differential form (1.15) or the integral form (1.17).

THEOREM 1.4 (General Itô's formula)
Let $x(t)$ be an $\mathcal{R}^n$-valued Itô process expressed by SDE (1.2), $V(x, t) \in \mathcal{C}^{2,1}(\mathcal{R}^n \times \mathcal{R}^+; \mathcal{R}^{p \times q})$, $V(x, t) = (v_{ij})_{p \times q}$. Then $V(x, t)$ is a matrix-valued Itô process whose component v_{ij} is given by

$$dv_{ij}(x, t) = \frac{\partial v_{ij}(x, t)}{\partial t} dt + \left(\frac{\partial v_{ij}(x, t)}{\partial x} \right)^T dx + \frac{1}{2} dx^T \frac{\partial^2 v_{ij}(x, t)}{\partial x^2} dx. \quad (1.18)$$

Example 1.3
In the *pth*-moment estimate, one often needs to take $V(x) = \|x\|^p$, $p \geq 2$, $x \in \mathcal{R}^n$. Obviously, $V(x) \in \mathcal{C}^2(\mathcal{R}^n; \mathcal{R}^+)$. It is easy to compute that

$$V_x(x) = p\|x\|^{p-2}x, \quad V_{xx} = p\|x\|^{p-2}I_{n \times n} + p(p-2)\|x\|^{p-4}xx^T.$$

Associated with SDE (1.2), by Itô's formula, we have

$$\begin{aligned}
\|x(t)\|^p = \|x_0\|^p + \int_0^t &\left\{ p\|x(s)\|^{p-2}x^T(s)f(x(s), s) \right. \\
&+ \frac{1}{2}\text{Trace}[g^T(x(s), s)[p\|x(s)\|^{p-2}I_{n \times n} \\
&+ p(p-2)\|x(s)\|^{p-4}x(s)x^T(s)]g(x(s), s)] \right\} ds \\
&+ \int_0^t p\|x(s)\|^{p-2}x^T(s)g(x(s), s) \, dB(s).
\end{aligned}$$

☐

Example 1.4
Let $x(t)$ and $y(t)$ be Itô processes [145] in $\mathcal{R}$; we now prove

$$d[x(t)y(t)] = y(t)dx(t) + x(t)dy(t) + dx(t) \cdot dy(t). \quad (1.19)$$

Set $z(t) = \begin{bmatrix} x(t) & y(t) \end{bmatrix}^T$, $g(z) = xy = \frac{1}{2}z^T \begin{bmatrix} 0 & 1 \\ 1 & 0 \end{bmatrix} z$. By Itô's formula, we have

$$d[x(t)y(t)] = dg(z(t)) = g_z^T dz + \frac{1}{2}(dz)^T g_{zz} dz$$

$$= \begin{bmatrix} y(t) & x(t) \end{bmatrix} \begin{bmatrix} dx(t) \\ dy(t) \end{bmatrix} + \frac{1}{2} \begin{bmatrix} dx(t) & dy(t) \end{bmatrix} \begin{bmatrix} 0 & 1 \\ 1 & 0 \end{bmatrix} \begin{bmatrix} dx(t) \\ dy(t) \end{bmatrix}$$
$$= y(t)dx(t) + x(t)dy(t) + dx(t) \cdot dy(t).$$

More generally, by (1.19), for $x(t), y(t) \in \mathcal{R}^n$, we still have

$$d[x(t)y^T(t)] = dx(t) \cdot y^T(t) + x(t) \cdot dy^T(t) + dx(t) \cdot dy^T(t). \tag{1.20}$$

Taking integration from 0 to t on the above equation, it yields the following general integration by parts formula:

$$\int_0^t x(s)dy^T(s) = x(t)y^T(t) - x(0)y^T(0) - \int_0^t [dx(s) \cdot y^T(s)] - \int_0^t [dx(s) \cdot dy^T(s)].$$

☐

Example 1.5

For the linear stochastic time-invariant system

$$\begin{cases} dx(t) = Fx(t)\, dt + Gx(t)dB(t), \\ x(0) = x_0 \in \mathcal{L}^2_{\mathcal{F}_0}(\Omega, \mathcal{R}^n), \ t \in [0, T], \end{cases} \tag{1.21}$$

where F and $G \in \mathcal{R}^{n \times n}$ are real constant matrices, $B(t)$ is the one-dimensional standard Brownian motion. By (1.20), we have

$$d[x(t)x^T(t)] = [dx(t)]x^T(t) + x(t)[dx^T(t)] + [dx(t)][dx^T(t)],$$

which yields that

$$x(t)x^T(t) = x_0x_0^T + \int_0^t [Fx(s)x^T(s) + x(s)x^T(s)F^T + Gx(s)x^T(s)G^T]\, ds$$
$$+ \int_0^T [Gx(s)x^T(s) + x(s)x^T(s)G^T]\, dB(s), \ a.s..$$

Letting $X(t) = \mathcal{E}[x(t)x^T(t)]$ and taking the mathematical expectation on both sides of the above equation lead to

$$X(t) = \mathcal{E}[x_0x_0^T] + \int_0^t [FX(s) + X(s)F^T + GX(s)G^T]\, ds.$$

Further, by taking the derivative with respect to t, a deterministic matrix differential equation on $X(t)$ is derived as follows:

$$\dot{X}(t) = FX(t) + X(t)F^T + GX(t)G^T, \ X(0) = \mathcal{E}[x_0x_0^T]. \tag{1.22}$$

☐

We should point out that the equation (1.22) plays an important role in stochastic spectrum analysis, which can be seen in subsequent sections. In addition, by means of (1.22), the study of the mean square stability of (1.21) can be transformed into that of asymptotic stability of an ordinary differential equation.

Example 1.6

Associated with SDE (1.2), if $V(x,t) > 0$ for $x \neq 0$, then by Itô's formula,

$$d\ln(V(x(t),t)) = \left\{\frac{1}{V}V_t + \frac{1}{V}V_x^T f + \frac{1}{2}\text{Trace}\left[g^T\left(-\frac{1}{V^2}V_x V_x^T + \frac{1}{V}V_{xx}\right)g\right]\right\} dt$$
$$+ \frac{1}{V}V_x^T g\, dB(t)$$

or its integral form

$$\ln(V(x(t),t)) = \ln(V(x(0),0) + \int_0^t \left(\frac{1}{V}V_t + \frac{1}{V}V_x^T f\right) ds$$
$$+ \int_0^t \left\{\frac{1}{2}\text{Trace}\left[g^T\left(-\frac{1}{V^2}V_x V_x^T + \frac{1}{V}V_{xx}\right)g\right]\right\} ds$$
$$+ \int_0^t \frac{1}{V}V_x^T g\, dB(s).$$

Examples 1.3 and 1.6 present, respectively, the Itô formulae for $V(x) = \|x\|^p$ and $\ln V(x,t)$ for $V(x,t) > 0$ except for $x = 0$, which are useful in studying stochastic exponential stability; see [130]. □

Example 1.7

It is key to compute V_t, V_x and V_{xx} in using Itô's formula, so some useful formulae should be remembered. For example, if $V(x) = \|x\|^{-1}$ for $x \neq 0$, $x \in \mathcal{R}^n$, then

$$V_x = -\|x\|^{-3}x^T, \quad V_{xx} = -\|x\|^{-3}I + 3\|x\|^{-5}xx^T.$$

□

REMARK 1.4   Because there is no essential difference whether x_0 is random or not in the stochastic H_2/H_∞ control to be discussed, for simplicity, we always assume $x_0 \in \mathcal{R}^n$ to be a deterministic vector from now on. ∎

REMARK 1.5   We also notice that in existing literature, the following model is often considered:

$$\begin{cases} dx(t) = f(x(t),t)\, dt + \sum_{i=1}^d g_i(x(t),t)dB_i(t), \\ x(0) = x_0 \in \mathcal{R}^n, \ t \in [0,T], \end{cases} \tag{1.23}$$

where $B_1, \cdots, B_d$ are independent one-dimensional Brownian motions, $g_i \in \mathcal{R}^n$, $i = 1, 2, \cdots, d$. Note that (1.2) is more general than (1.23), which can be seen by setting

$$g(x,t) = \left[g_1(x,t)\ g_2(x,t) \cdots g_d(x,t) \right]_{n \times d}, B(t) = \mathrm{Col}(B_1(t), B_2(t), \cdots, B_d(t)).$$

In most control problems, such as stochastic LQ control [2] and stochastic H_2/H_∞ control [36, 213], we can, without loss of generality, assume $B(t)$ to be a one-dimensional Brownian motion. So in what follows, we always assume $B(t)$ to be a one-dimensional Brownian motion. ∎

1.1.3   Various definitions of stability

It is well known that in infinite-time stochastic LQ control, H_2/H_∞ control and Kalman filtering, the mean square stability is an essential requirement, so we first introduce the following definition for linear stochastic time-invariant Itô systems.

DEFINITION 1.4   *The linear stochastic time-invariant Itô system*

$$dx(t) = Fx(t)\, dt + Gx(t)dB(t) \tag{1.24}$$

is said to be asymptotically stable in the mean square (ASMS) sense if for any $x_0 \in \mathcal{R}^n$, we have

$$\lim_{t \to \infty} \mathcal{E}\|x(t)\|^2 = 0.$$

When (1.24) is ASMS, we also call (F, G) stable for short.

THEOREM 1.5 [60]
System (1.24) is ASMS if and only if (iff) for any $Q > 0$, the following generalized Lyapunov equation (GLE)

$$PF + F^T P + G^T PG = -Q, \ Q > 0 \tag{1.25}$$

or the linear matrix inequality (LMI)

$$PF + F^T P + G^T PG < 0 \tag{1.26}$$

admits a solution $P > 0$.

Theorem 1.5 generalizes the classical Lyapunov theorem on the asymptotic stability of $\dot{x}(t) = Fx(t)$ to stochastic systems. GLE (1.25) can be viewed as an extension of Lyapunov equation

$$PF + F^T P = -Q, \ Q > 0. \tag{1.27}$$

We will generalize Theorem 1.5 to the case of $Q \geq 0$ by means of exact observability and detectability later.

Now, we define stability of a nonlinear stochastic time-invariant Itô system

$$dx(t) = f(x(t)) \, dt + g(x(t)) \, dB(t), x(0) = x_0 \in \mathcal{R}^n, \ f(0) = 0, \ g(0) = 0.$$
(1.28)

In this case, $x \equiv 0$ is called a trivial solution (corresponding to $x_0 = 0$) or the equilibrium point of (1.28).

DEFINITION 1.5

1) *The trivial solution $x \equiv 0$ of (1.28) is said to be stable in probability if for any $\epsilon > 0$,*

$$\lim_{x_0 \to 0} \mathcal{P}(\sup_{t \geq 0} \|x(t)\| > \epsilon) = 0. \tag{1.29}$$

2) *The trivial solution $x \equiv 0$ of (1.28) is said to be locally asymptotically stable in probability if (1.29) holds and*

$$\lim_{x_0 \to 0} \mathcal{P}(\lim_{t \to \infty} x(t) = 0) = 1.$$

3) *The trivial solution $x \equiv 0$ of (1.28) is said to be globally asymptotically stable in probability if it is stable in probability and*

$$\mathcal{P}(\lim_{t \to \infty} x(t) = 0) = 1.$$

4) *The trivial solution $x \equiv 0$ of (1.28) is said to be ASMS if*

$$\lim_{t \to \infty} \mathcal{E}\|x(t)\|^2 = 0.$$

5) *The trivial solution $x \equiv 0$ of (1.28) is said to be exponentially mean square stable if there exist ϱ and $\rho > 0$, such that*

$$\mathcal{E}\|x(t)\|^2 \leq \rho\|x_0\|^2 exp(-\varrho t).$$

In this book, since we are concerned with the stability of nonlinear stochastic time-invariant system (1.28) only, we refer the reader to [87, 130] for stability of general nonlinear time-varying system

$$dx(t) = f(x(t), t) \, dt + g(x(t), t) \, dB(t), \ f(0, t) \equiv 0, \ g(0, t) \equiv 0.$$

THEOREM 1.6

Let U be a neighborhood of the origin and $V(x)$ be positive definite on U with $V(x) \in \mathcal{C}^2(U; \mathcal{R}^+).$

(1) If

$$\mathcal{L}V(x) = V_x^T(x)f(x) + \frac{1}{2}\text{Trace}(g^T(x)V_{xx}(x)g(x)) \leq 0,$$

then the trivial solution $x \equiv 0$ of (1.28) is stable in probability, where $\mathcal{L}$ is said to be a differential operator associated with (1.28).

(2) If $\mathcal{L}V(x) < 0$ for $x \neq 0$, then the trivial solution $x \equiv 0$ of (1.28) is locally asymptotically stable in probability.

(3) If $V(x) \in \mathcal{C}^2(\mathcal{R}^n; \mathcal{R}^+)$, $\mathcal{L}V(x) < 0$ for $x \neq 0$, and $V(x)$ is radially unbounded, i.e.,

$$\lim_{\|x\|\to\infty} V(x) = \infty,$$

then the trivial solution $x \equiv 0$ of (1.28) is globally asymptotically stable in probability.

(4) If $V(x)$ satisfies

$$k_1\|x\|^2 \leq V(x) \leq k_2\|x\|^2$$

and

$$\mathcal{L}V(x) \leq -k_3\|x\|^2$$

for some constants $k_1 > 0$, $k_2 > 0$ and $k_3 > 0$, then the trivial solution $x \equiv 0$ of (1.28) is exponentially mean square stable.

REMARK 1.6   As noted in [130], the stability in the sense of Lyapunov implies that the trajectory behavior as $t \to \infty$ is independent of whether the initial state is deterministic or random. In fact, it is independent of the distribution of $x(0)$. Different from the Lyapunov stability, in recent years, finite-time stability has been paid much attention; see [8] and [210].   ∎

REMARK 1.7   In most cases, $\mathcal{L}V \leq 0(< 0)$ in stochastic stability plays a similar role to $\dot{V} \leq 0(< 0)$ in the stability test of deterministic systems.   ∎

1.2   Generalized Lyapunov Operators

It is well known that the time-invariant deterministic system

$$\dot{x}(t) = Fx(t), \ x(0) = x_0 \in \mathcal{R}^n \tag{1.30}$$

is asymptotically stable iff either one of the following conditions holds:

(i) The spectral set of the matrix F belongs to the open left-hand side of the complex plane, i.e., $\sigma(F) \subset \mathcal{C}^-$.

(ii) $PF + F^T P < 0$ admits a solution $P > 0$.

The above condition (ii) is derived by the Lyapunov direct method. Condition (i) is the so-called Hurwitz criterion or eigenvalue criterion, which is the basis of pole placement in modern control theory. In order to give a parallel result to the above (i) for stochastic system (1.24), we introduce a symmetric GLO as follows:

$$\mathcal{L}_{F,G} : Z \in \mathcal{S}_n \mapsto FZ + ZF^T + GZG^T \in \mathcal{S}_n,$$

where $\mathcal{S}_n$ is a set of all symmetric matrices.

DEFINITION 1.6 *If there exist $\lambda \in \mathcal{C}$ and a non-zero $Z \in \mathcal{S}_n$ such that $\mathcal{L}_{F,G}Z = \lambda Z$, then λ is called an eigenvalue and Z a corresponding eigenvector of $\mathcal{L}_{F,G}$.*

Example 1.8

If we take

$$F = \begin{bmatrix} -3 & \frac{1}{2} \\ -1 & -1 \end{bmatrix}, \ G = \begin{bmatrix} 2 & 0 \\ 0 & 0 \end{bmatrix}, \ Z = \begin{bmatrix} z_{11} & z_{12} \\ z_{12} & z_{22} \end{bmatrix},$$

then by solving the following characteristic equation

$$\mathcal{L}_{F,G}Z = FZ + ZF^T + GZG^T = \lambda Z, \ 0 \neq Z \in \mathcal{S}_2, \qquad (1.31)$$

we obtain $\sigma(\mathcal{L}_{F,G}) = \{-3 + i, -3 - i, -2\}$. $\quad \square$

REMARK 1.8 It is easily seen that the following operator

$$\mathcal{L}^*_{F,G} : Z \in \mathcal{S}_n \mapsto ZF + F^T Z + G^T ZG$$

is the adjoint operator of $\mathcal{L}_{F,G}$ with the inner product $< X, Y > = \text{Trace}(X^*Y)$ for any $X, Y \in \mathcal{S}_n$, where X^* is the conjugate transpose of X. $\quad \blacksquare$

Similar to $\mathcal{L}_{F,G}$, the operators $\mathcal{H}_{F,G}$ and $\mathcal{T}_{F,G}$ were respectively introduced in [56] and [71]:

$$\mathcal{H}_{F,G} : X \in \mathcal{H}_n \mapsto FX + XF^T + GXG^T \in \mathcal{H}_n,$$

$$\mathcal{T}_{F,G} : X \in \mathcal{C}_n \mapsto FX + XF^T + GXG^T \in \mathcal{C}_n,$$

where $\mathcal{H}_n$ and $\mathcal{C}_n$ are respectively the sets of all $n \times n$ Hermitian matrices and all $n \times n$ complex matrices. Note that the only difference among $\mathcal{L}_{F,G}$, $\mathcal{H}_{F,G}$ and $\mathcal{T}_{F,G}$ is that they are defined in different domains.

Example 1.9

In Example 1.8, it is easy to obtain that $\sigma(\mathcal{T}_{F,G}) = \{-3 + i, -3 - i, -2, -4\}$, $\sigma(\mathcal{H}_{F,G}) = \{-2, -4\}$. $\quad \square$

Generally speaking, for an n-dimensional SDE (1.24), $\sigma(\mathcal{L}_{F,G})$ contains $n(n+1)/2$ eigenvalues including repeated ones, $\sigma(\mathcal{T}_{F,G})$ contains total n^2 eigenvalues, and $\sigma(\mathcal{T}_{F,G}) = \sigma(F \otimes I + I \otimes F + G \otimes G)$. It is conjectured by [105] that $(F \otimes I + I \otimes F + G \otimes G)$ must have repeated eigenvalues, but Example 1.9 shows that this assertion is wrong. Obviously, $\sigma(\mathcal{L}_{F,G}) \subset \sigma(\mathcal{T}_{F,G})$, but there seems no inclusion between $\sigma(\mathcal{L}_{F,G})$ and $\sigma(\mathcal{H}_{F,G})$.

By means of the spectrum of $\mathcal{L}_{F,G}$, we are able to give a necessary and sufficient condition for the mean square stability of system (1.24). To this end, we introduce a special case of general $\mathcal{H}$-representation theory developed in [212]. For an $n \times n$ matrix $X = (x_{ij})_{n\times n}$, we use vec(X) to denote the vector formed by stacking the rows of X into one long vector, i.e., vec$(X) = [X^1, X^2, \cdots, X^n]^T$ with $X^i = \begin{bmatrix} x_{i1} & x_{i2} & \cdots & x_{in} \end{bmatrix}$ being the ith row of X. For example, if $X = (x_{ij})_{2\times 2} \in \mathcal{S}_2$, it is then easy to compute vec$(X) = \begin{bmatrix} x_{11} & x_{12} & x_{12} & x_{22} \end{bmatrix}^T$. In the space $\mathcal{S}_n$, we select a standard basis throughout this book as

$$\{E_{11}, E_{12}, \cdots, E_{1n}, E_{22}, \cdots, E_{2n}, \cdots, E_{nn}\} = \{E_{ij} : 1 \le i \le j \le n\},$$

where $E_{ij} = (e_{lk})_{n\times n}$ with $e_{ij} = e_{ji} = 1$ and all other entries being zero. Define an $\mathcal{H}$-representation matrix H_n as

$$H_n = \begin{bmatrix} \text{vec}(E_{11}), & \cdots, & \text{vec}(E_{1n}), & \text{vec}(E_{22}), & \cdots, & \text{vec}(E_{2n}), & \cdots, & \text{vec}(E_{nn}) \end{bmatrix}.$$
$$(1.32)$$

Example 1.10

In $\mathcal{S}_2$, its standard basis is

$$E_{11} = \begin{bmatrix} 1 & 0 \\ 0 & 0 \end{bmatrix}, \quad E_{12} = \begin{bmatrix} 0 & 1 \\ 1 & 0 \end{bmatrix}, \quad E_{22} = \begin{bmatrix} 0 & 0 \\ 0 & 1 \end{bmatrix}.$$

So

$$H_2 = \begin{bmatrix} 1 & 0 & 0 \\ 0 & 1 & 0 \\ 0 & 1 & 0 \\ 0 & 0 & 1 \end{bmatrix}.$$

□

LEMMA 1.1 [212]
For H_n, we have the following properties:

(i) *For any $X \in \mathcal{S}_n$, there is an $n^2 \times \frac{n(n+1)}{2}$ matrix H_n, independent of X, such that*

$$vec(X) = H_n \widetilde{X}, \quad \widetilde{X} = [x_{11}, x_{12}, \cdots, x_{1n}, x_{22}, \cdots, x_{2n}, \cdots, x_{nn}]^T$$

where $\widetilde{X}$ is an $\frac{n(n+1)}{2}$-dimensional vector that is derived by deleting the repeated elements of vec(X). Conversely, for any $\xi \in \mathcal{C}^{n(n+1)/2}$, there is $X \in \mathcal{S}_n$ such that vec$(X) = H_n\xi$.

(ii) $H_n^T H_n$ *is nonsingular, i.e.,* H_n *has full column rank.*

The following lemma is well known in matrix theory.

LEMMA 1.2
For any three matrices A, B and C of suitable dimension, $vec(ABC) = (A \otimes C^T)vec(B)$.

THEOREM 1.7
System (1.24) is ASMS iff $\sigma(\mathcal{L}_{F,G}) \subset \mathcal{C}^-$.

Proof. Set $X(t) = \mathcal{E}[x(t)x^T(t)]$, where $x(t)$ is the trajectory of (1.24). From Example 1.5, it follows that

$$\dot{X}(t) = \mathcal{L}_{F,G}X(t), \ X(0) = X_0 = x_0 x_0^T,$$

which is, by Lemma 1.2, equivalent to

$$vec[\dot{X}(t)] = (I \otimes F + F \otimes I + G \otimes G)vec[X(t)], \ vec(X_0) \in \mathcal{R}^{n^2}. \quad (1.33)$$

By Lemma 1.1-(i), (1.33) yields

$$H_n \dot{\tilde{X}}(t) = (F \otimes I + I \otimes F + G \otimes G)H_n \tilde{X}(t). \quad (1.34)$$

Pre-multiplying H_n^T on both sides of (1.34) and noting Lemma 1.1-(ii), the following is derived.

$$\begin{aligned} \dot{\tilde{X}}(t) &= (H_n^T H_n)^{-1} H_n^T (F \otimes I + I \otimes F + G \otimes G)H_n \tilde{X}(t) \\ &= H_n^+(F \otimes I + I \otimes F + G \otimes G)H_n \tilde{X}(t), \end{aligned} \quad (1.35)$$

where $H_n^+ := [H_n^T H_n]^{-1} H_n^T$ is the Moore–Penrose inverse of H_n. Observe that

$$\lim_{t \to \infty} \mathcal{E}\|x(t)\|^2 = 0 \Leftrightarrow \lim_{t \to \infty} X(t) = 0 \Leftrightarrow \lim_{t \to \infty} \tilde{X}(t) = 0.$$

Hence, system (1.24) is ASMS iff system (1.35) is ASMS. By the Hurwitz stability criterion, (1.35) is ASMS iff $\sigma(H_n^+(F \otimes I + I \otimes F + G \otimes G)H_n) \subset \mathcal{C}^-$. Below, we only need to show

$$\sigma(H_n^+(F \otimes I + I \otimes F + G \otimes G)H_n) = \sigma(\mathcal{L}_{F,G}).$$

We first show $\sigma(\mathcal{L}_{F,G}) \subseteq \sigma(H_n^+(F \otimes I + I \otimes F + G \otimes G)H_n)$. Suppose λ is any given eigenvalue of $\mathcal{L}_{F,G}$, and a non-zero $Z \in \mathcal{S}_n$ is the corresponding eigenvector, then from $\mathcal{L}_{F,G}Z = \lambda Z$, we have

$$FZ + ZF^T + GZG^T = \lambda Z$$

or equivalently

$$(F \otimes I + I \otimes F + G \otimes G) H_n \widetilde{Z} = \lambda H_n \widetilde{Z}. \qquad (1.36)$$

Pre-multiplying H_n^T on both sides of (1.36) and noting that $(H_n^T H_n)$ is nonsingular, (1.36) yields

$$H_n^+ (F \otimes I + I \otimes F + G \otimes G) H_n \widetilde{Z} = \lambda \widetilde{Z}$$

which shows that λ is an eigenvalue of $H_n^+ (F \otimes I + I \otimes F + G \otimes G) H_n$. So $\sigma(\mathcal{L}_{F,G}) \subseteq \sigma[H_n^+ (F \otimes I + I \otimes F + G \otimes G) H_n]$. In addition, because both $\sigma(\mathcal{L}_{F,G})$ and $\sigma[H_n^+ (F \otimes I + I \otimes F + G \otimes G) H_n]$ contain $n(n+1)/2$ eigenvalues, we must have $\sigma[H_n^+ (F \otimes I + I \otimes F + G \otimes G) H_n] = \sigma(\mathcal{L}_{F,G})$. The proof is complete. $\square$

REMARK 1.9　　(1.33) is not a standard linear system, because $\mathrm{vec}[X(t)]$ contains repeated or redundant components. Generally speaking, for such a non-standard linear system, the Hurwitz criterion does not hold, this is why we take great effort to transform (1.33) into (1.35). ∎

Example 1.11

$$\mathrm{vec}[\dot{X}(t)] = \begin{bmatrix} \dot{x}_{11}(t) \\ \dot{x}_{12}(t) \\ \dot{x}_{12}(t) \\ \dot{x}_{22}(t) \end{bmatrix} = \begin{bmatrix} -1 & 0 & 0 & 0 \\ 0 & -1 & 0 & 0 \\ 0 & -1 & 0 & 0 \\ 0 & 0 & 0 & -1 \end{bmatrix} \begin{bmatrix} x_{11}(t) \\ x_{12}(t) \\ x_{12}(t) \\ x_{22}(t) \end{bmatrix}.$$

Obviously, $\mathrm{vec}[X(t)] \to 0$ as $t \to \infty$, but the coefficient matrix contains one zero eigenvalue. ∎

REMARK 1.10　　In view of Example 1.11, we suspect the statement in [105] that system (1.24) is ASMS iff $\sigma(\mathcal{T}_{F,G}) \subset \mathcal{C}^-$. Considering Theorem 1.7, if the statement of [105] is right, then

$$\sigma(\mathcal{L}_{F,G}) \subset \mathcal{C}^- \Leftrightarrow \sigma(\mathcal{T}_{F,G}) \subset \mathcal{C}^-.$$

Otherwise, one can find F and G such that $\sigma(\mathcal{L}_{F,G}) \subset \mathcal{C}^-$, but $\sigma(\mathcal{T}_{F,G})$ is not the case. Example 1.11 cannot show that the statement in [105] is wrong, because we do not know whether there are matrices F and G such that

$$\begin{bmatrix} -1 & 0 & 0 & 0 \\ 0 & -1 & 0 & 0 \\ 0 & -1 & 0 & 0 \\ 0 & 0 & 0 & -1 \end{bmatrix} = F \otimes I + I \otimes F + G \otimes G.$$

∎

REMARK 1.11　　Note that the $\mathcal{H}$-representation technique developed in [212] is a powerful technique in studying continuous- and discrete-time stochastic moment stability; see [203] and [204]. It can also be used to study stochastic singular systems; see [215]. ∎

1.3 Basic Concepts of Stochastic Systems

It is well known that observability, controllability, detectability, and stabilizability are important structural properties of linear deterministic systems. This section aims to generalize these concepts to linear stochastic systems.

1.3.1 Exact observability

We first recall the complete observability of linear deterministic system

$$\dot{x}(t) = Fx(t), \ x(0) = x_0 \in \mathcal{R}^n, \ y(t) = Hx(t), \tag{1.37}$$

where $x(t)$ is the system state, $y(t) \in \mathcal{R}^m$ is the measurement output, and $F \in \mathcal{R}^{n \times n}$ and $H \in \mathcal{R}^{m \times n}$ are, respectively, the state matrix and measurement matrix.

DEFINITION 1.7 *For system (1.37), x_0 is an unobservable state if its corresponding output $y(t) \equiv 0$ on $[0, T]$ for any $T > 0$. If there exists no non-zero unobservable state x_0, system (1.37) or (F, H) is said to be completely observable.*

For linear deterministic systems, complete observability implies that by observing the measurement output $y(t)$ on a finite time interval, we are able to determine the system state completely, because once x_0 is determined, $x(t)$ for $t \geq 0$ can be determined. The following is the so-called Popov–Belevitch–Hautus (PBH) eigenvector test on complete observability.

THEOREM 1.8 (PBH eigenvector test)
(F, H) is completely observable iff there does not exist an eigenvector ζ (of course, $\zeta \neq 0$) of F satisfying

$$F\zeta = \lambda\zeta, \ H\zeta = 0, \ \lambda \in \mathcal{C}.$$

That is, (F, H) is not completely observable iff some eigenvectors of F are orthogonal with H.

In the following, we generalize complete observability of (F, H) to the linear time-invariant stochastic system

$$\begin{cases} dx(t) = Fx(t)\,dt + Gx(t)dB(t), \ x(0) = x_0 \in \mathcal{R}^n, \\ y(t) = Hx(t). \end{cases} \tag{1.38}$$

DEFINITION 1.8   *Consider system (1.38). $x(0) = x_0 \in \mathcal{R}^n$ is called an unobservable state if its corresponding output $y(t) \equiv 0, a.s.$ on $[0,T]$ for any $T > 0$. If (1.38) has no non-zero unobservable state x_0, then $(F, G|H)$ is said to be exactly observable.*

PROPOSITION 1.2
Let $\mathcal{U}_0$ be the set of all unobservable states of an unobservable system, then $\mathcal{U}_0$ is a linear vector space.

Proof. Obviously, $0 \in \mathcal{U}_0$. For clarity, we denote $x(t, x_0, 0)$ as the solution of (1.38) starting from the initial state x_0 at zero initial time. In addition, if $x_0^1, x_0^2 \in \mathcal{U}_0$, then $y_1(t) = Hx(t, x_0^1, 0) \equiv 0$, $y_2(t) = Hx(t, x_0^2, 0) \equiv 0$, a.s. on $[0, T]$ for any $T \geq 0$. By linearity, $x(t, x_0^1 + x_0^2, 0) = x(t, x_0^1, 0) + x(t, x_0^2, 0)$, so $y(t)|_{x(0) = x_0^1 + x_0^2} = Hx(t, x_0^1 + x_0^2, 0) = y_1(t) + y_2(t) \equiv 0$, a.s.. Hence, $x_0^1 + x_0^2 \in \mathcal{U}_0$. Similarly, for any $\alpha \in \mathcal{R}$ and $x_0 \in \mathcal{U}_0$, we have $\alpha x_0 \in \mathcal{U}_0$. Hence, $\mathcal{U}_0$ is a linear vector space.

We should remember that the set $\mathcal{O}_0$ of all observable states does not construct a linear vector space, because $\mathcal{O}_0$ does not contain the zero vector.

Another definition of observability of $(F, G|H)$ was first given in [216].

DEFINITION 1.9   *$(F, G|H)$ is observable if and only if there does not exist a non-trivial subspace $\mathcal{T}$, i.e., $\mathcal{T} \neq \{0\}$, such that*

$$F\mathcal{T} \subset \mathcal{T}, \ G\mathcal{T} \subset \mathcal{T}, \ \mathcal{T} \subset Ker(H), \tag{1.39}$$

where $Ker(H)$ represents the kernel space of the matrix H.

PROPOSITION 1.3
$(F, G|H)$ is exactly observable iff $(F, G|H)$ is observable in the sense of Definition 1.9. In other words, $(F, G|H)$ is exactly observable iff there does not exist a non-trivial subspace $\mathcal{T}$ satisfying (1.39).

Proof. Set

$$\mathcal{P}_0 = \left[H^T, \ F^T H^T, \ G^T H^T, \ F^T G^T H^T, \ G^T F^T H^T, \ (F^T)^2 H^T, \ (G^T)^2 H^T, \ \cdots \right]^T.$$

In [117], it is shown that $\mathcal{U}_0 = Ker(\mathcal{P}_0)$. If $(F, G|H)$ is exactly observable, then $\mathcal{U}_0 = \{0\}$, but (1.39) holds with a non-trivial subspace $\mathcal{T}$. By taking a non-zero $\xi \in \mathcal{T}$ and considering (1.39), it follows that

$$H\xi = 0, \ HF\xi = 0, \ HG\xi = 0.$$

By repeating the above procedure, we have

$$\mathcal{P}_0\xi = 0 \tag{1.40}$$

which yields a contradiction due to $Ker(\mathcal{P}_0) \neq \{0\} = \mathcal{U}_0$.

Conversely, if there does not exist a non-trivial subspace $\mathcal{T}$ satisfying (1.39), $(F, G|H)$ must be exactly observable. Otherwise, $\mathcal{U}_0 = \text{Ker}(\mathcal{P}_0)$ is a non-trivial subspace. Set $\mathcal{T} = \text{Ker}(\mathcal{P}_0)$, then $\mathcal{T}$ satisfies (1.39), leading to a contradiction. $\square$

We now introduce the following lemma.

LEMMA 1.3
Let $Z_1, Z_2 \in \mathcal{S}_n$. If $H_n^T vec(Z_1) = H_n^T vec(Z_2)$, where H_n is as defined in (1.32), then $vec(Z_1) = vec(Z_2)$.

Proof. By Lemma 1.1-(i), $\text{vec}(Z_1) = H_n\tilde{Z}_1$, $\text{vec}(Z_2) = H_n\tilde{Z}_2$. So from $H_n^T\text{vec}(Z_1) = H_n^T\text{vec}(Z_2)$, we have

$$H_n^T H_n \tilde{Z}_1 = H_n^T H_n \tilde{Z}_2.$$

By Lemma 1.1-(ii), the above leads to $\tilde{Z}_1 = \tilde{Z}_2$, which is equivalent to $\text{vec}(Z_1) = \text{vec}(Z_2)$.

Parallel to Theorem 1.8, we present a PBH eigenvector test for the exact observability of $(F, G|H)$.

THEOREM 1.9 (Stochastic PBH eigenvector test)
$(F, G|H)$ is exactly observable iff there does not exist a non-zero $Z \in \mathcal{S}_n$ such that

$$\mathcal{L}_{F,G}Z = \lambda Z, \ HZ = 0, \ \lambda \in \mathcal{C}. \tag{1.41}$$

Proof. Let $X(t) = \mathcal{E}[x(t)x^T(t)]$ and $Y(t) = \mathcal{E}[y(t)y^T(t)]$, where $x(t)$ is the solution of (1.38). As shown in Example 1.5, the following can be derived from (1.22).

$$\begin{cases} \dot{X}(t) = \mathcal{L}_{F,G}X(t), \ X(0) = X_0 = x_0x_0^T, \\ Y(t) = HX(t)H^T. \end{cases} \tag{1.42}$$

The first equation in (1.42) is, by Lemma 1.2, equivalent to

$$\text{vec}[\dot{X}(t)] = (I \otimes F + F \otimes I + G \otimes G)\text{vec}[X(t)], \ \text{vec}(X_0) \in \mathcal{R}^{n^2}. \tag{1.43}$$

We first show that (1.43) is equivalent to

$$\dot{\tilde{X}}(t) = (H_n^T H_n)^{-1}H_n^T(F \otimes I + I \otimes F + G \otimes G)H_n\tilde{X}(t). \tag{1.44}$$

(1.43) $\Rightarrow$ (1.44) can be seen from the proof of Theorem 1.7.

Conversely, from Lemma 1.1-(i), $\tilde{X}(t) = (H_n^T H_n)^{-1}H_n^T\text{vec}[X(t)]$. So it follows from (1.44) that

$$H_n^T\text{vec}[\dot{X}(t)] = H_n^T(F \otimes I + I \otimes F + G \otimes G)\text{vec}[X(t)]$$
$$= H_n^T\text{vec}[\mathcal{L}_{F,G}X(t)]. \tag{1.45}$$

By Lemma 1.3, (1.43) is derived.

Because $y(t) \equiv 0$ a.s. for $t \geq 0$ iff $\mathcal{E}\|y(t)\|^2 = 0$ or equivalently $Y_1 := HX(t) = 0$ for $t \geq 0$ due to $X(t) \geq 0$. Hence, $(F, G|H)$ is exactly observable iff for an arbitrary $X_0 = x_0 x_0^T \neq 0$, there exists a $\tilde{t} \geq 0$ such that

$$Y(\tilde{t}) = \mathcal{E}[y(\tilde{t})y^T(\tilde{t})] = HX(\tilde{t})H^T \neq 0, \tag{1.46}$$

or

$$Y_1(\tilde{t}) := HX(\tilde{t}) \neq 0 \tag{1.47}$$

which is equivalent to

$$\mathrm{vec}Y_1(\tilde{t}) = (H \otimes I)H_n \widetilde{X}(\tilde{t}) \neq 0. \tag{1.48}$$

So $(F, G|H)$ is exactly observable iff the linear deterministic system

$$\begin{cases} \dot{\widetilde{X}}(t) = (H_n^T H_n)^{-1} H_n^T (F \otimes I + I \otimes F + G \otimes G) H_n \widetilde{X}(t), \\ \mathrm{vec}Y_1(t) = (H \otimes I)H_n \widetilde{X}(t) \end{cases} \tag{1.49}$$

is completely observable. By Theorem 1.8, (1.49) is completely observable iff there does not exist $0 \neq \xi \in \mathcal{C}^{\frac{n(n+1)}{2}}$ such that

$$(H_n^T H_n)^{-1} H_n^T (F \otimes I + I \otimes F + G \otimes G) H_n \xi = \lambda \xi, \ (H \otimes I)H_n \xi = 0, \ \lambda \in \mathcal{C}.$$

By setting $\mathrm{vec}(Z) = H_n \xi \in \mathcal{C}^{n^2}$, $Z \in \mathcal{S}_n$ satisfies (1.41). The proof is completed.
□

Finally, we give some definitions and results on the observability of linear time-varying systems. Firstly, consider the following deterministic system

$$\dot{x}(t) = F(t)x(t), \ x(0) = x_0 \in \mathcal{R}^n, \ y(t) = H(t)x(t), \tag{1.50}$$

where $F(t)$ and $H(t)$ are piecewise continuous matrix-valued functions.

DEFINITION 1.10   *System (1.50) or $(F(t), H(t))$ is observable at $t_0 \in \mathcal{R}^+$, if there exists some finite time interval $[t_0, t_1]$ with $t_1 > t_0$, such that when $y(t) \equiv 0, \forall t \in [t_0, t_1]$, we have $x(t_0) = 0$. If the above statement is true for all $t_0 \in \mathcal{R}^+$, then system (1.50) is said to be completely observable.*

LEMMA 1.4 [20]
Let $\Phi_0(t, t_0)$ be the state transition matrix of $\dot{x}(t) = F(t)x(t)$. Then $(F(t), H(t))$ is observable at $t_0 \in \mathcal{R}^+$ iff one of the following holds.

(i) The Gramian matrix

$$W_0[t_0, t_1] = \int_{t_0}^{t_1} \Phi_0^T(t, t_0) H^T(t) H(t) \Phi_0(t, t_0) \, dt \tag{1.51}$$

is nonsingular, i.e., $\det W_0[t_0, t_1] \neq 0$.

(ii) $W_0[t_0, t_1]$ *is positive definite.*

Below, we define the observability of linear time-varying stochastic system

$$\begin{cases} dx(t) = F(t)x(t)\, dt + G(t)x(t)\, dB(t), \\ x(0) = x_0 \in \mathcal{R}^n, \\ y(t) = H(t)x(t), \end{cases} \tag{1.52}$$

where $F(t), G(t) \in \mathcal{R}^{n\times}$, and $H(t) \in \mathcal{R}^{m\times n}$.

DEFINITION 1.11   *System (1.52) or $(F(t), G(t)|H(t))$ is observable at $t_0 \in \mathcal{R}^+$, if there exists some finite time interval $[t_0, t_1]$ with $t_1 > t_0$, such that when $y(t) \equiv 0$ a.s., $\forall t \in [t_0, t_1]$, we have $x(t_0) = 0$ a.s.. If the above statement is true for all $t_0 \in \mathcal{R}^+$, then system (1.52) is said to be completely observable.*

The only difference between Definition 1.10 and Definition 1.11 is that $y(t) \equiv 0$ and $x(t_0) \equiv 0$ in Definition 1.10 are respectively replaced by their corresponding almost sure equality.

THEOREM 1.10
Set

$$N(t) = (H(t) \otimes I)H_n, \tag{1.53}$$

where H_n is defined in (1.32), and let $\Phi(t, t_0)$ be the state transition matrix with

$$\dot{\Phi}(t, t_0) = M(t)\Phi(t, t_0), \quad \Phi(t_0, t_0) = I,$$

where

$$M(t) = [H_n^T H_n]^{-1} H_n^T (F(t) \otimes I + I \otimes F(t) + G(t) \otimes G(t))H_n. \tag{1.54}$$

Then system (1.52) is observable at time $t_0 \in \mathcal{R}^+$ iff there exists some finite time interval $[t_0, t_1]$ with $t_1 > t_0$, such that one of the following holds:

(i) The Gramian matrix

$$W_1[t_0, t_1] = \int_{t_0}^{t_1} \Phi^T(t, t_0)N^T(t)N(t)\Phi(t, t_0)\, dt \tag{1.55}$$

is nonsingular, i.e., $\det W_1[t_0, t_1] \neq 0$.

(ii) $W_1[t_0, t_1]$ is positive definite.

Proof. As in the proof of Theorem 1.9, $(F(t), G(t)|H(t))$ is observable at $t_0 \in \mathcal{R}^+$ iff the following deterministic system

$$\begin{cases} \dot{\widetilde{X}}(t) = M(t)\widetilde{X}(t), \\ \mathrm{vec}Y_1(t) = N(t)\widetilde{X}(t) \end{cases} \tag{1.56}$$

is observable at t_0, which, following Lemma 1.4, is equivalent to one of the conditions (i)–(ii) holding. The proof is completed. $\square$

As in deterministic time-varying systems, it is not convenient to use Theorem 1.10 in practical applications, as it is not easy to compute $\Phi(t, t_0)$. To address this issue, we present a sufficient criterion on the observability of (1.52) that can be easily verified using Theorem 9.10 of [154].

THEOREM 1.11
Suppose there exists a positive integer q such that $F(t)$ and $G(t)$ are $(q-1)$-times continuously differentiable and $H(t)$ is q-times continuously differentiable for $t \in [t_0, t_f]$. Define a sequence of $(mn) \times n(n+1)/2$ matrices as

$$
\begin{aligned}
P_0(t) &= N(t), \\
P_1(t) &= P_0(t)M(t) + \dot{P}_0(t), \\
&\vdots \\
P_q(t) &= P_{q-1}(t)M(t) + \dot{P}_{q-1}(t),
\end{aligned}
\tag{1.57}
$$

where $M(t)$ and $N(t)$ are defined in Theorem 1.10. Then, $(F(t), G(t)|H(t))$ is observable at time $t_0 \in \mathcal{R}^+$ if there exists $t_1 > t_0$ satisfying

$$
\mathrm{rank} P(t_1) = \frac{n(n+1)}{2},
\tag{1.58}
$$

where

$$
P(t) = \left[P_0^T(t),\ P_1^T(t),\ \cdots,\ P_q^T(t) \right]^T.
$$

Example 1.12
In (1.52), we take

$$
F(t) = \begin{bmatrix} t & 0 \\ 1 & 2 \end{bmatrix}, \ G(t) = \begin{bmatrix} 1 & 0 \\ 0 & 1 \end{bmatrix}, \ H(t) = \begin{bmatrix} 1 & 3 \end{bmatrix}.
$$

It is easy to compute that

$$
\begin{aligned}
M(t) &= [H_2^T H_2]^{-1} H_2^T (F(t) \otimes I + I \otimes F(t) + G(t) \otimes G(t)) H_2 \\
&= \begin{bmatrix} 2t+1 & 0 & 0 \\ 1 & t+3 & 0 \\ 0 & 2 & 5 \end{bmatrix}, \\
N(t) &= (H(t) \otimes I) H_2 = \begin{bmatrix} 1 & 6 & 9 \end{bmatrix}.
\end{aligned}
$$

Hence,

$$
\begin{aligned}
P_0(t) &= N(t) = \begin{bmatrix} 1 & 6 & 9 \end{bmatrix}, \\
P_1(t) &= P_0(t)M(t) + \dot{P}_0(t) = \begin{bmatrix} 2t+7 & 6t+36 & 45 \end{bmatrix}, \\
P_2(t) &= P_1(t)M(t) + \dot{P}_1(t) = \begin{bmatrix} 4t^2+22t+45 & 6t^2+54t+204 & 225 \end{bmatrix}.
\end{aligned}
$$

Therefore,

$$P(t) = \begin{bmatrix} 1 & 6 & 9 \\ 2t+7 & 6t+36 & 45 \\ 4t^2+22t+45 & 6t^2+54t+204 & 225 \end{bmatrix}.$$

This leads to

$$\det(P(t)) = -108t^3 - 324t^2 - 324t - 108 = -108(t+1)^3 < 0$$

for $t > 0$, so $\text{rank}\, P(t) = 3$ whenever $t > 0$. By Theorem 1.11 and Definition 1.11, this system is completely observable. $\quad\Box$

1.3.2 Exact detectability

DEFINITION 1.12 *For system (1.37), if each unobservable state x_0 leads to*

$$\lim_{t\to\infty} \|x(t, x_0, 0)\| = 0,$$

then (1.37) or (F, H) is said to be completely detectable, where $x(t, x_0, 0)$ denotes the trajectory of (1.37) starting from $x(0) = x_0$.

REMARK 1.12 Complete observability tells us that $y(t) \equiv 0$ on $[0, T]$ for any $T > 0$ implies that $x_0 = 0$. However, complete detectability implies that, from $y(t) \equiv 0$ on $[0, T]$ for any $T > 0$, we only have $\lim_{t\to\infty} \|x(t, x_0, 0)\| = 0$. So complete detectability is weaker than complete observability. $\quad\blacksquare$

THEOREM 1.12 (PBH eigenvector test)
(F, H) is completely detectable iff there does not exist a non-zero eigenvector ζ of F satisfying

$$F\zeta = \lambda\zeta, \quad H\zeta = 0, \quad Re(\lambda) \geq 0.$$

Now, we define exact detectability for the system (1.38), which is weaker than exact observability.

DEFINITION 1.13 *System (1.38) or $(F, G|H)$ is said to be exactly detectable, if $y(t) = Hx(t) \equiv 0$ a.s., $t \in [0, T]$, $\forall T > 0$, implies*

$$\lim_{t\to\infty} \mathcal{E}\|x(t)\|^2 = 0.$$

Now, we are in a position to present a stochastic PBH eigenvector test for the exact detectability of $(F, G|H)$.

THEOREM 1.13 (Stochastic PBH eigenvector test)
$(F, G|H)$ *is exactly detectable iff there does not exist a non-zero $Z \in \mathcal{S}_n$ such that*

$$\mathcal{L}_{F,G}Z = \lambda Z, \quad HZ = 0, \quad Re\lambda \geq 0. \tag{1.59}$$

Proof. We note that

$$\lim_{t\to\infty} \mathcal{E}\|x(t)\|^2 = 0 \Leftrightarrow \lim_{t\to\infty} X(t) = 0 \Leftrightarrow \lim_{t\to\infty} \tilde{X}(t) = 0$$

where $X(t)$ and $\tilde{X}(t)$ are defined in (1.42) and (1.44), respectively. Hence, from the proof of Theorem 1.9, $(F, G|H)$ is exactly detectable iff (1.49) is completely detectable. Using Theorem 1.12 and repeating the same procedure as in the proof of Theorem 1.9, the result of the theorem can be established. $\square$

COROLLARY 1.1
(F, H) *is completely detectable iff there does not exist a non-zero $Z \in \mathcal{S}_n$ such that*

$$\mathcal{L}_{F,0}Z = FZ + ZF^T = \lambda Z, \quad HZ = 0, \quad Re\lambda \geq 0.$$

DEFINITION 1.14   *System (1.38) or $(F, G|H)$ is said to be stochastically detectable, if there is a constant matrix K such that*

$$dx(t) = (F + KH)x(t)\,dt + Gx(t)\,dB(t) \tag{1.60}$$

is ASMS, i.e., $\lim_{t\to\infty} \mathcal{E}\|x(t)\|^2 = 0$ for any initial state x_0, where $x(t)$ is the state trajectory of (1.60).

Obviously, when $G \equiv 0$, complete detectability, exact detectability and stochastic detectability are the same.

PROPOSITION 1.4

(i) $(F, G|H)$ *is stochastically detectable, then there does not exist a non-zero $Z \in \mathcal{S}_n$ such that*

$$\mathcal{L}_{F,G}Z = FZ + ZF^T + GZG^T = \lambda Z, \quad HZ = 0, \quad Re(\lambda) \geq 0. \tag{1.61}$$

(ii) $(F, G|H)$ *is stochastically detectable iff the following LMI*

$$\begin{bmatrix} F^T X + XF + H^T Y^T + YH & G^T X \\ XG & -X \end{bmatrix} < 0 \tag{1.62}$$

admits matrix-valued solutions $X > 0$ and Y of suitable dimension.

Proof. By Definition 1.14 and Theorem 1.7, $(F, G|H)$ is stochastically detectable iff there exists a constant matrix K such that $\sigma(\mathcal{L}_{F+KH,G}) \subset \mathcal{C}^-$. If (1.61) holds, then for any K,

$$\mathcal{L}_{F+KH,G}Z = (F + KH)Z + Z(F + KH)^T + GZG^T$$
$$= FZ + ZF^T + GZG^T = \lambda Z, \ Re(\lambda) \geq 0, \qquad (1.63)$$

which means that (1.60) is not ASMS according to Theorem 1.7. Hence, (i) is shown.

To prove (ii), we recall the well known Schur's complement lemma [24] which asserts that the following two matrix inequalities are equivalent.

$$M + NR^{-1}N^T < 0, R > 0 \ \Leftrightarrow \ \begin{bmatrix} M & N \\ N^T & -R \end{bmatrix} < 0.$$

By Theorem 1.5, (1.60) is ASMS iff

$$P(F + KH) + (F + KH)^T P + G^T PG$$
$$= P(F + KH) + (F + KH)^T P + G^T PP^{-1}PG < 0$$

admits a solution $P > 0$. Set $X = P, Y = PK$, using Schur's complement, (ii) can be proved. $\square$

THEOREM 1.14

If $(F, G|H)$ is stochastically detectable, then $(F, G|H)$ is exactly detectable and (F, H) is completely detectable.

Proof. By combining Proposition 1.4-(i) with Theorem 1.13, we know that stochastic detectability implies exact detectability. Because the ASMS of (1.60) implies that there exists a matrix K such that $F + KH$ is a stable matrix [87], the complete detectability of (F, H) is derived. $\square$

It should be pointed out that the converse of Theorem 1.14 is not true; see the following examples:

Example 1.13

Consider that

$$F = \begin{bmatrix} 0 & 0 \\ 1 & -1 \end{bmatrix}, \ G = \begin{bmatrix} 1 & 0 \\ -1 & 0 \end{bmatrix}, \ H = \begin{bmatrix} 0 & 1 \end{bmatrix}.$$

Then it is easy to verify that there is no $0 \neq Z \in \mathcal{S}_2$ and $Re(\lambda) \geq 0$ satisfying (1.59). Hence, $(F, G|H)$ is exactly detectable. However, by exploiting the LMI Control Toolbox [24, 34, 83], we can know that (1.62) does not have solutions $X > 0$ and Y. Hence, by Proposition 1.4-(ii), $(F, G|H)$ is not stochastically detectable. $\square$

Example 1.14

For the given system in Example 1.13, if we take $K = \begin{bmatrix} -1 & -1 \end{bmatrix}^T$, then $(F + KH)$ is stable. Hence, (F, H) is completely detectable. However, as shown in Example 1.13, $(F, G|H)$ is not stochastically detectable.   ⬜

The following example tells us that there is no any inclusion relation between exact detectability and complete detectability.

Example 1.15

Consider the system with

$$F = \begin{bmatrix} 1 & 0 \\ 0 & 0 \end{bmatrix}, \; G = \begin{bmatrix} 1 & 1 \\ 1 & 1 \end{bmatrix}, \; H = \begin{bmatrix} 0 & 1 \end{bmatrix}.$$

It is easy to verify that there does not exist a non-zero $Z \in \mathcal{S}_2$ and $Re(\lambda) \geq 0$ satisfying (1.59). By the stochastic PBH eigenvector test, $(F, G|H)$ is exactly detectable. However, (F, H) is not completely detectable, because $1 \in \sigma(F + KH)$ for any $K = \begin{bmatrix} k_1 & k_2 \end{bmatrix}^T$.

On the other hand, if we take

$$F = -I_{2\times2}, \; G = 2I_{2\times2}, \; H = \begin{bmatrix} 0 & 1 \end{bmatrix},$$

then it is easy to show that (F, H) is completely detectable but $(F, G|H)$ is not exactly detectable.   ⬜

1.3.3   Mean square stabilization

Consider the following control system

$$\begin{cases} dx(t) = [Fx(t) + F_1 u(t)] \, dt + [Gx(t) + G_1 u(t)] \, dB(t), \\ x(0) = x_0 \in \mathcal{R}^n, \end{cases} \qquad (1.64)$$

where $x(t) \in \mathcal{R}^n$ is the system state, $u(t) \in \mathcal{R}^m$ is the control input. We assume $u(t) \in \mathcal{L}^2_{\mathcal{F}}(\mathcal{R}^+, \mathcal{R}^m)$ which means that the system (1.64) has a unique strong solution.

DEFINITION 1.15   *System (1.64) or $(F, F_1; G, G_1)$ is called stabilizable in mean square sense if there exists a constant state feedback $u(t) = Kx(t)$ such that*

$$dx(t) = (F + F_1 K)x(t) \, dt + (G + G_1 K)x(t) \, dB(t) \qquad (1.65)$$

is ASMS.

THEOREM 1.15
$(F, F_1; G, G_1)$ *is stabilizable iff the LMI*

$$\begin{bmatrix} FX + XF^T + F_1Y + Y^T F_1^T & XG^T + Y^T G_1^T \\ GX + G_1Y & -X \end{bmatrix} < 0 \qquad (1.66)$$

or

$$\begin{bmatrix} FX + XF^T + F_1Y + Y^T F_1^T & GX + G_1Y \\ XG^T + Y^T G_1^T & -X \end{bmatrix} < 0 \qquad (1.67)$$

admits solutions $X > 0$ and Y.

Proof. By Definition 1.15 and Theorem 1.5, $(F, F_1; G, G_1)$ is stabilizable iff there exists K such that

$$P(F + F_1K) + (F + F_1K)^T P + (G + G_1K)^T P(G + G_1K) < 0 \qquad (1.68)$$

has a solution $P > 0$. Pre- and post-multiplying P^{-1} on both sides of (1.68) leads to

$$(F + F_1K)P^{-1} + P^{-1}(F + F_1K)^T + P^{-1}(G + G_1K)^T P(G + G_1K)P^{-1} < 0.$$

Let $X = P^{-1}$ and $Y = KP^{-1}$. By applying Schur's complement, (1.66) is immediately derived.

To prove (1.67), we note that (1.65) is ASMS iff its dual system

$$dx(t) = (F + F_1K)^T x(t)\, dt + (G + G_1K)^T x(t)\, dB(t)$$

is ASMS, which is equivalent to the matrix inequality

$$P(F + F_1K)^T + (F + F_1K)P + (G + G_1K)P(G + G_1K)^T < 0$$

having a solution $P > 0$. By setting $X = P$, $Y = KP$, and repeating the above procedure, (1.67) is derived. $\square$

REMARK 1.13 When $G = 0$ and $G_1 = 0$, the stabilizability of $(F, F_1; 0, 0)$ reduces to that of (F, F_1) in linear system theory. As is well known, Definition 1.12 is equivalent to its dual system

$$\dot{x}(t) = F^T x(t), \; x(0) = x_0 \in \mathcal{R}^n, \; y(t) = F_1^T x(t)$$

being completely detectable. ∎

REMARK 1.14 The stochastic detectability of system (1.38) is equivalent to the following system

$$dx(t) = [F^T x(t) + H^T u(t)]\, dt + G^T x(t)dB(t), \; x(0) = x_0 \in \mathcal{R}^n \quad (1.69)$$

being stabilizable. However, exact detectability is not dual to the stabilizability of (1.69), which reveals the complexity of detectability in stochastic systems; we refer the reader to [14] for the implication relationship about various detectabilities of stochastic systems. ∎

REMARK 1.15   We refer the reader to [152] for the spectral test for uniform exponential stability, observability and detectability of deterministic discrete time-varying systems, which needs further study on how to generalize the results of [152] to stochastic time-varying systems. ∎

1.4   Notes and References

There are many excellent books on SDEs. For instance, for stochastic stability, we refer the reader to [87, 104, 130]. Reference [145] is a very good introductory book for SDEs, while [106] gives a more thorough treatment of the general diffusion processes. The book [196] provides a summary of some recent development of stochastic maximum principle and dynamical programming. For the general theory of forward-backward SDEs, the reader is referred to [140].

The notion of exact observability of linear time-invariant SDEs was first independently introduced by [216] and [117], where the former defines the notion from a mathematical viewpoint while the latter from the viewpoint of state-measurement. On the other hand, the exact detectability was first defined in [206], which is weaker than the exact observability. The PBH criteria for exact observability and detectability were presented in [205] and [206], respectively, and have been extended to linear time-invariant Markov jump systems; see [118, 120, 97, 143, 161, 217]. Recently, the authors in [218] studied the uniform detectability, exact detectability and exact observability of linear time-varying stochastic systems with multiplicative noise, and extended some results of [10] to discrete-time stochastic time-varying systems. Other definitions of observability and detectability can be found in [41, 42, 50, 51, 71].

The mean square stability of linear time-invariant SDEs can be characterized via the spectra of GLOs [52, 53, 56, 57, 205, 211], based on which, we succeeded in presenting PBH criteria for the exact observability and detectability. Moreover, we can exactly describe the convergence rate based on the spectral technique; see [121, 211]. How to generalize the spectral technique to nonlinear time-invariant systems is an interesting research topic. The $\mathcal{H}$-representation technique was developed in [212], where its aim is to transform a non-standard linear system (the state vector includes repeated components) into a standard linear system for which many results in modern control theory [203, 204] can be applied.

2

Linear Continuous-Time Stochastic H_2/H_∞ Control

The aim of this chapter is to extend the classical H_2/H_∞ control theory to linear stochastic Itô systems. A relationship between the stochastic H_2/H_∞ control and two-person non-zero sum Nash game is revealed. New types of cross-coupled generalized differential Riccati equations (GDREs) and generalized algebraic Riccati equations (GAREs) are derived. The results of this chapter contribute to both H_∞ control theory and differential game theory.

2.1 Introduction

Since the seminal work [232], H_∞ control has become one of the most important robust control approaches and attracted a lot of interest in the past thirty years. H_∞ optimal control requires one to design a controller to attenuate the effect of external disturbance v on regulated output z as far as possible, while H_∞ suboptimal control only requires attenuating the effect of v on z by a prescribed level $\gamma > 0$. H_∞ optimal control can be computationally involved and the so-called H_∞ control is normally referred to as suboptimal H_∞ control. In the early development, H_∞ control was studied in the frequency domain, which is associated with the computation of the H_∞ norm of a transfer function. A well known breakthrough was made in [62], where it is shown that a solution to the H_∞ control problem can be obtained by solving two coupled algebraic Riccati equations (AREs) and that the H_∞ norm of a transfer function that is less than $\gamma > 0$ is equivalent to the $\mathcal{L}_2$-gain from v to z being less than γ. Such an important finding made it possible to generalize the H_∞ control theory of deterministic linear time-invariant systems to linear time-varying systems, nonlinear systems and stochastic systems, etc. We refer the reader to [69, 227] for the early development of H_∞ control along this direction.

By the 1990s, the deterministic H_∞ control theory had reached maturity for linear and nonlinear systems [92, 101, 111, 158, 173], where the theory of differential geometry is used to deal with nonlinear H_∞ control. Before 1998, there were few results on stochastic H_∞ control for Itô-type differential systems with multiplicative noises. In [150], stochastic H_2/H_∞ control for systems with additive noises is discussed. The references [84] and [169] reported almost at the same time results on

H_∞ control of linear Itô-type systems, in particular, the reference [84] obtained a stochastic bounded real lemma (SBRL) in terms of an LMI, which is very useful in designing an H_∞ filter [78, 191]. In [169], a very interesting example on a two-mass spring system is given. After 1998, stochastic H_∞ control has become one of the most popular research topics [53, 52, 146, 162], and has been applied to systems biology [38, 39], hard disk drives [55] and mathematical finance [139].

Note that there generally exists a family of suboptimal H_∞ controllers which offer the same level of disturbance attenuation. This freedom can be exploited to consider other desirable performances such as the H_2 performance. Therefore, the so-called mixed H_2/H_∞ control has attracted a lot of interest. The H_2/H_∞ control is more appealing than the pure H_∞ control from an engineering perspective [59, 181] and requires one to search for a controller u^* not only to attenuate the effect of distur-bance to some desirable level, but also to minimize an additional desired H_2 per-formance; see, for example, [21, 37, 113], where the H_2 performance optimization is defined in various senses. For example, in [21] and [37], the H_2 performance is optimized when the H_∞ disturbance $v \equiv 0$. On the other hand, a Nash game approach is adopted in [113] to deal with the H_2/H_∞ control problem, where the H_2 performance optimization is carried out under the worst-case H_∞ disturbance $v = v^*$, which, we believe, is more reasonable than assuming $v \equiv 0$. However, such an H_2/H_∞ controller design appears to be more complicated.

In this chapter, we study linear continuous-time stochastic H_2/H_∞ control of Itô-type systems based on the Nash game approach, which can be viewed as an exten-sion of the deterministic H_2/H_∞ control in [113]. To this end, we have to develop SBRLs, generalized Lyapunov-type theorems, and indefinite stochastic LQ theory. It turns out that the solvability of the finite-time (respectively, infinite-time) stochastic H_2/H_∞ control is equivalent to the solvability of some cross-coupled GDREs (re-spectively, GAREs). A relationship between the solvability of the stochastic H_2/H_∞ control and the existence of an equilibrium point of a two-person non-zero sum Nash game is revealed. A unified treatment for the finite-time H_2, H_∞ and mixed H_2/H_∞ control is presented, which shows that the pure H_2 or H_∞ control can be treated as special cases of the mixed H_2/H_∞ control. The results of this chapter contribute not only to the stochastic H_2/H_∞ control but also to multiple decision making [135] and differential game theory [136].

2.2   Finite Horizon H_2/H_∞ Control

In this section, we discuss the finite horizon stochastic H_2/H_∞ control problem. Consider the following stochastic linear system with state- and disturbance-dependent noises

$$\begin{cases} dx(t) = [A_1(t)x(t) + B_1(t)u(t) + C_1(t)v(t)]\,dt + [A_2(t)x(t) + C_2(t)v(t)]\,dB(t), \\ x(0) = x_0 \in \mathcal{R}^n, \\ z(t) = \begin{bmatrix} C(t)x(t) \\ D(t)u(t) \end{bmatrix}, \quad D^T(t)D(t) = I, \ t \in [0,T] \end{cases}$$

$$(2.1)$$

where all coefficients are matrix-valued continuous functions of time with suitable dimensions, $x(t)$, $u(t)$ and $v(t)$ are respectively the system state, control input and external disturbance. Without loss of generality, we assume $B(t)$ to be one-dimensional standard Brownian motion defined on the filtered probability space $(\Omega, \mathcal{F}, \{\mathcal{F}_t\}_{t \geq 0}, \mathcal{P})$ with $\mathcal{F}_t = \sigma(B(s) : 0 \leq s \leq t)$. $x_0 \in \mathcal{R}^n$ is the initial state which is assumed to be deterministic. By [106], for $(u, v, x_0) \in \mathcal{L}_{\mathcal{F}}^2([0,T], \mathcal{R}^{n_u}) \times \mathcal{L}_{\mathcal{F}}^2([0,T], \mathcal{R}^{n_v}) \times \mathcal{R}^n$, there exists a unique solution $x(t) \in \mathcal{L}_{\mathcal{F}}^2([0,T], \mathcal{R}^n)$ to (2.1). In (2.1), we take $z(t) = \begin{bmatrix} x^T(t)C^T(t) & u^T(t)D^T(t) \end{bmatrix}^T$ instead of $z(t) = Cx(t) + Du(t)$ only for simplicity and in order to avoid the presence of cross-term $x(t)u(t)$ in $\|z(t)\|^2$. In fact, if we take $z(t) = Cx(t) + Du(t)$, $\tilde{u} = u + D^T Cx(t)$, then the cross-term will disappear because

$$\begin{aligned} \|z(t)\|^2 &= \|Cx(t)\|^2 + \|u(t)\|^2 + 2x^T(t)C^T Du(t) \\ &= [u + D^T Cx(t)]^T [u + D^T Cx(t)] + x^T(t)(C^T C - C^T D D^T C)x(t) \\ &= \tilde{u}^T(t)\tilde{u}(t) + x^T(t)(C^T C - C^T D D^T C)x(t). \end{aligned}$$

For the same reason, $D^T(t)D(t) = I$ is also not an essential requirement.

2.2.1   Definitions and lemmas

The finite horizon stochastic H_2/H_∞ control problem can be stated as follows, which can be viewed as an extension of the H_2/H_∞ control for deterministic linear systems [113].

DEFINITION 2.1   *Consider system (2.1). For a given disturbance attenuation level $\gamma > 0$, $0 < T < \infty$, find a feedback control $u_T^*(t) \in \mathcal{L}_{\mathcal{F}}^2([0,T], \mathcal{R}^{n_u})$ such that*
1)

$$\begin{aligned} \|\mathcal{L}_T\| &= \sup_{v \in \mathcal{L}_{\mathcal{F}}^2([0,T], \mathcal{R}^{n_v}), v \neq 0, u = u_T^*, x_0 = 0} \frac{\|z\|_{[0,T]}}{\|v\|_{[0,T]}} \\ &:= \sup_{v \in \mathcal{L}_{\mathcal{F}}^2([0,T], \mathcal{R}^{n_v}), v \neq 0, u = u_T^*, x_0 = 0} \frac{\left\{ \mathcal{E} \int_0^T \|z(t)\|^2\,dt \right\}^{1/2}}{\left\{ \mathcal{E} \int_0^T \|v(t)\|^2\,dt \right\}^{1/2}} \\ &= \sup_{v \in \mathcal{L}_{\mathcal{F}}^2([0,T], \mathcal{R}^{n_v}), v \neq 0, u = u_T^*, x_0 = 0} \frac{\left\{ \mathcal{E} \int_0^T (\|C(t)x(t)\|^2 + \|u_T^*(t)\|^2)\,dt \right\}^{1/2}}{\left\{ \mathcal{E} \int_0^T \|v(t)\|^2\,dt \right\}^{1/2}} \\ &< \gamma \end{aligned}$$

where $\mathcal{L}_T$ is an operator associated with the system

$$\begin{cases} dx(t) = [A_1(t)x(t) + B_1(t)u_T^*(t) + C_1(t)v(t)]\, dt + [A_2(t)x(t) + C_2(t)v(t)]\, dB(t), \\ x(0) = 0, \\ z(t) = \begin{bmatrix} C(t)x(t) \\ D(t)u_T^*(t) \end{bmatrix}, \quad D^T(t)D(t) = I, \ t \in [0, T], \end{cases}$$

(2.2)

which is defined as

$$\mathcal{L}_T : \mathcal{L}_\mathcal{F}^2([0, T], \mathcal{R}^{n_v}) \mapsto \mathcal{L}_\mathcal{F}^2([0, T], \mathcal{R}^{n_z}), \quad \mathcal{L}_T(v(t)) = z(t)|_{x_0=0}, \ t \in [0, T].$$

2) When the worst-case disturbance*, $v_T^*(t) \in \mathcal{L}_\mathcal{F}^2([0, T], \mathcal{R}^{n_v})$, is applied to (2.1), u_T^* minimizes the output energy

$$J_{2,T}(u, v_T^*) := \|z(t)\|_{[0,T]}^2 = \mathcal{E} \int_0^T \|z(t)\|^2\, dt$$

$$= \mathcal{E} \int_0^T (\|C(t)x(t)\|^2 + \|u(t)\|^2)\, dt,$$

where v_T^* is defined as

$$v_T^* = \arg\min \left\{ J_{1,T}(u_T^*, v) := \mathcal{E} \int_0^T (\gamma^2 \|v(t)\|^2 - \|z(t)\|^2)\, dt \right\}.$$

If the above (u_T^*, v_T^*) exist, then we say that the finite horizon H_2/H_∞ control is solvable and has a pair of solutions (u_T^*, v_T^*).

REMARK 2.1   If $u_T^* \in \mathcal{L}_\mathcal{F}^2([0, T], \mathcal{R}^{n_u})$ only satisfies Definition 2.1-1) but does not necessarily satisfy Definition 2.1-2), then such a u_T^* is called an H_∞ control for system (2.1).   ∎

REMARK 2.2   When $x_0 \neq 0$, $\|\mathcal{L}_T\|$ is normally defined as

$$\sup_{v \in \mathcal{L}_\mathcal{F}^2([0,T],\mathcal{R}^{n_v}), v \neq 0, u = u_T^*} \frac{\left\{ \mathcal{E} \int_0^T (\|C(t)x(t)\|^2 + \|u_T^*(t)\|^2)\, dt \right\}^{1/2}}{\left\{ x_0^T \Pi x_0 + \mathcal{E} \int_0^T \|v(t)\|^2\, dt \right\}^{1/2}},$$

where $\Pi > 0$ is a measure of the uncertainty for the initial state relative to the uncertainty in $v(\cdot)$; see [189]. Because there is no essential difference between $x_0 = 0$ and $x_0 \neq 0$, we always assume $x_0 = 0$ in the definitions of H_∞ and mixed H_2/H_∞ control.   ∎

---

*$v_T^*(t)$ is the worst-case in the sense that it achieves the maximum possible gain from $v(t)$ to $z(t)$; see [113].

REMARK 2.3   We note in some references such as [37], the mixed H_2/H_∞ control problem is tackled by the decoupled H_∞ and H_2 problems, where the H_2 optimization is investigated under $v(t) \equiv 0$. In this case, the H_2/H_∞ design becomes easier.   ∎

As the stochastic H_2/H_∞ control problem can be formulated as a stochastic LQ two-person non-zero sum game [113], we recall the following definition.

DEFINITION 2.2 (u_T^*, v_T^*) *is called the Nash equilibrium point of a two-person non-zero sum LQ game corresponding to cost functions $J_{1,T}(u,v)$ and $J_{2,T}(u,v)$ if*

$$J_{1,T}(u_T^*, v_T^*) \le J_{1,T}(u_T^*, v), \quad J_{2,T}(u_T^*, v_T^*) \le J_{2,T}(u, v_T^*), \qquad (2.3)$$

$$(u,v) \in \mathcal{L}_\mathcal{F}^2([0,T], \mathcal{R}^{n_u}) \times \mathcal{L}_\mathcal{F}^2([0,T], \mathcal{R}^{n_v}).$$

Obviously, if the above Nash equilibrium point (u_T^*, v_T^*) exists, v_T^* is the worst-case disturbance, while u_T^* not only minimizes $J_{2,T}(u, v_T^*)$ but also makes $\|\mathcal{L}_T\| \le \gamma$ provided $J_{1,T}(u_T^*, v_T^*) \ge 0$ for $x_0 = 0$. In what follows, we will establish an essential relationship between the stochastic H_2/H_∞ control and the existence of Nash equilibrium point (u_T^*, v_T^*) of (2.3). In particular, the pure stochastic H_∞ control problem only needs to solve

$$J_{1,T}(u_T^*, v_T^*) \le J_{1,T}(u_T^*, v),$$

while the pure stochastic H_2 control only needs to solve

$$J_{2,T}(u_T^*, v_T^*) \le J_{2,T}(u, v_T^*).$$

Since Definitions 2.1 and 2.2 are closely related to each other, we call the stochastic H_2/H_∞ control in Definition 2.1 Nash game-based H_2/H_∞ control. To study the finite-time H_2/H_∞ control problem, we first need to establish a finite-time SBRL, which is key to developing the H_∞ theory. Consider the following stochastic perturbed system

$$\begin{cases} dx(t) = [A_{11}(t)x(t) + B_{11}(t)v(t)]\,dt + [A_{12}(t)x(t) + B_{12}(t)v(t)]\,dB(t), \\ x(0) = x_0, \\ z_1(t) = C_{11}(t)x(t), \ t \in [0,T], \end{cases}$$

$$(2.4)$$

where A_{11}, B_{11}, A_{12}, B_{12} and C_{11} are continuous matrix-valued functions of suitable dimensions. Associated with system (2.4), define the perturbation operator $\tilde{\mathcal{L}}_T : \mathcal{L}_\mathcal{F}^2([0,T], \mathcal{R}^{n_v}) \mapsto \mathcal{L}_\mathcal{F}^2([0,T], \mathcal{R}^{n_{z_1}})$ as

$$\tilde{\mathcal{L}}_T(v) = z_1|_{x_0=0} = C_{11}(t)x(t)|_{x_0=0}, \ v \in \mathcal{L}_\mathcal{F}^2([0,T], \mathcal{R}^{n_v}), \ t \in [0,T],$$

then

$$\|\tilde{\mathcal{L}}_T\| = \sup_{v \in \mathcal{L}^2_{\mathcal{F}}([0,T], \mathcal{R}^{n_v}), v \neq 0, x_0 = 0} \frac{\|z_1\|_{[0,T]}}{\|v\|_{[0,T]}}$$

$$= \sup_{v \in \mathcal{L}^2_{\mathcal{F}}([0,T], \mathcal{R}^{n_v}), v \neq 0, x_0 = 0} \frac{\left\{\mathcal{E} \int_0^T \|C_{11} x(t)\|^2 \, dt\right\}^{1/2}}{\left\{\mathcal{E} \int_0^T \|v(t)\|^2 \, dt\right\}^{1/2}}. \tag{2.5}$$

2.2.2 Finite horizon stochastic bounded real lemma (SBRL)

The following lemma is the so-called finite horizon SBRL, which establishes a relationship between the disturbance attenuation problem and the solvability of GDREs, and generalizes Lemma 2.2 of [113] to its stochastic counterpart and Lemma 7 of [36] to systems with state- and disturbance-dependent noises.

LEMMA 2.1
For stochastic system (2.4), $\|\tilde{\mathcal{L}}_T\| < \gamma$ for some $\gamma > 0$ iff the following GDRE (with the time argument t suppressed)

$$\begin{cases} \dot{P} + PA_{11} + A_{11}^T P + A_{12}^T P A_{12} - (PB_{11} + A_{12}^T P B_{12})(\gamma^2 I + B_{12}^T P B_{12})^{-1} \\ \qquad \cdot (B_{11}^T P + B_{12}^T P A_{12}) - C_{11}^T C_{11} = 0, \\ P(T) = 0, \\ \gamma^2 I + B_{12}^T P B_{12} > 0, \ \forall t \in [0, T] \end{cases}$$

$$\tag{2.6}$$

has a unique solution $P_T(t) \leq 0$ on $[0, T]$.

Proof. The sufficiency can be proved in the same way as Lemma 4.2 of [113]. Next, we prove the necessity, i.e., $\|\tilde{\mathcal{L}}_T\| < \gamma$ implies that (2.6) has a solution $P_T(t) \leq 0$ on $[0, T]$. Otherwise, by the standard theory of differential equations, there exists a unique solution $P_T(t)$ backwards in time on a maximal interval $(T_0, T]$, $T_0 \geq 0$, and as $t \to T_0$, $P_T(t)$ becomes unbounded, i.e., (2.6) exhibits the phenomenon of finite-time escape. We shall show that the existence of finite-time escape will lead to a contradiction. Take a sufficiently small $\varepsilon > 0$ with $0 < \varepsilon < T - T_0$, $x(T_0 + \varepsilon) := x_{T_0, \varepsilon} \in \mathcal{R}^n$. For $t \in [T_0 + \varepsilon, T]$, let

$$J_1^T(x, v; x_{T_0, \varepsilon}, T_0 + \varepsilon) := \mathcal{E} \int_{T_0 + \varepsilon}^T (\gamma^2 \|v\|^2 - \|z_1\|^2) \, dt$$

$$= \mathcal{E} \int_{T_0 + \varepsilon}^T (\gamma^2 \|v\|^2 - \|C_{11} x\|^2) \, dt,$$

$$\mathcal{D}(P) = \dot{P} + PA_{11} + A_{11}^T P + A_{12}^T P A_{12} - (PB_{11} + A_{12}^T P B_{12})$$

$$\cdot (\gamma^2 I + B_{12}^T P B_{12})^{-1} (B_{11}^T P + B_{12}^T P A_{12}) - C_{11}^T C_{11},$$

and
$$K_T = -(\gamma^2 I + B_{12}^T P_T B_{12})^{-1}(B_{11}^T P_T + B_{12}^T P_T A_{12}).$$

By Itô's formula together with the technique of completing squares, it follows that

$$
\begin{aligned}
J_1^T(x, v; x_{T_0,\varepsilon}, T_0 + \varepsilon) &= x_{T_0,\varepsilon}^T P_T(T_0 + \varepsilon)x_{T_0,\varepsilon} - \mathcal{E}[x^T(T)P_T(T)x(T)] \\
&\quad + \mathcal{E}\int_{T_0+\varepsilon}^T (\gamma^2\|v\|^2 - \|z_1\|^2)\,dt + \mathcal{E}\int_{T_0+\varepsilon}^T d(x^T P_T x) \\
&= x_{T_0,\varepsilon}^T P_T(T_0 + \epsilon)x_{T_0,\varepsilon} \\
&\quad + \mathcal{E}\int_{T_0+\varepsilon}^T (v - K_T x)^T(\gamma^2 I + B_{12}^T P_T B_{12})(v - K_T x)\,dt \\
&\quad + \mathcal{E}\int_{T_0+\varepsilon}^T x^T \mathcal{D}(P_T)x\,dt \\
&= \mathcal{E}\int_{T_0+\varepsilon}^T (v - K_T x)^T(\gamma^2 I + B_{12}^T P_T B_{12})(v - K_T x)\,dt \\
&\quad + x_{T_0,\varepsilon}^T P_T(T_0 + \varepsilon)x_{T_0,\varepsilon}.
\end{aligned}
$$

So

$$
\begin{aligned}
\min_{v \in \mathcal{L}_\mathcal{F}^2([T_0+\varepsilon, T], \mathcal{R}^{n_v})} J_1^T(x, v; x_{T_0,\varepsilon}, T_0 + \varepsilon) &= J_1^T(x, v^*; x_{T_0,\varepsilon}, T_0 + \varepsilon) \\
&= x_{T_0,\varepsilon}^T P_T(T_0 + \varepsilon)x_{T_0,\varepsilon} \le J_1^T(x, 0; x_{T_0,\varepsilon}, T_0 + \varepsilon) \\
&= \mathcal{E}\int_{T_0+\varepsilon}^T (-\|z_1\|^2)\,dt \le 0, \quad (2.7)
\end{aligned}
$$

where the corresponding optimal $v^*(t)$ is given by

$$
\begin{aligned}
v^*(t) &= K_T(t)x(t) \\
&= -[\gamma^2 I + B_{12}^T(t)P_T(t)B_{12}(t)]^{-1}[B_{11}^T(t)P_T(t) + B_{12}^T(t)P_T(t)A_{12}(t)]x(t).
\end{aligned}
$$

It follows from (2.7) that
$$P_T(T_0 + \varepsilon) \le 0 \qquad (2.8)$$

for any $0 < \varepsilon < T - T_0$. On the other hand, denote by $X_T(t)$ the solution of

$$
\begin{cases}
\dot{X}(t) + X(t)A_{11}(t) + A_{11}^T(t)X(t) + A_{12}^T(t)X(t)A_{12}(t) - C_{11}^T(t)C_{11}(t) = 0, \\
X(T) = 0.
\end{cases}
$$
$$(2.9)$$

For clarity, we denote the solution of (2.4) with $x(t_0) = x_{t_0}$ as $x(t, v, x_{t_0}, t_0)$. By linearity, the solution $x(t, v, x_{T_0,\varepsilon}, T_0 + \varepsilon)$ of system (2.4) satisfies

$$x(t, v, x_{T_0,\varepsilon}, T_0 + \varepsilon) = x(t, 0, x_{T_0,\varepsilon}, T_0 + \varepsilon) + x(t, v, 0, T_0 + \varepsilon),$$

where $x(t, 0, x_{T_0,\varepsilon}, T_0 + \varepsilon)$ is the trajectory of stochastic unperturbed system

$$
\begin{cases}
dx(t) = A_{11}(t)x(t)\,dt + A_{12}(t)x(t)\,dB(t), \\
x(T_0 + \varepsilon) = x_{T_0,\varepsilon}, \quad t \in [T_0 + \varepsilon, T],
\end{cases}
$$
$$(2.10)$$

while $x(t, v, 0, T_0 + \varepsilon)$ is the trajectory of the following stochastic perturbed system with zero initial state

$$\begin{cases} dx(t) = [A_{11}(t)x(t) + B_{11}(t)v(t)]\,dt + [A_{12}(t)x(t) + B_{12}(t)v(t)]\,dB(t), \\ x(T_0 + \epsilon) = x_{T_0,\epsilon} = 0, \ \ t \in [T_0 + \epsilon, T]. \end{cases}$$

It is easy to check that for $x_{T_0,\varepsilon} \in \mathcal{R}^n$,

$$J_1^T(x, v; x_{T_0,\varepsilon}, T_0 + \varepsilon) - J_1^T(x, v; 0, T_0 + \varepsilon)$$

$$= -\mathcal{E} \int_{T_0+\varepsilon}^T \|C_{11}(x(t, 0, x_{T_0,\varepsilon}, T_0 + \varepsilon) + x(t, v, 0, T_0 + \varepsilon))\|^2\,dt$$

$$+\mathcal{E} \int_{T_0+\varepsilon}^T \|C_{11}x(t, v, 0, T_0 + \varepsilon)\|^2\,dt$$

$$= -\mathcal{E} \int_{T_0+\varepsilon}^T \|C_{11}x(t, 0, x_{T_0,\varepsilon}, T_0 + \varepsilon)\|^2\,dt$$

$$-\mathcal{E} \int_{T_0+\varepsilon}^T x^T(t, v, 0, T_0 + \varepsilon)C_{11}^T C_{11}x(t, 0, x_{T_0,\varepsilon}, T_0 + \varepsilon)\,dt$$

$$-\mathcal{E} \int_{T_0+\varepsilon}^T x^T(t, 0, x_{T_0,\varepsilon}, T_0 + \varepsilon)C_{11}^T C_{11}x(t, v, 0, T_0 + \varepsilon)\,dt. \qquad (2.11)$$

Under the constraint of (2.10) and in view of (2.9), by applying Itô's formula, we have

$$-\mathcal{E} \int_{T_0+\varepsilon}^T \|C_{11}(x(t, 0, x_{T_0,\varepsilon}, T_0 + \varepsilon)\|^2\,dt$$

$$= -\mathcal{E} \int_{T_0+\varepsilon}^T \|C_{11}(x(t, 0, x_{T_0,\varepsilon}, T_0 + \varepsilon)\|^2\,dt$$

$$+\mathcal{E} \int_{T_0+\varepsilon}^T d(x^T(t, 0, x_{T_0,\varepsilon}, T_0 + \varepsilon)X_T(t)x(t, 0, x_{T_0,\varepsilon}, T_0 + \varepsilon))$$

$$+x_{T_0,\varepsilon}^T X_T(T_0 + \varepsilon)x_{T_0,\varepsilon} - \mathcal{E}[x^T(T, 0, x_{T_0,\varepsilon}, T_0 + \varepsilon)X_T(T)x(T, 0, x_{T_0,\varepsilon}, T_0 + \varepsilon)]$$

$$= \mathcal{E} \int_{T_0+\varepsilon}^T x^T(t, 0, x_{T_0,\varepsilon}, T_0 + \varepsilon)\mathcal{M}(X_T(t))x(t, 0, x_{T_0,\varepsilon}, T_0 + \varepsilon)\,dt$$

$$+x_{T_0,\varepsilon}^T X_T(T_0 + \varepsilon)x_{T_0,\varepsilon}$$

$$= x_{T_0,\varepsilon}^T X_T(T_0 + \varepsilon)x_{T_0,\varepsilon}, \qquad (2.12)$$

where

$$\mathcal{M}(X_T(t)) = \dot{X}(t) + X(t)A_{11}(t) + A_{11}^T(t)X(t) + A_{12}^T(t)X(t)A_{12}(t) - C_{11}^T(t)C_{11}(t).$$

Similarly, considering $x(T_0 + \varepsilon, v, 0, T_0 + \varepsilon) = 0$ and $X_T(T) = 0$, we have

$$-\mathcal{E} \int_{T_0+\varepsilon}^T x^T(t, v, 0, T_0 + \varepsilon)C_{11}^T C_{11}x(t, 0, x_{T_0,\varepsilon}, T_0 + \varepsilon)\,dt$$

$$= -\mathcal{E}\int_{T_0+\varepsilon}^{T} x^T(t,v,0,T_0+\varepsilon)C_{11}^T C_{11}x(t,0,x_{T_0,\varepsilon},T_0+\varepsilon)\,dt$$

$$+\mathcal{E}\int_{T_0+\varepsilon}^{T} d[x^T(t,v,0,T_0+\varepsilon)X_T(t)x(t,0,x_{T_0,\varepsilon},T_0+\varepsilon)]\,dt$$

$$= \mathcal{E}\int_{T_0+\varepsilon}^{T} x^T(t,v,0,T_0+\varepsilon)\mathcal{M}(X_T(t))x(t,0,x_{T_0,\varepsilon},T_0+\varepsilon)\,dt$$

$$+\mathcal{E}\int_{T_0+\varepsilon}^{T} v^T B_{11}^T X_T(t)x(t,0,x_{T_0,\varepsilon},T_0+\varepsilon)\,dt$$

$$+\mathcal{E}\int_{T_0+\varepsilon}^{T} v^T B_{12}^T X_T(t)A_{12}x(t,0,x_{T_0,\varepsilon},T_0+\varepsilon)\,dt$$

$$= \mathcal{E}\int_{T_0+\varepsilon}^{T} v^T B_{11}^T X_T(t)x(t,0,x_{T_0,\varepsilon},T_0+\varepsilon)\,dt$$

$$+\mathcal{E}\int_{T_0+\varepsilon}^{T} v^T B_{12}^T X_T(t)A_{12}x(t,0,x_{T_0,\varepsilon},T_0+\varepsilon)\,dt \qquad (2.13)$$

and

$$-\,\mathcal{E}\int_{T_0+\varepsilon}^{T} x^T(t,0,x_{T_0,\varepsilon},T_0+\varepsilon)C_{11}^T C_{11}x(t,v,0,T_0+\varepsilon)\,dt$$

$$= \mathcal{E}\int_{T_0+\varepsilon}^{T} x^T(t,0,x_{T_0,\varepsilon},T_0+\varepsilon)X_T(t)B_{11}v\,dt$$

$$+\mathcal{E}\int_{T_0+\varepsilon}^{T} x^T(t,0,x_{T_0,\varepsilon},T_0+\varepsilon)A_{12}^T X_T(t)B_{12}v\,dt. \qquad (2.14)$$

Substituting (2.12)–(2.14) into (2.11) yields

$$J_1^T(x,v;x_{T_0,\varepsilon},T_0+\varepsilon) - J_1^T(x,v;0,T_0+\varepsilon)$$

$$= x_{T_0,\varepsilon}^T X_T(T_0+\varepsilon)x_{T_0,\varepsilon} + \mathcal{E}\int_{T_0+\varepsilon}^{T} v^T B_{11}^T X_T(t)x(t,0,x_{T_0,\varepsilon},T_0+\varepsilon)\,dt$$

$$+\mathcal{E}\int_{T_0+\varepsilon}^{T} x^T(t,0,x_{T_0,\varepsilon},T_0+\varepsilon)X_T(t)B_{11}v\,dt$$

$$+\mathcal{E}\int_{T_0+\varepsilon}^{T} x^T(t,0,x_{T_0,\varepsilon},T_0+\varepsilon)X_T(t)B_{12}v\,dt$$

$$+\mathcal{E}\int_{T_0+\varepsilon}^{T} v^T B_{12}^T X_T(t)A_{12}x(t,0,x_{T_0,\varepsilon},T_0+\varepsilon)\,dt.$$

Take $0 < \epsilon^2 < \gamma^2 - \|\tilde{\mathcal{L}}_T\|^2$, then

$$J_1^T(x,v;0,T_0+\varepsilon) \geq \gamma^2\|\bar{v}\|_{[0,T]}^2 - \|z\|_{[0,T]}^2 \qquad (2.15)$$

$$\geq \epsilon^2\|\bar{v}\|_{[0,T]}^2 = \epsilon^2\|v\|_{[T_0+\varepsilon,T]}^2,$$

where $\bar{v}$ is the extension of v from $[T_0 + \varepsilon, T]$ to $[0, T]$ by setting $\bar{v}(t) = 0$, $\forall t \in [0, T_0 + \varepsilon)$. Therefore, as in [84],

$$J_1^T(x, v; x_{T_0,\varepsilon}, T_0 + \varepsilon)$$

$$\geq \mathcal{E} \int_{T_0+\varepsilon}^{T} [\varepsilon^2 \|v(t)\|^2 + v^T(t) B_{11}^T(t) X_T(t) x(t, 0, x_{T_0,\varepsilon}, T_0 + \varepsilon)$$

$$+ x^T(t, 0, x_{T_0,\varepsilon}, T_0 + \varepsilon) X_T(t) B_{11}(t) v(t)$$

$$+ v^T(t) B_{12}^T X_T(t) A_{12} x(t, 0, x_{T_0,\varepsilon}, T_0 + \varepsilon)$$

$$+ x^T(t, 0, x_{T_0,\varepsilon}, T_0 + \varepsilon) A_{12}^T X_T(t) B_{12}(t) v(t)] \, dt + x_{T_0,\varepsilon}^T X_T(T_0 + \varepsilon) x_{T_0,\varepsilon}$$

$$= \mathcal{E} \int_{T_0+\varepsilon}^{T} \left\| \frac{\varepsilon}{\sqrt{2}} v(t) + \frac{\sqrt{2}}{\varepsilon} B_{11}^T(t) X_T(t) x(t, 0, x_{T_0,\varepsilon}, T_0 + \varepsilon) \right\|^2 dt$$

$$- \mathcal{E} \int_{T_0+\varepsilon}^{T} \left\| \frac{\sqrt{2}}{\varepsilon} B_{11}^T(t) X_T(t) x(t, 0, x_{T_0,\varepsilon}, T_0 + \varepsilon) \right\|^2 dt$$

$$+ \mathcal{E} \int_{T_0+\varepsilon}^{T} \left\| \frac{\varepsilon}{\sqrt{2}} v(t) + \frac{\sqrt{2}}{\varepsilon} B_{12}^T(t) X_T(t) A_{12} x(t, 0, x_{T_0,\varepsilon}, T_0 + \varepsilon) \right\|^2 dt$$

$$- \mathcal{E} \int_{T_0+\varepsilon}^{T} \left\| \frac{\sqrt{2}}{\varepsilon} B_{12}^T(t) X_T(t) A_{12} x(t, 0, x_{T_0,\varepsilon}, T_0 + \varepsilon) \right\|^2 dt$$

$$+ x_{T_0,\varepsilon}^T X_T(T_0 + \varepsilon) x_{T_0,\varepsilon}$$

$$\geq x_{T_0,\varepsilon}^T X_T(T_0 + \varepsilon) x_{T_0,\varepsilon}$$

$$- \mathcal{E} \int_{T_0+\varepsilon}^{T} \left\| \frac{\sqrt{2}}{\varepsilon} B_{11}^T(t) X_T(t) x(t, 0, x_{T_0,\varepsilon}, T_0 + \varepsilon) \right\|^2 dt$$

$$- \mathcal{E} \int_{T_0+\varepsilon}^{T} \left\| \frac{\sqrt{2}}{\varepsilon} B_{12}^T(t) X_T(t) A_{12} x(t, 0, x_{T_0,\varepsilon}, T_0 + \varepsilon) \right\|^2 dt. \tag{2.16}$$

It is well known that there exists $\alpha > 0$ such that

$$\mathcal{E} \int_{T_0+\varepsilon}^{T} \|x(t, 0, x_{T_0,\varepsilon}, T_0 + \varepsilon)\|^2 \, dt \leq \alpha \|x_{T_0,\varepsilon}\|^2.$$

As seen from the above, there exist $\beta, \beta_1, \beta_2 > 0$ satisfying

$$x_{T_0,\varepsilon}^T X_T(T_0 + \varepsilon) x_{T_0,\varepsilon} = -\mathcal{E} \int_{T_0+\varepsilon}^{T} \|C_{11} x(t, 0, x_{T_0,\varepsilon}, T_0 + \varepsilon)\|^2 \, dt$$

$$\geq -\beta \|x_{T_0,\varepsilon}\|^2,$$

$$\mathcal{E} \int_{T_0+\varepsilon}^{T} \left\| \frac{\sqrt{2}}{\varepsilon} B_{11}^T X_T(t) x(t, 0, x_{T_0,\varepsilon}, T_0 + \varepsilon) \right\|^2 dt \leq \beta_1 \|x_{T_0,\varepsilon}\|^2,$$

$$\mathcal{E} \int_{T_0+\varepsilon}^{T} \left\| \frac{\sqrt{2}}{\epsilon} B_{12}^T X_T(t) A_{12} x(t, 0, x_{T_0,\varepsilon}, T_0 + \varepsilon) \right\|^2 dt \leq \beta_2 \|x_{T_0,\varepsilon}\|^2.$$

So from (2.16),

$$J_1^T(x, v; x_{T_0,\varepsilon}, T_0 + \varepsilon) \geq -(\beta + \beta_1 + \beta_2)\|x_{T_0,\varepsilon}\|^2 := -C\|x_{T_0,\varepsilon}\|^2.$$

The above inequality together with (2.7) and (2.8) yields, for any $0 < \varepsilon < T - T_0$,

$$-CI_{n\times n} \leq P_T(T_0 + \varepsilon) \leq 0.$$

So $P_T(T_0 + \varepsilon)$ cannot tend to ∞ as $\varepsilon \to 0$, showing that GDRE (2.6) has a unique solution $P_T(t) \leq 0$ on $[0, T]$. This lemma is proved. $\square$

As said before, Lemma 2.1 generalizes Lemma 7 of [36] to stochastic systems with both state- and disturbance-dependent noises and also generalizes Lemma 2.2 of [113] to the stochastic case. By checking the proof of Lemma 2.2 [113], it can be found that, in order to show that $\|\tilde{\mathcal{L}}_T\| < \gamma$ implies the existence of the solution $P_T(t) \leq 0$ of the differential Riccati equation (DRE)

$$\begin{cases} \dot{P}_T + P_T A_{11} + A_{11}^T P_T - \gamma^{-2} P_T B_{11} B_{11}^T P_T - C_{11}^T C_{11} = 0, \\ P_T(T) = 0, t \in [0, T], \end{cases}$$

the authors of [113] have to establish that the two-point boundary value problem

$$\begin{bmatrix} \dot{x} \\ \dot{\lambda} \end{bmatrix} = \begin{bmatrix} A_{11} & \gamma^{-2} B_{11} B_{11}^T \\ -C_{11}^T C_{11} & -A_{11} \end{bmatrix} \begin{bmatrix} x \\ \lambda \end{bmatrix}, \quad \begin{bmatrix} x(t^*) \\ \lambda(T) \end{bmatrix} = \begin{bmatrix} 0 \\ 0 \end{bmatrix}, \forall t^* \in [0, T] \quad (2.17)$$

has no conjugate points on $[0, T]$. This method, however, cannot be applied to prove Lemma 2.1, because we do not know what form (2.17) should take for stochastic system (2.4).

2.2.3 Finite horizon stochastic LQ control

LQ optimal control was pioneered by R. E. Kalman [107], and has played a central role in modern control theory. Note that the deterministic LQ control theory has been completely established; see the monograph [11]. On the other hand, stochastic LQ control for systems governed by Itô equations was initiated by Wonham [180]. It is well known that when the system state is perturbed by an additive Gaussian white noise, the corresponding quadratic optimal control problem is called "linear quadratic Gaussian (LQG) control," which received much attention in the 1960s [19]. Different from the deterministic LQ or LQG problem, even for indefinite state and control weighting matrices, the stochastic LQ optimal control may still be well posed, which was first found in [45], and has inspired a series of works; see [196] and the references therein.

We first recall the definition of pseudo matrix inverse [85].

DEFINITION 2.3   *Given a matrix $M \in \mathcal{R}^{m \times n}$, the pseudoinverse (Moore–Penrose inverse) of M is the unique matrix $M^+ \in \mathcal{R}^{n \times m}$ such that*

$$\begin{cases} MM^+M = M, \; M^+MM^+ = M^+, \\ (MM^+)^T = MM^+, \; (M^+M)^T = M^+M. \end{cases}$$

In particular, when $M \in \mathcal{S}_n$, the Moore–Penrose inverse has the following properties: (i) $M^+ = (M^+)^T$. (ii) $M \geq 0$ iff $M^+ \geq 0$. (iii) $MM^+ = M^+M$.

The finite horizon stochastic LQ optimal control can be stated as follows: Under the constraint of

$$\begin{cases} dx(t) = [A_{11}(t)x(t) + B_{11}(t)u(t)]\,dt + [A_{12}(t)x(t) + B_{12}(t)u(t)]\,dB(t), \\ x(0) = x_0 \in \mathcal{R}^n, \; t \in [0, T], \end{cases}$$

$$(2.18)$$

find $u \in \mathcal{L}_{\mathcal{F}}^2([0, T], \mathcal{R}^{n_u})$ that minimizes the quadratic performance

$$J_T(0, x_0; u) := \mathcal{E} \int_0^T [x^T(t)Q(t)x(t) + u^T(t)R(t)u(t)]\,dt, \qquad (2.19)$$

where $Q(t) \in \mathcal{C}([0, T]; \mathcal{S}_n)$ and $R(t) \in \mathcal{C}([0, T]; \mathcal{S}_{n_u})$ are continuous symmetric matrix-valued functions. Note that in (2.19), $Q(t)$ and $R(t)$ are indefinite matrices, which are different from the traditional assumptions [180, 23] that $Q(t) \geq 0$ and $R(t) > 0$ on $[0, T]$. Hence, the above is an indefinite stochastic LQ control problem that was first investigated in [45].

DEFINITION 2.4   *The LQ optimization problem (2.18)–(2.19) is called well posed if*

$$V_T(x_0) := \min_{u \in \mathcal{L}_{\mathcal{F}}^2([0,T],\mathcal{R}^{n_u})} J_T(0, x_0; u) > -\infty, \; \forall x_0 \in \mathcal{R}^n.$$

DEFINITION 2.5   *The LQ optimization problem (2.18)–(2.19) is called attainable if it is well posed, and there is $\bar{u}_T^* \in \mathcal{L}_{\mathcal{F}}^2([0, T], \mathcal{R}^{n_u})$ achieving $V(x_0)$, i.e.,*

$$V_T(x_0) = J_T(0, x_0; \bar{u}_T^*).$$

In this case, $\bar{u}_T^*(t)$ is called the optimal control, and the state $x^*(t)$ corresponding to $\bar{u}_T^*(t)$ is called the optimal trajectory.

The indefinite LQ optimal control is associated with the following GDRE (the time argument t is suppressed)

$$\begin{cases} \dot{P} + PA_{11} + A_{11}^T P + A_{12}^T P A_{12} + Q - (PB_{11} + A_{12}^T P B_{12})(R + B_{12}^T P B_{12})^+ \\ \qquad \cdot (B_{11}^T P + B_{12}^T P A_{12}) = 0, \\ P(T) = 0, \\ (R + B_{12}^T P B_{12})(R + B_{12}^T P B_{12})^+ (B_{11}^T P + B_{12}^T P A_{12}) - (B_{11}^T P + B_{12}^T P A_{12}) = 0, \\ R + B_{12}^T P B_{12} \geq 0, \; \text{a.e. } t \in [0, T]. \end{cases}$$

$$(2.20)$$

Note that when $R + B_{12}^T PB_{12} > 0$, GDRE (2.20) reduces to

$$\begin{cases} \dot{P} + PA_{11} + A_{11}^T P + A_{12}^T PA_{12} + Q - (PB_{11} + A_{12}^T PB_{12})(R + B_{12}^T PB_{12})^{-1} \\ \qquad \cdot (B_{11}^T P + B_{12}^T PA_{12}) = 0, \\ P(T) = 0, \\ R + B_{12}^T PB_{12} > 0, \text{a.e. } \forall t \in [0, T]. \end{cases}$$
(2.21)

We state the following well known results on indefinite and standard stochastic LQ controls.

LEMMA 2.2 [4]

(i) *If the GDRE (2.20) admits a real symmetric solution $P(\cdot) \in \mathcal{S}_n$, then the stochastic LQ control problem (2.18)–(2.19) is not only well posed but also attainable. In particular, the optimal cost $V_T(x_0) = x_0^T P(0)x_0$, but the optimal control is not unique, which is generally represented by*

$$\begin{aligned} \bar{u}_T^*(\mathcal{Y}, \mathcal{Z}; t) = &-[(R + B_{12}^T PB_{12})^+(B_{11}^T P + B_{12}^T PA_{12}) + \mathcal{Y}(t) \\ &- (R + B_{12}^T PB_{12})^+(R + B_{12}^T PB_{12})\mathcal{Y}(t)]x(t) \\ &+ \mathcal{Z}(t) - (R + B_{12}^T PB_{12})^+(R + B_{12}^T PB_{12})\mathcal{Z}(t) \end{aligned}$$

for any $\mathcal{Y}(t) \in \mathcal{L}_\mathcal{F}^2([0, T], \mathcal{R}^{n_u \times n})$ and $\mathcal{Z}(t) \in \mathcal{L}_\mathcal{F}^2([0, T], \mathcal{R}^{n_u})$.

(ii) *If there is an optimal linear feedback $\bar{u}_T^*(t) = K(t)x(t)$ to the LQ problem (2.18)–(2.19) with respect to $(0, x_0)$, then GDRE (2.20) must have a real solution $P(\cdot) \in \mathcal{S}_n$, and $\bar{u}_T^*(t) = \bar{u}_T^*(\mathcal{Y}, 0; t)$. Moreover, $\bar{u}_T^*(t) = \bar{u}_T^*(\mathcal{Y}, 0; t)$ is still optimal with respect to any initial value $x(s) = y$ for $(s, y) \in [0, T) \times \mathcal{R}^n$.*

LEMMA 2.3 [23]
For the standard stochastic LQ control, i.e., $Q(t) \geq 0$ and $R(t) > 0$ on $[0, T]$, GDRE (2.21) admits a unique global solution $P(t) \geq 0$ on $[0, T]$. The optimal cost and the unique optimal control law are respectively given by

$$V_T(x_0) = J_T(0, x_0; \bar{u}_T^*) = x_0^T P(0)x_0$$

and

$$\bar{u}_T^*(t) = -[R(t) + B_{12(t)}^T P(t)B_{12}(t)]^{-1}[B_{11}^T(t)P(t) + B_{12}^T(t)P(t)A_{12}(t)]x(t).$$

2.2.4 Conditions for the existence of Nash equilibrium strategies

For clarity, we consider the following system with only the state-dependent noise:

$$\begin{cases} dx(t) = [A_1(t)x(t) + B_1(t)u(t) + C_1(t)v(t)] dt + A_2(t)x(t) dB(t), \\ x(0) = x_0, \\ z(t) = \begin{bmatrix} C(t)x(t) \\ D(t)u(t) \end{bmatrix}, \quad D^T(t)D(t) = I, \, t \in [0, T]. \end{cases}$$
(2.22)

For system (2.22), we present a necessary and sufficient condition for the existence of two-person non-zero sum Nash equilibrium strategies.

LEMMA 2.4
For system (2.22), there exists a linear memoryless state feedback Nash equilibrium strategy

$$u_T^*(t) = K_{2,T}(t)x(t), \quad v_T^*(t) = K_{1,T}(t)x(t),$$

i.e.,

$$J_{1,T}(u_T^*, v_T^*) \leq J_{1,T}(u_T^*, v), \quad \forall v(t) \in \mathcal{L}_{\mathcal{F}}^2([0,T], \mathcal{R}^{n_v}) \qquad (2.23)$$

and

$$J_{2,T}(u_T^*, v_T^*) \leq J_{2,T}(u, v_T^*), \quad \forall u(t) \in \mathcal{L}_{\mathcal{F}}^2([0,T], \mathcal{R}^{n_u}), \qquad (2.24)$$

iff the coupled GDREs

$$\begin{cases} -\dot{P}_{1,T} = A_1^T P_{1,T} + P_{1,T} A_1 + A_2^T P_{1,T} A_2 - C^T C \\ \qquad - \begin{bmatrix} P_{1,T} & P_{2,T} \end{bmatrix} \begin{bmatrix} \gamma^{-2} C_1 C_1^T & B_1 B_1^T \\ B_1 B_1^T & B_1 B_1^T \end{bmatrix} \begin{bmatrix} P_{1,T} \\ P_{2,T} \end{bmatrix} \\ P_{1,T}(T) = 0 \end{cases} \qquad (2.25)$$

and

$$\begin{cases} -\dot{P}_{2,T} = A_1^T P_{2,T} + P_{2,T} A_1 + A_2^T P_{2,T} A_2 + C^T C \\ \qquad - \begin{bmatrix} P_{1,T} & P_{2,T} \end{bmatrix} \begin{bmatrix} 0 & \gamma^{-2} C_1 C_1^T \\ \gamma^{-2} C_1 C_1^T & B_1 B_1^T \end{bmatrix} \begin{bmatrix} P_{1,T} \\ P_{2,T} \end{bmatrix} \\ P_{2,T}(T) = 0 \end{cases} \qquad (2.26)$$

have a solution $(P_{1,T}(\cdot), P_{2,T}(\cdot))$ on $[0,T]$. If the solution of (2.25)–(2.26) exists, then

1)

$$u_T^*(t) = -B_1^T(t) P_{2,T}(t)x(t), \quad v_T^*(t) = -\gamma^{-2} C_1^T(t) P_{1,T}(t)x(t); \quad (2.27)$$

2)

$$J_{1,T}(u_T^*, v_T^*) = x_0^T P_{1,T}(0)x_0, \quad J_{2,T}(u_T^*, v_T^*) = x_0^T P_{2,T}(0)x_0;$$

3)

$$P_{1,T}(t) \leq 0, \quad P_{2,T}(t) \geq 0, \quad t \in [0,T].$$

Proof. Sufficiency: Applying the standard completion of squares argument and Itô's formula, under the constraint of (2.22), we have

$$J_{1,T}(u,v) = x_0^T P_{1,T}(0)x_0 - \mathcal{E}[x^T(T)P_{1,T}(T)x(T)]$$

$$+ \mathcal{E} \int_0^T [(\gamma^2 \|v(t)\|^2 - \|z(t)\|^2)\, dt + d(x^T(t)P_{1,T}(t)x(t))]$$

$$= x_0^T P_{1,T}(0)x_0 + \mathcal{E} \int_0^T \Big[(\gamma^2 \|v(t)\|^2 - \|z(t)\|^2)\, dt + x^T(t)\dot{P}_{1,T}(t)x(t)$$

$$+ dx^T(t)P_{1,T}(t)x(t) + x^T(t)P_{1,T}(t)dx(t) + dx^T(t)P_{1,T}(t)dx(t) \Big].$$

By a series of simple computations and considering (2.25), the above can be written as

$$
\begin{aligned}
J_{1,T}&(u,v) - x_0^T P_{1,T}(0)x_0 \\
&= \mathcal{E} \int_0^T \big[\gamma^2 \| v(t) - v_T^*(t) \|^2 - \| u(t) \|^2 + \| u_T^*(t) \|^2 \\
&\quad + 2x^T(t) P_{1,T}(t) B_1(t)(u(t) - u_T^*(t)) \big]\, dt,
\end{aligned}
\tag{2.28}
$$

where $u_T^*(t)$ and $v_T^*(t)$ are defined by (2.27)–(2.28) yields

$$
J_{1,T}(u_T^*, v) \geq J_{1,T}(u_T^*, v_T^*) = x_0^T P_{1,T}(0)x_0.
$$

Accordingly, the first Nash inequality (2.23) is derived.

In addition, when $v_T^*(t) = -\gamma^{-2} C_1^T(t) P_{1,T}(t)x(t)$ is implemented in (2.22), it becomes

$$
\begin{aligned}
dx(t) &= \big[(A_1(t) - \gamma^{-2} C_1(t) C_1^T(t) P_{1,T}(t))x(t) + B_1(t)u(t) \big]\, dt \\
&\quad + A_2(t)x(t)\, dB(t).
\end{aligned}
\tag{2.29}
$$

Under the constraint of (2.29), to minimize $J_{2,T}(u, v_T^*)$ is a standard stochastic LQ optimization problem. Considering Lemma 2.3 and GDRE (2.26), we have

$$
\begin{aligned}
&\min_{u \in \mathcal{L}_{\mathcal{F}}^2([0,T], \mathcal{R}^{n_u}),\, x(0)=x_0} \left\{ J_{2,T}(u, v_T^*) = \mathcal{E} \int_0^T \| z(t) \|^2\, dt \right\} \\
&= \min_{u \in \mathcal{L}_{\mathcal{F}}^2([0,T], \mathcal{R}^{n_u}),\, x(0)=x_0} \mathcal{E} \int_0^T \big(\| Cx(t) \|^2 + \| u(t) \|^2 \big)\, dt \\
&= J_{2,T}(u_T^*, v_T^*) = x_0^T P_{2,T}(0)x_0.
\end{aligned}
\tag{2.30}
$$

Hence, the second Nash inequality (2.24) is derived. The sufficiency is thus proved.

Necessity: Implement $u_T^*(t) = K_{2,T}(t)x(t)$ in (2.22), then

$$
\begin{cases}
dx(t) = \{ [A_1(t) + B_1(t)K_{2,T}(t)]x(t) + C_1(t)v(t) \}\, dt + A_2(t)x(t)\, dB(t), \\
x(0) = x_0, \\
z(t) = \begin{bmatrix} C(t)x(t) \\ D(t)K_{2,T}(t)x(t) \end{bmatrix}, \quad D^T(t)D(t) = I.
\end{cases}
\tag{2.31}
$$

By (2.23), $v_T^*(t) = K_{1,T}(t)x(t)$ minimizes $J_{1,T}(u_T^*, v)$. By means of Lemma 2.2-(ii), where $R(t) = \gamma^2 I$, $Q(t) = -[C^T(t)C(t) + K_{2,T}^T(t)K_{2,T}(t)]$, the following GDRE

$$
\begin{cases}
\dot{P} + P(A_1 + B_1 K_{2,T}) + (A_1 + B_1 K_{2,T})^T P + A_2^T P A_2 - \gamma^{-2} P C_1 C_1^T P \\
\quad - C^T C - K_{2,T}^T K_{2,T} = 0, \\
P(T) = 0
\end{cases}
\tag{2.32}
$$

has a solution $P_{1,T}(t)$ on $[0, T]$. Moreover,

$$v_T^*(t) = K_{1,T}(t)x(t) = -\gamma^{-2}C_1^T(t)P_{1,T}(t)x(t)$$

is as in (2.27), and

$$J_{1,T}(u_T^*, v_T^*) = x_0^T P_{1,T}(0)x_0. \tag{2.33}$$

Substituting $v = v_T^* = -\gamma^{-2}C_1^T(t)P_{1,T}(t)x(t)$ into (2.22) yields

$$\begin{cases} dx(t) = \left\{ [A_1(t) - \gamma^{-2}C_1(t)C_1^T(t)P_{1,T}(t)]x(t) + B_1(t)u(t) \right\} dt \\ \qquad\quad + A_2(t)x(t)\, dB(t), \\ x(0) = x_0, \\ z(t) = \begin{bmatrix} C(t)x(t) \\ D(t)u(t) \end{bmatrix}, \quad D^T(t)D(t) = I. \end{cases} \tag{2.34}$$

How to minimize $J_{2,T}(u, v_T^*)$ under the constraint of (2.34) is a standard optimal regulator problem. By Lemma 2.3, the following GDRE

$$\begin{cases} \dot{P} + P(A_1 - \gamma^{-2}C_1C_1^T P) + (A_1 - \gamma^{-2}C_1C_1^T P)^T P + A_2^T P A_2 - PB_1B_1^T P \\ \qquad + C^T C = 0, \\ P(T) = 0 \end{cases}$$
$$\tag{2.35}$$

has a positive semi-definite solution $P_{2,T}(t) \geq 0$ on $[0, T]$, which is the same with (2.26). Moreover,

$$u_T^*(t) = K_{2,T}(t)x(t) = -B_1^T(t)P_{2,T}(t)x(t)$$

and

$$J_{2,T}(u_T^*, v_T^*) = x_0^T P_{2,T}(0)x_0.$$

Substituting $K_{2,T}(t) = -B_1^T(t)P_{2,T}(t)$ into (2.32), (2.25) is obtained. The rest is to show $P_{1,T}(t) \leq 0$ on $[0, T]$, which can be seen from (2.33) and (2.23) that for any $x_0 \in \mathcal{R}^n$,

$$J_{1,T}(u_T^*, v_T^*) = x_0^T P_{1,T}(0)x_0 \leq J_{1,T}(u_T^*, 0) \leq 0.$$

The proof of this theorem is complete. □

2.2.5   Main results

We first consider system (2.22) and give its finite horizon H_2/H_∞ control as follows:

THEOREM 2.1
The following three statements are equivalent:

  (i) The finite horizon H_2/H_∞ control of (2.22) has a solution (u_T^*, v_T^*) as

$$u_T^*(t) = K_{2,T}(t)x(t)$$

    and

$$v_T^*(t) = K_{1,T}(t)x(t).$$

(ii) There exists a linear memoryless state feedback Nash equilibrium strategy $u_T^*(t)$ and $v_T^*(t)$ for (2.23)–(2.24).

(iii) The coupled GDREs (2.25)–(2.26) have a solution $(P_{1,T}(\cdot), P_{2,T}(\cdot))$ on $[0, T]$.

Proof. (ii) $\Leftrightarrow$ (iii) is shown in Lemma 2.4. So, to prove this theorem, we only need to show (i) $\Leftrightarrow$ (ii). From Lemma 2.4, (ii) implies that

$$\|\mathcal{L}_T\| \le \gamma, \quad J_{2,T}(u_T^*, v_T^*) \le J_{2,T}(u, v_T^*), \quad \forall u(t) \in \mathcal{L}_{\mathcal{F}}^2([0, T], \mathcal{R}^{n_u}), \qquad (2.36)$$

where u_T^* and v_T^* are defined by (2.27). In order to show (ii) $\Rightarrow$ (i), it suffices to show that $\|\mathcal{L}_T\| \le \gamma$ in (2.36) can be replaced by $\|\mathcal{L}_T\| < \gamma$.

Following a similar line of argument as in [113], define an operator

$$\mathcal{L}_0 : \mathcal{L}_{\mathcal{F}}^2([0, T], \mathcal{R}^{n_v}) \mapsto \mathcal{L}_{\mathcal{F}}^2([0, T], \mathcal{R}^{n_v})$$

as

$$\mathcal{L}_0 v(t) = v(t) - v_T^*(t)$$

with the realization

$$dx(t) = [(A_1 - B_1 B_1^T P_{2,T})x(t) + C_1 v(t)]\, dt + A_2 x(t) dB(t), \quad x(0) = 0$$

and

$$v(t) - v_T^*(t) = v(t) + \gamma^{-2} C_1^T P_{1,T} x(t).$$

Then $\mathcal{L}_0^{-1}$ exists, which is determined by

$$\begin{cases} dx(t) = (A - B_1 B_1^T P_{2,T} - \gamma^{-2} C_1 C_1^T P_{1,T})x(t)\, dt \\ \qquad\quad + C_1(v(t) - v_T^*(t))\, dt + A_2 x(t) dB(t), \\ x(0) = 0, \\ v(t) = -\gamma^{-2} C_1^T P_{1,T} x(t) + (v(t) - v_T^*(t)). \end{cases}$$

From (2.28), we have

$$\gamma^2 \|v\|_{[0,T]}^2 - \|z\|_{[0,T]}^2 = \gamma^2 \|\mathcal{L}_0 v\|_{[0,T]}^2 \ge \varepsilon \|v\|_{[0,T]}^2$$

for some sufficiently small $\varepsilon > 0$, which yields $\|\mathcal{L}_T\| < \gamma$. (ii) $\Rightarrow$ (i) is complete.

(i) $\Rightarrow$ (ii): By Definition 2.1 and keeping the definition of v_T^* in mind, (i) $\Rightarrow$ (ii) is obvious.

Summarizing the above, the proof is complete. $\square$

Theorem 2.1 reveals that the solvability of H_2/H_∞ control of system (2.22) is equivalent to the existence of a Nash equilibrium strategy (2.23)–(2.24) or the existence of solutions to coupled GDREs (2.25)–(2.26). Repeating the same procedure as in Theorem 2.1, we are able to obtain the following theorem for a slightly more general system with state- and control-dependent noises:

$$\begin{cases} dx(t) = [A_1(t)x(t) + B_1(t)u(t) + C_1(t)v(t)]\, dt + [A_2(t)x(t) + B_2(t)u(t)]\, dB(t), \\ z(t) = \begin{bmatrix} C(t)x(t) \\ D(t)u(t) \end{bmatrix}, \quad D^T(t)D(t) = I, \ t \in [0, T]. \end{cases}$$

$$(2.37)$$

REMARK 2.4   In view of Theorem 4.5 [113], we guess that (ii) of Theorem 2.1 can be replaced with a weaker condition. That is, we may only assume that $u_T^*(t) = u(t,x)$ and $v_T^*(t) = v(t,x)$ are continuous on $t \in [0,T]$ and have power series expansions in the neighborhood of $x = 0$, but may not necessarily be a linear memoryless state feedback Nash equilibrium strategy. This conjecture needs to be further verified. ∎

THEOREM 2.2

For system (2.37), the following three statements are equivalent:
(i) The finite horizon H_2/H_∞ control has a solution (u_T^*, v_T^*) with

$$u_T^* = K_{2,T}(t)x(t)$$

and

$$v_T^* = K_{1,T}(t)x(t).$$

(ii) There exists a linear memoryless state feedback Nash equilibrium strategy $(u_T^*(t), v_T^*(t))$ for (2.23)–(2.24).
(iii) The coupled GDREs

$$\begin{cases} -\dot{P}_{1,T} = P_{1,T}(A_1 + B_1 K_{2,T}) + (A_1 + B_1 K_{2,T})^T P_{1,T} + (A_2 + B_2 K_{2,T})^T P_{1,T} \\ \qquad \cdot (A_2 + B_2 K_{2,T}) - C^T C - K_{2,T}^T K_{2,T} - \gamma^{-2} P_{1,T} C_1 C_1^T P_{1,T}, \\ K_{2,T} = -(I + B_2^T P_{2,T} B_2)^{-1}(B_1^T P_{2,T} + B_2^T P_{2,T} A_2), \\ P_{1,T}(T) = 0 \end{cases}$$

$$(2.38)$$

and

$$\begin{cases} -\dot{P}_{2,T} = P_{2,T}(A_1 + C_1 K_{1,T}) + (A_1 + C_1 K_{1,T})^T P_{2,T} + C^T C \\ \qquad + A_2^T P_{2,T} A_2 - (P_{2,T} B_1 + A_2^T P_{2,T} B_2)(I + B_2^T P_{2,T} B_2)^{-1} \\ \qquad \cdot (B_1^T P_{2,T} + B_2^T P_{2,T} A_2), \\ K_{1,T} = -\gamma^{-2} C_1^T P_{1,T}, \\ P_{2,T}(T) = 0, \\ I + B_2^T P_{2,T} B_2 > 0, \ \forall t \in [0,T] \end{cases}$$

$$(2.39)$$

have a solution $(P_{1,T}(\cdot), P_{2,T}(\cdot))$ on $[0,T]$. If the solution of GDREs (2.38)–(2.39) exists, then
1.

$$u_T^*(t) = K_{2,T}(t)x(t), \quad v_T^*(t) = K_{1,T}(t)x(t);$$

2.

$$J_{1,T}(u_T^*, v_T^*) = x_0^T P_{1,T}(0)x_0, \quad J_{2,T}(u_T^*, v_T^*) = x_0^T P_{2,T}(0)x_0;$$

3.

$$P_{1,T}(t) \leq 0, \quad P_{2,T}(t) \geq 0, \quad t \in [0,T].$$

Proof. (Key sketch) One only needs to note that GDRE (2.38) corresponds to the indefinite LQ control problem

$$\min_{v \in \mathcal{L}_{\mathcal{F}}^2([0,T], \mathcal{R}^{n_v})} J_{1,T}(u_T^* = K_{2,T} x, v) \tag{2.40}$$

subject to

$$\begin{cases} dx(t) = [(A_1(t) + B_1(t)K_{2,T}(t))x(t) + C_1(t)v(t))]\, dt \\ \qquad + [A_2(t) + B_2(t)K_{2,T}(t)]x(t)\, dB(t), \\ x(0) = x_0, \\ z(t) = \begin{bmatrix} C(t)x(t) \\ D(t)K_{2,T}(t)x(t) \end{bmatrix}, \quad D^T(t)D(t) = I, \end{cases} \tag{2.41}$$

while GDRE (2.39) corresponds to the standard LQ control problem

$$\min_{u \in \mathcal{L}_{\mathcal{F}}^2([0,T], \mathcal{R}^{n_u})} J_{2,T}(u, v_T^* = K_{1,T} x) \tag{2.42}$$

subject to

$$\begin{cases} dx(t) = \{[A_1(t) + C_1(t)K_{1,T}(t)]x(t) + B_1(t)u(t)\}\, dt + [A_2(t)x(t) + B_2(t)u(t)]\, dB(t), \\ x(0) = x_0, \\ z(t) = \begin{bmatrix} C(t)x(t) \\ D(t)u(t) \end{bmatrix}, \quad D^T(t)D(t) = I. \end{cases} \tag{2.43}$$

The rest is similar to the proof of Theorem 2.1. □

Although for systems (2.22) and (2.37), we have (i)⇔ (ii) ⇔ (iii) in Theorems 2.1–2.2, it is not the case for system (2.1). It can be shown that, for (2.1), (i)⇔ (iii), but (i) and (iii) are not equivalent to (ii). Lemma 2.1 plays an important role in proving the following theorem.

THEOREM 2.3
For system (2.1), its finite horizon H_2/H_∞ control has solution (u_T^*, v_T^*) with

$$u_T^* = K_{2,T}(t)x(t)$$

and

$$v_T^* = K_{1,T}(t)x(t)$$

iff the coupled GDREs

$$\begin{cases} -\dot{P}_{1,T} = P_{1,T}(A_1 + B_1 K_{2,T}) + (A + B_1 K_{2,T})^T P_{1,T} + A_2^T P_{1,T} A_2 - C^T C \\ \quad - K_{2,T}^T K_{2,T} - (P_{1,T}C_1 + A_2^T P_{1,T} C_2)(\gamma^2 I + C_2^T P_{1,T} C_2)^{-1}(C_1^T P_{1,T} + C_2^T P_{1,T} A_2), \\ K_{2,T} = -B_1^T P_{2,T}, \\ P_{1,T}(T) = 0, \\ \gamma^2 I + C_2^T P_{1,T} C_2 > 0, \ \forall t \in [0,T] \end{cases} \tag{2.44}$$

and

$$\begin{cases} -\dot{P}_{2,T} = P_{2,T}(A_1 + C_1 K_{1,T}) + (A_1 + C_1 K_{1,T})^T P_{2,T} + C^T C \\ \quad + (A_2 + C_2 K_{1,T})^T P_{2,T}(A_2 + C_2 K_{1,T}) - P_{2,T} B_1 B_1^T P_{2,T}, \\ K_{1,T} = -(\gamma^2 I + C_2^T P_{1,T} C_2)^{-1}(C_1^T P_{1,T} + C_2^T P_{1,T} A_2), \\ P_{2,T}(T) = 0 \end{cases} \tag{2.45}$$

have a solution $(P_{1,T}(\cdot), P_{2,T}(\cdot))$ on $[0,T]$. If the solution of GDREs (2.44)–(2.45) exists, then

1.

$$u_T^*(t) = K_{2,T}(t)x(t), \quad v_T^*(t) = K_{1,T}(t)x(t);$$

2.

$$J_{1,T}(u_T^*, v_T^*) = x_0^T P_{1,T}(0)x_0,$$
$$J_{2,T}(u_T^*, v_T^*) = x_0^T P_{2,T}(0)x_0;$$

3.

$$P_{1,T}(t) \le 0, \quad P_{2,T}(t) \ge 0, \quad t \in [0,T].$$

Proof. Sufficiency: We note that GDRE (2.44) corresponds to the indefinite LQ control:

$$\min_{v \in \mathcal{L}_{\mathcal{F}}^2([0,T], \mathcal{R}^{n_v})} J_{1,T}(u_T^* = K_{2,T}x, v) \tag{2.46}$$

subject to

$$\begin{cases} dx(t) = \{[A_1(t) + B_1(t)K_{2,T}]x(t) + C_1(t)v(t)\}\, dt + [A_2(t)x(t) + C_2(t)v(t)]\, dB(t), \\ x(0) = x_0, \\ z(t) = \begin{bmatrix} C(t)x(t) \\ D(t)K_{2,T}(t)x(t) \end{bmatrix}, \quad D^T(t)D(t) = I. \end{cases}$$
$$\tag{2.47}$$

Using Lemma 2.2, for any $x_0 \in \mathcal{R}^n$, we have

$$x_0^T P_{1,T}(0)x_0 = J_{1,T}(u_T^*, v_T^*) \le J_{1,T}(u_T^*, v), \forall v \in \mathcal{L}_{\mathcal{F}}^2([0,T], \mathcal{R}^{n_v}). \tag{2.48}$$

Similarly, GDRE (2.45) corresponds to the standard LQ control problem

$$\min_{u \in \mathcal{L}_{\mathcal{F}}^2([0,T], \mathcal{R}^{n_u})} J_{2,T}(u, v_T^*) \tag{2.49}$$

subject to

$$\begin{cases} dx(t) = \{[A_1(t) + C_1(t)K_{1,T}]x(t) + B_1(t)u(t)\}\, dt + [A_2(t) + C_2(t)K_{1,T}]x(t)\, dB(t), \\ x(0) = x_0, \\ z(t) = \begin{bmatrix} C(t)x(t) \\ D(t)u(t) \end{bmatrix}, \quad D^T(t)D(t) = I. \end{cases}$$
$$\tag{2.50}$$

For the LQ optimization (2.49)–(2.50), using Lemma 2.3, it holds that

$$x_0^T P_{2,T}(0)x_0 = J_{2,T}(u_T^*, v_T^*) \le J_{2,T}(u, v_T^*), \forall u \in \mathcal{L}_{\mathcal{F}}^2([0,T], \mathcal{R}^{n_u}). \tag{2.51}$$

The inequalities (2.48) and (2.51) imply that (u_T^*, v_T^*) is a two-person non-zero sum Nash equilibrium point. To show that (u_T^*, v_T^*) solves the H_2/H_∞ control problem, it only needs to prove $\|\mathcal{L}_T\| < \gamma$, which can be shown along the line of Theorem 2.1.

Necessity: By part 1) of Definition 2.1, associated with system (2.47), $\|\mathcal{L}_T\| < \gamma$. By Lemma 2.1, the GDRE (2.44) admits a solution $P_{1,T}(t) \le 0$ on $[0,T]$. By Lemma 2.3, (2.45) must have a unique solution $P_{2,T}(t) \ge 0$ on $[0,T]$. $\square$

THEOREM 2.4

For system (2.1), there exists a linear memoryless state feedback Nash equilibrium strategy $(u_T^*(t), v_T^*(t))$ for (2.23)–(2.24) iff the following two coupled GDREs

$$
\begin{cases}
-\dot{P}_{1,T} = P_{1,T}(A_1 + B_1 K_{2,T}) + (A_1 + B_1 K_{2,T})^T P_{1,T} + A_2^T P_{1,T} A_2 - C^T C \\
\quad -K_{2,T}^T K_{2,T} - (P_{1,T} C_1 + A_2^T P_{1,T} C_2)(\gamma^2 I + C_2^T P_{1,T} C_2)^+ (C_1^T P_{1,T} + C_2^T P_{1,T} A_2), \\
(\gamma^2 I + C_2^T P_{1,T} C_2)(\gamma^2 I + C_2^T P_{1,T} C_2)^+ (C_1^T P_{1,T} + C_2^T P_{1,T} A_2) \\
\qquad\qquad\qquad\qquad\qquad\qquad\qquad\qquad -(C_1^T P_{1,T} + C_2^T P_{1,T} A_2) = 0, \\
K_{2,T} = -B_1^T P_{2,T}, \\
P_{1,T}(T) = 0, \\
\gamma^2 I + C_2^T P_{1,T} C_2 \geq 0, \forall t \in [0, T]
\end{cases}
\tag{2.52}
$$

and

$$
\begin{cases}
-\dot{P}_{2,T} = P_{2,T}(A_1 + C_1 K_{1,T}) + (A_1 + C_1 K_{1,T})^T P_{2,T} + C^T C \\
\quad + (A_2 + C_2 K_{1,T})^T P_{2,T}(A_2 + C_2 K_{1,T}) - P_{2,T} B_1 B_1^T P_{2,T}, \\
K_{1,T} = -(\gamma^2 I + C_2^T P_{1,T} C_2)^+ (C_1^T P_{1,T} + C_2^T P_{1,T} A_2), \\
P_{2,T} = 0
\end{cases}
\tag{2.53}
$$

have a solution $(P_{1,T}(\cdot), P_{2,T}(\cdot))$ on $[0, T]$.

Proof. By Lemma 2.2-(i), if (2.52) has a solution $P_{1,T}(\cdot)$, then $v_T^* = K_{1,T} x$ solves the indefinite stochastic LQ control (2.46)–(2.47). Hence, the first Nash inequality (2.23) holds. Additionally, by Lemma 2.3, $u_T^* = -B_1^T P_{2,T} x$ is an optimal control for the standard LQ (2.49)–(2.50). Hence, the second Nash inequality (2.24) is derived. The sufficiency part is complete.

Applying Lemma 2.2-(ii) and Lemma 2.3, the necessity can be proved. □

Obviously, when $\gamma^2 I + C_2^T(t) P_{1,T}(t) C_2(t) > 0$ on $[0, T]$, then GDREs (2.52)–(2.53) reduce to GDREs (2.44)–(2.45).

REMARK 2.5   From the above theorems, we know that, different from systems (2.22) and (2.37), the solvability of the H_2/H_∞ control problem is not equivalent to the existence of a Nash equilibrium point for system (2.1). This is due to the fact that v enters into the diffusion term. It can also be seen that when the diffusion term contains external disturbances, $\|\mathcal{L}_T\| < \gamma$ is not equivalent to $\|\mathcal{L}_T\| \leq \gamma$ or $\|z(t)\|_{[0,T]} \leq \gamma \|v(t)\|_{0,T]}$.   ∎

Repeating the same steps as above, we can even present a result for the H_2/H_∞ control of the following more general system with state-, control- and disturbance-dependent noises

$$
\begin{cases}
dx(t) = [A_1(t)x(t) + B_1(t)u(t) + C_1(t)v(t)]\, dt + [A_2(t)x(t) + B_2(t)u(t) \\
\qquad\qquad + C_2(t)v(t)]\, dB(t), \\
x(0) = x_0, \\
z(t) = \begin{bmatrix} C(t)x(t) \\ D(t)u(t) \end{bmatrix}, \ D^T(t)D(t) = I.
\end{cases}
\tag{2.54}
$$

Different from systems (2.1), (2.22) and (2.37), we should involve four GDREs instead of two in the H_2/H_∞ design for system (2.54). We list the following result but the proof is omitted.

THEOREM 2.5

For system (2.54), its finite horizon H_2/H_∞ control has a solution (u_T^*, v_T^*) with

$$u_T^* = K_{2,T}(t)x(t)$$

and

$$v_T^* = K_{1,T}(t)x(t)$$

iff the following four coupled GDREs

$$
\begin{cases}
-\dot{P}_{1,T} = P_{1,T}(A_1 + B_1 K_{2,T}) + (A_1 + B_1 K_{2,T})^T P_{1,T} + (A_2 + B_2 K_{2,T})^T P_{1,T} \\
\quad \cdot (A_2 + B_2 K_{2,T}) - C^T C - K_{2,T}^T K_{2,T} - [P_{1,T} C_1 + (A_2 + B_2 K_{2,T})^T P_{1,T} C_2] \\
\quad \cdot (\gamma^2 I + C_2^T P_{1,T} C_2)^{-1} [C_1^T P_{1,T} + C_2^T P_{1,T}(A_2 + B_2 K_{2,T})], \\
P_{1,T}(T) = 0, \\
\gamma^2 I + C_2^T P_{1,T} C_2 > 0, \ \forall t \in [0, T],
\end{cases}
$$

$$(2.55)$$

$$
\begin{cases}
-\dot{P}_{2,T} = P_{2,T}(A_1 + C_1 K_{1,T}) + (A_1 + C_1 K_{1,T})^T P_{2,T} + (A_2 + C_2 K_{1,T})^T P_{2,T} \\
\quad \cdot (A_2 + C_2 K_{1,T}) + C^T C - [P_{2,T} B_1 + (A_2 + C_2 K_{1,T})^T P_{2,T} B_2](I + B_2^T P_{2,T} B_2)^{-1} \\
\quad \cdot [B_1^T P_{2,T} + B_2^T P_{2,T}(A_2 + C_2 K_{1,T})], \\
P_{2,T}(T) = 0, \\
I + B_2^T P_{2,T} B_2 > 0, \ \forall t \in [0, T],
\end{cases}
$$

$$(2.56)$$

$$K_{1,T} = -(\gamma^2 I + C_2^T P_{1,T} C_2)^{-1} [C_1^T P_{1,T} + C_2^T P_{1,T}(A_2 + B_2 K_{2,T})], \quad (2.57)$$

and

$$K_{2,T} = -(I + B_2^T P_{2,T} B_2)^{-1} [B_1^T P_{2,T} + B_2^T P_{2,T}(A_2 + C_2 K_{1,T})] \quad (2.58)$$

have a solution $(P_{1,T} \leq 0, K_{1,T}; P_{2,T} \geq 0, K_{2,T})$.

REMARK 2.6   Strictly speaking, the equations (2.55)–(2.58) cannot be called coupled GDREs; this is because (2.57) and (2.58) are not differential equations. Here, we call the equations (2.55)–(2.58) four coupled GDREs in order to adopt the same name with the classical DRE. Maybe a more appropriate name to describe equations (2.55)–(2.58) is coupled differential-algebraic equations.   ∎

REMARK 2.7   We have to admit that it is not an easy task to solve the above cross-coupled GDREs. For the deterministic H_2/H_∞ control, i.e.,

$A_2 \equiv 0$ in (2.22), the cross-coupled GDREs (2.25)–(2.26) reduce to

$$
\begin{cases}
-\dot{P}_{1,T} = A_1^T P_{1,T} + P_{1,T} A_1 - C^T C \\
\qquad - \begin{bmatrix} P_{1,T} & P_{2,T} \end{bmatrix} \begin{bmatrix} \gamma^{-2} C_1 C_1^T & B_1 B_1^T \\ B_1 B_1^T & B_1 B_1^T \end{bmatrix} \begin{bmatrix} P_{1,T} \\ P_{2,T} \end{bmatrix}, \\
P_{1,T}(T) = 0
\end{cases}
\tag{2.59}
$$

and

$$
\begin{cases}
-\dot{P}_{2,T} = A_1^T P_{2,T} + P_{2,T} A_1 + C^T C \\
\qquad - \begin{bmatrix} P_{1,T} & P_{2,T} \end{bmatrix} \begin{bmatrix} 0 & \gamma^{-2} C_1 C_1^T \\ \gamma^{-2} C_1 C_1^T & B_1 B_1^T \end{bmatrix} \begin{bmatrix} P_{1,T} \\ P_{2,T} \end{bmatrix}, \\
P_{2,T}(T) = 0.
\end{cases}
\tag{2.60}
$$

The reference [113] presented a Runge–Kutta integration procedure to solve GDREs (2.59)–(2.60) for a scalar system. Of course, the Runge–Kutta integration procedure can also be used to solve GDREs (2.25)–(2.26), but it is by no means easy for high-order stochastic systems. Hence, it is necessary to search for valuable numerical algorithms for the solutions of the given cross-coupled GDREs (2.55)–(2.58), which is a key difficulty in designing H_2/H_∞ controllers. ∎

In the following, we present a discretization method to solve (2.55)–(2.58) approximately. Set $h = \frac{T}{n}$ for a natural number $n > 0$, and denote $t_i = ih$ with $i = 0, 1, 2, \cdots, n$. When n is sufficiently large, or equivalently, when h is sufficiently small, we may replace $\dot{P}_{1,T}(t_{i+1})$ and $\dot{P}_{2,T}(t_{i+1})$ with $\frac{P_{1,T}(t_i)-P_{1,T}(t_{i+1})}{-h}$ and $\frac{P_{2,T}(t_i)-P_{2,T}(t_{i+1})}{-h}$ in (2.55) and (2.56), respectively. Then a backward recursive algorithm can be given as follows:

(i) By solving (2.57) and (2.58), it follows that $K_{1,T}(T) = K_{2,T}(T) = 0$ from the given terminal condition $P_{1,T}(T) = P_{2,T}(T) = 0$.

(ii) Solving (2.55) and (2.56) yields $P_{1,T}(t_{n-1}) = P_{1,T}(T-h) = -hC^T(T)C(T) \le 0$ and $P_{2,T}(t_{n-1}) = P_{2,T}(T-h) = hC^T(T)C(T) \ge 0$.

(iii) Repeating the above steps (i)-(ii), $P_{1,T}(t_i)$ and $P_{2,T}(t_i)$ may be computed if $P_{1,T}(t_{i+1}) \le 0$ and $P_{2,T}(t_{i+1}) \ge 0$ are available with

$$\gamma^2 I + C_2^T(t_{i+1}) P_{1,T}(t_{i+1}) C_2(t_{i+1}) > 0$$

and

$$I + B_2^T(t_{i+1}) P_{2,T}(t_{i+1}) B_2(t_{i+1}) > 0, \ i = n, n-1, \cdots, 0.$$

The above recursions can be continued forever if $P_{1,T}(t_i) \le 0$, $P_{2,T}(t_i) \ge 0$, $\gamma^2 I + C_2^T(t_i) P_{1,T}(t_i) C_2(t_i) > 0$, $I + B_2^T(t_i) P_{2,T}(t_i) B_2(t_i) > 0$, $i = 1, 2, \cdots, n$. Because if the coupled GDREs (2.55)–(2.58) admit a pair of solutions ($\bar{P}_{1,T} \le$

$0, \bar{P}_{2,T} \geq 0$), then $\bar{P}_{1,T}(t)$ and $\bar{P}_{2,T}(t)$ must be uniformly continuous on $[0, T]$. Therefore, we have

$$\lim_{h \to 0} \max_{t_i \leq t < t_{i+1}, i=0,1,\cdots,n-1} \{|\bar{P}_{2,T}(t) - P_{2,T}(t_i)|, |\bar{P}_{1,T}(t) - P_{1,T}(t_i)|\} = 0.$$

For some special systems, we may solve (2.55)-(2.58) analytically; see the following example.

Example 2.1
We consider a scalar and time-invariant case of system (2.22):

$$\begin{cases} dx(t) = [2x(t) + 3u(t) + v(t)]\,dt + x(t)\,dB(t), \\ z(t) = \begin{bmatrix} 3x(t) \\ u(t) \end{bmatrix}, \quad t \in [0, 2]. \end{cases} \tag{2.61}$$

Note that the above state equation is a stochastic version of the equation (5.82) of [113] via adding "$x(t)\,dB(t)$." Then the coupled GDREs (2.25)–(2.26) become

$$\begin{cases} -\dot{p}_{1,2}(t) = 5p_{1,2}(t) - 9 - \gamma^{-2}p_{1,2}^2(t) - 9p_{2,2}^2(t) - 18p_{1,2}(t)p_{2,2}(t), \\ p_{1,2}(2) = 0 \end{cases} \tag{2.62}$$

and

$$\begin{cases} -\dot{p}_{2,2}(t) = 5p_{2,2}(t) + 9 - 9p_{2,2}^2(t) - 2\gamma^{-2}p_{1,2}(t)p_{2,2}(t), \\ p_{2,2}(2) = 0. \end{cases} \tag{2.63}$$

It is difficult to obtain the analytical solution $(p_{1,2}, p_{2,2})$ of (2.62)–(2.63), but we can obtain a numerical solution. Set $\gamma = 0.35$ and $\gamma = 0.4$; the solution trajectories are shown in Figure 2.1 and Figure 2.2, respectively. From Figure 2.1, we can see that both $p_{1,2}(t)$ and $p_{2,2}(t)$ are divergent for $\gamma = 0.35$, while Figure 2.2 indicates that $p_{1,2}(t)$ and $p_{2,2}(t)$ converge as $t \to \infty$ for $\gamma = 0.4$.

□

2.2.6 Unified treatment of H_2, H_∞ and mixed H_2/H_∞ control problems

In this section, we give a unified treatment for the H_2 optimal control, H_∞ control and mixed H_2/H_∞ control problems. For simplicity, we only consider system (2.22). Associated with (2.22), define

$$J_{1,T}(u, v) = \mathcal{E} \int_0^T (\gamma^2 \|v(t)\|^2 - \|z(t)\|^2)\,dt$$

and

$$J_{2,T}^\varrho(u, v) := \mathcal{E} \int_0^T [\|z(t)\|^2 - \varrho^2 \|v(t)\|^2]\,dt, \quad \varrho \geq 0.$$

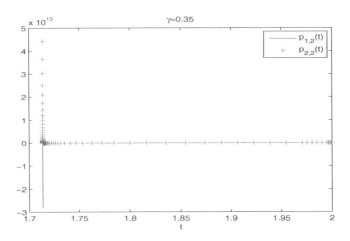

FIGURE 2.1

Trajectories of $p_{1,2}(t)$ and $p_{2,2}(t)$ for $\gamma = 0.35$.

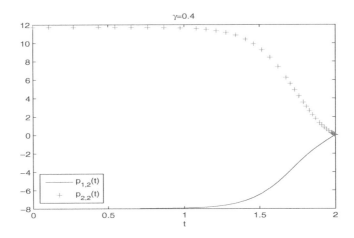

FIGURE 2.2

Trajectories of $p_{1,2}(t)$ and $p_{2,2}(t)$ for $\gamma = 0.4$.

In order to find a Nash equilibrium (u_T^*, v_T^*) satisfying

$$J_{1,T}(u_T^*, v_T^*) \leq J_{1,T}(u_T^*, v), \quad J_{2,T}^\varrho(u_T^*, v_T^*) \leq J_{2,T}^\varrho(u, v_T^*)$$

in the set Ω of linear memoryless state feedback laws, repeating the same procedure used in Section 2.2.5, we only need to solve the following coupled GDREs

$$\begin{cases} -\dot{Z}_{1,T} = A_1^T Z_{1,T} + Z_{1,T} A_1 + A_2^T Z_{1,T} A_2 - C^T C \\ \qquad - [\, Z_{1,T} \; Z_{2,T} \,] \begin{bmatrix} \gamma^{-2} C_1 C_1^T & B_1 B_1^T \\ B_1 B_1^T & B_1 B_1^T \end{bmatrix} \begin{bmatrix} Z_{1,T} \\ Z_{2,T} \end{bmatrix}, \\ Z_{1,T}(T) = 0 \end{cases} \qquad (2.64)$$

and

$$\begin{cases} -\dot{Z}_{2,T} = A_1^T Z_{2,T} + Z_{2,T} A_1 + A_2^T Z_{2,T} A_2 + C^T C \\ \qquad - [\, Z_{1,T} \; Z_{2,T} \,] \begin{bmatrix} \varrho^2 \gamma^{-4} C_1 C_1^T & \gamma^{-2} C_1 C_1^T \\ \gamma^{-2} C_1 C_1^T & B_1 B_1^T \end{bmatrix} \begin{bmatrix} Z_{1,T} \\ Z_{2,T} \end{bmatrix}, \\ Z_{2,T}(T) = 0 \end{cases} \qquad (2.65)$$

for $Z_{1,T}, Z_{2,T} \in \mathcal{S}_n$. In this case,

$$u_T^*(t) = -B_1^T(t) Z_{2,T}(t) x(t), \quad v_T^*(t) = -\gamma^{-2} C_1^T(t) Z_{1,T}(t) x(t).$$

1. The H_2 (LQ) optimal control problem: Find $\bar{u}_T^* \in \mathcal{L}_{\mathcal{F}}^2([0,T], \mathcal{R}^{n_u})$ such that it minimizes the cost

$$J_{2,T}(u,0) = \mathcal{E} \int_0^T \|z(t)\|^2 \, dt$$

subject to

$$\begin{cases} dx(t) = [A_1(t)x(t) + B_1(t)u(t)] \, dt + A_2(t)x(t) dB(t), \\ z(t) = \begin{bmatrix} C(t)x(t) \\ D(t)u(t) \end{bmatrix}, \quad D^T(t)D(t) = I. \end{cases} \qquad (2.66)$$

By setting $\gamma \to \infty$ and $\varrho = 0$ in (2.64) and (2.65), it can be seen that

$$Z_{1,T}(t) \to -P_T(t), \quad Z_{2,T}(t) \to P_T(t)$$

where $P_T(t)$ solves

$$\begin{cases} -\dot{P}_T(t) = A_1^T(t) P_T(t) + P_T(t) A_1(t) + A_2^T(t) P_T(t) A_2(t) + C^T(t) C(t) \\ \qquad - P_T(t) B_1(t) B_1^T(t) P_T(t), \\ P_T(T) = 0. \end{cases}$$

$$(2.67)$$

The H_2 optimal control is given by $\bar{u}_T^*(t) = -B_1^T(t) P_T(t) x(t)$.

2. H_∞ control: By setting $\varrho = \gamma$, it can be seen that

$$Z_{2,T} = -Z_{1,T} = P_{\infty,T},$$

where $P_{\infty,T}$ solves

$$\begin{cases} -\dot{P}_{\infty,T}(t) = A_1^T(t)P_{\infty,T}(t) + P_{\infty,T}(t)A_1(t) + A_2^T(t)P_{\infty,T}(t)A_2(t) \\ \qquad\qquad +C^T(t)C(t) - P_{\infty,T}(t)[B_1(t)B_1^T(t) - \gamma^{-2}C_1C_1^T]P_{\infty,T}(t), \\ P_{\infty,T}(T) = 0. \end{cases}$$

(2.68)

In this case, the Nash equilibrium point $(\tilde{u}_T^*, \tilde{v}_T^*)$ is given by

$$\tilde{u}_T^*(t) = -B_1^T(t)P_{\infty,T}(t)x(t), \quad \tilde{v}_T^*(t) = \gamma^{-2}C_1^T(t)P_{\infty,T}(t)x(t),$$

which is the saddle point of the two-person zero sum game problem:

$$J_{2,T}^\gamma(\tilde{u}_T^*, v) \le J_{2,T}^\gamma(\tilde{u}_T^*, \tilde{v}_T^*) \le J_{2,T}^\gamma(u, \tilde{v}_T^*).$$

The H_∞ control law for system (2.22) is obtained by $\tilde{u}_T^*(t) = -B_1^T(t)P_{\infty,T}(t)x(t)$, while $\tilde{v}_T^*(t) = \gamma^{-2}C_1^T(t)P_{\infty,T}(t)x(t)$ is the worst-case disturbance.

3. Mixed H_2/H_∞ control: Let $\varrho = 0$, then the mixed H_2/H_∞ control is retrieved. In this case, $P_{1,T} = Z_{1,T}$, $P_{2,T} = Z_{2,T}$.

2.3 Infinite Horizon H_2/H_∞ Control

In this section, we investigate the infinite horizon H_2/H_∞ control for the following stochastic time-invariant Itô system with state-, control- and disturbance-dependent noises

$$\begin{cases} dx(t) = [A_1x(t) + B_1u(t) + C_1v(t)]\,dt + [A_2x(t) + B_2u(t) \\ \qquad\qquad +C_2v(t)]\,dB(t), \\ x(0) = x_0 \in \mathcal{R}^n, \\ z(t) = \begin{bmatrix} Cx(t) \\ Du(t) \end{bmatrix}, \quad D^TD = I, \; t \in [0, \infty). \end{cases}$$

(2.69)

Firstly, we give the following definition for internal stabilization.

DEFINITION 2.6 *For a given $u_\infty^*(t) \in \mathcal{L}_\mathcal{F}^2(\mathcal{R}^+, \mathcal{R}^{n_u})$, system (2.69) is called internally stabilizable if it is ASMS via $u_\infty^*(t)$ in the absence of v, i.e.,*

$$dx(t) = [A_1x(t) + B_1u_\infty^*(t)]\,dt + [A_2x(t) + B_2u_\infty^*(t)]\,dB(t)$$

is ASMS.

The so-called infinite horizon stochastic H_2/H_∞ control is stated as follows:

DEFINITION 2.7 *The infinite horizon stochastic H_2/H_∞ control problem is to find $(u_\infty^*(t), v_\infty^*(t)) \in \mathcal{L}_\mathcal{F}^2(\mathcal{R}^+, \mathcal{R}^{n_u}) \times \mathcal{L}_\mathcal{F}^2(\mathcal{R}^+, \mathcal{R}^{n_v})$ such that:*

(i) When $u_\infty^*(t)$ is applied to (2.69), system (2.69) is internally stabilizable.

(ii) For a given disturbance attenuation $\gamma > 0$, under the constraint of

$$\begin{cases} dx(t) = [A_1 x(t) + B_1 u_\infty^*(t) + C_1 v(t)]\, dt + [A_2 x(t) + B_2 u_\infty^*(t) + C_2 v(t)]\, dB(t), \\ z(t) = \begin{bmatrix} Cx(t) \\ Du_\infty^*(t) \end{bmatrix}, \quad D^T D = I, \end{cases}$$

(2.70)

we have

$$\|\mathcal{L}_\infty\| = \sup_{v \in \mathcal{L}_{\mathcal{F}}^2(\mathcal{R}^+, \mathcal{R}^{n_v}), v \neq 0, u = u_\infty^*, x_0 = 0} \frac{\|z\|_{[0,\infty)}}{\|v\|_{[0,\infty)}}$$

$$:= \sup_{v \in \mathcal{L}_{\mathcal{F}}^2(\mathcal{R}^+, \mathcal{R}^{n_v}), v \neq 0, u = u_\infty^*, x_0 = 0} \frac{\left\{\mathcal{E} \int_0^\infty (\|C(t)x(t)\|^2 + \|u_\infty^*(t)\|^2)\, dt\right\}^{1/2}}{\left\{\mathcal{E} \int_0^\infty \|v(x(t))\|^2\, dt\right\}^{1/2}}$$

$$< \gamma.$$

(iii) When the worst-case $v_\infty^*(t) \in \mathcal{L}_{\mathcal{F}}^2(\mathcal{R}^+, \mathcal{R}^{n_v})$ is applied to (2.69), $u_\infty^*(t)$ minimizes the output energy

$$\|z(t)\|_{[0,\infty)}^2 = \mathcal{E} \int_0^\infty \|z(t)\|^2\, dt.$$

If the above (u_∞^*, v_∞^*) exist, then we say that the infinite horizon H_2/H_∞ control has a solution pair (u_∞^*, v_∞^*).

Similar to the finite horizon H_2/H_∞ case, if we set

$$J_{1,\infty}(u, v) = \mathcal{E} \int_0^\infty (\gamma^2 \|v(t)\|^2 - \|z(t)\|^2)\, dt$$

and

$$J_{2,\infty}(u, v) = \|z(t)\|_{[0,\infty)}^2,$$

then the above stochastic H_2/H_∞ control problem has a close relation to an infinite horizon stochastic LQ non-zero sum game. Roughly speaking, to find (u_∞^*, v_∞^*) only requires finding a Nash equilibrium point

$$J_{1,\infty}(u_\infty^*, v_\infty^*) \leq J_{1,\infty}(u_\infty^*, v), \quad J_{2,\infty}(u_\infty^*, v_\infty^*) \leq J_{2,\infty}(u, v_\infty^*), \qquad (2.71)$$

where the stochastic H_∞ control problem requires us to solve

$$J_{1,\infty}(u_\infty^*, v_\infty^*) \leq J_{1,\infty}(u_\infty^*, v)$$

and the stochastic H_2 control requires us to solve

$$J_{2,\infty}(u_\infty^*, v_\infty^*) \leq J_{2,\infty}(u, v_\infty^*).$$

Note that the infinite horizon H_2/H_∞ control requires us to establish stochastic stability, so it is more difficult to deal with than the finite horizon case. In the subsequent subsections, we will present some preliminaries.

2.3.1   Two Lyapunov-type theorems

Consider the following stochastic time-invariant system

$$\begin{cases} dx(t) = Ax(t)\, dt + A_1 x(t)\, dB(t), \ x(0) = x_0 \in \mathcal{R}^n, \\ y(t) = Cx(t) \end{cases} \tag{2.72}$$

together with the GLE

$$PA + A^T P + A_1^T P A_1 = -C^T C. \tag{2.73}$$

Our aim in this section is, under the assumptions of exact observability and detectability, to extend the classical Lyapunov theorem to stochastic system (2.72), which will be used in establishing our main results on the infinite horizon H_2/H_∞ control.

THEOREM 2.6
If $(A, A_1|C)$ is exactly observable, then (A, A_1) is stable iff the GLE (2.73) has a positive definite solution $P > 0$.

Theorem 2.6 extends Theorem 1.5 to the case $Q = C^T C \geq 0$.

Proof. If (A, A_1) is stable, then (2.73) has a solution $P \geq 0$; see [67]. For system (2.72), by Dynkin's formula [145], we have

$$0 \leq \mathcal{E}[x^T(t)Px(t)] = x_0^T Px_0 + \mathcal{E} \int_0^t x^T(s)(PA + A^T P + A_1^T P A_1)x(s)\, ds$$

$$= x_0^T Px_0 - \mathcal{E} \int_0^t x^T(s)C^T Cx(s)\, ds. \tag{2.74}$$

Now we show $P > 0$; otherwise, there exists $x_0 \neq 0$, such that $Px_0 = 0$. From (2.74), for any $T > 0$, we have

$$0 \leq \mathcal{E} \int_0^T x^T(s)C^T Cx(s)\, ds = -\mathcal{E}x^T(T)Px(T) \leq 0,$$

which implies $y(t) = Cx(t) \equiv 0, a.s., \forall t \in [0, T]$, but this is impossible because of exact observability, so $P > 0$.

If (2.73) has a positive definite solution $P > 0$, from (2.74), we know $V(x(t)) := \mathcal{E}[x^T(t)Px(t)]$ is monotonically decreasing and bounded from below with respect to t, so $\lim_{t \to \infty} V(x(t))$ exists. For any fixed T, if we let $t_n = nT$, then for $t \in [t_n, t_{n+1}]$, we have

$$V(x(t_{n+1})) \leq V(x(t)) \leq V(x(t_n)). \tag{2.75}$$

Again, by (2.74),

$$V(x(t_{n+1})) - V(x(t_n)) = -\mathcal{E} \int_{t_n}^{t_{(n+1)}} x^T(t)C^T Cx(t)\, dt.$$

By Lemma 2.3, we know the following backward differential equation

$$\begin{cases} -\dot{H}(t) = H(t)A + A^T H(t) + A_1^T H(t)A_1 + C^T C, \\ H(T) = 0 \end{cases} \qquad (2.76)$$

has a unique solution $H(\cdot) \geq 0$ on $[0, T]$. By completing the squares, we have

$$\begin{aligned} V(x(t_1)) - V(x(t_0)) &= -\mathcal{E}\int_0^T x^T(t)C^T Cx(t)\, dt \\ &= -x_0^T H(0)x_0 + \mathcal{E}x^T(T)H(T)x(T) \\ &\quad -\mathcal{E}\int_0^T x^T(t)C^T Cx(t)\, dt - \mathcal{E}\int_0^T d(x^T(t)H(t)x(t)) \\ &= -\mathcal{E}\int_0^T \Big\{ x^T(t)[\dot{H}(t) + H(t)A + A^T H(t) + A_1^T H(t)A_1 \\ &\quad + C^T C]x(t) \Big\}\, dt - x_0^T H(0)x_0 \\ &= -x_0^T H(0)x_0. \end{aligned}$$

As in the necessity proof, one has $H(0) > 0$ because of exact observability. Due to time-invariance of system (2.72), it is easy to prove

$$V(x(t_{n+1})) - V(x(t_n)) = -\mathcal{E}[x^T(t_n)H(0)x(t_n)].$$

Taking a limit in the above, we have

$$\lim_{n\to\infty} V(x(t_n)) = \lim_{n\to\infty} \mathcal{E}\|x(t_n)\|^2 = 0.$$

By (2.75) and $P > 0$,

$$\lim_{n\to\infty} \mathcal{E}\|x(t)\|^2 = \lim_{n\to\infty} V(x(t)) = 0.$$

Hence, (A, A_1) is stable, and the proof is complete. $\square$

THEOREM 2.7
If $(A, A_1|C)$ is exactly detectable, then (A, A_1) is stable iff the GLE (2.73) has a positive semi-definite solution $P \geq 0$.

Theorem 2.7 can be viewed as a corollary of Proposition 14 of [57] which is a very general and abstract result with an elementary proof given in [206]. Here, for the reader's convenience, we present two methods to prove Theorem 2.7. The first proof needs to use the following lemma:

LEMMA 2.5 [57]

(i) *The generalized Lyapunov operator $\mathcal{L}_{A,A_1}$ is resolvent positive.*

(ii) *If we set $\beta := \max_{\lambda_i \in \sigma(\mathcal{L}_{A,A_1})} Re(\lambda_i)$, then there exists a non-zero $X \geq 0$, such that $\mathcal{L}_{A,A_1}X = \beta X$.*

First proof of Theorem 2.7. Sufficiency: If (A, A_1) is not stable, then, by Theorem 1.7, $\beta = \max Re\sigma(\mathcal{L}_{A,A_1}) \geq 0$. According to Lemma 2.5, there exists a non-zero $X \geq 0$, such that $\mathcal{L}_{A,A_1}X = \beta X$. Note that

$$0 \geq < -C^T C, X >=< \mathcal{L}^*_{A,A_1}P, X >=< P, \mathcal{L}_{A,A_1}X >=< P, \beta X >\geq 0, \tag{2.77}$$

where $< A, B >:= \text{Trace}(A^T B)$, $\mathcal{L}^*_{A,A_1}$ is the adjoint operator of $\mathcal{L}_{A,A_1}$, $\mathcal{L}^*_{A,A_1}(P) = PA + A^T P + A_1^T P A_1$. It follows from inequality (2.77) that $\text{Trace}(C^T CX) = 0$, which implies $CX = 0$ due to $X \geq 0$. According to Theorem 1.13, $CX = 0$ together with $\mathcal{L}_{A,A_1}X = \beta X$ contradicts with the exact detectability of $(A, A_1|C)$. So (A, A_1) is stable.

Necessity: If (A, A_1) is stable, even without exact detectability of $(A, A_1|C)$, it is easy to derive that under the constraint of (2.72),

$$0 \leq \mathcal{E} \int_0^\infty x^T(t)C^T Cx(t)\, dt = \mathcal{E} \int_0^\infty x^T(t)C^T Cx(t)\, dt + x_0^T P x_0$$
$$- \lim_{t \to \infty} \mathcal{E}[x(t)^T Px(t)] + \mathcal{E} \int_0^\infty d(x^T(t)Px(t))$$
$$= x_0^T P x_0 \tag{2.78}$$

for any $x_0 \in \mathcal{R}^n$, which yields $P \geq 0$. Theorem 2.7 is shown. $\square$

The first proof of Theorem 2.7 is based on resolvent positive operator theory. In the following, we would like to present another proof, which is based on the matrix decomposition technique and could be useful in dealing with other control problems. The following lemma will be used in the second proof of Theorem 2.7.

LEMMA 2.6
If $P \geq 0$ solves (2.73) and $(A, A_1|C)$ is exactly detectable, then the unobservable subspace $\mathcal{N}_0$ of $(A, A_1|C)$ satisfies $\mathcal{N}_0 = \text{Ker}(P)$.

Proof. Under the constraint of (2.72), applying Itô's formula to $x^T(t)Px(t)$, it follows that for any $T \geq 0$,

$$0 \leq \mathcal{E} \int_0^T x^T(t)C^T Cx(t)\, dt$$
$$= \mathcal{E} \int_0^T x^T(t)C^T Cx(t)\, dt + \mathcal{E} \int_0^T d(x^T(t)Px(t)) - \mathcal{E}[x^T(T)Px(T)]$$
$$+ x_0' P x_0$$

$$= \mathcal{E} \int_0^T x^T (PA + A^T P + A_1^T PA_1 + C^T C)x \, dt + x_0^T Px_0 - \mathcal{E}[x^T(T)Px(T)]$$
$$= x_0^T Px_0 - \mathcal{E}[x^T(T)Px(T)]. \tag{2.79}$$

Obviously, for any $x_0 \in \mathrm{Ker}(P)$, (2.79) yields $y(t) = Cx(t) \equiv 0$ a.s. $\forall t \in [0, T]$, i.e., $x_0 \in \mathcal{N}_0$. Conversely, $\forall x_0 \in \mathcal{N}_0$, it follows that $\lim_{T \to \infty} \mathcal{E}[x^T(T)Px(T)] = 0$ from exact detectability. Taking limit in (2.79), we have

$$x_0^T Px_0 = \mathcal{E} \int_0^\infty \|y(t)\|^2 \, dt = 0,$$

which yields $x_0 \in \mathrm{Ker}(P)$ due to $P \geq 0$. In conclusion, $\mathcal{N}_0 = \mathrm{Ker}(P)$. □

Second proof of Theorem 2.7. The proof of the necessity is the same as in the first proof of Theorem 2.7. In the following, we only present an alternative proof for the sufficiency part.

If P is strictly positive definite, then, by Lemma 2.6, $(A, A_1|C)$ is exactly observable, while the stability of (A, A_1) follows from Theorem 2.6.

If $P \geq 0$ but not $P > 0$, then $\mathrm{Ker}(P)$ contains at least one non-zero vector. For any $\xi \neq 0$, $\xi \in \mathrm{Ker}(P)$, it is easy to test $\xi \in \mathrm{Ker}(C)$, i.e., $\mathrm{Ker}(P) \subset \mathrm{Ker}(C)$. Moreover, we can see that $\mathrm{Ker}(P)$ is an invariant subspace with respect to both A and A_1. Suppose S is an orthogonal matrix such that

$$S^T PS = \begin{bmatrix} 0 & 0 \\ 0 & P_2 \end{bmatrix}, \quad P_2 > 0$$

then

$$S^T C^T CS = \begin{bmatrix} 0 & 0 \\ 0 & Q_2 \end{bmatrix}, \quad S^T AS = \begin{bmatrix} A_{11} & A_{12} \\ 0 & A_2 \end{bmatrix},$$
$$S^T A_1 S = \begin{bmatrix} C_1 & C_{12} \\ 0 & C_2 \end{bmatrix}.$$

Pre- and post-multiplying both sides of (2.73) respectively by S^T and S, it follows that

$$S^T PS \cdot S^T AS + S^T A^T S \cdot S^T PS + S^T A_1^T S \cdot S^T PS \cdot S^T A_1 S = -S^T C^T CS,$$

which is equivalent to

$$P_2 A_2 + A_2^T P_2 + C_2^T P_2 C_2 = -Q_2, \quad Q_2 \geq 0. \tag{2.80}$$

In addition, applying Itô's formula to

$$\eta = \begin{bmatrix} \eta_1 \\ \eta_2 \end{bmatrix} = S^T x = \begin{bmatrix} S_{11} & S_{12} \\ S_{21} & S_{22} \end{bmatrix}^T x,$$

it follows that

$$\begin{cases} d\eta_1 = (A_{11}\eta_1 + A_{12}\eta_2) \, dt + (C_1\eta_1 + C_{12}\eta_2) \, dB, \\ d\eta_2 = A_2\eta_2 \, dt + C_2\eta_2 \, dB. \end{cases} \tag{2.81}$$

Obviously, that $y(t) = Cx(t) \equiv 0\, a.s.$ is equivalent to $Q_2\eta_2 \equiv 0\, a.s.$, for which a sufficient condition is $\eta_2 \equiv 0$. Hence, by Definition 1.13, we have that (A_{11}, C_1) is stable.

Below, we further show that (A_2, C_2) is stable. Since $(A, A_1|C)$ is exactly detectable, it is easy to see the exact detectability of $(A_2, C_2|Q_2^{1/2})$. In addition, (2.80) has a positive definite solution $P_2 > 0$. So, from Lemma 2.6, $(A_2, C_2|Q_2^{1/2})$ is exactly observable. By Theorem 2.6, the stability of (A_2, C_2) is derived.

Now, from the stability of (A_{11}, C_1) and (A_2, C_2), we have $\lim_{t\to\infty} \mathcal{E}\|\eta(t)\|^2 = 0$ [53, 84], which is equivalent to $\lim_{t\to\infty} \mathcal{E}\|x(t)\|^2 = 0$. The proof of this theorem is completed. $\square$

Using the method as in the first proof of Theorem 2.7, we are able to verify Conjecture 3.1 of [206], which is given in the following theorem:

THEOREM 2.8
Suppose $\sigma(\mathcal{L}_{A,A_1}) \subset \mathcal{C}^{-,0} := \{\lambda \in \mathcal{C} : Re(\lambda) \leq 0\}$ and $(A, A_1|C)$ is exactly detectable. If P is a real symmetric solution of (2.73), then $P \geq 0$ and (A, A_1) is stable.

Proof. Because $\sigma(\mathcal{L}_{A,A_1}) \subset \mathcal{C}^{-,0}$, the spectral abscissa $\beta = \max_{\lambda_i \in \sigma(\mathcal{L}_{A,A_1})} Re(\lambda_i) \leq 0$. If $\beta < 0$, then (A, A_1) is stable, which yields $P \geq 0$ by the result of [67]. If $\beta = 0$, by Lemma 2.5, there exists a non-zero $X \geq 0$, such that $\mathcal{L}_{A,A_1} X = 0$. Therefore, we have

$$0 \geq < -C^T C, X > = < \mathcal{L}^*_{A,A_1}(P), X > = < P, \mathcal{L}_{A,A_1}(X) > = < P, 0 > = 0. \tag{2.82}$$

It follows from inequality (2.82) that $CX = 0$, which together with $\mathcal{L}_{A,A_1} X = 0$ contradicts with the exact detectability of $(A, A_1|C)$. Therefore, we must have $\beta < 0$. This proof is complete. $\square$

2.3.2 Infinite horizon stochastic LQ control

Below, we revisit the infinite horizon stochastic LQ control of Itô systems. For simplicity, we only consider the standard LQ control, i.e., the state weighting matrix is positive semi-definite, and the control weighting matrix is positive definite. The infinite horizon stochastic LQ control can be considered in the two cases: free terminal state case and stable terminal state case.

Case 1 (Free terminal state). Under the constraint of the linear time-invariant Itô system

$$\begin{cases} dx(t) = [A_{11}x(t) + B_{11}u(t)]\,dt + [A_{12}x(t) + B_{12}u(t)]\,dB(t), \\ x(0) = x_0, \end{cases} \tag{2.83}$$

minimize the quadratic performance

$$\min_{u \in \mathcal{L}^2_{\mathcal{F}}(\mathcal{R}^+, \mathcal{R}^{n_u})} J^\infty(u), \tag{2.84}$$

where

$$J^\infty(u) := \mathcal{E} \int_0^\infty [x^T(t)Qx(t) + u^T(t)Ru(t)] \, dt, \ Q \geq 0, \ R > 0.$$

If there is $u_f^* \in \mathcal{L}_\mathcal{F}^2(\mathcal{R}^+, \mathcal{R}^{n_u})$ such that $\min_{u \in \mathcal{L}_\mathcal{F}^2(\mathcal{R}^+, \mathcal{R}^{n_u})} J^\infty(u) = J^\infty(u_f^*)$, then u_f^* is called an optimal stochastic LQ control with free terminal state.

   Case 2 (Stable terminal state). Under the constraint of (2.83), search for a $u_s^* \in \mathcal{U}_{ad}$ to minimize

$$\min_{u \in \mathcal{U}_{ad}} J^\infty(u), \tag{2.85}$$

where $\mathcal{U}_{ad}$ is defined as

$$\mathcal{U}_{ad} := \{u : u \in \mathcal{L}_\mathcal{F}^2(\mathcal{R}^+, \mathcal{R}^{n_u}), \lim_{t \to \infty} \mathcal{E}\|x(t)\|^2 = 0\}.$$

DEFINITION 2.8   *The infinite horizon stochastic LQ control problem (2.83)–(2.84) is said to be well posed if*

$$-\infty < \min_{u \in \mathcal{L}_\mathcal{F}^2(\mathcal{R}^+, \mathcal{R}^{n_u})} J^\infty(u) < \infty.$$

Similarly, the LQ control (2.83) and (2.85) is said to be well posed if

$$-\infty < \min_{u \in \mathcal{U}_{ad}} J^\infty(u) < \infty.$$

DEFINITION 2.9   *The LQ control problem (2.83)–(2.84) is said to be attainable if it is well posed, and there is $u_f^* \in \mathcal{L}_\mathcal{F}^2(\mathcal{R}^+, \mathcal{R}^{n_u})$ achieving*

$$\min_{u \in \mathcal{L}_\mathcal{F}^2(\mathcal{R}^+, \mathcal{R}^{n_u})} J^\infty(u),$$

i.e.,

$$J^\infty(u_f^*) = \min_{u \in \mathcal{L}_\mathcal{F}^2(\mathcal{R}^+, \mathcal{R}^{n_u})} J^\infty(u).$$

In this case, $u_f^*(t)$ is called the optimal control, and the state $x^*(t)$ corresponding to $u_f^*(t)$ is called the optimal trajectory.

   Similar definitions for attainability and u_s^* can be given for the LQ control (2.83) and (2.85).

THEOREM 2.9
If $(A_{11}, B_{11}; A_{12}, B_{12})$ is stabilizable, then, we have
   1)

$$\min_{u \in \mathcal{L}_\mathcal{F}^2(\mathcal{R}^+, \mathcal{R}^{n_u})} J^\infty(u) = x_0^T \bar{P} x_0$$

and the corresponding optimal control is given by

$$u_f^*(t) = -(R + B_{12}^T \bar{P} B_{12})^{-1}(B_{11}^T \bar{P} + B_{12}^T \bar{P} A_{12})x(t),$$

where $\bar{P} \geq 0$ is a solution of GARE

$$\begin{cases} PA_{11} + A_{11}^T P + A_{12}^T PA_{12} + Q - (PB_{11} + A_{12}^T PB_{12})(R + B_{12}^T PB_{12})^{-1} \\ \qquad \cdot (B_{11}^T P + B_{12}^T PA_{12}) = 0, \\ R + B_{12}^T PB_{12} > 0, \ Q \geq 0, \ R > 0. \end{cases}$$

(2.86)

2) $\bar{P}$ is the minimal positive semi-definite solution of GARE (2.86) denoted by $P_{\min}$, i.e., for any solution $P \geq 0$ of (2.86), $P \geq \bar{P}$.

Proof. To prove 1), we need the following facts:

Fact 1: By Lemma 2.3, we know

(i) The following GDRE

$$\begin{cases} \dot{P} + PA_{11} + A_{11}^T P + A_{12}^T PA_{12} + Q - (PB_{11} + A_{12}^T PB_{12})(R + B_{12}^T PB_{12})^{-1} \\ \qquad \cdot (B_{11}^T P + B_{12}^T PA_{12}) = 0, \\ P(T) = 0, \\ R + B_{12}^T PB_{12} > 0, \ \forall t \in [0, T], \\ Q \geq 0, \ R > 0 \end{cases}$$

(2.87)

has a solution $P(t) \geq 0$ on $[0, T]$ denoted by $P_T(t)$.

(ii) The optimal performance index and control are given by

$$\inf_{u \in \mathcal{L}_{\mathcal{F}}^2([0,T], \mathcal{R}^{n_u})} \mathcal{E} \int_0^T [x^T(t)Qx(t) + u^T(t)Ru(t)]\, dt = x_0^T P_T(0)x_0$$

and

$$\bar{u}_T^* = -(R + B_{12}^T P_T(t)B_{12})^{-1}(B_{11}^T P_T(t) + B_{12}^T P_T(t)A_{12})x(t)$$

respectively.

In addition, it is easy to prove the following properties of $P_T(t)$.

Fact 2. $P_T(t)$ is monotonically increasing with respect to T, i.e., $P_{T_2}(t) \geq P_{T_1}(t)$ for any $T_2 \geq T_1 \geq 0, t \in [0, T_1]$.

Fact 3. Under the condition of Theorem 2.9, or equivalently, under the stabilizability of $(A_{11}, B_{11}; A_{12}, B_{12})$, for any $t \geq 0$, $P_T(t)$ is uniformly bounded with respect to T for $T \geq t \geq 0$.

By applying Facts 1–3, 1) can be proved similarly to the deterministic LQ control as discussed in [11]. Moreover, $\bar{P} = \lim_{T \to \infty} P_T(0)$. The detail is omitted.

To prove 2), let

$$J^T(u) = \mathcal{E} \int_0^T [x^T(t)Qx(t) + u^T(t)Ru(t)]\, dt.$$

For any $P \in \mathcal{S}_n$, by applying Itô's formula to $x^T P x$ and completing the squares, the following identity holds:

$$J^T(u) = J^T(u) + \mathcal{E} \int_0^T d(x^T(t)Px(t)) + x_0^T Px_0 - \mathcal{E}[x^T(T)Px(T)]$$

$$= x_0^T P x_0 - \mathcal{E}[x^T(T)Px(T)] + \mathcal{E} \int_0^T x^T(t)\mathcal{P}(P)x(t)\, dt$$

$$+\mathcal{E} \int_0^T [u(t) - Kx(t)]^T (R + B_{12}^T PB_{12})[u(t) - Kx(t)]\, dt, \quad (2.88)$$

where

$$K = -(R + B_{12}^T PB_{12})^{-1}(B_{11}^T P + B_{12}^T PA_{12})$$

and

$$\begin{aligned}
\mathcal{P}(P) = {} & PA_{11} + A_{11}^T P + A_{12}^T PA_{12} + Q \\
& -(PB_{11} + A_{12}^T PB_{12})(R + B_{12}^T PB_{12})^{-1}(B_{11}^T P + B_{12}^T PA_{12}).
\end{aligned}$$

Suppose $\tilde{P} \geq 0$ is a solution of GARE (2.86) and let

$$\tilde{u}(t) = -(R + B_{12}^T \tilde{P}B_{12})^{-1}(B_{11}^T \tilde{P} + B_{12}^T \tilde{P}A_{12})x(t).$$

Then by taking $T \to \infty$ in (2.88), we first know $\tilde{u} \in \mathcal{L}_{\mathcal{F}}^2(\mathcal{R}^+, \mathcal{R}^{n_u})$ from

$$J^T(\tilde{u}) \leq x_0^T \tilde{P}x_0 < +\infty.$$

Secondly, by the principle of optimality, for any $x_0 \in \mathcal{R}^n$, we have

$$\min_{u \in \mathcal{L}_{\mathcal{F}}^2(\mathcal{R}^+, \mathcal{R}^{n_u})} J^\infty(u) = x_0^T \bar{P}x_0 \leq \mathcal{E} \int_0^\infty [x^T(t)Qx(t) + \tilde{u}^T(t)R\tilde{u}(t)]\, dt$$

$$\leq x_0^T \tilde{P}x_0 - \varliminf_{T \to \infty}\mathcal{E}[x^T(T)\tilde{P}x(T)]$$

$$\leq x_0^T \tilde{P}x_0. \quad (2.89)$$

Since x_0 is arbitrary, $\bar{P} \leq \tilde{P}$, therefore, $\bar{P} := P_{\min}$ is the minimal positive semi-definite solution of (2.86). □

The stabilizability of $(A_{11}, B_{11}; A_{12}, B_{12})$ is only a sufficient but not a necessary condition for the existence of a solution $\bar{P} \geq 0$ to GARE (2.86); see the following example:

Example 2.2

We consider a scalar ARE coming from deterministic LQ control as

$$2ap + q - b^2 r^{-1} p^2 = 0, \quad q \geq 0, \quad r > 0, \quad (2.90)$$

i.e., in (2.83)–(2.84), $A_{11} = a$, $B_{11} = b$, $Q = q$, $R = r$, $A_{12} = 0$, $B_{12} = 0$. If in (2.90), $a > 0$, $b = 0$, $q = 0$, then $(a, b) = (a, 0)$ is not stabilizable, but $\bar{p} = 0$ is the unique solution to (2.90). ☐

THEOREM 2.10

If $(A_{11}, B_{11}; A_{12}, B_{12})$ is stabilizable, then we have:

1) *GARE (2.86) has a maximal solution $P_{\max} \geq 0$, i.e., for any real symmetric solution P to (2.86), $P_{\max} \geq P$.*

2) *For the LQ optimal control problem (2.83) and (2.85) with stable terminal state, we have*

$$\min_{u \in \mathcal{U}_{ad}} J^\infty(u) = x_0^T P_{\max} x_0.$$

3) *If, in addition, $(A_{11}, A_{12}|Q^{1/2})$ is exactly observable, then $P_{\max} > 0$, which is the unique positive definite symmetric solution of GARE (2.86). In this case, the optimal control u_s^* exists, and*

$$\min_{u \in \mathcal{U}_{ad}} J^\infty(u) = J^\infty(u_s^*), \quad u_s^*(t) = K_{\max} x(t),$$

where

$$K_{\max} = -(R + B_{12}^T P_{\max} B_{12})^{-1}(B_{11}^T P_{\max} + B_{12}^T P_{\max} A_{12}).$$

To prove Theorem 2.10, we first show the following lemma:

LEMMA 2.7
If $(A_{11}, A_{12}|Q^{1/2})$ is exactly observable, so is

$$\left(A_{11} + B_{11} K_{\max}, A_{12} + B_{12} K_{\max} \middle| \begin{bmatrix} Q^{1/2} \\ R^{1/2} K_{\max} \end{bmatrix} \right).$$

Proof. If $\left(A_{11} + B_{11} K_{\max}, A_{12} + B_{12} K_{\max} \middle| \begin{bmatrix} Q^{1/2} \\ R^{1/2} K_{\max} \end{bmatrix} \right)$ is not exactly observable, then by Theorem 1.9, there exist a non-zero $Z \in \mathcal{S}_n$ and a $\lambda \in \mathcal{C}$, such that

$$(A_{11} + B_{11} K_{\max})Z + Z(A_{11} + B_{11} K_{\max})^T$$
$$+(A_{12} + B_{12} K_{\max})Z(A_{12} + B_{12} K_{\max})^T = \lambda Z \quad (2.91)$$

and

$$\begin{bmatrix} Q^{1/2} \\ R^{1/2} K_{\max} \end{bmatrix} Z = 0. \quad (2.92)$$

From (2.92), in view of $R > 0$, we have

$$K_{\max} Z = 0, \quad Q^{1/2} Z = 0. \quad (2.93)$$

Substituting $K_{\max} Z = 0$ into (2.91), it follows that

$$A_{11} Z + Z A_{11}^T + A_{12} Z A_{12}^T = \lambda Z,$$

which together with $Q^{1/2}Z = 0$ (the second equation of (2.93)) contradicts the exact observability of $(A_{11}, A_{12}|Q^{1/2})$ according to Theorem 1.9. □

Proof of Theorem 2.10. Corollary 4 and Theorem 12 of [2] immediately yield 1) and 2). As for 3), we note that if $P_{max} \geq 0$ is a solution of (2.86), then the first equation of (2.86) can also be rewritten as

$$
\begin{aligned}
&P_{max}(A_{11} + B_{11}K_{max}) + (A_{11} + B_{11}K_{max})^T P_{max} \\
&+ (A_{12} + B_{12}K_{max})^T P_{max}(A_{12} + B_{12}K_{max}) \\
&= -\begin{bmatrix} Q^{1/2} & K_{max}^T R^{1/2} \end{bmatrix} \begin{bmatrix} Q^{1/2} \\ R^{1/2}K_{max} \end{bmatrix}.
\end{aligned}
\tag{2.94}
$$

By Lemma 2.7, we know that

$$
\left(A_{11} + B_{11}K_{max}, A_{12} + B_{12}K_{max} \,\middle|\, \begin{bmatrix} Q^{1/2} \\ R^{1/2}K_{max} \end{bmatrix} \right)
$$

is exactly observable.

If $P_{max} \geq 0$ but is not strictly positive definite, then $\mathcal{T} := \mathrm{Ker}(P_{max})$ is not trivial. For any $\xi \in \mathcal{T}$, pre- and post-multiplying both sides of (2.94) by ξ^T and ξ, respectively, it follows that

$$
P_{max}(A_{12} + B_{12}K_{max})\xi = 0, \quad \begin{bmatrix} Q^{1/2} \\ R^{1/2}K_{max} \end{bmatrix} \xi = 0,
$$

which implies that

$$
(A_{12} + B_{12}K_{max})\mathcal{T} \subset \mathcal{T}, \quad \mathcal{T} \subset \mathrm{Ker}\left(\begin{bmatrix} Q^{1/2} \\ R^{1/2}K_{max} \end{bmatrix} \right).
\tag{2.95}
$$

Based on (2.95), by post-multiplying (2.94) with $\xi \in \mathcal{T}$, we further have

$$
(A_{11} + B_{11}K_{max})\mathcal{T} \subset \mathcal{T}.
\tag{2.96}
$$

By Proposition 1.3, (2.95) together with (2.96) contradicts the exact observability of

$$
\left(A_{11} + B_{11}K_{max}, A_{12} + B_{12}K_{max} \,\middle|\, \begin{bmatrix} Q^{1/2} \\ R^{1/2}K_{max} \end{bmatrix} \right).
$$

So $P_{max} > 0$.

By applying Theorem 2.6, $(A_{11} + B_{11}K_{max}, A_{12} + B_{12}K_{max})$ is stable. From (2.88),

$$
J^T(u_s^*) = x_0^T P_{max} x_0 - \mathcal{E}[x^T(T)P_{max}x(T)].
$$

Taking the limit on the above and in view of the stability of $(A_{11} + B_{11}K_{max}, A_{12} + B_{12}K_{max})$, it yields that

$$
J^\infty(u_s^*) = x_0^T P_{max} x_0 = \min_{u \in \mathcal{U}_{ad}} J^\infty(u).
$$

Therefore, u_s^* is an optimal control.

Finally, we show $P_{\max} > 0$ is the unique positive definite solution of GARE (2.86). By contradiction, assume that there is another positive definite solution $\hat{P} > 0$. Let

$$\hat{K} = -(R + B_{12}^T \hat{P} B_{12})^{-1}(B_{11}^T \hat{P} + B_{12}^T \hat{P} A_{12})x(t), \quad \hat{u}^*(t) = \hat{K}x(t).$$

Then, repeating the above discussion, we know that $(A_{11}+B_{11}K_{\max}, A_{12}+B_{12}K_{\max})$ is stable, and

$$J^\infty(\hat{u}^*) = x_0^T \hat{P} x_0.$$

By the optimality principle and noting that $P_{\max}$ is the maximal solution of GARE (2.86), the following is derived:

$$x_0^T P_{\max} x_0 = \min_{u \in \mathcal{U}_{ad}} J^\infty(u) \le J^\infty(\hat{u}^*) = x_0^T \hat{P} x_0,$$

which results in a contradiction as $P_{\max} \le \hat{P}$. So under the condition of exact observability, GARE (2.86) has at most one positive definition solution. This theorem is proved. $\square$

DEFINITION 2.10 *A real symmetric solution P to GARE (2.86) is called a feedback stabilizing solution, if $(A_{11} + B_{11}K, A_{12} + B_{12}K)$ is stable, where*

$$K = -(R + B_{12}^T P B_{12})^{-1}(B_{11}^T P + B_{12}^T P A_{12}).$$

Theorem 2.10 tells us that, under the stabilizability of $(A_{11}, B_{11}; A_{12}, B_{12})$ and exact observability of $(A_{11}, A_{12}|Q^{1/2})$, GARE (2.86) admits a unique feedback stabilizing solution.

Similar to Theorem 2.10 and Lemma 2.7, under exact detectability, we have the following results. The proofs are omitted.

LEMMA 2.8
If $(A_{11}, A_{12}|Q^{1/2})$ is exactly detectable, so is

$$\left(A_{11} + B_{11}K_{\max}, A_{12} + B_{12}K_{\max} \left| \begin{bmatrix} Q^{1/2} \\ R^{1/2}K_{\max} \end{bmatrix} \right.\right).$$

THEOREM 2.11
If $(A_{11}, B_{11}; A_{12}, B_{12})$ is stabilizable and $(A_{11}, A_{12}|Q^{1/2})$ is exactly detectable, then

(i) *$P_{\max} \ge 0$ is the unique positive semi-definite symmetric solution and a feedback stabilizing solution of GARE (2.86).*

(ii)

$$\min_{u \in \mathcal{U}_{ad}} J^\infty(u) = J^\infty(u_s^*) = x_0^T P_{\max} x_0, \quad u_s^* = K_{\max} x(t).$$

REMARK 2.8   The stochastic LQ control (2.83)–(2.84) with free terminal state only has theoretical interest. In practice, we are more interested in the LQ control with stable terminal state. ∎

REMARK 2.9   In [71], MS-stabilizability and MS-detectability were defined in order to study a class of linearly perturbed GAREs arising from stochastic control. It can be easily tested that Theorem 2.11 improves Theorem 4.1 of [71] when GARE (2.86) reduces to GARE (2.2) of [71], this is because MS-stabilizability and MS-detectability are stronger than stabilizability of $(A_{11}, B_{11}; A_{12}, B_{12})$ and exact detectability of $(A_{11}, A_{12}|Q^{1/2})$, respectively. ∎

The exact detectability of $(A_{11}, A_{12}|Q^{1/2})$ is only sufficient but not necessary for the existence of a feedback stabilizing solution of GARE (2.86). The reader can easily construct an example to verify this fact.

Below, we give a simple example to illustrate the difference between the free terminal and stable terminal stochastic LQ control problems.

Example 2.3

For simplicity, consider a scalar deterministic system

$$\dot{x}(t) = ax(t) + u(t), \quad x(0) = x_0 \neq 0 \tag{2.97}$$

with the performance index

$$J^\infty(u) = \int_0^\infty u^2(t)\, dt,$$

where the state weighting scalar $q = 0$. In this situation, (2.97) is stabilizable for any $a \in \mathcal{R}$. GARE (2.86) becomes $2aP - P^2 = 0$ which has two solutions $P_1 = 2a$ and $P_2 = 0$.

1) When $a > 0$, $(a, q) = (a, 0)$ is not detectable, $P_{max} = 2a$, $P_{min} = 0$, $\min_{u \in \mathcal{U}_{ad}} J^\infty(u) = 2ax_0^2$, $u_s^*(t) = -2ax(t)$; $\min_{u \in \mathcal{L}_{\mathcal{F}}^2(\mathcal{R}^+, \mathcal{R}^{n_u})} J^\infty(u) = 0$, $u_f^*(t) = 0$. $P_{max} > P_{min}$.

2) When $a < 0$, $(a, q) = (a, 0)$ is detectable, $P_{max} = P_{min} = 0$, and there is another solution $P = 2a < 0$. $\min_{u \in \mathcal{U}_{ad}} J^\infty(u) = \min_{u \in \mathcal{L}_{\mathcal{F}}^2(\mathcal{R}^+, \mathcal{R}^{n_u})} J^\infty(u) = 0$; $u_f^*(t) = u_s^*(t) = 0$.

3) When $a = 0$, $(a, q) = (0, 0)$ is not detectable, $P_{max} = P_{min} = 0$. So

$$\min_{u \in \mathcal{U}_{ad}} J^\infty(u) = \min_{u \in \mathcal{L}_{\mathcal{F}}^2(\mathcal{R}^+, \mathcal{R}^{n_u})} J^\infty(u) = 0, \quad u_f^* = 0$$

but u_s^* does not exist. $\Box$

REMARK 2.10   It can be seen from Example 2.3-3) that only under stabilizability of $(A_{11}, B_{11}; A_{12}, B_{12})$, although the LQ control with stable terminal state (2.83)–(2.85) is well posed, i.e., $-\infty < \min_{u \in \mathcal{U}_{ad}} J^\infty(u) < +\infty$, u_s^* does not necessarily exist.   ■

Finally, we give a modified Kleinman iteration to search for the feedback stabilizing solution $P_{\max}$ to (2.86) under stabilizability of $(A_{11}, B_{11}; A_{12}, B_{12})$ and exact detectability of $(A_{11}, A_{12}|Q^{1/2})$. By Theorem 2.11, the feedback stabilizing solution $P_{max} \geq 0$, hence, the constraint condition $R + B_{12}^T P_{\max} B_{12} > 0$ holds automatically. So we only need to search for the maximal solution $P_{\max}$ to

$$\begin{cases} PA_{11} + A_{11}^T P + A_{12}^T PA_{12} + Q - (PB_{11} + A_{12}^T PB_{12})(R + B_{12}^T PB_{12})^{-1} \\ \qquad \cdot (B_{11}^T P + B_{12}^T PA_{12}) = 0, \\ Q \geq 0, \ R > 0. \end{cases}$$
$$(2.98)$$

The equation (2.98) can be written as

$$P(A_{11} + B_{11}K) + (A_{11} + B_{11}K)^T P + (A_{12} + B_{12}K)^T P(A_{12} + B_{12}K)$$
$$= -Q - K^T RK, \qquad (2.99)$$

where $K = -(R + B_{12}^T PB_{12})^{-1}(B_{11}^T P + B_{12}^T PA_{12})$. Because $(A_{11}, B_{11}; A_{12}, B_{12})$ is stabilizable, there exists a matrix K_0 such that $(A_{11} + B_{11}K_0, A_{12} + B_{12}K_0)$ is stable. By Theorem 2.7, there is a solution $P_1 \geq 0$ to GARE

$$P(A_{11} + B_{11}K_0) + (A_{11} + B_{11}K_0)^T P + (A_{12} + B_{12}K_0)^T P(A_{12} + B_{12}K_0)$$
$$= -Q - K_0^T RK_0, \qquad (2.100)$$

and then the second iteration gain K_1 taken as

$$K_1 = -(R + B_{12}^T P_1 B_{12})^{-1}(B_{11}^T P_1 + B_{12}^T P_1 A_{12})$$

can be derived. Summarizing the above analysis, we name it Step 1 and Step 2.
   Step 1: Find K_0 such that $(A_{11} + B_{11}K_0, A_{12} + B_{12}K_0)$ is stable.
   Step 2: P_1 and K_1 are in turn obtained, i.e.,

$$K_0 \to P_1 \to K_1.$$

We assert that $(A_{11} + B_{11}K_1, A_{12} + B_{12}K_1)$ is also stable. It is easy to test that

$$P_1(A_{11} + B_{11}K_1) + (A_{11} + B_{11}K_1)^T P_1 + (A_{12} + B_{12}K_1)^T P_1(A_{12} + B_{12}K_1)$$
$$= -Q - K_1^T RK_1 - (K_1 - K_0)^T (R + B_{12}^T P_1 B_{12})(K_1 - K_0). \qquad (2.101)$$

By Theorem 1.13, $(A_{11} + B_{11}K_1, A_{12} + B_{12}K_1|\tilde{H})$ is exactly detectable, where

$$\tilde{H} = Q + K_1^T K_1 + (K_1 - K_0)^T (R + B_{12}^T P_1 B_{12})(K_1 - K_0).$$

Again, by Theorem 2.7, it yields the stability of $(A_{11} + B_{11}K_1, A_{12} + B_{12}K_1)$. Repeating Steps 1 and 2, P_2 and K_2 are obtained. In general, we have the following general iteration formulae:

$$P_{k+1}(A_{11} + B_{11}K_k) + (A_{11} + B_{11}K_k)^T P_{k+1}$$
$$+(A_{12} + B_{12}K_k)^T P_{k+1}(A_{12} + B_{12}K_k) = -Q - K_k^T R K_k, \quad (2.102)$$
$$K_k = -(R + B_{12}^T P_k B_{12})^{-1}(B_{11}^T P_k + B_{12}^T P_k A_{12}), \quad (2.103)$$

$$P_k(A_{11} + B_{11}K_k) + (A_{11} + B_{11}K_k)^T P_k + (A_{12} + B_{12}K_k)^T P_k(A_{12} + B_{12}K_k)$$
$$= -Q - K_k^T R K_k - (K_k - K_{k-1})^T(R + B_{12}^T P_k B_{12})(K_k - K_{k-1}). \quad (2.104)$$

Hence, Steps 1 and 2 can proceed forever. A general iteration algorithm is taken as Step 3.

Step 3: $K_0 \to P_1 \to K_1 \to P_2 \to K_2 \to \cdots$.

If we subtract (2.102) from (2.104), then we have

$$(P_k - P_{k+1})(A_{11} + B_{11}K_k) + (A_{11} + B_{11}K_k)^T(P_k - P_{k+1})$$
$$+(A_{12} + B_{12}K_k)^T(P_k - P_{k+1})(A_{12} + B_{12}K_k)$$
$$= -(K_k - K_{k-1})^T(R + B_{12}^T P_k B_{12})(K_k - K_{k-1}), \quad (2.105)$$

which deduces $P_1 \geq P_2 \geq \ldots \geq 0$, i.e., $\{P_i \geq 0\}_{i \geq 1}$ is a sequence that is monotonically decreasing and bounded from below, so $\bar{P} = \lim_{k \to \infty} P_k \geq 0$ exists, which is a feedback stabilizing solution to (2.86). In [216], it is further shown that the sequence $\{P_k\}_{k \geq 1}$ converges in a quadratic rate to $P_{\max}$.

2.3.3 Infinite horizon SBRL

Consider the following stochastic time-invariant system with state- and disturbance-dependent noises:

$$\begin{cases} dx(t) = [A_{11}x(t) + B_{11}v(t)]\, dt + [A_{12}x(t) + B_{12}v(t)]\, dB(t), \\ z_1(t) = C_{11}x(t), \ x(0) = x_0. \end{cases} \quad (2.106)$$

Associated with system (2.106), define the perturbation operator $\tilde{\mathcal{L}}_\infty : \mathcal{L}_{\mathcal{F}}^2(\mathcal{R}^+, \mathcal{R}^{n_v}) \mapsto \mathcal{L}_{\mathcal{F}}^2(\mathcal{R}^+, \mathcal{R}^{n_{z_1}})$ as

$$\tilde{\mathcal{L}}_\infty(v) = z_1|_{x_0=0} = C_{11}x(t)|_{x_0=0}, \ t \geq 0, \ v \in \mathcal{L}_{\mathcal{F}}^2(\mathcal{R}^+, \mathcal{R}^{n_v}).$$

Then,

$$\|\tilde{\mathcal{L}}_\infty\| = \sup_{v \in \mathcal{L}_{\mathcal{F}}^2(\mathcal{R}^+, \mathcal{R}^{n_v}), v \neq 0, x_0=0} \frac{\|z_1\|_{[0,\infty)}}{\|v\|_{[0,\infty)}}$$

$$= \sup_{v \in \mathcal{L}_{\mathcal{F}}^2(\mathcal{R}^+, \mathcal{R}^{n_v}), v \neq 0, x_0=0} \frac{\left\{\mathcal{E} \int_0^\infty \|C_{11}x(t)\|^2\, dt\right\}^{1/2}}{\left\{\mathcal{E} \int_0^\infty \|v(t)\|^2\, dt\right\}^{1/2}}.$$

The following lemma is the infinite horizon SBRL, which is a very important result in H_∞ analysis and control, and can be viewed as an extension of Lemma 4 of [62].

THEOREM 2.12
System (2.106) is internally stable and $\|\tilde{\mathcal{L}}_\infty\| < \gamma$ iff the GARE

$$\begin{cases} PA_{11} + A_{11}^T P + A_{12}^T PA_{12} - (PB_{11} + A_{12}^T PB_{12})(\gamma^2 I + B_{12}^T PB_{12})^{-1} \\ \qquad \cdot (B_{11}^T P + B_{12}^T PA_{12}) - C_{11}^T C_{11} = 0, \\ \gamma^2 I + B_{12}^T PB_{12} > 0 \end{cases}$$

(2.107)

has a feedback stabilizing solution $P_1 \leq 0$. More specifically, the following two statements are equivalent:

(i) *System (2.106) is internally stable (i.e., (A_{11}, A_{12}) is stable) and $\|\tilde{\mathcal{L}}_\infty\| < \gamma$ for some $\gamma > 0$.*

(ii) *GARE (2.107) has a feedback stabilizing solution $P_1 \leq 0$, i.e., $(A_{11} + B_{11}\tilde{K}, A_{12} + B_{12}\tilde{K})$ is stable, where*

$$\tilde{K} = -(\gamma^2 I + B_{12}^T P_1 B_{12})^{-1}(B_{11}^T P_1 + B_{12}^T P_1 A_{12}).$$

REMARK 2.11   Comparing GARE (2.107) with GARE (2.86), it can be found that, by letting $Q = -C_{11}^T C_{11} \leq 0$ and $R = \gamma^2 I > 0$, (2.86) leads to (2.107). In other words, the GARE arising from SBRL is simply a GARE from indefinite stochastic LQ control.   ∎

REMARK 2.12   In Lemma 5 of [36], it was shown that (i) $\Rightarrow$ (ii) for the case $B_{12} = 0$ in (2.106). In Lemma 4.3 of [206], under the condition of internal stability of the system (2.106), we showed that $\|\tilde{\mathcal{L}}_\infty\| < \gamma$ is equivalent to the above (ii). So Theorem 2.12 improves Lemma 5 of [36] and Lemma 4.3 of [206].   ∎

Proof. If (i) holds, by Corollary 2.14 of [84], there exists a sufficiently small $\delta > 0$, such that the following GARE

$$\begin{cases} PA_{11} + A_{11}^T P + A_{12}^T PA_{12} - \delta^2 I - (PB_{11} + A_{12}^T PB_{12})(\gamma^2 I + B_{12}^T PB_{12})^{-1} \\ \qquad \cdot (B_{11}^T P + B_{12}^T PA_{12}) - C_{11}^T C_{11} = 0, \\ \gamma^2 I + B_{12}^T PB_{12} > 0 \end{cases}$$

(2.108)

has a solution $P_\delta < 0$. By Theorem 5 of [205], which is called a comparison theorem on GAREs, there exists a solution P_1 to (2.107) with $P_1 \geq P_\delta$. Furthermore, from the proof of Theorem 5 of [205], it can be seen that $P_1 > P_\delta$. In addition, P_1 satisfies

$$\min_{v \in \mathcal{U}_{ad}} \mathcal{E} \int_0^\infty (\gamma^2 \|v(t)\|^2 - \|z_1(t)\|^2)\, dt = x_0^T P_1 x_0$$

$$\leq J_1^\infty(x, 0; x_0, 0) \leq 0, \tag{2.109}$$

where

$$J_1^\infty(x, v; x(t_0), t_0) = \mathcal{E} \int_{t_0}^\infty (\gamma^2 \|v(t)\|^2 - \|z_1(t)\|^2) \, dt.$$

Because x_0 is arbitrary, $P_1 \leq 0$.

The rest is to prove that $(A_{11} + B_{11}\tilde{K}, A_{12} + B_{12}\tilde{K})$ is stable. Denote $\hat{A}_{11} = A_{11} + B_{11}\tilde{K}$, $\hat{A}_{12} = A_{12} + B_{12}\tilde{K}$, $\hat{A}_{11}^\delta = A_{11} + B_{11}K_\delta$, and $\hat{A}_{12}^\delta = A_{12} + B_{12}K_\delta$, where

$$K_\delta = -(\gamma^2 I + B_{12}^T P_\delta B_{12})^{-1}(B_{11}^T P_\delta + B_{12}^T P_\delta A_{12}).$$

Note that the first equations in (2.107) and (2.108) can be rewritten as

$$P_1 \hat{A}_{11} + \hat{A}_{11}^T P_1 + \hat{A}_{12}^T P_1 \hat{A}_{12} + \gamma^2 \tilde{K}^T \tilde{K} - C_{11}^T C_{11} = 0 \tag{2.110}$$

and

$$P_\delta \hat{A}_{11}^\delta + (\hat{A}_{11}^\delta)^T P_\delta + (\hat{A}_{12}^\delta)^T P_\delta \hat{A}_{12}^\delta + \gamma^2 K_\delta^T K_\delta - C_{11}^T C_{11} - \delta^2 I = 0, \tag{2.111}$$

respectively. Subtracting (2.111) from (2.110), by a series of computations, we have

$$\begin{aligned}
&(P_1 - P_\delta)\hat{A}_{11} + \hat{A}_{11}^T(P_1 - P_\delta) + \hat{A}_{12}^T(P_1 - P_\delta)\hat{A}_{12} \\
&= -[(P_\delta B_{11} + A_{12}^T P_\delta B_{12}) + \tilde{K}^T(\gamma^2 I + B_{12}^T P_\delta B_{12})](\gamma^2 I + B_{12}^T P_\delta B_{12})^{-1} \\
&\quad \cdot [(P_\delta B_{11} + A_{12}^T P_\delta B_{12}) + \tilde{K}^T(\gamma^2 I + B_{12}^T P_\delta B_{12})]^T - \delta^2 I \\
&< 0,
\end{aligned}$$

which means that the GLI

$$P\hat{A}_{11} + \hat{A}_{11}^T P + \hat{A}_{12}^T P \hat{A}_{12} < 0$$

admits a positive definite solution $P := P_1 - P_\delta > 0$. Applying Theorem 1.5, $(\hat{A}_{11}, \hat{A}_{12})$ is stable. So (i)$\Rightarrow$(ii).

Conversely, if GARE (2.107) has a feedback stabilizing solution $P_1 \leq 0$, i.e., $(A_{11} + B_{11}\tilde{K}, A_{12} + B_{12}\tilde{K})$ is stable, which implies that $(A_{11}, B_{11}; A_{12}, B_{12})$ is stabilizable. By the results of [2], there is a P_0 such that

$$\mathcal{M}(P_0) := \begin{bmatrix} P_0 A_{11} + A_{11}^T P_0 + A_{12}^T P_0 A_{12} - C_{11}^T C_{11} & P_0 B_{11} + A_{12}^T P_0 B_{12} \\ B_{11}^T P_0 + B_{12}^T P_0 A_{12} & \gamma^2 I + B_{12}^T P_0 B_{12} \end{bmatrix} > 0,$$

which implies

$$\mathcal{M}(P_0, \epsilon) := \begin{bmatrix} P_0 A_{11} + A_{11}^T P_0 + A_{12}^T P_0 A_{12} - C_{11}^T C_{11} - \epsilon I & P_0 B_{11} + A_{12}^T P_0 B_{12} \\ B_{11}^T P_0 + B_{12}^T P_0 A_{12} & \gamma^2 I + B_{12}^T P_0 B_{12} \end{bmatrix} > 0$$

for sufficiently small $\epsilon > 0$. Again, by [2],

$$\begin{cases} PA_{11} + A_{11}^T P + A_{12}^T P A_{12} - (PB_{11} + A_{12}^T P B_{12})(\gamma^2 I + B_{12}^T P B_{12})^{-1} \\ \qquad \cdot (B_{11}^T P + B_{12}^T P A_{12}) - C_{11}^T C_{11} - \epsilon I = 0, \\ \gamma^2 I + B_{12}^T P B_{12} > 0 \end{cases}$$

$$\tag{2.112}$$

admits a solution $P_\epsilon \le P_1 \le 0$. We now assert that $P_\epsilon < 0$. Otherwise, there exists a non-zero $x_0 \in \text{Ker}(P_\epsilon)$ such that $P_\epsilon x_0 = 0$. Pre- and post-multiplying both sides of (2.112) by x_0^T and x_0, respectively, it follows that

$$
\begin{aligned}
0 &\ge x_0^T A_{12}^T P_\epsilon A_{12} x_0 \\
&= x_0^T [C_{11}^T C_{11} + \epsilon I + (A_{12}^T P_\epsilon B_{12})(\gamma^2 I + B_{12}^T P_\epsilon B_{12})^{-1}(B_{12}^T P_\epsilon A_{12})] x_0 \\
&> 0,
\end{aligned}
$$

which is a contradiction. Therefore, $P_\epsilon < 0$. GARE (2.112) shows that $\bar{P}_\epsilon = -P_\epsilon > 0$ solves the GLI

$$
\bar{P}_\epsilon A_{11} + A_{11}^T \bar{P}_\epsilon + A_{12}^T \bar{P}_\epsilon A_{12} < 0
$$

from which the internal stability of (2.106) is derived by applying Theorem 1.5.

To prove $\|\tilde{\mathcal{L}}_\infty\| < \gamma$, define a linear operator $\mathcal{L}_1$ as follows:

$$
\mathcal{L}_1 : \mathcal{L}_{\mathcal{F}}^2(\mathcal{R}^+, \mathcal{R}^{n_v}) \mapsto \mathcal{L}_{\mathcal{F}}^2(\mathcal{R}^+, \mathcal{R}^{n_v}), \quad \mathcal{L}_1 v = v - v_s^*, \quad v_s^*(t) = \tilde{K} x(t)
$$

associated with

$$
dx = (A_{11}x(t) + B_{11}v_s^*(t))\, dt + (A_{12}x + B_{12}v_s^*(t))\, dB(t).
$$

Then $\mathcal{L}_1$ is invertible, so there exists $\varrho > 0$, such that $\|\mathcal{L}_1\|^2 \ge \varrho$. By completing the squares, we have

$$
\begin{aligned}
\|z_1\|_{[0,\infty)}^2 - \gamma^2 \|v\|_{[0,\infty)}^2 &= -\mathcal{E} \int_0^\infty \|v - v_s^*\|_{(\gamma^2 I + B_{12}^T P_1 B_{12})}^2 \, dt \\
&\le -\rho \|v - v_s^*\|_{[0,\infty)}^2 \\
&= -\rho \|\mathcal{L}_1 v\|_{[0,\infty)}^2 \le -\rho\varrho \|v\|_{[0,\infty)}^2, \quad (2.113)
\end{aligned}
$$

where $\|Z\|_{(\gamma^2 I + B_{12}^T P_1 B_{12})}^2 := Z^T(\gamma^2 I + B_{12}^T P_1 B_{12})Z$ and ρ is such that $\gamma^2 I + B_{12}^T P_1 B_{12} > \rho I$. (2.113) leads to $\|\tilde{\mathcal{L}}_\infty\| < \gamma$. This completes (ii)$\Rightarrow$ (i). Summarizing the above, this theorem is proved. $\square$

Theorem 2.12 is an infinite horizon SBRL based on GARE (2.107). We note that in Theorem 2.8 of [84], another form of SBRL for stochastic perturbed systems with state- and disturbance-dependent independent noises was given based on LMIs. Following a similar procedure as in [84], we are able to obtain an SBRL similarly to Theorem 2.8 of [84].

THEOREM 2.13

For system (2.106), the following two statements are equivalent:

(i) The system (2.106) is internally stable and $\|\tilde{\mathcal{L}}_\infty\| < \gamma$.

(ii) There exists $P < 0$ such that

$$
\mathcal{M}(P) = \begin{bmatrix} PA_{11} + A_{11}^T P + A_{12}^T P A_{12} - C_{11}^T C_{11} & PB_{11} + A_{12}^T PB_{12} \\ B_{11}^T P + B_{12}^T P A_{12} & \gamma^2 I + B_{12}^T PB_{12} \end{bmatrix} > 0.
$$

$$(2.114)$$

REMARK 2.13 Theorem 2.12 and Theorem 2.13 are equivalent, and the conditions given can be easily tested using MATLAB Toolbox [24, 83]. The existence of a feedback stabilizing solution P_1 of GARE (2.107) is equivalent to the existence of an optimal solution P_1 to the semi-definite programming problem:

$$\max_{\text{subject to } \mathcal{M}(P)>0,\ P\leq 0} \text{Trace}(P).$$

However, in order to use the Nash game approach to study the H_2/H_∞ control design, we have to use Theorem 2.12 instead of Theorem 2.13. ∎

COROLLARY 2.1
GARE (2.107) has a feedback stabilizing solution $P_1 \leq 0$ iff LMI (2.114) has a solution $P < 0$.

REMARK 2.14 We note that Theorem 10 of [53] gives similar results as Theorems 2.12–2.13 for stochastic Markovian systems based on a finite horizon SBRL and stochastic moment estimation. In the above, we have presented very simple proofs for various forms of SBRLs. ∎

In the following, we give an application of SBRL in Theorem 2.12 to obtain a stochastic small gain theorem, which generalizes Theorem 1 of [54] to systems with both state- and control-dependent noise.

LEMMA 2.9 [85]
For any given matrices L, M and N of appropriate size, the matrix equation

$$LXM = N \tag{2.115}$$

admits a solution X iff
$$LL^+NM^+M = N.$$

In particular, any solution to (2.115) may be represented by

$$X = L^+NM^+ + S - L^+LSMM^+$$

with S being a matrix of suitable dimension.

Consider the following two linear stochastic controlled systems with state and control-dependent noise

$$\begin{cases} dx_1(t) = [A_{10}x_1(t) + B_{10}u_1(t)]\,dt + [A_{20}x_1(t) + B_{20}u_1(t)]\,dB(t), \\ y_1(t) = C_{10}x_1(t) + D_{10}u_1(t) \end{cases} \tag{2.116}$$

and

$$\begin{cases} dx_2(t) = [A_{20}x_2(t) + B_{20}u_2(t)]\,dt + [A_{21}x_2(t) + B_{21}u_2(t)]\,dB(t), \\ y_2(t) = C_{20}x_2(t). \end{cases} \tag{2.117}$$

THEOREM 2.14

Let T_1 and T_2 be the perturbed operators of (2.116) and (2.117), respectively. In addition, we assume both (2.116) and (2.117) are internally stable. If for some $\gamma > 0$, $\|T_1\| < \gamma, \|T_2\| < \frac{1}{\gamma}$, then the system

$$\begin{bmatrix} dx_1 \\ dx_2 \end{bmatrix} = \begin{bmatrix} A_{10} & B_{10}C_{20} \\ B_{20}C_{10} & A_{20} + B_{20}D_{10}C_{20} \end{bmatrix} \begin{bmatrix} x_1 \\ x_2 \end{bmatrix} dt$$
$$+ \begin{bmatrix} A_{20} & B_{20}C_{20} \\ B_{21}C_{10} & A_{21} + B_{21}D_{10}C_{20} \end{bmatrix} \begin{bmatrix} x_1 \\ x_2 \end{bmatrix} dB(t) \qquad (2.118)$$

is ASMS. System (2.118) is a combination of (2.116) and (2.117) for $u_1 = y_2$ and $u_2 = y_1$.

To prove Theorem 2.14, we first present the following lemma.

LEMMA 2.10

If system (2.106) is internally stable and $\|\tilde{\mathcal{L}}_\infty\| < 1$, then $I - \tilde{\mathcal{L}}_\infty$ is invertible with its realization being internally stable.

Proof. It is easy to verify that the realization of $(I - \tilde{\mathcal{L}}_\infty)^{-1}$ takes the form of

$$\begin{cases} dx(t) = [(A_{11} + B_{11}C_{11})x(t) + B_{11}v(t)] \, dt + [(A_{12} + B_{12}C_{11})x(t) + B_{12}v(t)] \, dB, \\ z_1(t) = C_{11}x(t) + v(t). \end{cases}$$
$$(2.119)$$

Obviously, we only need to prove that $(A_{11} + B_{11}C_{11}, A_{12} + B_{12}C_{11})$ is stable. By Theorem 2.12, the following GARE

$$\begin{cases} PA_{11} + A_{11}^T P + A_{12}^T PA_{12} + (PB_{11} + A_{12}^T PB_{12})(I - B_{12}^T PB_{12})^{-1} \\ \quad \cdot (B_{11}^T P + B_{12}^T PA_{12}) + C_{11}^T C_{11} = 0, \\ I - B_{12}^T PB_{12} > 0 \end{cases}$$
$$(2.120)$$

admits a feedback stabilizing solution $P \geq 0$, and $(A_{11} + B_{11}\tilde{K}, A_{12} + B_{12}\tilde{K})$ is stable with $\tilde{K} = (I - B_{12}^T PB_{12})^{-1}(B_{11}^T P + B_{12}^T PA_{12})$. Note that (2.120) may be written as

$$P\tilde{A}_{11} + \tilde{A}_{11}^T P + \tilde{A}_{12}^T P\tilde{A}_{12} = -\tilde{C}_{11}^T(I - B_{12}^T PB_{12})\tilde{C}_{11}, \quad I - B_{12}^T PB_{12} > 0,$$
$$(2.121)$$

where

$$\tilde{A}_{11} = A_{11} + B_{11}C_{11}, \tilde{A}_{12} = A_{12} + B_{12}C_{11}, \tilde{C}_{11} = \tilde{K} - C_{11}.$$

We first assert that $(\tilde{A}_{11}, \tilde{A}_{12}|\tilde{C}_{11})$ is exactly detectable. Otherwise, by Theorem 1.13, there exists non-zero $Z \in \mathcal{S}_n$ such that

$$\tilde{A}_{11}Z + Z\tilde{A}_{11}^T + \tilde{A}_{12}Z\tilde{A}_{12}^T = \lambda Z, \quad \tilde{C}_{11}Z = 0, \quad \mathrm{Re}\lambda \geq 0. \qquad (2.122)$$

From the second equation of (2.122), it follows that

$$\tilde{K}Z = C_{11}Z. \qquad (2.123)$$

Applying Lemma 2.9, we have

$$C_{11} = \tilde{K}ZZ^+ + S - SZZ^+,$$

which, by means of Definition 2.3, yields

$$
\begin{aligned}
B_{12}C_{11}ZC_{11}^T B_{12}^T &= B_{12}(\tilde{K}ZZ^+ + S - SZZ^+)Z(\tilde{K}ZZ^+ + S - SZZ^+)^T B_{12}^T \\
&= B_{12}(\tilde{K}ZZ^+Z + SZ - SZZ^+Z)(\tilde{K}ZZ^+ + S - SZZ^+)^T B_{12}^T \\
&= B_{12}(\tilde{K}Z)(Z^+Z\tilde{K}^T + S^T - Z^+ZS^T)B_{12}^T \\
&= B_{12}(\tilde{K}ZZ^+Z\tilde{K}^T)B_{12} = B_{12}\tilde{K}Z\tilde{K}^T B_{12}. \qquad (2.124)
\end{aligned}
$$

In view of (2.123) and (2.124), we have

$$
\begin{aligned}
(A_{11} + B_{11}\tilde{K})Z &+ Z(A_{11} + B_{11}\tilde{K})^T + (A_{12} + B_{12}\tilde{K})Z(A_{12} + B_{12}\tilde{K})^T \\
&= (A_{11} + B_{11}C_{11})Z + Z(A_{11} + B_{11}C_{11})^T \\
&\quad + (A_{12} + B_{12}C_{11})Z(A_{12} + B_{12}C_{11})^T \\
&= \lambda Z, \; Re\lambda \geq 0, \qquad (2.125)
\end{aligned}
$$

which contradicts the stability of $(A_{11} + B_{11}\tilde{K}, A_{12} + B_{12}\tilde{K})$ according to Theorem 1.7. Therefore, $(\tilde{A}_{11}, \tilde{A}_{12}|\tilde{C}_{11})$ is exactly detectable. Applying Theorem 2.7 to (2.121), the stability of $(A_{11} + B_{11}C_{11}, A_{12} + B_{12}C_{11})$ is derived. $\square$

Proof of Theorem 2.14. Clearly, the realization of T_1T_2 is as follows:

$$
\begin{cases}
dx_1(t) = [A_{10}x_1(t) + B_{10}C_{20}x_2(t)]\,dt + [A_{20}x_1(t) + B_{20}C_{20}x_2(t)]\,dB(t), \\
dx_2(t) = [A_{20}x_2(t) + B_{20}u_2(t)]\,dt + [A_{21}x_2(t) + B_{21}u_2(t)]\,dB(t), \\
y_1(t) = C_{10}x_1(t) + D_{10}C_{20}x_2(t).
\end{cases}
$$

$$(2.126)$$

By assumptions, both (2.116) and (2.117) are internally stable, so (2.126) is also internally stable; see [84]. Moreover, $\|T_1T_2\| \leq \|T_1\| \cdot \|T_2\| < 1$. So this theorem can be proved by a direct application of Lemma 2.10. $\square$

2.3.4   Stochastic H_2/H_∞ control

Based on the preliminaries of the last sections, we shall present the main results of the H_2/H_∞ control in this section.

Case 1: H_2/H_∞ control with state-dependent noise

For clarity, we first discuss the stochastic H_2/H_∞ control for the system with only state-dependent noise:

$$
\begin{cases}
dx(t) = [A_1x(t) + B_1u(t) + C_1(t)v(t)]\,dt + A_2x(t)\,dB(t), \\
x(0) = x_0 \in \mathcal{R}^n, \\
z(t) = \begin{bmatrix} Cx(t) \\ Du(t) \end{bmatrix}, \quad D^T D = I,
\end{cases}
$$

$$(2.127)$$

where x, u and z are, respectively, the state, control input and regulated output.

THEOREM 2.15
Consider system (2.127). Assume that the following coupled GAREs

$$A_1^T P_{1,\infty} + P_{1,\infty} A_1 + A_2^T P_{1,\infty} A_2 - C^T C$$
$$- \begin{bmatrix} P_{1,\infty} & P_{2,\infty} \end{bmatrix} \begin{bmatrix} \gamma^{-2} C_1 C_1^T & B_1 B_1^T \\ B_1 B_1^T & B_1 B_1^T \end{bmatrix} \begin{bmatrix} P_{1,\infty} \\ P_{2,\infty} \end{bmatrix} = 0 \quad (2.128)$$

and

$$A_1^T P_{2,\infty} + P_{2,\infty} A_1 + A_2^T P_{2,\infty} A_2 + C^T C$$
$$- \begin{bmatrix} P_{1,\infty} & P_{2,\infty} \end{bmatrix} \begin{bmatrix} 0 & \gamma^{-2} C_1 C_1^T \\ \gamma^{-2} C_1 C_1^T & B_1 B_1^T \end{bmatrix} \begin{bmatrix} P_{1,\infty} \\ P_{2,\infty} \end{bmatrix} = 0 \quad (2.129)$$

have a pair of solutions $(P_{1,\infty}, P_{2,\infty})$ with $P_{1,\infty} < 0$ and $P_{2,\infty} > 0$. Additionally, if $(A_1, A_2|C)$ and $(A_1 - \gamma^{-2} C_1 C_1^T P_{1,\infty}, A_2|C)$ are exactly observable, then the infinite horizon stochastic H_2/H_∞ control problem admits a pair of solutions $(u_\infty^*(t), v_\infty^*(t))$ with

$$u_\infty^*(t) = -B_1^T P_{2,\infty} x(t), \quad v_\infty^*(t) = -\gamma^{-2} C_1^T P_{1,\infty} x(t).$$

In other words, we have
i) *$(A_1 - B_1 B_1^T P_{2,\infty}, A_2)$ is stable.*
ii) *$\|\mathcal{L}_\infty\| < \gamma$.*
iii) *u_∞^* minimizes the output energy $\|z\|_{[0,\infty)}^2$ when v_∞^* is applied to (2.127).*

To prove Theorem 2.15, we need the following lemma:

LEMMA 2.11
Assume $\gamma \neq 0$ and let

$$\tilde{A}_2 = \begin{bmatrix} C \\ \gamma^{-1} C_1^T P_{1,\infty} \\ B_1^T P_{2,\infty} \end{bmatrix}, \quad \tilde{A}_3 = \begin{bmatrix} C \\ B_1^T P_{2,\infty} \end{bmatrix}.$$

(i) *If $(A_1, A_2|C)$ is exactly observable (respectively, exactly detectable), so is $(A_1 - B_1 B_1^T P_{2,\infty}, A_2|\tilde{A}_2)$.*

(ii) *If $(A_1 - \gamma^{-2} C_1 C_1^T P_{1,\infty}, A_2|C)$ is exactly observable (respectively, exactly detectable), so is*

$$(A_1 - \gamma^{-2} C_1 C_1^T P_{1,\infty} - B_1 B_1^T P_{2,\infty}, A_2|\tilde{A}_3).$$

Proof. If $(A_1 - B_1 B_1^T P_{2,\infty}, A_2|\tilde{A}_2)$ is not exactly observable, then by Theorem 1.9, there exists a non-zero $Z \in \mathcal{S}_n$ such that

$$(A_1 - B_1 B_1^T P_{2,\infty}) Z + Z(A_1 - B_1 B_1^T P_{2,\infty})^T + A_2 Z A_2^T = \lambda Z, \lambda \in \mathcal{C} \quad (2.130)$$

and

$$\tilde{A}_2 Z = \begin{bmatrix} C \\ \gamma^{-1} C_1^T P_{1,\infty} \\ B_1^T P_{2,\infty} \end{bmatrix} Z = 0. \tag{2.131}$$

From (2.131), we have

$$B_1^T P_{2,\infty} Z = 0, \quad CZ = 0. \tag{2.132}$$

Substituting the first equation of (2.132) into (2.130), it follows that

$$A_1 Z + Z A_1^T + A_2 Z A_2^T = \lambda Z, \quad \lambda \in \mathcal{C},$$

which together with $CZ = 0$ (the second equation of (2.132)) contradicts the exact observability of $(A_1, A_2 | C)$ according to Theorem 1.9. (i) is proved. Repeating the same procedure as in (i), (ii) can be shown. Using Theorem 1.13 and repeating the above procedure, the exact detectability can be established. $\square$

Proof of Theorem 2.15. Note that (2.128) and (2.129) can be rearranged as

$$P_{1,\infty}(A_1 - B_1 B_1^T P_{2,\infty}) + (A_1 - B_1 B_1^T P_{2,\infty})^T P_{1,\infty} + A_2^T P_{1,\infty} A_2 = \tilde{A}_2^T \tilde{A}_2 \tag{2.133}$$

and

$$P_{2,\infty}(A_1 - B_1 B_1^T P_{2,\infty} - \gamma^{-2} C_1 C_1^T P_{1,\infty})$$
$$+ (A_1 - B_1 B_1^T P_{2,\infty} - \gamma^{-2} C_1 C_1^T P_{1,\infty})^T P_{2,\infty} + A_2^T P_{2,\infty} A_2$$
$$= -\tilde{A}_3^T \tilde{A}_3, \tag{2.134}$$

respectively, where $\tilde{A}_2$ and $\tilde{A}_3$ are defined in Lemma 2.11. By Lemma 2.11-(ii), $(A_1 - B_1 B_1^T P_{2,\infty} - \gamma^{-2} C_1 C_1^T P_{1,\infty}, A_2 | \tilde{A}_3)$ is exactly observable. Using Theorem 2.6, it is known that $(A_1 - B_1 B_1^T P_{2,\infty} - \gamma^{-2} C_1 C_1^T P_{1,\infty}, A_2)$ is stable. Hence, $(u_\infty^*, v_\infty^*) \in \mathcal{L}_\mathcal{F}^2(\mathcal{R}^+, \mathcal{R}^{n_u}) \times \mathcal{L}_\mathcal{F}^2(\mathcal{R}^+, \mathcal{R}^{n_v})$. Next, by Lemma 2.11-(i) and Theorem 2.6, (2.133) implies that $(A_1 - B_1 B_1^T P_{2,\infty}, A_2)$ is stable. Hence, i) of Theorem 2.15 is proved.

As for ii), substituting $u(t) = u_\infty^*(t) = -B_1^T P_{2,\infty} x(t)$ into (2.127) gives

$$\begin{cases} dx(t) = [(A_1 - B_1 B_1^T P_{2,\infty}) x(t) + C_1 v(t)] \, dt + A_2 x(t) dB(t), \\ x(0) = x_0 \in \mathcal{R}^n, \\ z(t) = \begin{bmatrix} Cx(t) \\ -DB_1^T P_{2,\infty} x(t) \end{bmatrix}, \quad D^T D = I. \end{cases} \tag{2.135}$$

Since $(A - B_1 B_1^T P_{2,\infty}, A_2)$ is stable, $v \in \mathcal{L}_\mathcal{F}^2(\mathcal{R}^+, \mathcal{R}^{n_v})$, we have $x(t) \in \mathcal{L}_\mathcal{F}^2(\mathcal{R}^+, \mathcal{R}^n)$ from Remark 2.6 of [84], where $x(t)$ is the solution of (2.135). Consider (2.128) and (2.135) and apply Itô's formula to $x^T P_{1,\infty} x$, by completing the squares, we have

$$J_{1,\infty}(u_\infty^*, v) = \mathcal{E} \int_0^\infty (\gamma^2 \|v(t)\|^2 - \|z(t)\|^2) \, dt$$
$$= x_0^T P_{1,\infty} x_0 + \gamma^2 \mathcal{E} \int_0^\infty \|(v(t) - v_\infty^*(t)\|^2 \, dt$$
$$\geq J_{1,\infty}(u_\infty^*, v_\infty^*) = x_0^T P_{1,\infty} x_0. \tag{2.136}$$

From (2.136), it can be seen that the worst-case disturbance corresponding to u_∞^* is $v_\infty^* = -\gamma^{-2}C_1^T P_{1,\infty}x$. Following the line of [113], define an operator

$$\mathcal{L}_1 : \mathcal{L}_{\mathcal{F}}^2(\mathcal{R}^+, \mathcal{R}^{n_v}) \mapsto \mathcal{L}_{\mathcal{F}}^2(\mathcal{R}^+, \mathcal{R}^{n_v})$$

as

$$\mathcal{L}_1 v(t) = v(t) - v_\infty^*(t)$$

with the realization

$$dx(t) = [(A_1 - B_1 B_1^T P_{2,\infty})x(t) + C_1 v(t)]\, dt + A_2 x(t)dB(t),\ x(0) = 0$$

and

$$v(t) - v_\infty^*(t) = v(t) + \gamma^{-2}C_1^T P_{1,\infty}x(t).$$

Then $\mathcal{L}_1^{-1}$ exists, which is determined by

$$\begin{cases} dx(t) = (A - B_1 B_1^T P_{2,\infty} - \gamma^{-2}C_1 C_1^T P_{1,\infty})x(t)\,dt \\ \qquad + C_1(v(t) - v_\infty^*(t))\,dt + A_2 x(t)dB(t), \\ x(0) = 0, \\ v(t) = -\gamma^{-2}C_1^T P_{1,\infty}x(t) + (v(t) - v_\infty^*(t)). \end{cases}$$

From (2.136), we have

$$\gamma^2\|v\|_{[0,\infty)}^2 - \|z\|_{[0,\infty)}^2 = \gamma^2\|\mathcal{L}_1 v\|_{[0,\infty)}^2 \geq \varepsilon\|v\|_{[0,\infty)}^2 > 0$$

for some sufficiently small $\varepsilon > 0$, which yields $\|\mathcal{L}_\infty\| < \gamma$. ii) of this theorem is proved.

Finally, when the worst-case disturbance $v = v_\infty^*(t) = -\gamma^{-2}C_1^T P_{1,\infty}x(t)$ is implemented in system (2.127), we have

$$\begin{cases} dx(t) = [(A_1 - \gamma^{-2}C_1 C_1^T P_{1,\infty})x(t) + B_1 u(t)]\,dt + A_2 x(t)\,dB(t), \\ x(0) = x_0, \\ z(t) = \begin{bmatrix} Cx(t) \\ Du(t) \end{bmatrix}, \quad D^T D = I. \end{cases} \tag{2.137}$$

Now, the H_2 optimization becomes a standard stochastic LQ optimal control problem:

$$\min_{u\in\mathcal{U}_{ad}} J_{2,\infty}(u, v_\infty^*)$$

under the constraint of (2.137). Because $(A_1 - B_1 B_1^T P_{2,\infty} - \gamma^{-2}C_1 C_1^T P_{1,\infty}, A_2)$ is stable, we conclude that $(A_1 - \gamma^{-2}C_1 C_1^T P_{1,\infty}, B_1; A_2)$ is stabilizable. In addition, (2.129) can be written as

$$P_{2,\infty}(A_1 - \gamma^{-2}C_1 C_1^T P_{1,\infty}) + (A_1 - \gamma^{-2}C_1 C_1^T P_{1,\infty})^T P_{2,\infty} + A_2^T P_{2,\infty}A_2 \\ - P_{2,\infty}B_1 B_1^T P_{2,\infty} + C^T C = 0.$$

By taking into account that $(A_1 - \gamma^{-2}C_1C_1^T P_{1,\infty}, A_2|C)$ is exactly observable from the assumption, it follows from Theorem 2.10 immediately that

$$\min_{u\in\mathcal{U}_{ad}} J_{2,\infty}(u, v_\infty^*) = J_{2,\infty}(u_\infty^*, v_\infty^*) = x_0^T P_{2,\infty} x_0.$$

Accordingly, iii) is proved, and the proof of the theorem is complete. $\square$

By using Theorem 2.11 instead of Theorem 2.10 and repeating the same procedure as above, the conditions of Theorem 2.15 can be weakened.

THEOREM 2.16

Theorem 2.15 still holds under the following weaker conditions:

1. *The cross-coupled GAREs (2.128)–(2.129) have a pair of solutions $(P_{1,\infty} \leq 0, P_{2,\infty} \geq 0)$.*

2. *$(A_1, A_2|C)$ and $(A_1 - \gamma^{-2}C_1C_1^T P_{1,\infty}, A_2|C)$ are exactly detectable.*

By Theorem 2.10-3) and Theorem 2.11-(ii), under the conditions of exact observability and exact detectability, the cross-coupled GAREs (2.128)–(2.129) have at most one solution pair $(P_{1,\infty} \leq 0, P_{2,\infty} \geq 0)$.

THEOREM 2.17

Suppose the stochastic H_2/H_∞ control problem has a pair of solutions $(u_\infty^*(t), v_\infty^*(t))$ with

$$u_\infty^*(t) = K_2 x(t), \quad v_\infty^*(t) = K_1 x(t),$$

where $u_\infty^*(t)$ and $v_\infty^*(t)$ are time-invariant feedback laws. If $(A_1 + C_1 K_1, A_2|C)$ is exactly observable (respectively, exactly detectable), then the coupled GAREs (2.128)–(2.129) admit solutions $P_{1,\infty} \leq 0$ and $P_{2,\infty} > 0$ (respectively, $P_{2,\infty} \geq 0$). Moreover,

$$u_\infty^*(t) = -B_1^T P_{2,\infty} x(t), \quad v_\infty^*(t) = -\gamma^{-2}C_1^T P_{1,\infty} x(t).$$

Proof. By means of Theorem 2.12, this theorem can be proved following the line of Theorem 3.1 [113]. For the reader's convenience, we give the detail as follows. Implementing $u(t) = u_\infty^*(t) = K_2 x(t)$ in (2.127) gives

$$dx(t) = [(A_1 + B_1 K_2)x(t)\, dt + C_1 v(t)]\, dt + A_2 x(t)\, dB(t), \quad x(0) = x_0 \quad (2.138)$$

$$z(t) = \begin{bmatrix} Cx(t) \\ DK_2 x(t) \end{bmatrix}. \quad (2.139)$$

By Definition 2.7, $(A_1 + B_1 K_2, A_2)$ is stable and $\|\mathcal{L}_\infty\| < \gamma$. By Theorem 2.12, the following equation

$$P(A_1 + B_1 K_2) + (A_1 + B_1 K_2)^T P + A_2^T P A_2 - C^T C - K_2^T K_2 - \gamma^{-2} P C_1 C_1^T P = 0 \quad (2.140)$$

has a solution $P_{1,\infty} \leq 0$, and $(A_1 + B_1 K_2 - \gamma^{-2} C_1 C_1^T P_{1,\infty}, A_2)$ is stable.

Since $(A_1 + B_1 K_2, A_2)$ is stable, for any $v \in \mathcal{L}_\mathcal{F}^2(\mathcal{R}^+, \mathcal{R}^{n_v})$, by Remark 2.6 of [84], $x(t) \in \mathcal{L}_\mathcal{F}^2(\mathcal{R}^+, \mathcal{R}^n)$. Consider system (2.138) and GARE (2.140), by using Itô's formula and completing the squares, we have that

$$J_{1,\infty}(u_\infty^*, v) = \mathcal{E} \int_0^\infty (\gamma^2 \|v(t)\|^2 - \|z(t)\|^2)\, dt$$

$$= \mathcal{E} \int_0^\infty (\gamma^2 \|v(t)\|^2 - \|z(t)\|^2)\, dt$$

$$+ \mathcal{E} \int_0^\infty d(x^T P_{1,\infty} x(t))$$

$$+ x_0^T P_{1,\infty} x_0 - \lim_{t \to \infty} \mathcal{E}[x^T(t) P_{1,\infty} x(t)]$$

$$= x_0^T P_{1,\infty} x_0 + \gamma^2 \mathcal{E} \int_0^\infty \|v(t) + \gamma^{-2} C_1^T P_{1,\infty} x(t)\|^2\, dt$$

$$\geq x_0^T P_{1,\infty} x_0 = J_{1,\infty}(u_\infty^*, v_\infty^*),$$

where $v_\infty^*(t) = -\gamma^{-2} C_1^T P_{1,\infty} x(t) \in \mathcal{L}_\mathcal{F}^2(\mathcal{R}^+, \mathcal{R}^{n_v})$. So

$$K_1 = -\gamma^{-2} C_1^T P_{1,\infty}.$$

Implementing $v(t) = v_\infty^*(t) = -\gamma^{-2} C_1^T P_{1,\infty} x(t)$ in system (2.127) yields

$$\begin{cases} dx(t) = [(A - \gamma^{-2} C_1 C_1^T P_{1,\infty}) x(t) + B_1 u(t)]\, dt + A_2 x(t)\, dB(t), \\ x(0) = x_0. \end{cases} \quad (2.141)$$

By Theorem 2.12, $(A_1 + B_1 K_2 - \gamma^{-2} C_1 C_1^T P_{1,\infty}, A_2)$ is stable, so $(A_1 - \gamma^{-2} C_1 C_1^T P_{1,\infty}, B_1; A_2)$ is stabilizable. Further, noting the assumption that $(A_1 + C_1 K_1, A_2 | C) = (A_1 - \gamma^{-2} C_1 C_1^T P_{1,\infty}, A_2 | C)$ is exactly observable (respectively, exactly detectable), it follows from Theorem 2.10 (respectively, Theorem 2.11) that the following GARE

$$P_{2,\infty}(A_1 - \gamma^{-2} C_1 C_1^T P_{1,\infty}) + (A_1 - \gamma^{-2} C_1 C_1^T P_{1,\infty})^T P_{2,\infty} + A_2^T P_{2,\infty} A_2$$
$$- P_{2,\infty} B_1 B_1^T P_{2,\infty} + C^T C = 0 \quad (2.142)$$

has a unique solution $P_{2,\infty} > 0$ (respectively, $P_{2,\infty} \geq 0$), which satisfies that

$$\min_{u \in \mathcal{U}_{ad}} \mathcal{E} \int_0^\infty \|z(t)\|^2\, dt = x_0^T P_{2,\infty} x_0,$$

$$u_\infty^*(t) = -B_1^T P_{2,\infty} x(t).$$

So $K_2 = -B_1^T P_{2,\infty}$. It is easy to test that equation (2.142) is the same as (2.129). Substituting $K_2 = -B_1^T P_{2,\infty}$ into (2.140) yields GARE (2.128). This ends the proof of Theorem 2.17. $\square$

Theorems 2.16–2.17 extend Theorem 3.1 of [113] to stochastic systems.

REMARK 2.15   In Theorems 2.15–2.17, if $C^T C > 0$, then exact observability and exact detectability are automatically satisfied. The mixed H_2/H_∞ control can be decomposed into two problems: one is about a standard stochastic LQ control, and the other one is SBRL. ∎

Case 2: H_2/H_∞ control with (x, u)-dependent noise

As a special case of (2.69) but a more general case than (2.127), we consider the following system with state- and control-dependent noise ((x, u)-dependent noise for short):

$$\begin{cases} dx(t) = [A_1 x(t) + B_1 u(t) + C_1 v(t)]\, dt + [A_2 x(t) + B_2 u(t)]\, dB(t), \\ x(0) = x_0, \\ z(t) = \begin{bmatrix} Cx(t) \\ Du(t) \end{bmatrix}, \ D^T D = I. \end{cases} \quad (2.143)$$

Following the proofs of Theorems 2.15–2.17, it is easy to obtain the following theorem for stochastic H_2/H_∞ control with (x, u)-dependent noise.

THEOREM 2.18

For system (2.143), assume that $(A_1, A_2|C)$ and $(A_1 + C_1 K_1, A_2|C)$ are exactly observable (respectively, exactly detectable) for a constant matrix K_1. Then the stochastic H_2/H_∞ control admits a pair of solutions (u_∞^*, v_∞^*) with $u_\infty^*(t) = K_2 x(t)$ and $v_\infty^*(t) = K_1 x(t)$ iff the following two coupled GAREs

$$\begin{cases} P_{1,\infty}(A_1 + B_1 K_2) + (A_1 + B_1 K_2)^T P_{1,\infty} + (A_2 + B_2 K_2)^T P_{1,\infty}(A_2 + B_2 K_2) \\ \quad -C^T C - K_2^T K_2 - \gamma^{-2} P_{1,\infty} C_1 C_1^T P_{1,\infty} = 0, \\ K_2 = -(I + B_2^T P_{2,\infty} B_2)^{-1}(B_1^T P_{2,\infty} + B_2^T P_{2,\infty} A_2) \end{cases} \quad (2.144)$$

and

$$\begin{cases} P_{2,\infty}(A_1 + C_1 K_1) + (A_1 + C_1 K_1)^T P_{2,\infty} + A_2^T P_{2,\infty} A_2 + C^T C \\ \quad -(P_{2,\infty} B_1 + A_2^T P_{2,\infty} B_2)(I + B_2^T P_{2,\infty} B_2)^{-1}(B_1^T P_{2,\infty} + B_2^T P_{2,\infty} A_2) = 0, \\ I + B_2^T P_{2,\infty} B_2 > 0, \\ K_1 = -\gamma^{-2} C_1^T P_{1,\infty} \end{cases} \quad (2.145)$$

have a solution pair $(P_{1,\infty} < 0, P_{2,\infty} > 0)$ (respectively, $(P_{1,\infty} \le 0, P_{2,\infty} \ge 0)$).

Case 3: H_2/H_∞ control with (x, v)-dependent noise

Similarly, for the following system with state- and disturbance-dependent noise ((x, v)-dependent noise for short)

$$\begin{cases} dx(t) = [A_1 x(t) + B_1 u(t) + C_1(t) v(t)]\, dt + [A_2 x(t) + C_2 v(t)]\, dB(t), \\ x(0) = x_0 \in \mathcal{R}^n, \\ z(t) = \begin{bmatrix} Cx(t) \\ Du(t) \end{bmatrix}, \ D^T D = I, \end{cases}$$

$$(2.146)$$

we have the following mixed H_2/H_∞ control result.

THEOREM 2.19
For system (2.146), assume $(A_1, A_2|C)$ and $(A_1 + C_1 K_1, A_2 + C_2 K_1|C)$ are exactly observable (respectively, exactly detectable) for a constant matrix K_1. Then the stochastic H_2/H_∞ control admits a pair of memoryless state feedback solutions (u_∞^*, v_∞^*) with $u_\infty^*(t) = K_2 x(t)$ and $v_\infty^*(t) = K_1 x(t)$ iff the following two coupled GAREs

$$\begin{cases} P_{1,\infty}(A_1 + B_1 K_2) + (A_1 + B_1 K_2)^T P_{1,\infty} + A_2^T P_{1,\infty} A_2 - C^T C - K_2^T K_2 \\ \quad -(P_{1,\infty} C_1 + A_2^T P_{1,\infty} C_2)(\gamma^2 I + C_2^T P_{1,\infty} C_2)^{-1}(C_1^T P_{1,\infty} + C_2^T P_{1,\infty} A_2) = 0, \\ \gamma^2 I + C_2^T P_{1,\infty} C_2 > 0, \\ K_2 = -B_1^T P_{2,\infty} \end{cases}$$

(2.147)

and

$$\begin{cases} P_{2,\infty}(A_1 + C_1 K_1) + (A_1 + C_1 K_1)^T P_{2,\infty} + (A_2 + C_2 K_1)^T P_{2,\infty}(A_2 + C_2 K_1) \\ \quad + C^T C - P_{2,\infty} B_1 B_1^T P_{2,\infty} = 0, \\ K_1 = -(\gamma^2 I + C_2^T P_{1,\infty} C_2)^{-1}(C_1^T P_{1,\infty} + C_2^T P_{1,\infty} A_2) \end{cases}$$

(2.148)

have a solution $(P_{1,\infty} < 0, P_{2,\infty} > 0)$ (respectively, $(P_{1,\infty} \le 0, P_{2,\infty} \ge 0)$).

Case 4: H_2/H_∞ control with (x, u, v)-dependent noise
System (2.69) is more general since it has not only state-dependent noise, but also control- and disturbance-dependent noise $((x, u, v)$-dependent noise for short). For system (2.69), we have:

THEOREM 2.20
For system (2.69), assume that $(A_1, A_2|C)$ and $(A_1 + C_1 K_1, A_2 + C_2 K_1|C)$ are exactly observable (resp. exactly detectable) for a constant matrix K_1. Then the stochastic H_2/H_∞ control admits a pair of constant state feedback solutions (u_∞^*, v_∞^*) with $u_\infty^*(t) = K_2 x(t)$ and $v_\infty^*(t) = K_1 x(t)$ iff the following four coupled matrix-valued equations

$$\begin{cases} P_{1,\infty}(A_1 + B_1 K_2) + (A_1 + B_1 K_2)^T P_{1,\infty} + (A_2 + B_2 K_2)^T P_{1,\infty}(A_2 + B_2 K_2) \\ \quad -C^T C - K_2^T K_2 - [P_{1,\infty} C_1 + (A_2 + B_2 K_2)^T P_{1,\infty} C_2] \\ \quad \cdot (\gamma^2 I + C_2^T P_{1,\infty} C_2)^{-1}[C_1^T P_{1,\infty} + C_2^T P_{1,\infty}(A_2 + B_2 K_2)] = 0, \\ \gamma^2 I + C_2^T P_{1,\infty} C_2 > 0, \end{cases}$$

(2.149)

$$\begin{cases} P_{2,\infty}(A_1 + C_1 K_1) + (A_1 + C_1 K_1)^T P_{2,\infty} + (A_2 + C_2 K_1)^T P_{2,\infty}(A_2 + C_2 K_1) \\ \quad + C^T C - [P_{2,\infty} B_1 + (A_2 + C_2 K_1)^T P_{2,\infty} B_2](I + B_2^T P_{2,\infty} B_2)^{-1} \\ \quad \cdot [B_1^T P_{2,\infty} + B_2^T P_{2,\infty}(A_2 + C_2 K_1)] = 0, \\ I + B_2^T P_{2,\infty} B_2 > 0, \end{cases}$$

(2.150)

$$K_1 = -(\gamma^2 I + C_2^T P_{1,\infty} C_2)^{-1}[C_1^T P_{1,\infty} + C_2^T P_{1,\infty}(A_2 + B_2 K_2)], \quad (2.151)$$

$$K_2 = -(I + B_2^T P_{2,\infty} B_2)^{-1}[B_1^T P_{2,\infty} + B_2^T P_{2,\infty}(A_2 + C_2 K_1)] \quad (2.152)$$

have a solution $(P_{1,\infty} < 0, K_1; P_{2,\infty} > 0, K_2)$ *(resp.*$(P_{1,\infty} \leq 0, K_1; P_{2,\infty} \geq 0, K_2)$).

REMARK 2.16 From Theorem 2.20, it can be found that it is very difficult to design a general H_2/H_∞ controller with (x, u, v)-dependent noise, which requires us to solve four coupled matrix-valued equations. Up to now, we have no efficient method to solve (2.149)–(2.152), which deserves further study. Because of the appearances of (2.151) and (2.152), it is not appropriate to call (2.149)–(2.152) coupled GAREs. ∎

2.4 Relationship between Stochastic H_2/H_∞ and Nash Game

Similar to the finite horizon H_2/H_∞ control, in this section, we further discuss the relationship between the infinite horizon stochastic H_2/H_∞ control and the existence of a two-person non-zero sum Nash equilibrium point $(u_\infty^* = K_2 x, v_\infty^* = K_1 x)$, which satisfies

$$J_{1,\infty}(u_\infty^*, v_\infty^*) \leq J_{1,\infty}(u_\infty^*, v), \quad J_{2,\infty}(u_\infty^*, v_\infty^*) \leq J_{2,\infty}(u, v_\infty^*) \quad (2.153)$$

with $(u, v) \in \mathcal{U}_{ad}(v_\infty^*) \times \mathcal{V}_{ad}(u_\infty^*)$, where $\mathcal{U}_{ad}(v_\infty^*) \subset \mathcal{L}_\mathcal{F}^2(\mathcal{R}^+, \mathcal{R}^{n_u})$ denotes the set of all $u \in \mathcal{L}_\mathcal{F}^2(\mathcal{R}^+, \mathcal{R}^{n_u})$ which make the closed-loop system

$$dx(t) = [(A_1 + C_1 K_1)x(t) + B_1 u(t)]\, dt + [(A_2 + C_2 K_1)x(t) + B_2 u(t)]\, dB(t)$$

ASMS. Similarly, $\mathcal{V}_{ad}(u_\infty^*)$ can be defined.

THEOREM 2.21
Consider system (2.69). There exists a Nash equilibrium strategy $u_\infty^*(t) = K_2 x(t) \in \mathcal{U}_{ad}(v_\infty^*)$ *and* $v_\infty^*(t) = K_1 x(t) \in \mathcal{V}_{ad}(u_\infty^*)$ *with K_1 and K_2 being constant matrices iff the following four coupled matrix-valued equations*

$$\begin{cases} P_{1,\infty}(A_1 + B_1 K_2) + (A_1 + B_1 K_2)^T P_{1,\infty} + (A_2 + B_2 K_2)^T P_{1,\infty}(A_2 + B_2 K_2) \\ \quad -C^T C - K_2^T K_2 - [P_{1,\infty} C_1 + (A_2 + B_2 K_2)^T P_{1,\infty} C_2] \\ \quad \cdot(\gamma^2 I + C_2^T P_{1,\infty} C_2)^+[C_1^T P_{1,\infty} + C_2^T P_{1,\infty}(A_2 + B_2 K_2)] = 0, \\ [I - (\gamma^2 I + C_2^T P_{1,\infty} C_2)(\gamma^2 I + C_2^T P_{1,\infty} C_2)^+]\, [C_1^T P_{1,\infty} + C_2^T P_{1,\infty}(A_2 + B_2 K_2)] \\ = 0, \\ \gamma^2 I + C_2^T P_{1,\infty} C_2 \geq 0, \end{cases}$$

$$(2.154)$$

$$\begin{cases} P_{2,\infty}(A_1 + C_1 K_1) + (A_1 + C_1 K_1)^T P_{2,\infty} + (A_2 + C_2 K_1)^T P_{2,\infty}(A_2 + C_2 K_1) \\ + C^T C - [P_{2,\infty} B_1 + (A_2 + C_2 K_1)^T P_{2,\infty} B_2](I + B_2^T P_{2,\infty} B_2)^{-1} \\ \cdot [B_1^T P_{2,\infty} + B_2^T P_{2,\infty}(A_2 + C_2 K_1)] = 0, \\ I + B_2^T P_{2,\infty} B_2 > 0, \end{cases}$$

$$(2.155)$$

$$K_1 = -(\gamma^2 I + C_2^T P_{1,\infty} C_2)^+ [C_1^T P_{1,\infty} + C_2^T P_{1,\infty}(A_2 + B_2 K_2)], \qquad (2.156)$$

and

$$K_2 = -(I + B_2^T P_{2,\infty} B_2)^{-1}[B_1^T P_{2,\infty} + B_2^T P_{2,\infty}(A_2 + C_2 K_1)] \qquad (2.157)$$

have a solution $(P_{1,\infty} \leq 0, K_1; P_{2,\infty} \geq 0, K_2)$.

Proof. Sufficiency: If the equations (2.154)–(2.157) have a solution $(P_{1,\infty} \leq 0, K_1; P_{2,\infty} \geq 0, K_2)$, substituting $u = u_\infty^* = K_2 x$ into (2.69) with K_2 defined by (2.157), it follows that

$$\begin{cases} dx(t) = [(A_1 + B_1 K_2)x(t) + C_1(t)v(t)]\,dt + [(A_2 + B_2 K_2)x(t) + C_2 v(t)]\,dB(t), \\ x(0) = x_0 \in \mathcal{R}^n, \\ z(t) = \begin{bmatrix} Cx(t) \\ DK_2 x(t) \end{bmatrix}, \quad D^T D = I, \; t \in [0, \infty). \end{cases}$$

$$(2.158)$$

Under the constraint of

$$\begin{cases} dx(t) = [(A_1 + B_1 K_2)x(t) + C_1(t)v(t)]\,dt + [(A_2 + B_2 K_2)x(t) + C_2 v(t)]\,dB(t), \\ x(0) = x_0 \in \mathcal{R}^n, \end{cases}$$

$$(2.159)$$

to minimize

$$\min_{v \in \mathcal{V}_{ad}(u_\infty^*)} J_{1,\infty}(u_\infty^*, v)$$

is an indefinite LQ problem with the control weighting matrix $R := \gamma^2 I$, while the state weighting matrix $Q := -(C^T C + K_2^T K_2)$. Because of $v_\infty^*(t) = K_1 x(t) \in \mathcal{V}_{ad}(u_\infty^*)$ with K_1 defined by (2.156), using Theorem 2.1 of [6] and (2.154), $v_\infty^* = K_1 x$ is an optimal control law (note: the optimal controller is not unique) for this LQ optimal control. Hence, the first inequality of (2.153) holds. Similarly, substituting $v_\infty^* = K_1 x$ into (2.69), under the constraint of

$$\begin{cases} dx(t) = [(A_1 + C_1 K_1)x(t) + B_1(t)u(t)]\,dt + [(A_2 + C_2 K_1)x(t) + B_2 u(t)]\,dB(t), \\ x(0) = x_0 \in \mathcal{R}^n, \end{cases}$$

$$(2.160)$$

to minimize

$$\min_{u \in \mathcal{U}_{ad}(v_\infty^*)} J_{2,\infty}(u, v_\infty^*)$$

is a standard LQ control problem with the control weighting matrix $R := I$ and the state weighting matrix $Q := C^T C$. Considering (2.155) and $u_\infty^*(t) = K_2 x(t) \in$

$\mathcal{U}_{ad}(v_\infty^*)$, re-applying Theorem 2.1 of [6], the second inequality of (2.153) can be derived.

Necessity: This is a direct corollary of Theorem 4.1 of [6]. $\square$

When $C_2 = 0$, the second and third constraints of (2.154) hold automatically. The following theorem reveals the relationship between the H_2/H_∞ control and the existence of a Nash equilibrium strategy.

COROLLARY 2.2

For system (2.143), the following are equivalent:

(i) *There exists a Nash equilibrium strategy pair (u_∞^*, v_∞^*) with*

$$u_\infty^*(t) = K_2 x(t) \in \mathcal{U}_{ad}(v_\infty^*), \quad v_\infty^*(t) = K_1 x(t) \in \mathcal{V}_{ad}(u_\infty^*)$$

(ii) *The cross-coupled GAREs (2.144) and (2.145) have feedback stabilizing solutions $P_{1,\infty} \le 0$ and $P_{2,\infty} \ge 0$, respectively, i.e., $(A_1 + B_1 K_2 + C_1 K_1, A_2 + B_2 K_2)$ is stable with*

$$K_1 = -\gamma^{-2} C_1^T P_{1,\infty}, \quad K_2 = -(I + B_2^T P_{2,\infty} B_2)^{-1}(B_1^T P_{2,\infty} + B_2^T P_{2,\infty} A_2).$$

(iii) *The stochastic H_2/H_∞ control is solvable with*

$$u_\infty^*(t) = K_2 x(t) \in \mathcal{U}_{ad}(v_\infty^*), \ v_\infty^*(t) = K_1 x(t) \in \mathcal{V}_{ad}(u_\infty^*).$$

Proof. (i)$\Leftrightarrow$ (ii) is a corollary of Theorem 2.21. In fact, for $C_2 = 0$, (2.154) and (2.155) reduce to (2.144) and (2.145), respectively.

(iii)$\Rightarrow$ (ii): Following the proof of Theorem 2.17, if the stochastic H_2/H_∞ control is solvable with $u_\infty^*(t) = K_2 x(t) \in \mathcal{U}_{ad}(v_\infty^*)$, $v_\infty^*(t) = K_1 x(t) \in \mathcal{V}_{ad}(u_\infty^*)$, then the cross-coupled GAREs (2.144) and (2.145) have the feedback stabilizing solution $(P_{1,\infty} \le 0, P_{2,\infty} \ge 0)$ without the assumption of exact observability/exact detectability.

(ii) $\Rightarrow$ (iii): Because (2.144) has a feedback stabilizing solution $P_{1,\infty} \le 0$, by Theorem 2.12, system (2.143) is internally stable, and $\|\mathcal{L}_\infty\| < \gamma$. In addition, GARE (2.145) having a feedback stabilizing solution $P_{2,\infty} \ge 0$ leads to the fact that $u_\infty^*(t) = K_2 x(t)$ minimizes $J_{2,\infty}(u, v_\infty^*)$ under the constraint of

$$\begin{cases} dx(t) = [(A_1 + C_1 K_1)x(t) + B_1 u(t)] \, dt + [A_2 x(t) + B_2 u(t)] \, dB(t), \\ x(0) = x_0. \end{cases}$$

The proof is complete. $\square$

However, when $C_2 \ne 0$, Corollary 2.2 does not hold, because the existence of Nash equilibrium strategy does not guarantee $\gamma^2 I + C_2^T P_{1,\infty} C_2 > 0$. In fact, to search for a Nash equilibrium point (u_∞^*, v_∞^*), we need to solve two LQ optimization problems, one is an indefinite LQ optimization problem and the other is a standard one.

2.5   Algorithm for Solving Coupled GAREs

Some well-known algorithms for solving coupled GAREs can be found in [61, 108, 160]. In this section, we present a convex optimization algorithm to solve the two crossed-coupled GAREs arising from the stochastic H_2/H_∞ control for systems with state-, or (x, u)-, or (x, v)-dependent noise. For simplicity, we only consider the coupled GAREs (2.128)–(2.129) and assume $C^T C > 0$. Under the assumption $C^T C > 0$, $P_{1,\infty} < 0$ and $P_{2,\infty} > 0$, while the exact observability and detectability are automatically satisfied. Let

$$\mathcal{H}_1(P_{1,\infty}, P_{2,\infty}) := A_1^T P_{1,\infty} + P_{1,\infty} A_1 + A_2^T P_{1,\infty} A_2 - C^T C$$
$$- [P_{1,\infty}, P_{2,\infty}] \begin{bmatrix} \gamma^{-2} C_1 C_1^T & B_1 B_1^T \\ B_1 B_1^T & B_1 B_1^T \end{bmatrix} \begin{bmatrix} P_{1,\infty} \\ P_{2,\infty} \end{bmatrix}$$
$$= A_1^T P_{1,\infty} + P_{1,\infty} A_1 + A_2^T P_{1,\infty} A_2 - C^T C - P_{1,\infty} B_1 B_1^T P_{2,\infty}$$
$$- \gamma^{-2} P_{1,\infty} C_1 C_1^T P_{1,\infty} - P_{2,\infty} B_1 B_1^T P_{2,\infty} - P_{2,\infty} B_1 B_1^T P_{1,\infty}$$

and

$$\mathcal{H}_2(P_{1,\infty}, P_{2,\infty}) := A_1^T P_{2,\infty} + P_{2,\infty} A_1 + A_2^T P_{2,\infty} A_2 + C^T C$$
$$- [P_{1,\infty}, P_{2,\infty}] \begin{bmatrix} 0 & \gamma^{-2} C_1 C_1^T \\ \gamma^{-2} C_1 C_1^T & B_1 B_1^T \end{bmatrix} \begin{bmatrix} P_{1,\infty} \\ P_{2,\infty} \end{bmatrix}$$
$$= A_1^T P_{2,\infty} + P_{2,\infty} A_1 + A_2^T P_{2,\infty} A_2 + C^T C - \gamma^{-2} P_{1,\infty} C_1 C_1^T P_{2,\infty}$$
$$- \gamma^{-2} P_{2,\infty} C_1 C_1^T P_{1,\infty} - P_{2,\infty} B_1 B_1^T P_{2,\infty}.$$

By Theorem 10 of [2], $(P_{1,\infty}, P_{2,\infty})$ is the optimal solution to

$$\max_{\mathcal{H}_1(P_{1,\infty},P_{2,\infty})>0, \, \mathcal{H}_2(P_{1,\infty},P_{2,\infty})>0, P_{1,\infty}<0, \, P_{2,\infty}>0} \mathrm{Trace}(P_{1,\infty} + P_{2,\infty}).$$

We note that $\mathcal{H}_1(P_{1,\infty}, P_{2,\infty}) > 0$ and $\mathcal{H}_2(P_{1,\infty}, P_{2,\infty}) > 0$ only if

$$\bar{\mathcal{H}}_1(P_{1,\infty}, P_{2,\infty}) := A_1^T P_{1,\infty} + P_{1,\infty} A_1 + A_2^T P_{1,\infty} A_2 - C^T C$$
$$- P_{1,\infty}(\gamma^{-2} C_1 C_1^T + B_1 B_1^T) P_{1,\infty} - 2 P_{2,\infty} B_1 B_1^T P_{2,\infty} > 0$$

and

$$\bar{\mathcal{H}}_2(P_{1,\infty}, P_{2,\infty}) := A_1^T P_{2,\infty} + P_{2,\infty} A_1 + A_2^T P_{2,\infty} A_2 + C^T C$$
$$- P_{2,\infty}(\gamma^{-2} C_1 C_1^T + B_1 B_1^T) P_{2,\infty} - \gamma^{-2} P_{1,\infty} C_1 C_1^T P_{1,\infty} > 0,$$

respectively. By the well known Schur's complement lemma, $\bar{\mathcal{H}}_1(P_{1,\infty}, P_{2,\infty}) > 0$ and $\bar{\mathcal{H}}_2(P_{1,\infty}, P_{2,\infty}) > 0$ are respectively equivalent to

$$\bar{M}_1(P_{1,\infty}, P_{2,\infty}) := \begin{bmatrix} M_{11} & M_{12} & \sqrt{2}P_{2,\infty}B_1 \\ M_{12}^T & I & 0 \\ \sqrt{2}B_1^T P_{2,\infty} & 0 & I \end{bmatrix} > 0 \qquad (2.161)$$

and

$$\bar{N}_1(P_{1,\infty}, P_{2,\infty}) := \begin{bmatrix} N_{11} & N_{12} & \gamma^{-1}P_{2,\infty}C_1 \\ N_{12}^T & I & 0 \\ \gamma^{-1}C_1^T P_{2,\infty} & 0 & I \end{bmatrix} > 0, \qquad (2.162)$$

where

$$M_{11} = A_1^T P_{1,\infty} + P_{1,\infty}A_1 + A_2^T P_{1,\infty}A_2 - C^T C,$$

$$M_{12} = P_{1,\infty}(\gamma^{-2}C_1 C_1^T + B_1 B_1^T)^{1/2},$$

$$N_{11} = A_1^T P_{2,\infty} + P_{2,\infty}A_1 + A_2^T P_{2,\infty}A_2 + C^T C,$$

$$N_{12} = P_{2,\infty}(\gamma^{-2}C_1 C_1^T + B_1 B_1^T)^{1/2}.$$

(2.161) and (2.162) are LMIs, so a suboptimal solution to GAREs (2.128)–(2.129) can be obtained by solving the following convex optimization problem:

$$\max_{\bar{M}_1(P_1,P_2)>0,\ \bar{N}_1(P_1,P_2)>0,\ P_1<0,\ P_2>0} \text{Trace}(P_1 + P_2).$$

2.6   Notes and References

For linear stochastic H_∞ control of Itô systems, we refer the reader to the good references [84, 169]. The reference [36] first used the Nash game approach to study the stochastic H_2/H_∞ control design, which can be viewed as an extension of [113]. Since [36], a series of related works have appeared; see [90, 134, 139, 207, 208, 213, 214].

As far as the linear stochastic H_2/H_∞ control for Itô systems is concerned, there are many problems that remain unsolved. Firstly, although we can use convex optimization techniques to design H_∞ or H_2 controllers [2, 58, 84], and apply the iterative algorithm to solve the LQ zero-sum games [73], as mentioned above, currently, there is no efficient method to solve the four cross-coupled GDREs and the four cross-coupled GAREs. Therefore, there is a need to search for efficient algorithms in this regard. Secondly, in this chapter, we assume that all the states are available for feedback, that is, we only discuss the state feedback H_2/H_∞ control. However, in practical applications, we usually only know partial state information via direct measurement, in this case, we have to consider the output feedback H_2/H_∞ design as in [1, 84, 188, 79] for the H_∞ control and [35] for the H_2 one. Thirdly, because time delay widely exists in the real world, the study of time delay systems has received a great deal of attention [156, 202]. How to extend the results of this chapter to linear stochastic time-delay systems deserves further study. Finally, as in [33], we can consider combining the stochastic H_2/H_∞ design with the spectral placement constraints based on the spectrum technique of GLOs [211].

3

Linear Discrete-Time Stochastic H_2/H_∞ Control

In this chapter we continue to study the H_2/H_∞ control for discrete-time stochastic systems with multiplicative noise, which have many applications, for example, in networked control [187, 199, 201] and power control in CDMA systems [153]. Compared with discrete-time stochastic systems with additive noise [138], the H_2/H_∞ control of systems with multiplicative noise is more challenging. The study of the H_∞ control for discrete-time systems with state- and disturbance-dependent noise seems to start from [65], where a very useful SBRL was given in terms of LMIs, which has played an important role in H_∞ filter design [78]. For the system dealt with in [65], the finite and infinite horizon mixed H_2/H_∞ control problems were investigated in [208] and [207], respectively. Similar to linear continuous-time Itô systems, the existence of mixed H_2/H_∞ controllers for general discrete-time systems with multiplicative noise is equivalent to the solvability of four coupled difference matrix-valued equations. However, they differ in that the four coupled difference equations can be solved recursively.

3.1 Finite Horizon H_2/H_∞ Control

In this section, we will consider finite horizon H_2/H_∞ control for the following general linear discrete-time stochastic system with multiplicative noise

$$\begin{cases} x_{k+1} = A_1^k x_k + B_1^k u_k + C_1^k v_k + (A_2^k x_k + B_2^k u_k + C_2^k v_k)w_k, \ x_0 \in \mathcal{R}^n, \\ z_k = \begin{bmatrix} C_k x_k \\ D_k u_k \end{bmatrix}, \ D_k^T D_k = I, \ k \in N_T := \{0, 1, 2, \cdots, T\}, \end{cases}$$

$$(3.1)$$

where $x_k \in \mathcal{R}^n$, $u_k \in \mathcal{R}^{n_u}$ and $v_k \in \mathcal{R}^{n_v}$ are, as in Chapter 2, called respectively the system state, control input and external disturbance. $\{w_k, k \in N_T\}$ is a sequence of real independent random variables defined on a complete probability space $\{\Omega, \mathcal{F}, \mathcal{P}\}$, which is a wide sense stationary, second-order moment process with $\mathcal{E}(w_k) = 0$ and $\mathcal{E}(w_k w_s) = \delta_{sk}$, where δ_{sk} is a Kronecker function. We denote $\mathcal{F}_k$ as the σ-field generated by $w_s, s = 0, 1, \cdots, k$, i.e., $\mathcal{F}_k = \sigma(w_s : s \in N_k)$. Let $\mathcal{L}_{\mathcal{F}_i}^2(\Omega, \mathcal{R}^k)$ represent the space of $\mathcal{R}^k$-valued $\mathcal{F}_i$-measurable random vectors $\zeta(\omega)$ with $\mathcal{E}\|\zeta(\omega)\|^2 < \infty$, and $l_w^2(N_T, \mathcal{R}^k)$ consists of all finite sequences $y = \{y_i : y_i \in$

$\mathcal{R}^k\}_{i \in N_\mathcal{T}} = \{y_0, y_1, \cdots, y_\mathcal{T}\}$, such that $y_i \in \mathcal{L}_{\mathcal{F}_{i-1}}^2(\Omega, \mathcal{R}^k)$ for $i \in N_\mathcal{T}$, where we define $\mathcal{F}_{-1} = \{\phi, \Omega\}$, i.e, y_0 is a constant. The l^2-norm of $l_w^2(N_\mathcal{T}, \mathcal{R}^k)$ is defined by

$$\| y \|_{l_w^2(N_\mathcal{T}, \mathcal{R}^k)} = \left(\sum_{i=0}^{\mathcal{T}} \mathcal{E} \| y_i \|^2 \right)^{\frac{1}{2}}.$$

Obviously, for any $\mathcal{T} \in \mathcal{N} := \{0, 1, 2, \cdots, \}$ and $(x_0, u, v) \in \mathcal{R}^n \times l_w^2(N_\mathcal{T}, \mathcal{R}^{n_u}) \times l_w^2(N_\mathcal{T}, \mathcal{R}^{n_v})$, there exists a unique solution $\{x_{k;x_0}\} \in l_w^2(N_{\mathcal{T}+1}, \mathcal{R}^n)$ of (3.1) with initial value x_0. System (3.1) is very general in the sense that it is with (x, u, v)-dependent noise, which includes the (x, v)-dependent noise system [207, 208]

$$\begin{cases} x_{k+1} = (A_1^k x_k + B_1^k u_k + C_1^k v_k) + (A_2^k x_k + C_2^k v_k) w_k, \\ z_k = \begin{bmatrix} C_k x_k \\ D_k u_k \end{bmatrix}, \ D_k^T D_k = I, \ k \in N_\mathcal{T}, \end{cases} \tag{3.2}$$

and the (x, u)-dependent noise system

$$\begin{cases} x_{k+1} = (A_1^k x_k + B_1^k u_k + C_1^k v_k) + (A_2^k x_k + B_2^k u_k) w_k, \\ z_k = \begin{bmatrix} C_k x_k \\ D_k u_k \end{bmatrix}, \ D_k^T D_k = I, \ k \in N_\mathcal{T} \end{cases} \tag{3.3}$$

as special cases.

3.1.1   Definitions

We first define the so-called finite horizon mixed H_2/H_∞ control as follows.

DEFINITION 3.1 (Finite horizon mixed H_2/H_∞ control)
Consider system (3.1). Given a prescribed level of disturbance attenuation level $\gamma > 0$, and $0 < \mathcal{T} < \infty$, find, if it exists, a state feedback control $u = u_\mathcal{T}^* = \{u_k = u_{\mathcal{T},k}^* = K_{2,\mathcal{T}}^k x_k\}_{k \in N_\mathcal{T}} \in l_w^2(N_\mathcal{T}, \mathcal{R}^{n_u})$ such that
i) For the closed-loop system

$$\begin{cases} x_{k+1} = (A_1^k + B_1^k K_{2,\mathcal{T}}^k)x_k + C_1^k v_k + [(A_2^k + B_2^k K_{2,\mathcal{T}}^k)x_k + C_2^k v_k]w_k, \\ z_k = \begin{bmatrix} C_k x_k \\ D_k K_{2,\mathcal{T}}^k x_k \end{bmatrix}, \ D_k^T D_k = I, \ k \in N_\mathcal{T}, \end{cases} \tag{3.4}$$

the following holds:

$$\| \mathcal{L}_\mathcal{T} \| = \sup_{v \in l_w^2(N_\mathcal{T}, \mathcal{R}^{n_v}), v \neq 0, u = u_\mathcal{T}^*, x_0 = 0} \frac{\| z \|_{l_w^2(N_\mathcal{T}, \mathcal{R}^{n_z})}}{\| v \|_{l_w^2(N_\mathcal{T}, \mathcal{R}^{n_v})}}$$

$$= \sup_{v \in l_w^2(N_\mathcal{T}, \mathcal{R}^{n_v}), v \neq 0, u = u_\mathcal{T}^*, x_0 = 0} \frac{U}{V} < \gamma,$$

where

$$U = \left\{ \sum_{k=0}^{\mathcal{T}} \mathcal{E}[x_k^T C_k^T C_k x_k + x_k^T (K_{2,\mathcal{T}}^k)^T K_{2,\mathcal{T}}^k x_k] \right\}^{\frac{1}{2}},$$

$$V = \left[\sum_{k=0}^{\mathcal{T}} \mathcal{E}(v_k^T v_k) \right]^{\frac{1}{2}}.$$

ii) When the worst-case disturbance $v = v_{\mathcal{T}}^* = \{v_k = v_{\mathcal{T},k}^*, k \in N_{\mathcal{T}}\} \in l_w^2(N_{\mathcal{T}}, \mathcal{R}^{n_v})$, *if it exists, is implemented in (3.1),* $u_{\mathcal{T}}^*$ *simultaneously minimizes the output energy*

$$J_{2,\mathcal{T}}(u, v_{\mathcal{T}}^*) := \|z\|_{l_w^2(N_{\mathcal{T}}, \mathcal{R}^{n_z})}^2 = \sum_{k=0}^{\mathcal{T}} \mathcal{E}(x_k^T C_k^T C_k x_k + u_k^T u_k).$$

When the above $(u_{\mathcal{T}}^*, v_{\mathcal{T}}^*)$ exist, we say that the finite horizon H_2/H_∞ control is solvable. It should be noted that in the discrete-time case, x_k, u_k and v_k are all $\mathcal{F}_{k-1}$-adapted, but not $\mathcal{F}_k$-adapted as in Itô systems.

If we further define

$$J_{1,\mathcal{T}}(u, v) := \gamma^2 \|v\|_{l_w^2(N_{\mathcal{T}}, \mathcal{R}^{n_v})}^2 - \|z\|_{l_w^2(N_{\mathcal{T}}, \mathcal{R}^{n_z})}^2$$

associated with the system (3.1), then, as will be seen later, different from continuous-time Itô systems, even for the simplest system with only state-dependent noise

$$\begin{cases} x_{k+1} = (A_1^k x_k + B_1^k u_k + C_1^k v_k) + A_2^k x_k w_k, \\ z_k = \begin{bmatrix} C_k x_k \\ D_k u_k \end{bmatrix}, \ D_k^T D_k = I, \ k \in N_{\mathcal{T}}, \end{cases} \tag{3.5}$$

the solvability of the finite horizon discrete-time stochastic H_2/H_∞ control has no close relation with the existence of the following Nash equilibria $(u_{\mathcal{T}}^*, v_{\mathcal{T}}^*) \in l_w^2(N_{\mathcal{T}}, \mathcal{R}^{n_u}) \times l_w^2(N_{\mathcal{T}}, \mathcal{R}^{n_v})$ defined by

$$J_{1,\mathcal{T}}(u_{\mathcal{T}}^*, v_{\mathcal{T}}^*) \le J_{1,\mathcal{T}}(u_{\mathcal{T}}^*, v), \ J_{2,\mathcal{T}}(u_{\mathcal{T}}^*, v_{\mathcal{T}}^*) \le J_{2,\mathcal{T}}(u, v_{\mathcal{T}}^*). \tag{3.6}$$

In fact, by Definition 3.1, if the finite horizon H_2/H_∞ control is solvable, $(u_{\mathcal{T}}^*, v_{\mathcal{T}}^*)$ satisfy not only (3.6) but also

$$J_{1,\mathcal{T}}(u_{\mathcal{T}}^*, v_{\mathcal{T}}^*) > 0 \text{ for } x_0 = 0, (u_{\mathcal{T}}^*, v_{\mathcal{T}}^*) \in l_w^2(N_{\mathcal{T}}, \mathcal{R}^{n_u}) \times l_w^2(N_{\mathcal{T}}, \mathcal{R}^{n_v}). \tag{3.7}$$

Even so, we still call Definition 3.1 a Nash game-based definition for the H_2/H_∞ control in order to differentiate it from other previous definitions; see [37, 78, 79, 81, 112].

It is well known that SBRL plays a central role in the study of stochastic H_2/H_∞ control. To develop a finite horizon SBRL, we consider the following discrete-time linear time-varying stochastic system

$$\begin{cases} x_{k+1} = (A_{11}^k x_k + B_{11}^k v_k) + (A_{12}^k x_k + B_{12}^k v_k) w_k, \\ z_1^k = C_{11}^k x_k, \ k \in N_{\mathcal{T}}. \end{cases} \tag{3.8}$$

DEFINITION 3.2   *The perturbed operator of system (3.8) is defined by*

$$\tilde{\mathcal{L}}_T : l_w^2(N_T, \mathcal{R}^{n_v}) \mapsto l_w^2(N_T, \mathcal{R}^{n_{z_1}}),$$

$$\tilde{\mathcal{L}}_T(v_k) = z_1^k|_{x_0=0} = C_{11}^k x_k|_{x_0=0}, \ \forall v = \{v_i : i \in N_T\} \in l_w^2(N_T, \mathcal{R}^{n_v})$$

with its norm defined by

$$\|\tilde{\mathcal{L}}_T\| := \sup_{v \in l_w^2(N_T, \mathcal{R}^{n_v}), v \neq 0, x_0 = 0} \frac{\| z_1 \|_{l_w^2(N_T, \mathcal{R}^{n_{z_1}})}}{\| v \|_{l_w^2(N_T, \mathcal{R}^{n_v})}}$$

$$= \sup_{v \in l_w^2(N_T, \mathcal{R}^{n_v}), v \neq 0, x_0 = 0} \frac{\left(\displaystyle\sum_{k=0}^{T} \mathcal{E} \| C_{11}^k x_k \|^2 \right)^{\frac{1}{2}}}{\left(\displaystyle\sum_{k=0}^{T} \mathcal{E} \| v_k \|^2 \right)^{\frac{1}{2}}}.$$

3.1.2 Two identities

In the following, we give two identities that are necessary for the proofs of our main results.

LEMMA 3.1
In system (3.8), suppose $T \in \mathcal{N}$ is given and $P_0, P_1, \cdots, P_{T+1}$ belong to $\mathcal{S}_n$, then for any $x_0 \in \mathcal{R}^n$, we have

$$\sum_{k=0}^{T} \mathcal{E} \begin{bmatrix} x_k \\ v_k \end{bmatrix}^T Q(P_k) \begin{bmatrix} x_k \\ v_k \end{bmatrix} = \mathcal{E}(x_{T+1}^T P_{T+1} x_{T+1}) - x_0^T P_0 x_0, \qquad (3.9)$$

where

$$Q(P_k) = \begin{bmatrix} \Pi_{11} & (A_{11}^k)^T P_{k+1} B_{11}^k + (A_{12}^k)^T P_{k+1} B_{12}^k \\ \Pi_{21} & (B_{11}^k)^T P_{k+1} B_{11}^k + (B_{12}^k)^T P_{k+1} B_{12}^k \end{bmatrix},$$

$$\Pi_{11} = -P_k + (A_{11}^k)^T P_{k+1} A_{11}^k + (A_{12}^k)^T P_{k+1} A_{12}^k,$$

$$\Pi_{21} = (B_{11}^k)^T P_{k+1} A_{11}^k + (B_{12}^k)^T P_{k+1} A_{12}^k.$$

Proof. Since $x_0 \in \mathcal{R}^n$ is deterministic, $A_{11}^k x_k + B_{11}^k v_k$ and $A_{12}^k x_k + B_{12}^k v_k$ are independent of w_k, which are $\mathcal{F}_{k-1}$ measurable. In view of $\mathcal{E} w_k = 0$, we have

$$\mathcal{E} \left[(A_{11}^k x_k + B_{11}^k v_k)^T P_{k+1} (A_{12}^k x_k + B_{12}^k v_k) w_k \right]$$
$$= \mathcal{E} \left[(A_{12}^k x_k + B_{12}^k v_k)^T P_{k+1} (A_{11}^k x_k + B_{11}^k v_k) w_k \right]$$
$$= 0.$$

In addition, in view of $\mathcal{E} w_k^2 = 1$, it follows that

$$
\begin{aligned}
&\mathcal{E}(x_{k+1}^T P_{k+1} x_{k+1} - x_k^T P_k x_k) \\
&= \mathcal{E}\left[(A_{11}^k x_k + B_{11}^k v_k)^T P_{k+1}(A_{11}^k x_k + B_{11}^k v_k)\right] \\
&\quad + \mathcal{E}[(A_{12}^k x_k + B_{12}^k v_k)^T P_{k+1}(A_{12}^k x_k + B_{12}^k v_k)] - \mathcal{E}(x_k^T P_k x_k) \\
&= \mathcal{E}\begin{bmatrix} x_k \\ v_k \end{bmatrix}^T Q(P_k)\begin{bmatrix} x_k \\ v_k \end{bmatrix}.
\end{aligned}
$$

By taking summation of the above equality over $k = 0, 1, \cdots, \mathcal{T}$, we obtain (3.9) and the proof is complete. $\square$

LEMMA 3.2

In system (3.8), suppose $\mathcal{T} \in \mathcal{N}$ is given and $P_0, P_1, \cdots, P_{\mathcal{T}+1} \in \mathcal{S}_n$, then for any $x_0 \in \mathcal{R}^n, v \in l_w^2(N_{\mathcal{T}}, \mathcal{R}^{n_v})$, we have

$$
\sum_{k=0}^{\mathcal{T}} \mathcal{E}(\gamma^2 \parallel v_k \parallel^2 - \parallel z_1^k \parallel^2) = x_0^T P_0 x_0 - \mathcal{E}(x_{\mathcal{T}+1}^T P_{\mathcal{T}+1} x_{\mathcal{T}+1})
$$

$$
+ \sum_{k=0}^{\mathcal{T}} \mathcal{E}\begin{bmatrix} x_k \\ v_k \end{bmatrix}^T M(P_k)\begin{bmatrix} x_k \\ v_k \end{bmatrix}, \quad (3.10)
$$

where

$$
\begin{aligned}
M(P_k) &= \begin{bmatrix} \Delta_{11} & \Pi_{21}^T \\ \Pi_{21} & \gamma^2 I + (B_{11}^k)^T P_{k+1} B_{11}^k + (B_{12}^k)^T P_{k+1} B_{12}^k \end{bmatrix} \\
&= Q(P_k) + diag\{-(C_{11}^k)^T C_{11}^k, \gamma^2 I\},
\end{aligned}
$$

$$
\begin{aligned}
\Delta_{11} &= \Pi_{11} - (C_{11}^k)^T C_{11}^k \\
&= -P_k + (A_{11}^k)^T P_{k+1} A_{11}^k + (A_{12}^k)^T P_{k+1} A_{12}^k - (C_{11}^k)^T C_{11}^k,
\end{aligned}
$$

and Π_{21} is defined in Lemma 3.1.

Proof. We denote

$$
J_1^{\mathcal{T}}(x, v; \nu, k_0) = J_1^{\mathcal{T}}(x, v; x_{k_0} = \nu, k_0) := \sum_{k=k_0}^{\mathcal{T}} \mathcal{E}(\gamma^2 \parallel v_k \parallel^2 - \parallel z_1^k \parallel^2)
$$

associated with the system

$$
\begin{cases}
x_{k+1} = (A_{11}^k x_k + B_{11}^k v_k) + (A_{12}^k x_k + B_{12}^k v_k) w_k, \\
x_{k_0} = \nu \in \mathcal{R}^n, \\
z_1^k = C_{11}^k x_k, \ k \in \{k_0, k_0 + 1, \cdots, \mathcal{T}\}.
\end{cases}
$$

From Lemma 3.1, we have

$$J_1^{\mathcal{T}}(x, v; x_0, 0) = \sum_{k=0}^{\mathcal{T}} \mathcal{E}(\gamma^2 \parallel v_k \parallel^2 - \parallel z_1^k \parallel^2)$$

$$= \sum_{k=0}^{\mathcal{T}} \mathcal{E}[\gamma^2 v_k^T v_k - x_k^T (C_{11}^k)^T C_{11}^k x_k] + \sum_{k=0}^{\mathcal{T}} \mathcal{E} \begin{bmatrix} x_k \\ v_k \end{bmatrix}^T Q(P_k) \begin{bmatrix} x_k \\ v_k \end{bmatrix}$$

$$+ x_0^T P_0 x_0 - \mathcal{E}(x_{\mathcal{T}+1}^T P_{\mathcal{T}+1} x_{\mathcal{T}+1})$$

$$= x_0^T P_0 x_0 - \mathcal{E}(x_{\mathcal{T}+1}^T P_{\mathcal{T}+1} x_{\mathcal{T}+1}) + \sum_{k=0}^{\mathcal{T}} \mathcal{E} \begin{bmatrix} x_k \\ v_k \end{bmatrix}^T M(P_k) \begin{bmatrix} x_k \\ v_k \end{bmatrix},$$

which ends the proof. □

3.1.3 Finite horizon SBRL

For convenience, we adopt the following notations:

$$L(P_{k+1}) := (A_{11}^k)^T P_{k+1} A_{11}^k + (A_{12}^k)^T P_{k+1} A_{12}^k - (C_{11}^k)^T C_{11}^k,$$

$$K(P_{k+1}) := (A_{11}^k)^T P_{k+1} B_{11}^k + (A_{12}^k)^T P_{k+1} B_{12}^k,$$

$$H(P_{k+1}) := \gamma^2 I + (B_{11}^k)^T P_{k+1} B_{11}^k + (B_{12}^k)^T P_{k+1} B_{12}^k.$$

Hence, $M(P_k)$ can be simply written as

$$M(P_k) = \begin{bmatrix} -P_k + L(P_{k+1}) & K(P_{k+1}) \\ K(P_{k+1})^T & H(P_{k+1}) \end{bmatrix}.$$

The following lemma is the so-called finite horizon discrete SBRL.

LEMMA 3.3 (Finite horizon SBRL)
For stochastic system (3.8), $\parallel \tilde{\mathcal{L}}_{\mathcal{T}} \parallel < \gamma$ for some $\gamma > 0$ iff the following constrained backward difference equation

$$\begin{cases} P_k = L(P_{k+1}) - K(P_{k+1}) H(P_{k+1})^{-1} K(P_{k+1})^T, \\ P_{\mathcal{T}+1} = 0, \\ H(P_{k+1}) > 0, \ k \in N_{\mathcal{T}} \end{cases} \tag{3.11}$$

has a unique solution $P_{\mathcal{T}}^k \leq 0, k \in N_{\mathcal{T}}$.

Lemma 3.3 is in parallel to Lemma 2.1 of Chapter 2.
Proof. Sufficiency: From Lemma 3.2 and the backward difference equation (3.11), by using the completing squares method, we obtain for any $\{v_k : k \in N_{\mathcal{T}}\} \in$

$l_w^2(N_{\mathcal{T}}, \mathcal{R}^{n_v})$ and $x_0 = 0$,

$$J_1^{\mathcal{T}}(x, v; 0, 0) = \sum_{k=0}^{\mathcal{T}} \mathcal{E} \begin{bmatrix} x_k \\ v_k \end{bmatrix}^T M(P_{\mathcal{T}}^k) \begin{bmatrix} x_k \\ v_k \end{bmatrix}$$

$$= \sum_{k=0}^{\mathcal{T}} \mathcal{E}\{x_k^T[-P_{\mathcal{T}}^k + L(P_{\mathcal{T}}^{k+1}) - K(P_{\mathcal{T}}^{k+1})H(P_{\mathcal{T}}^{k+1})^{-1}K(P_{\mathcal{T}}^{k+1})^T]x_k\}$$

$$+ \sum_{k=0}^{\mathcal{T}} \mathcal{E}\{[v_k + H(P_{\mathcal{T}}^{k+1})^{-1}K(P_{\mathcal{T}}^{k+1})^T x_k]^T H(P_{\mathcal{T}}^{k+1})$$

$$\cdot [v_k + H(P_{\mathcal{T}}^{k+1})^{-1}K(P_{\mathcal{T}}^{k+1})^T x_k]\}$$

$$= \sum_{k=0}^{\mathcal{T}} \mathcal{E}\left[(v_k - v_k^*)^T H(P_{\mathcal{T}}^{k+1})(v_k - v_k^*)\right]$$

$$\geq 0,$$

where $v_k^* = -H(P_{\mathcal{T}}^{k+1})^{-1}K(P_{\mathcal{T}}^{k+1})^T x_k$. So we have $\| \tilde{\mathcal{L}}_{\mathcal{T}} \| \leq \gamma$. To show $\|\tilde{\mathcal{L}}_{\mathcal{T}}\| < \gamma$, we define the operator

$$L_1 : l_w^2(N_{\mathcal{T}}, \mathcal{R}^{n_v}) \mapsto l_w^2(N_{\mathcal{T}}, \mathcal{R}^{n_v}), \ L_1 v_k = v_k - v_k^*$$

with its realization

$$\begin{cases} x_{k+1} = A_{11}^k x_k + A_{12}^k x_k w_k + B_{11}^k v_k + B_{12}^k v_k w_k, \\ x_0 = 0, \\ v_k - v_k^* = v_k + H(P_{\mathcal{T}}^{k+1})^{-1}K(P_{\mathcal{T}}^{k+1})^T x_k. \end{cases}$$

Then L_1^{-1} exists, which is determined by

$$\begin{cases} x_{k+1} = [A_{11}^k - B_{11}^k H(P_{\mathcal{T}}^{k+1})^{-1}K(P_{\mathcal{T}}^{k+1})^T]x_k \\ \qquad + [A_{12}^k - B_{12}^k H(P_{\mathcal{T}}^{k+1})^{-1}K(P_{\mathcal{T}}^{k+1})^T]x_k w_k \\ \qquad + B_{11}^k(v_k - v_k^*) + B_{12}^k(v_k - v_k^*)w_k, \\ x_0 = 0, \end{cases}$$

$$v_k = -H(P_{\mathcal{T}}^{k+1})^{-1}K(P_{\mathcal{T}}^{k+1})^T x_k + (v_k - v_k^*).$$

We assume $\max_{k \in N_{\mathcal{T}}} H(P_{\mathcal{T}}^{k+1}) \geq \varepsilon I, \varepsilon > 0$, so there exists a sufficiently small constant $c > 0$ such that

$$J_1^{\mathcal{T}}(x, v; 0, 0) = \sum_{k=0}^{\mathcal{T}} \mathcal{E}[(L_1 v_k)^T H(P_{\mathcal{T}}^{k+1})(L_1 v_k)] \geq \varepsilon \| L_1 v_k \|_{l_w^2(N_{\mathcal{T}}, \mathcal{R}^{n_v})}^2$$

$$\geq c \| v_k \|_{l_w^2(N_{\mathcal{T}}, \mathcal{R}^{n_v})}^2 > 0,$$

i.e., $\|\tilde{\mathcal{L}}_{\mathcal{T}}\| < \gamma$.

Necessity: We first prove that $\|\tilde{\mathcal{L}}_T\| < \gamma$ implies the existence of a solution P_T^k of (3.11) on N_T. Obviously, there always exists a solution to (3.11) at $k = T$ due to $H(P_T^{T+1}) = \gamma^2 I > 0$, i.e.,

$$P_T^T = L(P_T^{T+1}) - \gamma^{-2} K(P_T^{T+1}) K(P_T^{T+1})^T.$$

If (3.11) does not admit a solution P_T^k for $k \in N_T$, then there must exist a minimum number $\mathcal{T}_0 \in N_T$, $0 < \mathcal{T}_0 \leq T$, such that (3.11) can be solved backward up to $k = \mathcal{T}_0$, which means that $P_T^{\mathcal{T}_0}, P_T^{\mathcal{T}_0+1}, \cdots, P_T^{T+1}$ satisfy (3.11), but $P_T^{\mathcal{T}_0-1}$ does not exist, or, equivalently, $H(P_T^{\mathcal{T}_0})$ is not a positive definite symmetric matrix. Let $\overline{F}_T^k = -H(P_T^{k+1})^{-1} K(P_T^{k+1})^T$, $k = \mathcal{T}_0, \mathcal{T}_0 + 1, \cdots, T$; then $\overline{F}_T^k$ is well defined. Again, let

$$\overline{F}_k = \begin{cases} 0, & k = 0, 1, \cdots, \mathcal{T}_0 - 1, \\ \overline{F}_T^k, & k = \mathcal{T}_0, \mathcal{T}_0 + 1, \cdots, T. \end{cases}$$

Consider the following backward matrix-valued equation

$$\begin{cases} P_k = L(P_{k+1}) + K(P_{k+1})\overline{F}_k + \overline{F}_k^T K(P_{k+1})^T + \overline{F}_k^T H(P_{k+1})\overline{F}_k, \\ P_{T+1} = 0. \end{cases} \quad (3.12)$$

The equation (3.12) has a unique solution P_2^k on N_{T+1} satisfying the terminal condition $P_2^{T+1} = 0$. Comparing (3.12) with (3.11), we see that $P_T^k = P_2^k, k = \mathcal{T}_0, \mathcal{T}_0+1, \cdots, T$. Following the line of Lemma 2.13 in [65] with almost no modification, we can assert that for stochastic system (3.8), if $\|\tilde{\mathcal{L}}_T\| < \gamma$, then $H(P_2^{k+1}) > 0$ for $k \in N_T$. In particular, $H(P_T^{\mathcal{T}_0}) = H(P_2^{\mathcal{T}_0}) > 0$, which contradicts the nonpositiveness of $H(P_T^{\mathcal{T}_0})$. Therefore, (3.11) admits a unique solution P_T^k on N_T.

Next, we prove $P_T^k \leq 0$ on N_T. For any $k_0 \in N_T$, and any deterministic vector $\nu \in \mathcal{R}^n$, let $x_{k_0} = \nu$ and consider the system

$$\begin{cases} x_{k+1} = (A_{11}^k x_k + B_{11}^k v_k) + (A_{12}^k x_k + B_{12}^k v_k) w_k, \\ x_{k_0} = \nu \in \mathcal{R}^n, \\ z_1^k = C_{11}^k x_k, \ k \in \{k_0, k_0 + 1, \cdots, T\}. \end{cases} \quad (3.13)$$

By Lemmas 3.1–3.2 and (3.11), using the completing squares method, we have

$$J_1^T(x, v; \nu, k_0) = \sum_{k=k_0}^{T} \mathcal{E}(\gamma^2 \| v_k \|^2 - \| z_1^k \|^2)$$

$$= \nu^T P_T^{k_0} \nu + \sum_{k=k_0}^{T} \mathcal{E}\left\{ \begin{bmatrix} x_k \\ v_k \end{bmatrix}^T M(P_T^k) \begin{bmatrix} x_k \\ v_k \end{bmatrix} \right\}$$

$$= \nu^T P_T^{k_0} \nu + \sum_{k=k_0}^{T} \mathcal{E}\left\{ (v_k - v_k^*)^T H(P_T^{k+1})(v_k - v_k^*) \right\}.$$

Obviously,

$$\min_{v \in l_w^2(N_T, \mathcal{R}^{n_v})} J_1^T(x, v; \nu, k_0) = J_1^T(x, v^*; \nu, k_0) = \nu^T P_T^{k_0} \nu$$

$$\leq J_1^T(x, 0; \nu, k_0) = -\sum_{k=k_0}^{T} \parallel z_1^k \parallel^2$$

$$\leq 0$$

for arbitrary $\nu \in \mathcal{R}^n$. So $P_T^k \leq 0, k \in N_T$. This lemma is proved. $\square$

3.1.4 Discrete-time stochastic LQ control

Consider the following discrete-time LQ optimal control problem [7]:

$$V_T(x_0) = \min_{u \in l_w^2(N_T, \mathcal{R}^{n_u})} J_T(0, x_0; u_0, \cdots, u_T)$$

$$= \min_{u \in l_w^2(N_T, \mathcal{R}^{n_u})} \sum_{k=0}^{T} \mathcal{E}(x_k^T Q_k x_k + u_k^T R_k u_k) \quad (3.14)$$

subject to

$$\begin{cases} x_{k+1} = (A_{11}^k x_k + B_{11}^k u_k) + (A_{12}^k x_k + B_{12}^k u_k)w_k, \\ x_0 \in \mathcal{R}^n, \ k \in N_T, \end{cases} \quad (3.15)$$

where in (3.14), Q_k and R_k are indefinite real symmetric matrices on N_T. The indefinite LQ optimal control (3.14)–(3.15) differs from the traditional LQ control definition [18] where it is assumed that $Q_k \geq 0$ and $R_k > 0$ for $k \in N_T$.

DEFINITION 3.3 *The LQ control problem (3.14)–(3.15) is well posed if*

$$V_T(x_0) > -\infty.$$

It is called attainable if it is well posed and in addition, there exist $u_0^*, \cdots, u_T^*$ such that

$$V_T(x_0) = J_T(0, x_0; u_0^*, \cdots, u_T^*).$$

For simplicity, we introduce the following notations:

$$\bar{L}(P_{k+1}) := (A_{11}^k)^T P_{k+1} A_{11}^k + (A_{12}^k)^T P_{k+1} A_{12}^k + Q_k,$$
$$K(P_{k+1}) := (A_{11}^k)^T P_{k+1} B_{11}^k + (A_{12}^k)^T P_{k+1} B_{12}^k,$$
$$\bar{H}(P_{k+1}) := R_k + (B_{11}^k)^T P_{k+1} B_{11}^k + (B_{12}^k)^T P_{k+1} B_{12}^k,$$

and the following generalized difference Riccati equation (GDRE)

$$\begin{cases} P_k = \bar{L}(P_{k+1}) - K(P_{k+1})\bar{H}^+(P_{k+1})K(P_{k+1})^T, \\ P_{T+1} = 0, \\ \bar{H}(P_{k+1})\bar{H}^+(P_{k+1})K(P_{k+1})^T - K(P_{k+1})^T = 0, \\ \bar{H}(P_{k+1}) \geq 0, \ k \in N_T. \end{cases} \quad (3.16)$$

LEMMA 3.4 [7]

The LQ control problem (3.14)–(3.15) is well posed iff there exists a symmetric matrix sequence $\{P_k\}_{k\in N_T}$ solving the GDRE (3.16). In this case, the optimal cost and the optimal control are respectively given by

$$V_T(x_0) = x_0^T P_0 x_0$$

and

$$
\begin{aligned}
u_{T,k}^*(\mathcal{Y},\mathcal{Z}) = &-[\bar{H}^+(P_{k+1})K(P_{k+1})^T + \mathcal{Y}_k \\
&-\bar{H}^+(P_{k+1})\bar{H}(P_{k+1})\mathcal{Y}_k]x_k \\
&+\mathcal{Z}_k - \bar{H}^+(P_{k+1})\bar{H}(P_{k+1})\mathcal{Z}_k
\end{aligned}
$$

for $k \in N_T$, where $\mathcal{Y}_k \in l_w^2(N_T,\mathcal{R}^{n_u \times n})$ and $\mathcal{Z}_k \in l_w^2(N_T,\mathcal{R}^{n_u})$.

Lemma 3.4 shows that for the LQ control problem (3.14)–(3.15), its well posedness is equivalent to its attainability, which differs from that in the case of Itô systems. Moreover, the optimal control is unique iff $\bar{H}^+(P_{k+1}) > 0$ on N_T. In this case, the unique optimal control is given by $u_{T,k}^* = -\bar{H}^{-1}(P_{k+1})K(P_{k+1})^T x_k$ for $k \in N_T$.

COROLLARY 3.1 [18]

In Lemma 3.4, if $Q_k \geq 0$, $R_k > 0$, $k \in N_T$, then GDRE

$$
\begin{cases}
P_k = \bar{L}(P_{k+1}) - K(P_{k+1})\bar{H}(P_{k+1})^{-1}K(P_{k+1})^T, \\
P_{T+1} = 0, \\
\bar{H}(P_{k+1}) > 0, \; k \in N_T
\end{cases}
\tag{3.17}
$$

has a unique solution sequence $\{P_k \geq 0\}_{k\in N_T}$ with the unique optimal control $u_{T,k}^* = -\bar{H}(P_{k+1})^{-1}K(P_{k+1})^T x_k$ for $k \in N_T$.

3.1.5 Finite horizon H_2/H_∞ with (x,v)-dependent noise

For simplicity and clarity, in this section, we first study the H_2/H_∞ control for system (3.2).

THEOREM 3.1

For a given disturbance attenuation level $\gamma > 0$, the finite horizon H_2/H_∞ control for system (3.2) has a solution (u_T^*, v_T^*) of the form

$$u_{T,k}^* = K_{2,T}^k x_k, \quad v_{T,k}^* = K_{1,T}^k x_k$$

with $K_{2,T}^k \in \mathcal{R}^{n_u \times n}$, $K_{1,T}^k \in \mathcal{R}^{n_v \times n}$ and $k \in N_T$ being discrete-time matrix-valued functions, iff the following four coupled matrix-valued equations have *a solution $(P_{1,\mathcal{T}}^k, K_{1,\mathcal{T}}^k; P_{2,\mathcal{T}}^k, K_{2,\mathcal{T}}^k)$ with $P_{1,\mathcal{T}}^k \leq 0$ and $P_{2,\mathcal{T}}^k \geq 0$ on $k \in N_{\mathcal{T}}$.*

$$\begin{cases} P_{1,\mathcal{T}}^k = (A_1^k + B_1^k K_{2,\mathcal{T}}^k)^T P_{1,\mathcal{T}}^{k+1}(A_1^k + B_1^k K_{2,\mathcal{T}}^k) - K_{3,\mathcal{T}}^k H_1(P_{1,\mathcal{T}}^{k+1})^{-1}(K_{3,\mathcal{T}}^k)^T \\ \qquad + (A_2^k)^T P_{1,\mathcal{T}}^{k+1} A_2^k - C_k^T C_k - (K_{2,\mathcal{T}}^k)^T K_{2,\mathcal{T}}^k, \\ P_{1,\mathcal{T}}^{\mathcal{T}+1} = 0, \\ H_1(P_{1,\mathcal{T}}^{k+1}) > 0, \end{cases}$$

$$\tag{3.18}$$

$$K_{1,\mathcal{T}}^k = -H_1(P_{1,\mathcal{T}}^{k+1})^{-1}(K_{3,\mathcal{T}}^k)^T, \tag{3.19}$$

$$\begin{cases} P_{2,\mathcal{T}}^k = \left(A_1^k + C_1^k K_{1,\mathcal{T}}^k\right)^T P_{2,\mathcal{T}}^{k+1}\left(A_1^k + C_1^k K_{1,\mathcal{T}}^k\right) \\ \qquad + (A_2^k + C_2^k K_{1,\mathcal{T}}^k)^T P_{2,\mathcal{T}}^{k+1}(A_2^k + C_2^k K_{1,\mathcal{T}}^k) \\ \qquad + C_k^T C_k - K_{4,\mathcal{T}}^k H_2(P_{2,\mathcal{T}}^{k+1})^{-1}(K_{4,\mathcal{T}}^k)^T, \\ H_2(P_{2,\mathcal{T}}^{k+1}) > 0, \\ P_{2,\mathcal{T}}^{\mathcal{T}+1} = 0, \end{cases}$$

$$\tag{3.20}$$

and

$$K_{2,\mathcal{T}}^k = -H_2(P_{2,\mathcal{T}}^{k+1})^{-1}(K_{4,\mathcal{T}}^k)^T, \tag{3.21}$$

where in (3.18)–(3.21),

$$K_{3,\mathcal{T}}^k = (A_1^k + B_1^k K_{2,\mathcal{T}}^k)^T P_{1,\mathcal{T}}^{k+1} C_1^k + (A_2^k)^T P_{1,\mathcal{T}}^{k+1} C_2^k,$$
$$K_{4,\mathcal{T}}^k = (A_1^k + C_1^k K_{1,\mathcal{T}}^k)^T P_{2,\mathcal{T}}^{k+1} B_1^k,$$
$$H_1(P_{1,\mathcal{T}}^{k+1}) = \gamma^2 I + (C_1^k)^T P_{1,\mathcal{T}}^{k+1} C_1^k + (C_2^k)^T P_{1,\mathcal{T}}^{k+1} C_2^k,$$
$$H_2(P_{2,\mathcal{T}}^{k+1}) = I + (B_1^k)^T P_{2,\mathcal{T}}^{k+1} B_1^k.$$

Proof. Necessity: Implementing $u_{\mathcal{T},k}^* = K_{2,\mathcal{T}}^k x_k$ in (3.2), we obtain

$$\begin{cases} x_{k+1} = (A_1^k + B_1^k K_{2,\mathcal{T}}^k)x_k + C_1^k v_k + (A_2^k x_k + C_2^k v_k)w_k, \\ x_0 \in \mathcal{R}^n, \\ z_k = \begin{bmatrix} C_k x_k \\ D_k K_{2,\mathcal{T}}^k x_k \end{bmatrix}, \; D_k^T D_k = I, \; k \in N_{\mathcal{T}}. \end{cases} \tag{3.22}$$

By Lemma 3.3, (3.18) has a unique solution $P_{1,\mathcal{T}}^k \leq 0$ on $N_{\mathcal{T}}$. From the sufficiency proof of Lemma 3.3, it can be seen that the worst-case disturbance $\{v_{\mathcal{T},k}^*\}_{k \in N_{\mathcal{T}}}$ is given by

$$v_{\mathcal{T},k}^* = K_{1,\mathcal{T}}^k x_k = -H_1(P_{1,\mathcal{T}}^{k+1})^{-1} K_{3,\mathcal{T}}^T x_k.$$

Substituting $v_k = v_{\mathcal{T},k}^*$ into (3.2), we have

$$\begin{cases} x_{k+1} = (A_1^k + C_1^k K_{1,\mathcal{T}}^k)x_k + (A_2^k + C_2^k K_{1,\mathcal{T}}^k)x_k w_k + B_1^k u_k, \\ z_k = \begin{bmatrix} C_k x_k \\ D_k x_k \end{bmatrix}, \\ D_k^T D_k = I, \; k \in N_{\mathcal{T}}. \end{cases} \tag{3.23}$$

On the other hand, the optimization problem

$$
\begin{cases}
\displaystyle\min_{u\in l^2_w(N_T,\mathcal{R}^{n_u})} J_{2,T}(u,v_T^*) \\
\text{subject to } (3.23)
\end{cases}
\tag{3.24}
$$

is a standard discrete-time LQ optimal control problem. By applying Corollary 3.1, (3.20) admits a unique solution $P^k_{2,T} \geq 0$ on N_T. Furthermore,

$$
\min_{u\in l^2_w(N_T,\mathcal{R}^{n_u})} J_{2,T}(u,v_T^*) = J_{2,T}(u_T^*,v_T^*) = x_0^T P^0_{2,T} x_0,
$$

where $u^*_{T,k} = K^k_{2,T} x_k$ is as in (3.21).

Sufficiency: Applying $u_k = u^*_{T,k} = K^k_{2,T} x_k$ to (3.2) yields (3.22). From (3.18) and Lemma 3.3, we have $\| \mathcal{L}_T \| < \gamma$. By Lemma 3.2 and (3.18), using the completing squares method, we immediately have

$$
\begin{aligned}
J_{1,T}(u_T^*,v) &= \sum_{k=0}^{T} \mathcal{E}(\gamma^2 \| v_k \|^2 - \| z_k \|^2) \\
&= x_0^T P^0_{1,T} x_0 + \sum_{k=0}^{T} \mathcal{E}\begin{bmatrix} x_k \\ v_k \end{bmatrix}^T M_1(P^k_{1,T})\begin{bmatrix} x_k \\ v_k \end{bmatrix} \\
&= x_0^T P^0_{1,T} x_0 + \sum_{k=0}^{T} \mathcal{E}[(v_k - v^*_{T,k})^T H_1(P^{k+1}_{1,T})(v_k - v^*_{T,k})] \\
&\geq J_{1,T}(u_T^*,v_T^*) = x_0^T P^0_{1,T} x_0,
\end{aligned}
\tag{3.25}
$$

where

$$
M_1(P^k_{1,T}) = \begin{bmatrix} H_0(P^{k+1}_{1,T}) & K^k_{3,T} \\ (K^k_{3,T})^T & H_1(P^{k+1}_{1,T}) \end{bmatrix},
$$

$$
\begin{aligned}
H_0(P^{k+1}_{1,T}) = &-P^k_{1,T} + (A^k_1 + B^k_1 K^k_{2,T})^T P^{k+1}_{1,T}(A^k_1 + B^k_1 K^k_{2,T}) + (A^k_2)^T P^{k+1}_{1,T} A^k_2 \\
&- C^T_k C_k - (K^k_{2,T})^T K^k_{2,T}.
\end{aligned}
$$

From (3.25), we see that $v^*_{T,k} = K^k_{1,T} x_k$ is the worst-case disturbance. Similarly, we have

$$
\begin{aligned}
J_{2,T}(u,v_T^*) &= \sum_{k=0}^{T} \mathcal{E}\| z_k \|^2 \\
&= x_0^T P^0_{2,T} x_0 + \sum_{k=0}^{T} \mathcal{E}\left[(u_k - u^*_{T,k})^T H_2(P^{k+1}_{2,T})(u_k - u^*_{T,k})\right] \\
&\geq J_{2,T}(u_T^*,v_T^*) = x_0^T P^0_{2,T} x_0.
\end{aligned}
$$

So (u_T^*,v_T^*) solves the finite horizon H_2/H_∞ control problem of system (3.2), and this theorem is proved. □

REMARK 3.1   Although for simplicity, we assume that w_k is a scalar white noise, we can, in fact, extend Theorem 3.1 to the following system with multiple noises without any difficulty.

$$\begin{cases} x_{k+1} = A_1^k x_k + B_1^k u_k + C_1^k v_k + \sum_{i=1}^{r} A_2^{k,i} x_k w_k^i + \sum_{i=1}^{r} C_2^{k,i} v_k w_k^i, \\ x_0 \in \mathcal{R}^n, \\ z_k = \begin{bmatrix} C_k x_k \\ D_k x_k \end{bmatrix}, \\ D_k^T D_k = I, k \in N_{\mathcal{T}}, \end{cases} \quad (3.26)$$

where

$$\mathcal{E}[w_k^i] = 0, \ \mathcal{E}[w_k^i w_k^j] = \delta_{ij}, \ i, j \in \{1, 2, \cdots, r\}, \ k \in N_{\mathcal{T}},$$

i.e., $w_k = Col(w_k^1, w_k^2, \cdots, w_k^r) \in \mathcal{R}^r, k \in N_{\mathcal{T}}$ is an r-dimensional stationary process consisting of uncorrelated random vectors with zero mean and covariance matrix I_r. In this case, Theorem 3.1 still holds if we replace

$$(A_2^k)^T P_{1,\mathcal{T}}^{k+1} A_2^k, \quad (A_2^k + C_2^k K_{1,\mathcal{T}}^k)^T P_{2,\mathcal{T}}^{k+1} (A_2^k + C_2^k K_{1,\mathcal{T}}^k),$$

$$(A_2^k)^T P_{1,\mathcal{T}}^{k+1} C_2^k, \quad (C_2^k)^T P_{1,\mathcal{T}}^{k+1} C_2^k$$

by

$$\sum_{i=1}^{r} (A_2^{k,i})^T P_{1,\mathcal{T}}^{k+1} A_2^{k,i}, \quad \sum_{i=1}^{r} (A_2^{k,i} + C_2^{k,i} K_{1,\mathcal{T}}^k)^T P_{2,\mathcal{T}}^{k+1} (A_2^{k,i} + C_2^{k,i} K_{1,\mathcal{T}}^k),$$

$$\sum_{i=1}^{r} (A_2^{k,i})^T P_{1,\mathcal{T}}^{k+1} C_2^{k,i}, \quad \sum_{i=1}^{r} (C_2^{k,i})^T P_{1,\mathcal{T}}^{k+1} C_2^{k,i},$$

respectively.   ∎

REMARK 3.2   Note that (3.18)–(3.21) are coupled backward matrix-valued equations. Hence, their solution

$$(P_{1,\mathcal{T}}^k \le 0, K_{1,\mathcal{T}}^k; P_{2,\mathcal{T}}^k \ge 0, K_{2,\mathcal{T}}^k),$$

if it exists, must be unique.   ∎

3.1.6   Unified treatment of H_2, H_∞ and H_2/H_∞ control

Similar to the continuous-time case in Chapter 2, we can give a unified treatment of H_2, H_∞ and H_2/H_∞ controls if we introduce an additional parameterized performance

$$J_{2,\mathcal{T}}^\varrho(u, v) = \sum_{k=0}^{\mathcal{T}} \mathcal{E}(\| z_k \|^2 - \varrho^2 \| v_k \|^2)$$

associated with system (3.2), where $\varrho \geq 0$. With a similar discussion as in Theorem 3.1, we can obtain the following result.

THEOREM 3.2

For system (3.2), if the following four coupled matrix-valued equations

$$\begin{cases} Z_{1,\mathcal{T}}^k = (A_1^k + B_1^k K_{2,\mathcal{T}}^k)^T Z_{1,\mathcal{T}}^{k+1} (A_1^k + B_1^k K_{2,\mathcal{T}}^k) - K_{3,\mathcal{T}}^k H_1 (Z_{1,\mathcal{T}}^{k+1})^{-1} (K_{3,\mathcal{T}}^k)^T \\ \quad + (A_2^k)^T Z_{1,\mathcal{T}}^{k+1} A_2^k - C_k^T C_k - (K_{2,\mathcal{T}}^k)^T K_{2,\mathcal{T}}^k, \\ Z_{1,\mathcal{T}}^{\mathcal{T}+1} = 0, \\ H_1(Z_{1,\mathcal{T}}^{k+1}) > 0, \end{cases}$$

$$\text{(3.27)}$$

$$K_{1,\mathcal{T}}^k = -H_1(Z_{1,\mathcal{T}}^{k+1})^{-1} (K_{3,\mathcal{T}}^k)^T, \qquad\qquad \text{(3.28)}$$

$$\begin{cases} Z_{2,\mathcal{T}}^k = (A_1^k + C_1^k K_{1,\mathcal{T}}^k)^T Z_{2,\mathcal{T}}^{k+1} (A_1^k + C_1^k K_{1,\mathcal{T}}^k) - K_{4,\mathcal{T}}^k H_2 (Z_{2,\mathcal{T}}^{k+1})^{-1} (K_{4,\mathcal{T}}^k)^T \\ \quad + (A_2^k + C_2^k K_{1,\mathcal{T}}^k)^T Z_{2,\mathcal{T}}^{k+1} (A_2^k + C_2^k K_{1,\mathcal{T}}^k) + C_k^T C_k - \varrho^2 (K_{1,\mathcal{T}}^k)^T K_{1,\mathcal{T}}^k, \\ Z_{2,\mathcal{T}}^{\mathcal{T}+1} = 0, \\ H_2(Z_{2,\mathcal{T}}^{k+1}) > 0, \end{cases}$$

$$\text{(3.29)}$$

and

$$K_{2,\mathcal{T}}^k = -H_2(Z_{2,\mathcal{T}}^{k+1})^{-1} (K_{4,\mathcal{T}}^k)^T \qquad\qquad \text{(3.30)}$$

have a solution $(Z_{1,\mathcal{T}}^k, K_{1,\mathcal{T}}^k; Z_{2,\mathcal{T}}^k, K_{2,\mathcal{T}}^k)$ *with* $Z_{1,\mathcal{T}}^k \leq 0$, $Z_{2,\mathcal{T}}^k \geq 0, k \in N_{\mathcal{T}}$, *then*

$$u_{\mathcal{T},k}^* = K_{2,\mathcal{T}}^k x_k, \quad v_{\mathcal{T},k}^* = K_{1,\mathcal{T}}^k x_k$$

solve the H_2/H_∞ *control associated with the performances* $J_{1,\mathcal{T}}(u,v)$ *and* $J_2^\varrho(u,v)$, *i.e.,*

$$\|\mathcal{L}_{\mathcal{T}}\| < \gamma, \ J_{2,\mathcal{T}}^\varrho(u_{\mathcal{T}}^*, v_{\mathcal{T}}^*) \leq J_{2,\mathcal{T}}^\varrho(u, v_{\mathcal{T}}^*).$$

REMARK 3.3   By comparing equations (3.27)–(3.30) with equations (3.18)-(3.21), it can be found that only (3.29) has a different form than (3.20).

∎

Similar to continuous-time Itô systems dealt with in Section 2.2.6, based on Theorem 3.2, we can give a unified treatment of H_2, H_∞ and mixed H_2/H_∞ controls as follows:

(i) H_2 control: Consider the following quadratic optimization problem

$$\min_{u \in l_w^2(N_{\mathcal{T}}, \mathcal{R}^{n_u})} J_{2,\mathcal{T}}(u, 0) = \sum_{k=0}^{\mathcal{T}} \mathcal{E} \| z_k \|^2$$

subject to

$$\begin{cases} x_{k+1} = A_1^k x_k + B_1^k u_k + A_2^k x_k w_k, \ x_0 \in \mathcal{R}^n, \\ z_k = \begin{bmatrix} C_k x_k \\ D_k x_k \end{bmatrix}, \\ D_k^T D_k = I, k \in N_{\mathcal{T}}. \end{cases}$$

Let $\varrho = 0, \gamma \to \infty$, we have

$$H_1(Z_{1,\mathcal{T}}^{k+1})^{-1} \to 0, \quad K_{1,\mathcal{T}}^k \to 0, \quad K_{4,\mathcal{T}}^k \to (A_1^k)^T P_{\mathcal{T}}^{k+1} B_1^k,$$

$$K_{2,\mathcal{T}}^k \to \bar{K}_2^k = -[I + (B_1^k)^T P_{\mathcal{T}}^{k+1} B_1^k]^{-1} (B_1^k)^T P_{\mathcal{T}}^{k+1} A_1^k,$$

where $P_{\mathcal{T}}^k, k = 1, 2, \cdots, N$, are determined by

$$
\begin{cases}
P_{\mathcal{T}}^k = (A_1^k)^T P_{\mathcal{T}}^{k+1} A_1^k + (A_2^k)^T P_{\mathcal{T}}^{k+1} A_2^k + C_k^T C_k \\
\quad -(A_1^k)^T P_{\mathcal{T}}^{k+1} B_1^k [I + (B_1^k)^T P_{\mathcal{T}}^{k+1} B_1^k]^{-1} (B_1^k)^T P_{\mathcal{T}}^{k+1} A_1^k, \\
I + (B_1^k)^T P_{\mathcal{T}}^{k+1} B_1^k > 0, \\
P_{\mathcal{T}}^{\mathcal{T}+1} = 0,
\end{cases}
\tag{3.31}
$$

which is the limiting form of (3.29). It is easy to see that when $\rho = 0, \gamma \to \infty$, the solution $(Z_{1,\mathcal{T}}^k, K_{1,\mathcal{T}}^k; Z_{2,\mathcal{T}}^k, K_{2,\mathcal{T}}^k)$ of (3.27)–(3.30) approaches $(-P_{\mathcal{T}}^k, 0; P_{\mathcal{T}}^k, \bar{K}_2^k)$. The optimal control and the optimal performance index are respectively given by

$$\bar{u}_{\mathcal{T},k}^* = \bar{K}_2^k x_k \tag{3.32}$$

and

$$\min_{u \in l_w^2(N_{\mathcal{T}}, \mathcal{R}^{n_u})} J_{2,\mathcal{T}}(u, 0) = J_{2,\mathcal{T}}(\bar{u}_{\mathcal{T}}^*, 0) = x_0^T P_{\mathcal{T}}^0 x_0. \tag{3.33}$$

Hence, the existing discrete stochastic LQ optimal control is recovered; see [7, 18, 97].

(ii) H_∞ control: If we set $\varrho = \gamma$ in (3.27)–(3.30), it is easy to see that $Z_{2,\mathcal{T}}^k = -Z_{1,\mathcal{T}}^k = P_{\infty,\mathcal{T}}^k$, while

$$\tilde{u}_{\mathcal{T},k}^* = -[I + (B_1^k)^T P_{\infty,\mathcal{T}}^{k+1} B_1^k]^{-1} (B_1^k)^T P_{\infty,\mathcal{T}}^{k+1} (A_1^k + C_1^k K_{1,\mathcal{T}}^k) x_k$$

and

$$\tilde{v}_{\mathcal{T},k}^* = [\gamma^2 I - (C_1^k)^T P_{\infty,\mathcal{T}}^{k+1} C_1^k - (C_2^k)^T P_{\infty,\mathcal{T}}^{k+1} C_2^k]^{-1}$$
$$\cdot [(C_1^k)^T P_{\infty,\mathcal{T}}^{k+1} (A_1^k + B_1^k K_{2,\mathcal{T}}^k) + (C_2^k)^T P_{\infty,\mathcal{T}}^{k+1} A_2^k] x_k$$

are respectively the H_∞ control and the corresponding worst-case disturbance. The result of [65] is retrieved.

(iii) Mixed H_2/H_∞: If $\varrho = 0$, Theorem 3.1 is retrieved.

REMARK 3.4 Different from classical linear quadratic Gaussian control, the system state is assumed to be measurable and the state feedback H_2/H_∞ control is solved. Otherwise, we must study the output feedback H_2/H_∞ control problem which, however, remains unsolved. Note that the output feedback H_2 and H_∞ control were addressed in [79, 188]. ∎

3.1.7 A numerical example

A key step to design an H_2/H_∞ controller is to solve the coupled equations (3.18)–(3.21). We note in Remark 2.7 that for the continuous-time stochastic H_2/H_∞ control, it is very difficult to solve the related GDREs (2.55)–(2.58). It can be found that for the discrete-time stochastic H_2/H_∞ control, if it is solvable, then the related four coupled equations (3.18)–(3.21) can be solved recursively as follows:

(i) Set $k = \mathcal{T}$, then

$$H_1(P_{1,\mathcal{T}}^{\mathcal{T}+1}),\ H_1(P_{1,\mathcal{T}}^{\mathcal{T}+1})^{-1}, H_2(P_{2,\mathcal{T}}^{\mathcal{T}+1}), H_2(P_{2,\mathcal{T}}^{\mathcal{T}+1})^{-1}$$

can be computed.

(ii) Solving the following linear matrix equations

$$\begin{cases} K_{1,\mathcal{T}}^{\mathcal{T}} = -H_1(P_{1,\mathcal{T}}^{\mathcal{T}+1})^{-1}\left[(C_1^{\mathcal{T}})^T P_{1,\mathcal{T}}^{\mathcal{T}+1}(A_1^{\mathcal{T}} + B_1^{\mathcal{T}} K_{2,\mathcal{T}}^{\mathcal{T}}) + (C_2^{\mathcal{T}})^T P_{1,\mathcal{T}}^{\mathcal{T}+1} A_2^{\mathcal{T}}\right], \\ K_{2,\mathcal{T}}^{\mathcal{T}} = -H_2(P_{2,\mathcal{T}}^{\mathcal{T}+1})^{-1}[(B_1^{\mathcal{T}})^T P_{2,\mathcal{T}}^{\mathcal{T}+1}(A_1^{\mathcal{T}} + C_1^{\mathcal{T}} K_{1,\mathcal{T}}^{\mathcal{T}})] \end{cases}$$

$$(3.34)$$

to obtain $K_{1,\mathcal{T}}^{\mathcal{T}}$ and $K_{2,\mathcal{T}}^{\mathcal{T}}$.

(iii) By solving (3.18) and (3.20), $(P_{1,\mathcal{T}}^{\mathcal{T}} \le 0, P_{2,\mathcal{T}}^{\mathcal{T}} \ge 0)$ are obtained.

(iv) Repeating the steps (i)–(iii), we can compute $(P_{1,\mathcal{T}}^{k}, K_{1,\mathcal{T}}^{k}; P_{2,\mathcal{T}}^{k}, K_{2,\mathcal{T}}^{k})$ for $k = \mathcal{T} - 1, \mathcal{T} - 2, \cdots, 0$, recursively.

The following numerical example illustrates the above procedure in finding the solution of (3.18)–(3.21).

Example 3.1

In system (3.2), take $\mathcal{T} = 2$ and let
$k = 0$:

$$A_1^0 = 0.8500,\ A_2^0 = 0.4000,\ C_1^0 = 0.4500,\ C_2^0 = 0.3000,$$

$$B_1^0 = 0.7000,\ C_0 = 0.5000,\ D_0 = 1.0000.$$

$k = 1$:

$$A_1^1 = 0.9000,\ A_2^1 = 0.5500,\ C_1^1 = 0.5000,\ C_2^1 = 0.6000,$$

$$B_1^1 = 0.6500,\ C_1 = 0.4500,\ D_1 = 1.0000.$$

$k = 2$:

$$A_1^2 = 0.8000,\ A_2^2 = 0.4000,\ C_1^2 = 0.6500,\ C_2^2 = 0.5000,$$

$$B_1^2 = 0.7500,\ C_2 = 0.4000,\ D_2 = 1.0000.$$

Set $\gamma = 0.8$; we solve the coupled matrix-valued equations (3.18)–(3.21) according to the above steps (i)–(iv).

Step 1. $k = 2$. We then have

$$H_1(P_{1,2}^3) = 0.6400,\ H_2(P_{2,2}^3) = 1.0000,$$

$$K_{1,2}^2 = K_{2,2}^2 = 0,\ P_{1,2}^2 = -0.5625,\ P_{2,2}^2 = 0.5625.$$

Thus
$$(P_{1,2}^2, K_{1,2}^2; P_{2,2}^2, K_{2,2}^2) = (-0.5625, 0; 0.5625, 0).$$

Step 2. $k = 1$. We have

$$H_1(P_{1,2}^2) = 0.2969, \quad H_2(P_{2,2}^2) = 1.2377,$$

$$H_1(P_{1,2}^2)^{-1} = 3.3600, \quad H_2(P_{2,2}^2)^{-1} = 0.8000,$$

$$\begin{cases} K_{1,2}^1 = 1.4742 + 0.6132K_{2,2}^1, \\ K_{2,2}^1 = -0.2633 - 0.1463K_{1,2}^1. \end{cases}$$

Solving the above equations, we have $(K_{1,2}^1, K_{2,2}^1) = (1.2047, -0.4394)$. Hence,

$$(P_{1,2}^1, P_{2,2}^1) = (-1.2097, 2.1420)$$

and
$$(P_{1,2}^1, K_{1,2}^1; P_{2,2}^1, K_{2,2}^1) = (-1.2097, 1.2047; 2.1420, -0.4394).$$

Step 3. $k = 0$. It is easy to compute

$$H_1(P_{1,2}^1) = 0.2826, \quad H_2(P_{2,2}^1) = 2.0496,$$

$$H_1(P_{1,2}^1)^{-1} = 3.4941, \quad H_2(P_{2,2}^1)^{-1} = 0.4879,$$

$$\begin{cases} K_{1,2}^0 = 2.1241 + 1.3316K_{2,2}^0, \\ K_{2,2}^0 = -0.6218 - 0.3292K_{1,2}^0. \end{cases}$$

Solving the above equation yields

$$(K_{1,2}^0, K_{2,2}^0) = (0.9010, -0.9185)$$

and accordingly
$$(P_{1,2}^0, P_{2,2}^0) = (-1.5714, 4.3794),$$

$$(P_{1,2}^0, K_{1,2}^0; P_{2,2}^0, K_{2,2}^0) = (-1.5714, 0.9010; 4.3794, -0.9185).$$

▢

3.1.8 H_2/H_∞ control of systems with (x, u)- and (x, u, v)-dependent noise

Repeating the same procedure as in Theorem 3.1, we are in a position to present the following theorems whose proofs are omitted in order to avoid repetition.

THEOREM 3.3

For a given disturbance attenuation level $\gamma > 0$, the finite horizon H_2/H_∞ control for system (3.3) has a solution (u_T^*, v_T^*) of the form

$$u_{T,k}^* = K_{2,T}^k x_k, \quad v_{T,k}^* = K_{1,T}^k x_k$$

with $K_{2,\mathcal{T}}^k \in \mathcal{R}^{n_u \times n}$ and $K_{1,\mathcal{T}}^k \in \mathcal{R}^{n_v \times n}, k \in N_{\mathcal{T}}$ *being discrete-time matrix-valued functions, iff the following four coupled matrix-valued equations have a solution* $(P_{1,\mathcal{T}}^k, K_{1,\mathcal{T}}^k; P_{2,\mathcal{T}}^k, K_{2,\mathcal{T}}^k)$ *with* $P_{1,\mathcal{T}}^k \le 0$ *and* $P_{2,\mathcal{T}}^k \ge 0$ *on* $k \in N_{\mathcal{T}}$.

$$
\begin{cases}
P_{1,\mathcal{T}}^k = (A_1^k + B_1^k K_{2,\mathcal{T}}^k)^T P_{1,\mathcal{T}}^{k+1}(A_1^k + B_1^k K_{2,\mathcal{T}}^k) - C_k^T C_k - (K_{2,\mathcal{T}}^k)^T K_{2,\mathcal{T}}^k \\
\quad + (A_2^k + B_2^k K_{2,\mathcal{T}}^k)^T P_{1,\mathcal{T}}^{k+1}(A_2^k + B_2^k K_{2,\mathcal{T}}^k) \\
\quad - K_{3,\mathcal{T}}^k H_1(P_{1,\mathcal{T}}^{k+1})^{-1}(K_{3,\mathcal{T}}^k)^T, \\
P_{1,\mathcal{T}}^{\mathcal{T}+1} = 0, \\
H_1(P_{1,\mathcal{T}}^{k+1}) > 0,
\end{cases}
$$

$$\text{(3.35)}$$

$$
K_{1,\mathcal{T}}^k = -H_1(P_{1,\mathcal{T}}^{k+1})^{-1}(K_{3,\mathcal{T}}^k)^T, \tag{3.36}
$$

$$
\begin{cases}
P_{2,\mathcal{T}}^k = \left(A_1^k + C_1^k K_{1,\mathcal{T}}^k\right)^T P_{2,\mathcal{T}}^{k+1}\left(A_1^k + C_1^k K_{1,\mathcal{T}}^k\right) + (A_2^k)^T P_{2,\mathcal{T}}^{k+1} A_2^k \\
\quad + C_k^T C_k - K_{4,\mathcal{T}}^k H_2(P_{2,\mathcal{T}}^{k+1})^{-1}(K_{4,\mathcal{T}}^k)^T, \\
P_{2,\mathcal{T}}^{\mathcal{T}+1} = 0, \\
H_2(P_{2,\mathcal{T}}^{k+1}) > 0,
\end{cases}
\tag{3.37}
$$

and

$$
K_{2,\mathcal{T}}^k = -H_2(P_{2,\mathcal{T}}^{k+1})^{-1}(K_{4,\mathcal{T}}^k)^T, \tag{3.38}
$$

where in (3.35)–(3.38),

$$
\begin{aligned}
K_{3,\mathcal{T}}^k &= (A_1^k + B_1^k K_{2,\mathcal{T}}^k)^T P_{1,\mathcal{T}}^{k+1} C_1^k, \\
K_{4,\mathcal{T}}^k &= (A_1^k + C_1^k K_{1,\mathcal{T}}^k)^T P_{2,\mathcal{T}}^{k+1} B_1^k + (A_2^k)^T P_{2,\mathcal{T}}^{k+1} B_2^k, \\
H_1(P_{1,\mathcal{T}}^{k+1}) &= \gamma^2 I + (C_1^k)^T P_{1,\mathcal{T}}^{k+1} C_1^k, \\
H_2(P_{2,\mathcal{T}}^{k+1}) &= I + (B_1^k)^T P_{2,\mathcal{T}}^{k+1} B_1^k + (B_2^k)^T P_{2,\mathcal{T}}^{k+1} B_2^k.
\end{aligned}
$$

THEOREM 3.4

For a given disturbance attenuation level $\gamma > 0$, the finite horizon H_2/H_∞ control for system (3.1) has a solution $(u_{\mathcal{T}}^*, v_{\mathcal{T}}^*)$ given by

$$
u_{\mathcal{T},k}^* = K_{2,\mathcal{T}}^k x_k, \quad v_{\mathcal{T},k}^* = K_{1,\mathcal{T}}^k x_k
$$

with $K_{2,\mathcal{T}}^k \in \mathcal{R}^{n_u \times n}$ and $K_{1,\mathcal{T}}^k \in \mathcal{R}^{n_v \times n}, k \in N_{\mathcal{T}}$ being discrete-time matrix-valued functions, iff the following four coupled matrix-valued equations have solutions $(P_{1,\mathcal{T}}^k, K_{1,\mathcal{T}}^k; P_{2,\mathcal{T}}^k, K_{2,\mathcal{T}}^k)$ with $P_{1,\mathcal{T}}^k \le 0$ and $P_{2,\mathcal{T}}^k \ge 0$ on $k \in N_{\mathcal{T}}$.

$$
\begin{cases}
P_{1,\mathcal{T}}^k = (A_1^k + B_1^k K_{2,\mathcal{T}}^k)^T P_{1,\mathcal{T}}^{k+1}(A_1^k + B_1^k K_{2,\mathcal{T}}^k) - C_k^T C_k - (K_{2,\mathcal{T}}^k)^T K_{2,\mathcal{T}}^k \\
\quad + (A_2^k + B_2^k K_{2,\mathcal{T}}^k)^T P_{1,\mathcal{T}}^{k+1}(A_2^k + B_2^k K_{2,\mathcal{T}}^k) - K_{3,\mathcal{T}}^k H_1(P_{1,\mathcal{T}}^{k+1})^{-1}(K_{3,\mathcal{T}}^k)^T, \\
H_1(P_{1,\mathcal{T}}^{k+1}) > 0, \\
P_{1,\mathcal{T}}^{\mathcal{T}+1} = 0,
\end{cases}
$$

$$\text{(3.39)}$$

$$
K_{1,\mathcal{T}}^k = -H_1(P_{1,\mathcal{T}}^{k+1})^{-1}(K_{3,\mathcal{T}}^k)^T, \tag{3.40}
$$

$$
\begin{cases}
P_{2,\mathcal{T}}^{k} = \left(A_1^k + C_1^k K_{1,\mathcal{T}}^k\right)^T P_{2,\mathcal{T}}^{k+1} \left(A_1^k + C_1^k K_{1,\mathcal{T}}^k\right) + C_k^T C_k \\
\quad + (A_2^k + C_2^k K_{1,\mathcal{T}}^k)^T P_{2,\mathcal{T}}^{k+1} (A_2^k + C_2^k K_{1,\mathcal{T}}^k) \\
\quad - K_{4,\mathcal{T}}^k H_2 (P_{2,\mathcal{T}}^{k+1})^{-1} (K_{4,\mathcal{T}}^k)^T, \\
H_2(P_{2,\mathcal{T}}^{k+1}) > 0, \\
P_{2,\mathcal{T}}^{\mathcal{T}+1} = 0,
\end{cases}
\tag{3.41}
$$

and

$$
K_{2,\mathcal{T}}^k = -H_2(P_{2,\mathcal{T}}^{k+1})^{-1}(K_{4,\mathcal{T}}^k)^T,
\tag{3.42}
$$

where in (3.39)–(3.42),

$$
\begin{aligned}
K_{3,\mathcal{T}}^k &= (A_1^k + B_1^k K_{2,\mathcal{T}}^k)^T P_{1,\mathcal{T}}^{k+1} C_1^k + (A_2^k + B_2^k K_{2,\mathcal{T}}^k)^T P_{1,\mathcal{T}}^{k+1} C_2^k, \\
K_{4,\mathcal{T}}^k &= (A_1^k + C_1^k K_{1,\mathcal{T}}^k)^T P_{2,\mathcal{T}}^{k+1} B_1^k + (A_2^k + C_2^k K_{1,\mathcal{T}}^k)^T P_{2,\mathcal{T}}^{k+1} B_2^k, \\
H_1(P_{1,\mathcal{T}}^{k+1}) &= \gamma^2 I + (C_1^k)^T P_{1,\mathcal{T}}^{k+1} C_1^k + (C_2^k)^T P_{1,\mathcal{T}}^{k+1} C_2^k, \\
H_2(P_{2,\mathcal{T}}^{k+1}) &= I + (B_1^k)^T P_{2,\mathcal{T}}^{k+1} B_1^k + (B_2^k)^T P_{2,\mathcal{T}}^{k+1} B_2^k.
\end{aligned}
$$

REMARK 3.5 Our solution to the finite horizon H_2/H_∞ control depends heavily on the SBRL (Lemma 3.3) and the results on standard stochastic LQ control [7, 18]. Different from linear Itô systems, in the presence of (x,u)- or (x,v)- or (x,u,v)-dependent noise, the discrete-time stochastic H_2/H_∞ control is always associated with four coupled matrix-valued difference equations which can be easily solved iteratively. ∎

3.2 Two-Person Non-Zero Sum Nash Game

Consider system (3.1) together with

$$
J_{1,\mathcal{T}}(u,v) = \gamma^2 \|v\|_{l_w^2(N_{\mathcal{T}}, \mathcal{R}^{n_v})}^2 - \|z\|_{l_w^2(N_{\mathcal{T}}, \mathcal{R}^{n_z})}^2
$$

and

$$
J_{2,\mathcal{T}}(u,v) = \|z\|_{l_w^2(N_{\mathcal{T}}, \mathcal{R}^{n_z})}^2.
$$

DEFINITION 3.4 $(u_{\mathcal{T}}^*, v_{\mathcal{T}}^*) \in l_w^2(N_{\mathcal{T}}, \mathcal{R}^{n_u}) \times l_w^2(N_{\mathcal{T}}, \mathcal{R}^{n_v})$ *is called a two-person non-zero sum Nash game strategy (i.e., equilibrium point) associated with the costs $J_{1,\mathcal{T}}(u,v)$ and $J_{2,\mathcal{T}}(u,v)$ if for any $(u,v) \in l_w^2(N_{\mathcal{T}}, \mathcal{R}^{n_u}) \times l_w^2(N_{\mathcal{T}}, \mathcal{R}^{n_v})$, we have*

$$
J_{1,\mathcal{T}}(u_{\mathcal{T}}^*, v_{\mathcal{T}}^*) \le J_{1,\mathcal{T}}(u_{\mathcal{T}}^*, v), \quad J_{2,\mathcal{T}}(u_{\mathcal{T}}^*, v_{\mathcal{T}}^*) \le J_{2,\mathcal{T}}(u, v_{\mathcal{T}}^*).
\tag{3.43}
$$

THEOREM 3.5

For system (3.1), there exists a linear memoryless state feedback Nash equilibrium strategy $(u_\mathcal{T}^*, v_\mathcal{T}^*)$ iff the following four coupled matrix-valued equations

$$
\begin{cases}
P_{1,\mathcal{T}}^k = (A_1^k + B_1^k K_{2,\mathcal{T}}^k)^T P_{1,\mathcal{T}}^{k+1}(A_1^k + B_1^k K_{2,\mathcal{T}}^k) - C_k^T C_k - (K_{2,\mathcal{T}}^k)^T K_{2,\mathcal{T}}^k \\
\quad + (A_2^k + B_2^k K_{2,\mathcal{T}}^k)^T P_{1,\mathcal{T}}^{k+1}(A_2^k + B_2^k K_{2,\mathcal{T}}^k) - K_{3,\mathcal{T}}^k H_1(P_{1,\mathcal{T}}^{k+1})^+(K_{3,\mathcal{T}}^k)^T, \\
P_{1,\mathcal{T}}^{\mathcal{T}+1} = 0, \\
H_1(P_{1,\mathcal{T}}^{k+1})H_1^+(P_{1,\mathcal{T}}^{k+1})(K_{3,\mathcal{T}}^k)^T - (K_{3,\mathcal{T}}^k)^T = 0, \\
H_1(P_{1,\mathcal{T}}^{k+1}) \geq 0, \ k \in N_\mathcal{T},
\end{cases}
$$

$$(3.44)$$

$$K_{1,\mathcal{T}}^k = -H_1(P_{1,\mathcal{T}}^{k+1})^+(K_{3,\mathcal{T}}^k)^T, \tag{3.45}$$

$$
\begin{cases}
P_{2,\mathcal{T}}^k = (A_1^k + C_1^k K_{1,\mathcal{T}}^k)^T P_{2,\mathcal{T}}^{k+1}(A_1^k + C_1^k K_{1,\mathcal{T}}^k) \\
\quad + (A_2^k + C_2^k K_{1,\mathcal{T}}^k)^T P_{2,\mathcal{T}}^{k+1}(A_2^k + C_2^k K_{1,\mathcal{T}}^k) \\
\quad + C_k^T C_k - K_{4,\mathcal{T}}^k H_2(P_{2,\mathcal{T}}^{k+1})^{-1}(K_{4,\mathcal{T}}^k)^T, \\
H_2(P_{2,\mathcal{T}}^{k+1}) > 0, \\
P_{2,\mathcal{T}}^{\mathcal{T}+1} = 0,
\end{cases}
$$

$$(3.46)$$

and

$$K_{2,\mathcal{T}}^k = -H_2(P_{2,\mathcal{T}}^{k+1})^{-1}(K_{4,\mathcal{T}}^k)^T \tag{3.47}$$

have a solution $(P_{1,\mathcal{T}}^k \leq 0, K_{1,\mathcal{T}}^k; P_{2,\mathcal{T}}^k \geq 0, K_{2,\mathcal{T}}^k)$ on $k \in N_\mathcal{T}$, where $K_{3,\mathcal{T}}^k$, $H_1(P_{1,\mathcal{T}}^{k+1})$, $K_{4,\mathcal{T}}^k$ and $H_2(P_{2,\mathcal{T}}^{k+1})$ are as defined in Theorem 3.4.

Proof. We only give a sketch of the proof. Necessity: If

$$(u_{\mathcal{T},k}^* = K_{2,\mathcal{T}}^k x_k, v_{\mathcal{T},k}^* = K_{1,\mathcal{T}}^k x_k)$$

is a solution to (3.43), then the first inequality of (3.43) implies that $J_{1,\mathcal{T}}(u_\mathcal{T}^*, v)$ achieves its infimum value at $v_k = v_{\mathcal{T},k}^* = K_{1,\mathcal{T}}^k x_k$, i.e., $v_{\mathcal{T},k}^* = K_{1,\mathcal{T}}^k x_k$ is an optimal control for the following LQ control problem:

$$\min_{v \in l_w^2(N_\mathcal{T}, \mathcal{R}^{n_v})} J_{1,\mathcal{T}}(u_\mathcal{T}^*, v) \tag{3.48}$$

subject to

$$
\begin{cases}
x_{k+1} = (A_1^k + B_1^k K_{2,\mathcal{T}}^k)x_k + C_1^k v_k + (A_2^k + B_2^k K_{2,\mathcal{T}}^k)x_k w_k + C_2^k v_k w_k, \\
x_0 \in \mathcal{R}^n, \ k \in N_\mathcal{T}.
\end{cases}
$$

$$(3.49)$$

By Lemma 3.4, (3.44) has a solution $P_{1,\mathcal{T}}^k$ on $k \in N_\mathcal{T}$, and the optimal feedback gain $K_{1,\mathcal{T}}^k$ is given by (3.45). We can further show $P_{1,\mathcal{T}}^k \leq 0$ on $k \in N_\mathcal{T}$ from the relation $J_{1,\mathcal{T}}(u_\mathcal{T}^*, v_\mathcal{T}^*) = x_0^T P_{1,\mathcal{T}}^0 x_0 \leq J_{1,\mathcal{T}}(u_\mathcal{T}^*, 0) \leq 0$ for any $x_0 \in \mathcal{R}^n$.

Similarly, the second inequality of (3.43) implies that $J_{2,\mathcal{T}}(u, v_\mathcal{T}^*)$ achieves its minimal value at $u_k = u_{\mathcal{T},k}^* = K_{2,\mathcal{T}}^k x_k$, i.e., $u_{\mathcal{T},k}^* = K_{2,\mathcal{T}}^k x_k$ is an optimal control for the following LQ control problem:

$$\min_{u \in l_w^2(N_\mathcal{T}, \mathcal{R}^{n_u})} J_{2,\mathcal{T}}(u, v_\mathcal{T}^*) \tag{3.50}$$

subject to

$$\begin{cases} x_{k+1} = (A_1^k + C_1^k K_{1,T}^k)x_k + B_1^k u_k + (A_2^k + C_2^k K_{1,T}^k)x_k w_k + B_2^k u_k w_k, \\ x_0 \in \mathcal{R}^n, \ k \in N_T. \end{cases}$$
$$(3.51)$$

Again, by Corollary 3.1, (3.46) must have a solution $P_{2,T}^k \geq 0$ on $k \in N_T$, and the optimal control $u_{T,k}^* = K_{2,T}^k x_k$ with $K_{2,T}^k$ given by (3.47).

At the end of this section, we give some comments on the relationship between the solvability of the finite horizon H_2/H_∞ control and the existence of a two-person non-zero sum Nash equilibrium point defined in (3.43).

REMARK 3.6   Obviously, only when $H_1(P_{1,T}^{k+1}) > 0$, the solvability of the discrete-time H_2/H_∞ control is equivalent to the existence of a two-person non-zero sum Nash equilibrium point. Hence, in the discrete-time case, the above two problems are generally not equivalent, which differs from continuous-time Itô systems as described in Theorems 2.1–2.2. A study on H_2/H_∞ control is based on an SBRL and a standard LQ regulator, while the problem of searching for a Nash equilibrium point (u_T^*, v_T^*) can be changed into that of solving two LQ regulators: one is indefinite and the other one is standard.   ∎

Different from the mixed H_2/H_∞ control that is closely related to a two-person non-zero sum Nash game, the pure H_∞ control is related to a minimax design problem that is a two-person zero sum game problem [31, 169]. There are other studies on the H_2/H_∞ control which are based on game theory; see [44, 134].

3.3   Infinite Horizon H_2/H_∞ Control

This section will study the infinite horizon mixed H_2/H_∞ control for discrete time-invariant stochastic systems with not only the state- but also the disturbance- and control-dependent noise of the form

$$\begin{cases} x_{k+1} = (A_1 x_k + B_1 u_k + C_1 v_k) + (A_2 x_k + B_2 u_k + C_2 v_k)w_k, \\ x_0 \in \mathcal{R}^n, \\ z_k = \begin{bmatrix} C x_k \\ D u_k \end{bmatrix}, \ D^T D = I, \ k \in \mathcal{N} := \{0, 1, 2, \cdots, \}. \end{cases}$$
$$(3.52)$$

We denote $l_w^2(\mathcal{N}, \mathcal{R}^k)$ as the set of all non-anticipative square summable stochastic processes

$$y = \{y_k : y_k \in \mathcal{L}_{\mathcal{F}_{k-1}}^2(\Omega, \mathcal{R}^k)\}_{k \in \mathcal{N}}.$$

The l^2-norm of $y \in l_w^2(\mathcal{N}, \mathcal{R}^k)$ is defined by

$$\| y \|_{l_w^2(\mathcal{N}, \mathcal{R}^k)} = \left(\sum_{k=0}^{\infty} \mathcal{E} \| y_k \|^2 \right)^{\frac{1}{2}}.$$

Obviously, for any $\mathcal{T} \in \mathcal{N}$ and $(x_0, u, v) \in \mathcal{R}^n \times l_w^2(N_\mathcal{T}, \mathcal{R}^{n_u}) \times l_w^2(N_\mathcal{T}, \mathcal{R}^{n_v})$, there exists a unique solution $x \in l_w^2(N_{\mathcal{T}+1}, \mathcal{R}^n)$ to (3.52).

As special cases of (3.52), the results of the H_2/H_∞ control for the system with state- and control-dependent noise

$$\begin{cases} x_{k+1} = (A_1 x_k + B_1 u_k + C_1 v_k) + (A_2 x_k + B_2 u_k) w_k, \\ x_0 \in \mathcal{R}^n, \\ z_k = \begin{bmatrix} C x_k \\ D u_k \end{bmatrix}, \ D^T D = I, \ k \in \mathcal{N} \end{cases} \tag{3.53}$$

and the system with state- and exogenous disturbance-dependent noise

$$\begin{cases} x_{k+1} = (A_1 x_k + B_1 u_k + C_1 v_k) + (A_2 x_k + C_2 v_k) w_k, \\ x_0 \in \mathcal{R}^n, \\ z_k = \begin{bmatrix} C x_k \\ D u_k \end{bmatrix}, \ D^T D = I, \ k \in \mathcal{N} \end{cases} \tag{3.54}$$

are obtained. Different from the finite horizon case in Section 3.1, to discuss the infinite horizon H_2/H_∞ control, the internal stability of the closed-loop system is to be established, which is non-trivial. Similar to linear Itô systems, we have to introduce exact observability and detectability for linear discrete-time systems, and then establish the Lyapunov-type theorems and an infinite horizon SBRL, which by itself has theoretical importance. Based on the SBRL, it is shown that under the conditions of exact observability and detectability, the existence of a static state feedback H_2/H_∞ controller is equivalent to the solvability of four coupled matrix-valued equations. A suboptimal H_2/H_∞ controller design is given based on a convex optimization approach and an iterative algorithm is proposed to solve the four coupled matrix-valued equations. We refer the reader to [97, 118, 161, 206, 207] for the content of this section.

3.3.1   Preliminaries

Consider the following discrete time-invariant stochastic system

$$x_{k+1} = A_1 x_k + A_2 x_k w_k, \ x_0 \in \mathcal{R}^n, \ k \in \mathcal{N}. \tag{3.55}$$

DEFINITION 3.5   *The system (3.55) is said to be ASMS if for any* $x_0 \in \mathcal{R}^n$*, the corresponding state satisfies*

$$\lim_{k \to \infty} \mathcal{E} \| x_k \|^2 = 0.$$

In this case, we say that (A_1, A_2) is Schur stable in short.

DEFINITION 3.6   *The time-invariant control system*

$$x_{k+1} = A_1 x_k + B_1 u_k + (A_2 x_k + B_2 u_k) w_k, \ k \in \mathcal{N} \qquad (3.56)$$

is said to be stabilizable in the mean square sense if there exists a feedback control law $u_k = K x_k$ with K a constant gain matrix, such that for any $x_0 \in \mathcal{R}^n$, the closed-loop system

$$x_{k+1} = (A_1 + B_1 K) x_k + (A_2 + B_2 K) x_k w_k \qquad (3.57)$$

is ASMS. In this situation, we say that $(A_1, B_1; A_2, B_2)$ is stabilizable in short.

A necessary and sufficient condition for the stabilizability of $(A_1, B_1; A_2, B_2)$ is that there are matrices $P > 0$ and U solving the LMI [65]

$$\begin{bmatrix} -P & A_1 P + B_1 U & A_2 P + B_2 U \\ P A_1^T + U^T B_1^T & -P & 0 \\ P A_2^T + U^T B_2^T & 0 & -P \end{bmatrix} < 0.$$

REMARK 3.7   From Proposition 2.2 in [65], the stabilizability of $(A_1, B_1; A_2, B_2)$ implies that of (A_1, B_1), and the Schur stability of (A_1, A_2) implies that of A_1. As is known, a matrix A_1 is called Schur stable if all eigenvalues of A_1 are located in the open unit circle $\mathcal{D}(0,1) := \{\lambda \in \mathcal{C} : |\lambda| < 1\}$. ∎

We now introduce the notions of exact observability and detectability of discrete-time stochastic systems, which are discrete versions of Definition 1.8 and Definition 1.13.

DEFINITION 3.7   *The system*

$$\begin{cases} x_{k+1} = A_1 x_k + A_2 x_k w_k, & x_0 \in \mathcal{R}^n, \\ y_k = C x_k, \ k \in \mathcal{N} \end{cases} \qquad (3.58)$$

or $(A_1, A_2|C)$ is said to be exactly observable if

$$y_k \equiv 0, a.s., \forall k \in \mathcal{N} \Rightarrow x_0 = 0.$$

$(A_1, A_2|C)$ is said to be exactly detectable if

$$y_k \equiv 0, a.s., \forall k \in \mathcal{N} \Rightarrow \lim_{k \to \infty} \mathcal{E} \|x_k\|^2 = 0.$$

Obviously, exact detectability is weaker than exact observability. Observe that when $w_k \equiv 0$, Definition 3.7 reduces to the complete observability and detectability of linear deterministic discrete-time systems.

Associated with the system

$$x_{k+1} = A_1 x_k + A_2 x_k w_k,$$

we introduce the following GLO $\mathcal{D}_{A_1,A_2}$:

$$\mathcal{D}_{A_1,A_2} X = A_1 X A_1^T + A_2 X A_2^T, \forall X \in \mathcal{S}_n.$$

Obviously, the adjoint operator $\mathcal{D}_{A_1,A_2}^*$ of $\mathcal{D}_{A_1,A_2}$ is given by

$$\mathcal{D}_{A_1,A_2}^* X = A_1^T X A_1 + A_2^T X A_2, \forall X \in \mathcal{S}_n.$$

In (3.58), let $X_k = \mathcal{E}[x_k x_k^T]$, $Y_k = \mathcal{E}[y_k y_k^T]$, then X_k and Y_k obey

$$\begin{cases} X_{k+1} = \mathcal{D}_{A_1,A_2} X_k, & X_0 = x_0 x_0^T, \\ Y_k = C X_k C^T, & k \in \mathcal{N}. \end{cases} \tag{3.59}$$

Similar to the proof of Theorem 1.7 and noticing the well known fact that a constant matrix A is Schur stable iff $\sigma(A) \subset \mathcal{D}(0,1)$, the following is obvious.

THEOREM 3.6
(A_1, A_2) *is Schur stable iff* $\sigma(\mathcal{D}_{A_1,A_2}) \subset \mathcal{D}(0,1)$.

Parallel to Theorem 1.9 and Theorem 1.13, it is easy to derive the following theorem for exact observability and detectability of discrete-time stochastic systems.

THEOREM 3.7 (Stochastic PBH eigenvector test)

(i) $(A_1, A_2|C)$ *is exactly observable iff there does not exist a non-zero* $Z \in \mathcal{S}_n$ *such that*

$$\mathcal{D}_{A_1,A_2} Z = \lambda Z, \ CZ = 0, \ \lambda \in \mathcal{C}. \tag{3.60}$$

(ii) $(A_1, A_2|C)$ *is exactly detectable iff there does not exist a non-zero* $Z \in \mathcal{S}_n$ *such that*

$$\mathcal{D}_{A_1,A_2} Z = \lambda Z, \ CZ = 0, \ |\lambda| \geq 1. \tag{3.61}$$

The following lemma generalizes Theorems 2.6–2.7 to consider a discrete-time Lyapunov-type equation, which also generalizes Proposition 2.2 of [65] to the case $C^T C \geq 0$.

LEMMA 3.5
The following hold:

(i) *If (A_1, A_2) is Schur stable, then the following discrete Lyapunov-type equation (i.e., GLE)*

$$-P + A_1^T P A_1 + A_2^T P A_2 + C^T C = 0 \qquad (3.62)$$

has a unique solution $P \geq 0$.

(ii) *If $(A_1, A_2|C)$ is exactly observable, then (A_1, A_2) is Schur stable iff (3.62) has a unique solution $P > 0$.*

(iii) *If $(A_1, A_2|C)$ is exactly detectable, then (A_1, A_2) is Schur stable iff (3.62) has a unique solution $P \geq 0$.*

(iv) *Suppose that $\sigma(\mathcal{D}_{A_1, A_2}) \subset \bar{\mathcal{D}}(0, 1) := \{\lambda \in \mathcal{C} : |\lambda| \leq 1\}$ and $(A_1, A_2|C)$ is exactly detectable. If P is a real symmetric solution of (3.62), then $P \geq 0$ and (A_1, A_2) is stable.*

Proof. (i) If (A_1, A_2) is stable, by Proposition 2.2 [65], (3.62) has a unique solution $P \geq 0$.

(ii) Necessity. In view of (i), we only need to show $P > 0$. Suppose by contradiction, this is not the case. Then, there exists $x_0 \neq 0$, such that $P x_0 = 0$. Using Lemma 3.1 with $v_k = 0$, A_{11} and A_{12} replaced by A_1 and A_2, respectively, we obtain that for any $\mathcal{T} \in \mathcal{N}$,

$$
\begin{aligned}
0 \leq \sum_{k=0}^{\mathcal{T}} \mathcal{E} \|C x_k\|^2 &= \sum_{k=0}^{\mathcal{T}} \mathcal{E}[x_k^T (-P + A_1^T P A_1 + A_2^T P A_2 + C^T C) x_k] \\
&\quad + x_0^T P x_0 - \mathcal{E}(x_{\mathcal{T}+1}^T P x_{\mathcal{T}+1}) \\
&= -\mathcal{E}(x_{\mathcal{T}+1}^T P x_{\mathcal{T}+1}) \leq 0, \qquad (3.63)
\end{aligned}
$$

which implies that $y_k = C x_k \equiv 0$ along the state trajectory of $x_{k+1} = A_1 x_k + A_2 x_k w_k$. According to Definition 3.7, we must have $x_0 = 0$, which is a contradiction. So $P > 0$.

Sufficiency. If $P > 0$ is a solution to (3.62), set $V(x_k) := \mathcal{E}(x_k^T P x_k)$, where x_k is the state trajectory of the system (3.58). From (3.63), we have

$$V(x_k) = x_0^T P x_0 - \sum_{i=0}^{k-1} \mathcal{E} \|C x_i\|^2,$$

which indicates that $V(x_k)$ is monotonically decreasing and bounded from below with respect to k. Therefore, $\lim_{k \to \infty} V(x_k)$ exists. It is easy to see that the following difference equation

$$
\begin{cases}
-H_k + A_1^T H_{k+1} A_1 + A_2^T H_{k+1} A_2 + C^T C = 0, \\
H_{\mathcal{T}+1} = 0, k \in N_{\mathcal{T}}
\end{cases}
\qquad (3.64)
$$

always admits a unique solution $H_k \geq 0$ for $k \in N_T$. Similar to (3.63), we have

$$\sum_{k=0}^{T} \mathcal{E}\|Cx_k\|^2 = x_0^T H_0 x_0 - \mathcal{E}(x_{T+1}^T H_{T+1} x_{T+1}) = x_0^T H_0 x_0.$$

So

$$V(x_{T+1}) - V(x_0) = -x_0^T H_0 x_0. \qquad (3.65)$$

It can be shown, as in the necessity part, that $H_0 > 0$ because of the exact observability of $(A_1, A_2|C)$. By the time-invariance of the system (3.58), the generalization of (3.65) still holds for any $k \in \mathcal{N}$, i.e.,

$$V(x_{T+k+1}) - V(x_k) = -\mathcal{E}(x_k^T H_0 x_k). \qquad (3.66)$$

Taking $k \to \infty$ and considering $H_0 > 0$, it follows that $\lim_{k \to \infty} \mathcal{E}\|x_k\|^2 = 0$, which completes the proof of the sufficiency.

(iii) By (i), it suffices to prove the sufficiency part. Let $\rho(\mathcal{D}_{A_1,A_2})$ denote the spectral radius of $\mathcal{D}_{A_1,A_2}$, by the finite dimensional Krein–Rutman theorem [164], there exists a non-zero $X_0 \geq 0$ such that $\mathcal{D}_{A_1,A_2} X_0 = \rho(\mathcal{D}_{A_1,A_2}) X_0$. If (A_1, A_2) is not Schur stable, then $\rho(\mathcal{D}_{A_1,A_2}) \geq 1$.

$$\begin{aligned}
0 \geq\; &< -C^T C, X_0 > = < -P + \mathcal{D}_{A_1,A_2}^*(P), X_0 > \\
&= < -P, X_0 > + < P, \mathcal{D}_{A,A_1}(X_0) > \\
&= < -P, X_0 > + < P, \rho(\mathcal{D}_{A_1,A_2}) X_0 > \\
&\geq 0. \qquad (3.67)
\end{aligned}$$

The inequality (3.67) implies $CX_0 = 0$ due to $X_0 \geq 0$. According to Theorem 3.7-(ii), $CX_0 = 0$ together with $\mathcal{D}_{A_1,A_2}(X_0) = \rho(\mathcal{D}_{A_1,A_2}) X_0$, $\rho(\mathcal{D}_{A_1,A_2}) \geq 1$, contradicts the exact detectability of $(A_1, A_2|C)$. So (A_1, A_2) is Schur stable.

(iv) If $\rho(\mathcal{D}_{A_1,A_2}) < 1$, then (A_1, A_2) is Schur stable by Theorem 3.6, which yields $P \geq 0$ according to (iii). If $\rho(\mathcal{D}_{A_1,A_2}) = 1$, then by Krein–Rutman theorem and repeating the same procedure as in (iii), (iv) can be shown. $\square$

For system (3.56), we define the admissible control set

$$\mathcal{U}_{ad} := \{u \in l_w^2(\mathcal{N}, \mathcal{R}^l) : \{u_k\}_{k \in \mathcal{N}} \text{ is a mean square stabilizing control sequence}\}$$

as well as the cost functional

$$J^\infty(u) := \sum_{k=0}^{\infty} \mathcal{E}(x_k^T Q x_k + u_k^T R x_k), \qquad (3.68)$$

where $Q \geq 0, R > 0$. The infinite horizon stochastic LQ optimal control with stable terminal states is to find a control $u_s^* = \in \mathcal{U}_{ad}$, if it exists, such that

$$J^\infty(u_s^*) = \Theta(x_0) := \inf_{u \in \mathcal{U}_{ad}} J^\infty(u). \qquad (3.69)$$

In this case, u_s^* is called the optimal control. We call $\{x_k^*\}_{k\in\mathcal{N}}$ corresponding to $\{u_{s,k}^*\}_{k\in\mathcal{N}}$ the optimal trajectory, and $\Theta(x_0)$ the optimal cost value.

Quite often, u_s^* does not necessarily exist. So we introduce the following definition:

DEFINITION 3.8   *The LQ control problem (3.56) and (3.69) is called well posed, if*

$$-\infty < \Theta(x_0) < \infty.$$

If, moreover, u_s^* exists, the LQ control problem (3.69) under the constraint of (3.56) is called attainable.

We will not extensively discuss the discrete-time stochastic LQ control associated with (3.56) and (3.69) with free terminal states since it is not relevant when the H_2/H_∞ control is concerned.

3.3.2   Standard LQ control result

The following lemma is about the infinite horizon stochastic LQ optimal control problem [97], which is needed later.

LEMMA 3.6
Consider the LQ control problem (3.56) and (3.69). If $(A_1, B_1; A_2, B_2)$ is stabilizable and $(A_1, A_2|Q^{1/2})$ is exactly observable (respectively, exactly detectable), then the following hold.

(i) The GARE

$$\begin{cases} A_1^T P A_1 + A_2^T P A_2 + Q - (A_1^T P B_1 + A_2^T P B_2)(R + B_1^T P B_1 + B_2^T P B_2)^{-1} \\ \quad \cdot (B_1^T P A_1 + B_2^T P A_2) = P, \\ R + B_1^T P B_1 + B_2^T P B_2 > 0, R > 0, Q \ge 0 \end{cases}$$

$$(3.70)$$

has a solution $P > 0$ (respectively, $P \ge 0$), which is also a feedback stabilizing solution, i.e., $(A_1 + B_1 K, A_2 + B_2 K)$ is Schur stable, where

$$K = -(R + B_1^T P B_1 + B_2^T P B_2)^{-1}(B_1^T P A_1 + B_2^T P A_2).$$

(ii) The optimal cost and optimal control sequence $u_s^* := \{u_{s,k}^*\}_{k\in\mathcal{N}}$ are given respectively by

$$\Theta(x_0) = x_0^T P x_0, \qquad (3.71)$$

and

$$u_{s,k}^* = -(R + B_1^T P B_1 + B_2^T P B_2)^{-1}(B_1^T P A_1 + B_2^T P A_2)x_k. \quad (3.72)$$

(iii) GARE (3.70) admits a unique positive definite solution (respectively, positive semi-definite solution).

Proof. In order to avoid repetition, we only prove this lemma under exact observability. From [7, 18], for any $\mathcal{T} \in \mathcal{N}$, the following finite horizon difference equation

$$
\begin{cases}
A_1^T P_{k+1} A_1 + A_2^T P_{k+1} A_2 + Q - (A_1^T P_{k+1} B_1 + A_2^T P_{k+1} B_2) \\
\quad \cdot (R + B_1^T P_{k+1} B_1 + B_2^T P_{k+1} B_2)^{-1} (B_1^T P_{k+1} A_1 + B_2^T P_{k+1} A_2) = P_k, \\
R + B_1^T P_{k+1} B_1 + B_2^T P_{k+1} B_2 > 0, k \in N_\mathcal{T}, \\
P_{\mathcal{T}+1} = 0
\end{cases}
\tag{3.73}
$$

admits a unique solution $P_\mathcal{T}^k \geq 0$ on $k \in N_\mathcal{T}$. Moreover, subject to the system (3.56), the finite horizon LQ optimal control leads to

$$
\min_{u \in l_w^2(N_\mathcal{T}, \mathcal{R}^{n_u})} \sum_{k=0}^{\mathcal{T}} \mathcal{E}(x_k^T Q x_k + u_k^T R u_k) = x_0^T P_\mathcal{T}^0 x_0.
$$

Similar to the discussion in [11], it is easy to show that under the stabilizability of $(A_1, B_1; A_2, B_2)$, $\lim_{T \to \infty} P_\mathcal{T}^0 = P \geq 0$, which is a solution to GARE (3.70). To show $P > 0$, we note that (3.70) can be written as

$$
-P + (A_1 + B_1 K)^T P (A_1 + B_1 K) + (A_2 + B_2 K)^T P (A_2 + B_2 K)
$$
$$
= -Q - K^T R K = - \left[Q^{1/2} \ K^T R^{1/2} \right] \begin{bmatrix} Q^{1/2} \\ R^{1/2} K \end{bmatrix} = 0.
\tag{3.74}
$$

Because $(A_1, A_2 | Q^{1/2})$ is exactly observable, by Theorem 3.7-(i), we know

$$
\left(A_1 + B_1 K, A_2 + B_2 K \middle| \begin{bmatrix} Q^{1/2} \\ R^{1/2} K \end{bmatrix} \right)
$$

is also exactly observable. If $P \geq 0$ but not $P > 0$, then there exists a non-zero $\xi \in \mathcal{R}^n$ such that $P\xi = 0$. Under the constraint of

$$
\begin{cases}
x_{k+1} = (A_1 + B_1 K) x_k + (A_2 + B_2 K) x_k w_k, \\
x_0 = \xi \in \mathcal{R}^n, \\
y_k = \begin{bmatrix} Q^{1/2} \\ R^{1/2} K \end{bmatrix} x_k, \ k \in \mathcal{N}
\end{cases}
\tag{3.75}
$$

and as in (3.63), we have that for any $\mathcal{T} \in \mathcal{N}$,

$$
0 \leq \sum_{k=0}^{\mathcal{T}} \mathcal{E} \|y_k\|^2 = \sum_{k=0}^{\mathcal{T}} \mathcal{E}\{ x_k^T [-P + (A_1 + B_1 K)^T P (A_1 + B_1 K)
$$
$$
+ (A_2 + B_2 K)^T P (A_2 + B_2 K) + Q + K^T R K] x_k \}
$$
$$
+ \xi^T P \xi - \mathcal{E}(x_{\mathcal{T}+1}^T P x_{\mathcal{T}+1})
$$
$$
= -\mathcal{E}(x_{\mathcal{T}+1}^T P x_{\mathcal{T}+1}) \leq 0.
\tag{3.76}
$$

The above implies $y_k \equiv 0, a.s., \forall k \in \mathcal{N}$ but $x_0 = \xi \neq 0$, which contradicts the exact observability of

$$
\left(A_1 + B_1 K, A_2 + B_2 K \middle| \begin{bmatrix} Q^{1/2} \\ R^{1/2} K \end{bmatrix} \right).
$$

Hence, we must have $P > 0$. On the other hand, $P > 0$ is a feedback stabilizing solution due to Lemma 3.5-(ii).

As for (ii), we show first that $\Theta(x_0)$ is bounded from below by $x_0^T P x_0$. Indeed, for any $u \in \mathcal{U}_{ad}$, using Lemma 3.1, we obtain

$$\sum_{k=0}^{\mathcal{T}} \mathcal{E}(x_k^T Q x_k + u_k^T R u_k) = \sum_{k=0}^{\mathcal{T}} \mathcal{E}(x_k^T Q x_k + u_k^T R u_k)$$

$$+ \sum_{k=0}^{\mathcal{T}} \mathcal{E} \begin{bmatrix} x_k \\ u_k \end{bmatrix}^T \begin{bmatrix} -P + A_1^T P A_1 + A_2^T P A_2 & A_1^T P B_1 + A_2^T P B_2 \\ B_1^T P A_1 + B_2^T P A_2 & B_1^T P B_1 + B_2^T P B_2 \end{bmatrix} \begin{bmatrix} x_k \\ u_k \end{bmatrix}$$

$$+ x_0^T P x_0 - \mathcal{E}(x_{\mathcal{T}+1}^T P x_{\mathcal{T}+1})$$

$$= \sum_{k=0}^{\mathcal{T}} \mathcal{E}\{x_k^T[-P + A_1^T P A_1 + A_2^T P A_2 + Q - (A_1^T P B_1 + A_2^T P B_2)$$

$$\cdot (R + B_1^T P B_1 + B_2^T P B_2)^{-1}(A_1^T P B_1 + A_2^T P B_2)^T]x_k\}$$

$$+ \sum_{k=0}^{\mathcal{T}} \mathcal{E}[(u_k - K x_k)^T (R + B_1^T P B_1 + B_2^T P B_2)(u_k - K x_k)]$$

$$+ x_0^T P x_0 - \mathcal{E}(x_{\mathcal{T}+1}^T P x_{\mathcal{T}+1}).$$

Set $\mathcal{T} \to \infty$, then for any $u \in \mathcal{U}_{ad}$, it follows that

$$J^\infty(u) = \sum_{k=0}^{\infty} \mathcal{E}[(u - K x_k)^T (R + B_1^T P B_1 + B_2^T P B_2)(u - K x_k)] + x_0^T P x_0, \quad (3.77)$$

which yields $J^\infty(u) \geq x_0^T P x_0$ because of $(R + B_1^T P B_1 + B_2^T P B_2) > 0$, i.e., $\Theta(x_0) \geq x_0^T P x_0$. Secondly, because $u_{s,k}^* = K x_k$ is a feedback stabilizing control law, so $u_s^* \in \mathcal{U}_{ad}$. From (3.77), we have $J^\infty(u_s^*) = x_0^T P x_0$, so u_s^* is the desired optimal control, and the optimal cost value is given by $\Theta(x_0) = x_0^T P x_0$. (ii) is proved.

We finally show that GARE (3.70) admits a unique positive definite solution $P > 0$. Otherwise, if $\tilde{P} > 0$ is another solution to GARE (3.70), then, by repeating the above procedure, we know

$$u_{1,k}^* = K_1 x_k = -(R + B_1^T \tilde{P} B_1 + B_2^T \tilde{P} B_2)^{-1}(B_1^T \tilde{P} A_1 + B_2^T \tilde{P} A_2)x_k$$

is another optimal control, i.e., $\Theta(x_0) = x_0^T \tilde{P} x_0$. Because the optimal cost value is unique, so

$$x_0^T \tilde{P} x_0 = x_0^T P x_0, \forall x_0 \in \mathcal{R}^n,$$

which implies $\tilde{P} = P$, (iii) is proved. $\square$

REMARK 3.8 Similar to the discussions in Section 2.3.2, it can be shown that the above $P > 0$ is also a maximal solution. Generally speaking, a feedback stabilizing solution must be a maximal solution, but the converse is not true. ∎

3.3.3   An SBRL

In this section we shall develop a discrete-time version of SBRL which will play a central role in the study of stochastic H_∞ control and estimation. To this end, we consider the following perturbed system

$$\begin{cases} x_{k+1} = A_{11}x_k + B_{11}v_k + (A_{12}x_k + B_{12}v_k)w_k, \\ x_0 \in \mathcal{R}^n, \\ z_1^k = C_{11}x_k, \ k \in \mathcal{N}, \end{cases} \tag{3.78}$$

where $z_1 \in \mathcal{R}^{n_{z1}}$ is the controlled output.

DEFINITION 3.9   *In system (3.78), if the disturbance input $v \in l_w^2$ $(\mathcal{N}, \mathcal{R}^{n_v})$ and the controlled output $z_1 \in l_w^2(\mathcal{N}, \mathcal{R}^{n_{z1}})$, then the perturbed operator $\tilde{\mathcal{L}}_\infty : l_w^2(\mathcal{N}, \mathcal{R}^{n_v}) \mapsto l_w^2(\mathcal{N}, \mathcal{R}^{n_{z1}})$ is defined by*

$$\tilde{\mathcal{L}}_\infty v_k := z_1^k|_{x_0=0} = C_{11}x_k|_{x_0=0}$$

with its norm

$$\| \tilde{\mathcal{L}}_\infty \| = \sup_{v \in l_w^2(\mathcal{N}, \mathcal{R}^{n_v}), v \neq 0, x_0=0} \frac{\| z_1 \|_{l_w^2(\mathcal{N}, \mathcal{R}^{n_{z1}})}}{\| v \|_{l_w^2(\mathcal{N}, \mathcal{R}^{n_v})}}$$

$$= \sup_{v \in l_w^2(\mathcal{N}, \mathcal{R}^{n_v}), v \neq 0, x_0=0} \frac{\left(\displaystyle\sum_{k=0}^{\infty} \mathcal{E} \| C_{11}x_k \|^2 \right)^{\frac{1}{2}}}{\left(\displaystyle\sum_{k=0}^{\infty} \mathcal{E} \| v_k \|^2 \right)^{\frac{1}{2}}}.$$

DEFINITION 3.10   *The system (3.78) is said to be internally stable if it is ASMS in the absence of v, i.e., $v_k \equiv 0$ for $k \in \mathcal{N}$.*

The following SBRL can be viewed as the discrete version of Theorem 2.12. It should be pointed out that a discrete-time bounded real lemma for deterministic systems can be found in [47].

LEMMA 3.7
If the system (3.78) is internally stable and $\| \tilde{\mathcal{L}}_\infty \| < \gamma$ for a given $\gamma > 0$, then there exists a stabilizing solution $P \leq 0$ to the following GARE

$$\begin{cases} -P + A_{11}^T P A_{11} + A_{12}^T P A_{12} - C_{11}^T C_{11} - (A_{11}^T P B_{11} + A_{12}^T P B_{12}) \\ \cdot (\gamma^2 I + B_{12}^T P B_{12} + B_{11}^T P B_{11})^{-1}(A_{11}^T P B_{11} + A_{12}^T P B_{12})^T = 0, \\ \gamma^2 I + B_{12}^T P B_{12} + B_{11}^T P B_{11} > 0, \end{cases} \tag{3.79}$$

i.e., $(A_{11} + B_{11}K, A_{12} + B_{12}K)$ is Schur stable with

$$K = -(\gamma^2 I + B_{12}^T P B_{12} + B_{11}^T P B_{11})^{-1}(A_{11}^T P B_{11} + A_{12}^T P B_{12})^T. \tag{3.80}$$

Conversely, if (3.79) has a stabilizing solution $P \leq 0$, then $\| \tilde{\mathcal{L}}_\infty \| < \gamma$.

Proof. We consider the following associated finite horizon GDRE

$$\begin{cases} -P_k + A_{11}^T P_{k+1} A_{11} + A_{12}^T P_{k+1} A_{12} - C_{11}^T C_{11} \\ -(A_{11}^T P_{k+1} B_{11} + A_{12}^T P_{k+1} B_{12})(\gamma^2 I + B_{12}^T P_{k+1} B_{12} + B_{11}^T P_{k+1} B_{11})^{-1} \\ \quad \cdot (A_{11}^T P_{k+1} B_{11} + A_{12}^T P_{k+1} B_{12})^T = 0, \\ \gamma^2 I + B_{12}^T P_{k+1} B_{12} + B_{11}^T P_{k+1} B_{11} > 0, \; k \in N_{\mathcal{T}}, \\ P_{\mathcal{T}+1} = 0 \end{cases} \tag{3.81}$$

and the corresponding quadratic cost functional

$$J_{\mathcal{T}}(x_0, v) := \sum_{k=0}^{\mathcal{T}} \mathcal{E}(\gamma^2 \| v_k \|^2 - \| z_1^k \|^2) = \sum_{k=0}^{\mathcal{T}} \mathcal{E}(\gamma^2 v_k^T v_k - x_k^T C_{11}^T C_{11} x_k),$$

where x_k is the solution of (3.78) on $N_{\mathcal{T}+1}$ with initial value x_0, z_1^k is the corresponding controlled output signal on $N_{\mathcal{T}}$. In addition, we assume $v = \{v_k\}_{k \in N_{\mathcal{T}}} \in l_w^2(N_{\mathcal{T}}, \mathcal{R}^{n_v})$. From Proposition 2.14 [65], we immediately obtain that GDRE (3.81) has a unique solution $P_{\mathcal{T}}^k \leq 0$ on $k \in N_{\mathcal{T}+1}$, where $P_{\mathcal{T}}^{\mathcal{T}+1} := P_{\mathcal{T}+1} = 0$. Moreover, the minimal cost is given by

$$\min_{v \in l_w^2(N_{\mathcal{T}}, \mathcal{R}^{n_v})} J_{\mathcal{T}}(x_0, v) = x_0^T P_{\mathcal{T}}^0 x_0.$$

Replacing C_{11} by $C_\delta = \begin{bmatrix} C_{11} \\ \delta I \end{bmatrix}$ and z_1^k by $z_\delta^k = C_\delta x_k$, we obtain the corresponding perturbation operator $L_\delta : l_w^2(\mathcal{N}, \mathcal{R}^{n_v}) \to l_w^2(\mathcal{N}, \mathcal{R}^{n_{z\delta}})$, defined by

$$L_\delta v_k = z_\delta^k|_{x_0=0} = C_\delta x_k|_{x_0=0}$$

and the cost

$$J_{\mathcal{T},\delta}(x_0, v) = \sum_{k=0}^{\mathcal{T}} \mathcal{E}(\gamma^2 \| v_k \|^2 - \| z_\delta^k \|^2)$$

$$= \sum_{k=0}^{\mathcal{T}} \mathcal{E}(\gamma^2 v_k^T v_k - x_k^T C_{11}^T C_{11} x_k - \delta^2 x_k^T x_k).$$

Since (3.78) is internally stable, from Remark 2.9 [65], the state trajectory of (3.78) belongs to $l_w^2(\mathcal{N}, \mathcal{R}^n)$ for every $v \in l_w^2(\mathcal{N}, \mathcal{R}^{n_v})$, so $\|L_\delta\| < \gamma$ for sufficiently small $\delta > 0$. Applying Proposition 2.14 [65] to the modified data, we find that GDRE

$$\begin{cases} -P_k + A_{11}^T P_{k+1} A_{11} + A_{12}^T P_{k+1} A_{12} - C_{11}^T C_{11} - \delta^2 I \\ -(A_{11}^T P_{k+1} B_{11} + A_{12}^T P_{k+1} B_{12})(\gamma^2 I + B_{12}^T P_{k+1} B_{12} + B_{11}^T P_{k+1} B_{11})^{-1} \\ \quad \cdot (A_{11}^T P_{k+1} B_{11} + A_{12}^T P_{k+1} B_{12})^T = 0, \\ \gamma^2 I + B_{12}^T P_{k+1} B_{12} + B_{11}^T P_{k+1} B_{11} > 0, k \in N_{\mathcal{T}}, \\ P_{\mathcal{T}+1} = 0 \end{cases} \tag{3.82}$$

admits a unique solution $P_{T,\delta}^k \leq 0$ on N_{T+1}, and

$$\min_{v \in l_w^2(N_T, \mathcal{R}^{n_v})} J_{T,\delta}(x_0, v) = x_0^T P_{T,\delta}^0 x_0.$$

By the time invariance of P_T^k and $P_{T,\delta}^k$ on N_{T+1}, i.e., $P_T^k = P_{T-k}^0$, $P_{T,\delta}^k = P_{T-k,\delta}^0$, $0 \leq t \leq T$, we have for any $x_0 \in \mathcal{R}^n$,

$$
\begin{aligned}
x_0^T P_{T,\delta}^k x_0 = x_0^T P_{T-k,\delta}^0 x_0 &= \min_{v \in l_w^2(N_{T-k}, \mathcal{R}^{n_v})} J_{T-k,\delta}(x_0, v) \\
&\leq \min_{v \in l_w^2(N_{T-k}, \mathcal{R}^{n_v})} J_{T-k}(x_0, v) \\
&= x_0^T P_{T-k}^0 x_0 = x_0^T P_T^k x_0
\end{aligned}
$$

i.e., $P_{T,\delta}^k \leq P_T^k$, $k \in N_T$. From Lemmas 2.12 and 2.15 [65], we can easily see that P_T^k and $P_{T,\delta}^k$ are bounded from below, and decrease as T increases, which implies that

$$\lim_{T \to \infty} P_T^k = \lim_{T \to \infty} P_{T-k}^0 = P, \quad \lim_{T \to \infty} P_{T,\delta}^k = \lim_{T \to \infty} P_{T-k,\delta}^0 = P_\delta$$

exist and $P \geq P_\delta$. Moreover, P and P_δ satisfy GARE (3.79) and the following GARE

$$
\begin{cases}
-P_\delta + A_{11}^T P_\delta A_{11} + A_{12}^T P_\delta A_{12} - C_{11}^T C_{11} - \delta^2 I - (A_{11}^T P_\delta B_{11} + A_{12}^T P_\delta B_{12}) \\
\cdot(\gamma^2 I + B_{12}^T P_\delta B_{12} + B_{11}^T P_\delta B_{11})^{-1}(A_{11}^T P_\delta B_{11} + A_{12}^T P_\delta B_{12})^T = 0, \\
\gamma^2 I + B_{12}^T P_\delta B_{12} + B_{11}^T P_\delta B_{11} > 0,
\end{cases}
$$
$$(3.83)$$

respectively. The rest is to prove that $(A_{11} + B_{11}K, A_{12} + B_{12}K)$ is stable. To this end, we note that the first equalities of (3.79) and (3.83) can be written as

$$
\begin{aligned}
-P + (A_{11} + B_{11}K)^T P(A_{11} + B_{11}K) + (A_{12} + B_{12}K)^T P(A_{12} + B_{12}K) \\
- C_{11}^T C_{11} + \gamma^2 K^T K = 0 \qquad (3.84)
\end{aligned}
$$

and

$$
\begin{aligned}
(A_{11} + B_{11}K_\delta)^T P_\delta(A_{11} + B_{11}K_\delta) + (A_{12} + B_{12}K_\delta)^T P_\delta(A_{12} + B_{12}K_\delta) \\
-C_{11}^T C_{11} - \delta^2 I + \gamma^2 K_\delta^T K_\delta - P_\delta = 0, \qquad (3.85)
\end{aligned}
$$

respectively, where

$$K_\delta = -(\gamma^2 I + B_{12}^T P_\delta B_{12} + B_{11}^T P_\delta B_{11})^{-1}(A_{11}^T P_\delta B_{11} + A_{12}^T P_\delta B_{12})^T.$$

Subtracting (3.85) from (3.84), we obtain

$$
\begin{aligned}
-(P - P_\delta) + (A_{11} + B_{11}K)^T (P - P_\delta)(A_{11} + B_{11}K) + (A_{12} + B_{12}K)^T(P - P_\delta) \\
\cdot(A_{12} + B_{12}K) \\
= -(A_{11} + B_{11}K)^T P_\delta(A_{11} + B_{11}K) - (A_{12} + B_{12}K)^T P_\delta(A_{12} + B_{12}K)
\end{aligned}
$$

$$+(A_{11} + B_{11}K_\delta)^T P_\delta(A_{11} + B_{11}K_\delta) + (A_{12} + B_{12}K_\delta)^T P_\delta(A_{12} + B_{12}K_\delta)$$
$$-\gamma^2(K^T K - K_\delta K_\delta) - \delta^2 I$$
$$= -K^T B_{11}^T P_\delta A_{11} + K_\delta^T B_{11}^T P_\delta A_{11} - A_{11}^T P_\delta B_{11}K + A_{11}^T P_\delta B_{11}K_\delta - K^T B_{12}^T P_\delta A_{12}$$
$$-A_{12}^T P_\delta B_{12}K + A_{12}^T P_\delta B_{12}K_\delta$$
$$-K^T(\gamma^2 I + B_{11}^T P_\delta B_{11} + B_{12}^T P_\delta B_{12})K$$
$$+K_\delta^T B_{12}^T P_\delta A_{12} + K_\delta^T(\gamma^2 I + B_{11}^T P_\delta B_{11} + B_{12}^T P_\delta B_{12})K_\delta - \delta^2 I$$
$$= -K^T(\gamma^2 I + B_{11}^T P_\delta B_{11} + B_{12}^T P_\delta B_{12})K - K^T(B_{11}^T P_\delta A_{11} + B_{12}^T P_\delta A_2)$$
$$-(A_{11}^T P_\delta B_{11} + A_{12}^T P_\delta B_{12})K - K_\delta^T(\gamma^2 I + B_{11}^T P_\delta A_{11} + B_{12}^T P_\delta A_{12}) - \delta^2 I$$
$$= -[K + (\gamma^2 I + B_{11}^T P_\delta B_{11} + B_{12}^T P_\delta B_{12})^{-1}(B_{11}^T P_\delta A_{11} + B_{12}^T P_\delta A_{12})]^T$$
$$\cdot(\gamma^2 I + B_{11}^T P_\delta B_{11} + B_{12}^T P_\delta B_{12})$$
$$\cdot[K + (\gamma^2 I + B_{11}^T P_\delta B_{11} + B_{12}^T P_\delta B_{12})^{-1}(B_{11}^T P_\delta A_{11} + B_{12}^T P_\delta A_{12})] - \delta^2 I.$$

So

$$\begin{aligned} &-(P - P_\delta) + (A_{11} + B_{11}K)^T(P - P_\delta)(A_{11} + B_{11}K) \\ &+(A_{12} + B_{12}K)^T(P - P_\delta)(A_{12} + B_{12}K) \\ &= -(K - K_\delta)^T(\gamma^2 I + B_{11}^T P_\delta B_{11} + B_{12}^T P_\delta B_{12})(K - K_\delta) - \delta^2 I. \end{aligned} \quad (3.86)$$

We assert that $P - P_\delta$ must be strictly positive definite. Otherwise, there exists a real vector $\xi \neq 0$ such that $(P - P_\delta)\xi = 0$ due to $P \geq P_\delta$. Multiplying ξ^T from the left-hand side and ξ from the right-hand side in (3.86) yields

$$\begin{aligned} 0 \leq\ &\xi^T(A_{11} + B_{11}K)^T(P - P_\delta)(A_{11} + B_{11}K)\xi \\ &+\xi^T(A_{12} + B_{12}K)^T(P - P_\delta)(A_{12} + B_{12}K)\xi \\ &= -\xi^T(K - K_\delta)^T(\gamma^2 I + B_{11}^T P_\delta B_{11} + B_{12}^T P_\delta B_{12})(K - K_\delta)\xi - \delta^2 \xi^T \xi \\ &< 0, \end{aligned}$$

which is a contradiction, so $P - P_\delta > 0$. In view of (3.86), $\bar{P} := P - P_\delta > 0$ is the solution to the Lyapunov-type inequality

$$-\bar{P} + (A_{11} + B_{11}K)^T \bar{P}(A_{11} + B_{11}K) + (A_{12} + B_{12}K)^T \bar{P}(A_{12} + B_{12}K) < 0$$

which implies the stability of $(A_{11} + B_{11}K, A_{12} + B_{12}K)$ by Proposition 2.2 [65]. The first part of this lemma is proved.

Conversely, considering (3.78), (3.79) and applying Lemma 3.1, we have

$$J_T(x_0, v) = \sum_{k=0}^{T} \mathcal{E}(\gamma^2 \parallel v_k \parallel^2 - \parallel z_1^k \parallel^2)$$

$$= \sum_{k=0}^{T} \mathcal{E}(\gamma^2 v_k^T v_k - x_k^T C_{11}^T C_{11} x_k)$$

$$= -\mathcal{E}(x_{T+1}^T P x_{T+1}) + x_0^T P x_0 + \sum_{k=0}^{T} \mathcal{E} \begin{bmatrix} x_k \\ v_k \end{bmatrix}^T \mathcal{Q}(P) \begin{bmatrix} x_k \\ v_k \end{bmatrix}$$

$$= x_0^T P x_0 - \mathcal{E}(x_{\mathcal{T}+1}^T P x_{\mathcal{T}+1}) + \sum_{k=0}^{\mathcal{T}} \mathcal{E}[(v_k - K x_k)^T H(P)(v_k - K x_k)],$$

$$(3.87)$$

where

$$H(P) = B_{11}^T P B_{11} + B_{12}^T P B_{12} + \gamma^2 I$$

and

$$\mathcal{Q}(P) = \begin{bmatrix} -P + A_{11}^T P A_{11} + A_{12}^T P A_{12} - C_{11}^T C_{11} & A_{11}^T P B_{11} + A_{12}^T P B_{12} \\ B_{11}^T P A_{11} + B_{12}^T P A_{12} & H(P) \end{bmatrix}.$$

Because P is a feedback stabilizing solution, taking the limit in (3.87) leads to

$$J_\infty(x_0, v) \geq J_\infty(x_0, v = Kx) = x_0^T P x_0.$$

When $x_0 = 0$, $J_\infty(0, v) \geq 0$, which is equivalent to $\| \tilde{\mathcal{L}}_\infty \| \leq \gamma$. To show $\| \tilde{\mathcal{L}}_\infty \| < \gamma$, we define an operator $L_1 : l_w^2(\mathcal{N}, \mathcal{R}^{n_v}) \mapsto l_w^2(\mathcal{N}, \mathcal{R}^{n_v})$, $L_1 v_k = v_k - K x_k$ with its realization

$$\begin{cases} x_{k+1} = A_{11} x_k + B_{11} v_k + (A_{12} x_k + B_{12} v_k) w_k, \\ x_0 = 0, \ k \in \mathcal{N}, \\ v_k - K x_k = v_k + H(P)^{-1}(B_{11}^T P A_{11} + B_{12}^T P A_{12}) x_k. \end{cases}$$

Then L_1^{-1} exists, which is determined by

$$\begin{cases} x_{k+1} = [A_{11} - B_{11} H(P)^{-1}(B_{11}^T P A_{11} + B_{12}^T P A_{12})] x_k \\ \qquad + B_{11}(v_k - K x_k) + [A_{12} - B_{12} H(P)^{-1}(B_{11}^T P A_{11} \\ \qquad + B_{12}^T P A_{12})] x_k w_k + B_{12}(v_k - K x_k) w_k, \\ x_0 = 0, k \in \mathcal{N}, \\ v_k = -H(P)^{-1}(B_{11}^T P A_{11} + B_{12}^T P A_{12}) x_k + (v_k - K x_k). \end{cases}$$

Since $H(P) > 0$, $H(P) \geq \varepsilon I$ for some $\varepsilon > 0$, there exists a constant $c > 0$ such that (with $x_0 = 0$ in mind)

$$J_\infty(0, v) = \sum_{k=0}^{\infty} \mathcal{E}\left[(L_1 v_k)^T H(P)(L_1 v_k)\right]$$

$$\geq \varepsilon \| L_1 v_k \|_{l_w^2(\mathcal{N}, \mathcal{R}^{n_v})}^2 \geq c \| v_k \|_{l_w^2(\mathcal{N}, \mathcal{R}^{n_v})}^2$$

$$> 0,$$

i.e., $\| \tilde{\mathcal{L}}_\infty \| < \gamma$. The proof of this lemma is completed. $\square$

In the next section, we shall apply the SBRL to the stochastic mixed H_2/H_∞ control.

REMARK 3.9   Lemma 3.7 shows that the internal stability of (3.78) together with $\| \tilde{\mathcal{L}}_\infty \| < \gamma$ is equivalent to (3.79) having a stabilizing solution

$P \leq 0$. Another SBRL can also be found in Theorem 2.5 of [65] which says that the internal stability of (3.78) and $\| \tilde{\mathcal{L}}_\infty \| < \gamma$ is equivalent to the solvability of some LMIs. ∎

REMARK 3.10   Lemma 3.7 improves Lemma 4 of [207], where it was asserted that a combination of the internal stability of (3.79) and the existence of a feedback stabilizing solution of (3.79) implies $\| \tilde{\mathcal{L}}_\infty \| < \gamma$. ∎

3.3.4 H_2/H_∞ control with (x, v)-dependent noise

Consider the system (3.52) together with a disturbance attenuation level $\gamma > 0$, and define two associated performance indices

$$J_{1,\infty}(u, v) = \sum_{k=0}^{\infty} \mathcal{E} \left(\gamma^2 \| v_k \|^2 - \| z_k \|^2 \right) \tag{3.88}$$

and

$$J_{2,\infty}(u, v) = \sum_{k=0}^{\infty} \mathcal{E} \| z_k \|^2 . \tag{3.89}$$

The infinite horizon stochastic H_2/H_∞ control of system (3.52) is stated as follows:

DEFINITION 3.11   *Given a scalar $\gamma > 0$, find, if possible, a control law $u_\infty^* = \{u_{\infty,k}^*\}_{k\in\mathcal{N}} \in l_w^2(\mathcal{N}, \mathcal{R}^{n_u})$ such that*

  (i) u_∞^* stabilizes system (3.52) internally, i.e., when $v_k \equiv 0, u_k = u_{\infty,k}^*$, the state trajectory of (3.52) with any initial value x_0 satisfies

$$\lim_{k \to \infty} \mathcal{E} \| x_k \|^2 = 0.$$

(ii)

$$\| \mathcal{L}_\infty \| = \sup_{v \in l_w^2(\mathcal{N}, \mathcal{R}^{n_v}), v \neq 0, x_0 = 0} \frac{\left(\sum_{k=0}^{\infty} \mathcal{E} \left(\| Cx_k \|^2 + \| u_{\infty,k}^* \|^2 \right) \right)^{\frac{1}{2}}}{\left(\sum_{k=0}^{\infty} \mathcal{E} \| v_k \|^2 \right)^{\frac{1}{2}}} < \gamma.$$

(iii) When the worst-case disturbance $v_\infty^* = \{v_{\infty,k}^*\}_{k\in\mathcal{N}} \in l_w^2(\mathcal{N}, \mathcal{R}^{n_v})$, if it exists, is implemented in (3.52), u_∞^* minimizes the output energy

$$J_{2,\infty}(u, v_\infty^*) = \sum_{k=0}^{\infty} \mathcal{E} \| z_k \|^2 .$$

If the above (u_∞^*, v_∞^*) exist, we say that the infinite horizon stochastic H_2/H_∞ control problem is solvable.

In this section, we shall first present a solution to the stochastic H_2/H_∞ control of (3.54) based on four coupled matrix-valued equations.

THEOREM 3.8
For system (3.54), suppose the following four coupled matrix-valued equations have a solution $(P_{1,\infty}, K_1; P_{2,\infty}, K_2)$ with $P_{1,\infty} \leq 0$ and $P_{2,\infty} \geq 0$.

$$\begin{cases} -P_{1,\infty} + (A_1 + B_1 K_2)^T P_{1,\infty}(A_1 + B_1 K_2) + A_2^T P_{1,\infty} A_2 - C^T C \\ \quad -K_2^T K_2 - K_3 H_1(P_{1,\infty})^{-1} K_3^T = 0, \\ H_1(P_{1,\infty}) > 0, \end{cases} \tag{3.90}$$

$$K_1 = -H_1(P_{1,\infty})^{-1} K_3^T, \tag{3.91}$$

$$\begin{cases} -P_{2,\infty} + (A_1 + C_1 K_1)^T P_{2,\infty}(A_1 + C_1 K_1) + (A_2 + C_2 K_1)^T P_{2,\infty}(A_2 + C_2 K_1) \\ \quad +C^T C - K_4 H_2(P_{2,\infty})^{-1} K_4^T = 0, \\ H_2(P_{2,\infty}) > 0, \end{cases}$$
$$\tag{3.92}$$

and

$$K_2 = -H_2(P_{2,\infty})^{-1} K_4^T, \tag{3.93}$$

where

$$K_3 = (A_1 + B_1 K_2)^T P_{1,\infty} C_1 + A_2^T P_{1,\infty} C_2,$$
$$H_1(P_{1,\infty}) = \gamma^2 I + C_2^T P_{1,\infty} C_2 + C_1^T P_{1,\infty} C_1,$$
$$K_4 = (A_1 + C_1 K_1)^T P_{2,\infty} B_1, \quad H_2(P_{2,\infty}) = I + B_1^T P_{2,\infty} B_1.$$

If $(A_1, A_2|C)$ and $(A_1 + C_1 K_1, A_2 + C_2 K_1|C)$ are exactly detectable, then the H_2/H_∞ control problem has a pair of solutions

$$u_{\infty,k}^* = K_2 x_k, \quad v_{\infty,k}^* = K_1 x_k.$$

REMARK 3.11   Because exact observability is stronger than exact detectability, Theorem 3.8 certainly holds under exact observability, which is Theorem 1 of [207]. Of course, if $C^T C > 0$, $(A_1, A_2|C)$ and $(A_1 + C_1 K_1, A_2 + C_2 K_1|C)$ are not only exactly detectable, but also exactly observable.   ∎

We first give the following lemma which will be used in the proof of Theorem 3.8.

LEMMA 3.8
Let K_1, K_2, K_3, $H_1(P_{1,\infty})$ be as defined in Theorem 3.8 and denote

$$\tilde{C}_1 = \begin{bmatrix} C \\ K_2 \end{bmatrix}, \quad \tilde{C}_2 = \begin{bmatrix} C \\ K_2 \\ H_1(P_{1,\infty})^{-1/2} K_3^T \end{bmatrix}, \tag{3.94}$$

then we have the following statements:

(i) If $(A_1, A_2|C)$ is exactly detectable, so is $(A_1 + B_1K_2, A_2|\tilde{C}_2)$.

(ii) If $(A_1 + C_1K_1, A_2 + C_2K_1|C)$ is exactly detectable, so is $(A_1 + C_1K_1 + B_1K_2, A_2 + C_2K_1|\tilde{C}_1)$.

Proof. (i) Suppose $(A_1, A_2|C)$ is exactly detectable but $(A_1 + B_1K_2, A_2|\tilde{C}_2)$ is not. By Theorem 3.7-(ii), there exists a non-zero symmetric matrix $Z \in \mathcal{S}_n$ such that

$$\mathcal{D}_{A_1+B_1K_2,A_2}Z = \lambda Z, \quad \tilde{C}_2 Z = 0, \quad |\lambda| \geq 1 \qquad (3.95)$$

which yields

$$(A_1 + B_1K_2)Z(A_1 + B_1K_2)^T + A_2 Z A_2^T = \lambda Z, \quad |\lambda| \geq 1,$$

$$CZ = 0, \quad K_2 Z = 0, \quad K_3^T Z = 0,$$

or

$$\mathcal{D}_{A_1,A_2}Z = \lambda Z, \quad CZ = 0, \quad |\lambda| \geq 1.$$

The above contradicts the exact detectability of $(A_1, A_2|C)$, (i) is proved.

(ii) Repeating the same procedure as above, the exact detectability of $(A_1 + C_1K_1 + B_1K_2, A_2 + C_2K_1|\tilde{C}_1)$ can be derived. $\square$

REMARK 3.12   Lemma 3.8 still holds under the assumption of exact observability, which can be verified following the same line as in Lemma 3.8.
▌

Proof of Theorem 3.8. We split the proof of Theorem 3.8 into three steps as follows:

Step 1: $(u_\infty^*, v_\infty^*) \in l_w^2(\mathcal{N}, \mathcal{R}^{n_u}) \times l_w^2(\mathcal{N}, \mathcal{R}^{n_v})$; $(A_1 + B_1K_2, A_2)$ is Schur stable.

To this end, we note that (3.90) and (3.92) can be rewritten as

$$-P_{1,\infty} + (A_1 + B_1K_2)^T P_{1,\infty}(A_1 + B_1K_2) + A_2^T P_{1,\infty}A_2 - \tilde{C}_2^T\tilde{C}_2 = 0 \qquad (3.96)$$

and

$$-P_{2,\infty} + (A_1 + C_1K_1 + B_1K_2)^T P_{2,\infty}(A_1 + C_1K_1 + B_1K_2)$$
$$+(A_2 + C_2K_1)^T P_{2,\infty}(A_2 + C_2K_1) + \tilde{C}_1^T\tilde{C}_1 = 0, \qquad (3.97)$$

respectively, where $\tilde{C}_1$ and $\tilde{C}_2$ are defined in (3.94). From Lemma 3.8, $(A_1 + C_1K_1 + B_1K_2, A_2 + C_2K_1|\tilde{C}_1)$ is also exactly detectable for the reason that $(A_1 + C_1K_1, A_2 + C_2K_1|C)$ is exactly detectable, which implies that $(A_1 + C_1K_1 + B_1K_2, A_2 + C_2K_1)$ is Schur stable by Lemma 3.5-(iii) and (3.97). Hence,

$$u_{\infty,k}^* = K_2 x_k \in l_w^2(\mathcal{N}, \mathcal{R}^{n_u}), \quad v_{\infty,k}^* = K_1 x_k \in l_w^2(\mathcal{N}, \mathcal{R}^{n_v}).$$

From Lemmas 3.5–3.8 and (3.96), the Schur stability of $(A_1 + B_1 K_2, A_2)$ follows, i.e., the system (3.54) is internally stabilized by $u^*_{\infty,k} = K_2 x_k$.

Step 2: $\| \mathcal{L}_\infty \| < \gamma$.

Substituting $u_k = u^*_{\infty,k} = K_2 x_k$ into (3.54) yields

$$\begin{cases} x_{k+1} = (A_1 + B_1 K_2) x_k + C_1 v_k + (A_2 x_k + C_2 v_k) w_k, \\ x_0 \in \mathcal{R}^n, \\ z_k = \begin{bmatrix} C x_k \\ D K_2 x_k \end{bmatrix}, \quad D^T D = I. \end{cases} \tag{3.98}$$

As shown in Step 1, $P_{1,\infty} \le 0$ is a feedback stabilizing solution of (3.90) and a direct application of Lemma 3.7 yields $\| \mathcal{L}_\infty \| < \gamma$.

Step 3: u^*_∞ also minimizes the output energy when the worst-case disturbance v^*_∞ is applied in the system (3.54).

We first show that $v^*_\infty = \{ v^*_{\infty,k} = K_1 x_k \}_{k \in \mathcal{N}}$ is the worst-case disturbance. Similar to the proof of Lemma 3.7, in view of (3.90), (3.98) and Lemma 3.1, we have

$$\begin{aligned} J_{1,\mathcal{T}}(u^*_\infty, v) &= \sum_{k=0}^{\mathcal{T}} \mathcal{E}(\gamma^2 \| v_k \|^2 - \| z_k \|^2) \\ &= \sum_{k=0}^{\mathcal{T}} \mathcal{E}[\gamma^2 v_k^T v_k - x_k^T (C^T C + K_2^T K_2) x_k] \\ &= x_0^T P_{1,\infty} x_0 - \mathcal{E}(x_{\mathcal{T}+1}^T P_{1,\infty} x_{\mathcal{T}+1}) \\ &\quad + \sum_{k=0}^{\mathcal{T}} \mathcal{E} \begin{bmatrix} x_k \\ v_k \end{bmatrix}^T Q_1(P_{1,\infty}) \begin{bmatrix} x_k \\ v_k \end{bmatrix} \\ &= x_0^T P_{1,\infty} x_0 - \mathcal{E}(x_{\mathcal{T}+1}^T P_{1,\infty} x_{\mathcal{T}+1}) \\ &\quad + \sum_{k=0}^{\mathcal{T}} \mathcal{E}(v_k - K_1 x_k)^T H_1(P_{1,\infty})(v_k - K_1 x_k), \end{aligned}$$

where

$$Q_1(P_{1,\infty}) =$$
$$\begin{bmatrix} \{(A_1 + B_1 K_2)^T P_{1,\infty}(A_1 + B_1 K_2) & (A_1 + B_1 K_2)^T P_{1,\infty} C_1 + A_2^T P_{1,\infty} C_2 \\ -P_{1,\infty} + A_2^T P_{1,\infty} A_2 - C^T C - K_2^T K_2 \} & \\ C_1^T P_{1,\infty}(A_1 + B_1 K_2) + C_2^T P_{1,\infty} A_2 & C_1^T P_{1,\infty} C_1 + C_2^T P_{1,\infty} C_2 + \gamma^2 I \end{bmatrix}.$$

Since $\lim_{\mathcal{T} \to \infty} \mathcal{E} \| x_{\mathcal{T}} \|^2 = 0$, we have

$$J_{1,\infty}(u^*_\infty, v) = x_0^T P_{1,\infty} x_0 + \sum_{k=0}^{\infty} \mathcal{E}(v_k - K_1 x_k)^T H_1(P_{1,\infty})(v_k - K_1 x_k)$$

$$\ge J_{1,\infty}(u^*_\infty, K_1 x) = x_0^T P_{1,\infty} x_0.$$

So $v^*_\infty \in l^2_w(\mathcal{N}, \mathcal{R}^{n_v})$ is the worst-case disturbance corresponding to u^*_∞.

When $v_k = v^*_{\infty,k} = K_1 x_k$ is implemented in the system (3.54), we obtain

$$\begin{cases} x_{k+1} = (A_1 + C_1 K_1) x_k + B_1 u_k + (A_2 + C_2 K_1) x_k w_k, \\ x_0 \in \mathcal{R}^n, \\ z_k = \begin{bmatrix} C x_k \\ D u_k \end{bmatrix}, \ D^T D = I, \ k \in \mathcal{N}. \end{cases} \tag{3.99}$$

The optimization problem

$$\begin{cases} \min_{u \in \mathcal{U}_{ad}} J_{2,\infty}(u, v^*_\infty) \\ \text{subject to} \quad (3.99) \end{cases} \tag{3.100}$$

is a standard LQ optimal control problem. Note that $(A_1 + C_1 K_1, B_1; A_2 + C_2 K_1, 0)$ is stabilizable by virtue of the stability of $(A_1 + C_1 K_1 + B_1 K_2, A_2 + C_2 K_1)$, while $(A_1 + C_1 K_1, A_2 + C_2 K_1 | C)$ is exactly detectable by assumptions. Applying Lemma 3.6, we immediately have

$$\min_{u \in \mathcal{U}_{ad}} J_{2,\infty}(u, v^*_\infty) = J_{2,\infty}(u^*_\infty, v^*_\infty) = x_0^T P_{2,\infty} x_0.$$

This theorem is proved. $\square$

It is very interesting to note that in [138] the linear discrete-time stochastic H_2/H_∞ control for systems with additive noise, unlike systems with multiplicative noise considered in this chapter, is associated with three instead of four coupled matrix-valued equations in Theorem 3.8. In addition, compared with Theorem 2.19, the discrete-time stochastic H_2/H_∞ control with (x, v)-dependent noise is associated with four coupled equations instead of two.

COROLLARY 3.2
Suppose $P_{1,\infty}$ and $P_{2,\infty}$ are respectively the solutions of (3.90) and (3.92) in Theorem 3.8. Then, $P_{1,\infty} + P_{2,\infty} \geq 0$.

Proof. We note that (3.90) and (3.92) can be written as

$$\begin{cases} -P_{1,\infty} + (A_1 + B_1 K_2 + C_1 K_1)^T P_{1,\infty} (A_1 + B_1 K_2 + C_1 K_1) + (A_2 + C_2 K_1)^T P_{1,\infty} \\ \qquad \cdot (A_2 + C_2 K_1) = C^T C + K_2^T K_2 - \gamma^2 K_1^T K_1, \\ H_1(P_{1,\infty}) > 0 \end{cases} \tag{3.101}$$

and

$$\begin{cases} -P_{2,\infty} + (A_1 + C_1 K_1 + B_1 K_2)^T P_{2,\infty} (A_1 + C_1 K_1 + B_1 K_2) + (A_2 + C_2 K_1)^T P_{2,\infty} \\ \qquad \cdot (A_2 + C_2 K_1) = -C^T C - K_2^T K_2, \end{cases} \tag{3.102}$$

respectively. Summing (3.101) with (3.102) yields

$$\begin{aligned} -(P_{1,\infty} + P_{2,\infty}) &+ (A_1 + C_1 K_1 + B_1 K_2)^T (P_{1,\infty} + P_{2,\infty})(A_1 + C_1 K_1 + B_1 K_2) \\ &+ (A_2 + C_2 K_1)^T (P_{1,\infty} + P_{2,\infty})(A_2 + C_2 K_1) \\ &= -\gamma^2 K_1^T K_1. \end{aligned} \tag{3.103}$$

Applying Lemma 3.5-(i) to (3.103), $P_{1,\infty} + P_{2,\infty} \geq 0$ is derived.

The following corollary shows that when $C_1 = 0$ in the system (3.54), the four coupled equations (3.90)–(3.93) reduce to two.

COROLLARY 3.3

Consider the case of $C_1 = 0$ in (3.54). Suppose the following two coupled matrix-valued equations

$$\begin{cases} -P_{1,\infty} + (A_1 + B_1 K_2)^T P_{1,\infty}(A_1 + B_1 K_2) + A_2^T P_{1,\infty} A_2 - C^T C - K_2^T K_2 \\ \qquad -K_3 H_1(P_{1,\infty})^{-1} K_3^T = 0, \\ \gamma^2 I + C_2^T P_{1,\infty} C_2 > 0 \end{cases}$$

(3.104)

and

$$-P_{2,\infty} + A_1^T P_{2,\infty} A_1 + (A_2 + C_2 K_1)^T P_{2,\infty}(A_2 + C_2 K_1)$$
$$+ C^T C - K_4 H_2(P_{2,\infty})^{-1} K_4^T = 0 \qquad (3.105)$$

admit a pair of solutions $(P_{1,\infty} \leq 0, P_{2,\infty} \geq 0)$, where

$$K_1 = -(\gamma^2 I + C_2^T P_{1,\infty} C_2)^{-1} C_2^T P_{1,\infty} A_2,$$
$$K_2 = -(I + B_1^T P_{2,\infty} B_1)^{-1} B_1^T P_{2,\infty} A_1,$$
$$K_3 = A_2^T P_{1,\infty} C_2, \quad K_4 = A_1^T P_{2,\infty} B_1,$$
$$H_1(P_{1,\infty}) = \gamma^2 I + C_2^T P_{1,\infty} C_2, \quad H_2(P_{2,\infty}) = I + B_1^T P_{2,\infty} B_1.$$

If $(A_1, A_2 | C)$ and $(A_1, A_2 + C_2 K_1 | C)$ are exactly detectable, then the H_2/H_∞ control problem has a solution (u_∞^*, v_∞^*) with

$$u_{\infty,k}^* = K_2 x_k, \quad v_{\infty,k}^* = K_1 x_k.$$

REMARK 3.13   From Corollary 3.3, we see that when $A_1 = 0$, $C_1 = 0$, (3.104) and (3.105) will be decoupled, then $u_\infty^* = 0$ and v_∞^* is only determined by (3.104). ∎

For convenience, we assume $C^T C > 0$ in the following corollaries, in this case, both $(A_1, A_2 | C)$ and $(A_1 + C_1 K_1, A_2 + C_2 K_1 | C)$ are not only exactly detectable, but also exactly observable. Similar to the finite horizon case, we can derive the following result for the stochastic H_∞ control by setting $P_{2,\infty} = -P_{1,\infty}$ in Theorem 3.8.

COROLLARY 3.4 (Stochastic H_∞ control)

Consider system (3.54). If the following three coupled matrix-valued equations

$$\begin{cases} -P_{1,\infty} + (A_1 + B_1 K_2)^T P_{1,\infty}(A_1 + B_1 K_2) + A_2^T P_{1,\infty} A_2 - C^T C - K_2^T K_2 \\ \qquad -K_3 H_1(P_{1,\infty})^{-1} K_3^T = 0, \\ H_1(P_{1,\infty}) > 0, \end{cases}$$

(3.106)

$$K_1 = -H_1(P_{1,\infty})^{-1}K_3^T$$
$$= -H_1(P_{1,\infty})^{-1}[C_2^T P_{1,\infty} A_2 + C_1^T P_{1,\infty}(A_1 + B_1 K_2)] \quad (3.107)$$

and

$$K_2 = -H_2(-P_{1,\infty})^{-1}K_4^T = H_2(-P_{1,\infty})^{-1}B_1^T P_{1,\infty}(A_1 + C_1 K_1) \quad (3.108)$$

have a solution $(P_{1,\infty} \le 0, K_1, K_2)$, then $\tilde{u}_{\infty,k}^* = K_2 x_k$ is the desired H_∞ control, while $\tilde{v}_{\infty,k}^* = K_1 x_k$ is the corresponding worst-case disturbance.

In Theorem 3.8, if we take $A_2 = 0, C_2 = 0$, then a result about deterministic discrete-time H_2/H_∞ control is derived.

COROLLARY 3.5 (Deterministic H_2/H_∞ control)
Consider the following deterministic discrete-time system

$$\begin{cases} x_{k+1} = A_1 x_k + B_1 x_k + C_1 v_k, \\ x_0 \in \mathcal{R}^n, \\ z_k = \begin{bmatrix} C x_k \\ D x_k \end{bmatrix}, \quad D^T D = I, \ k \in \mathcal{N}. \end{cases} \quad (3.109)$$

If the following four coupled matrix-valued equations

$$\begin{cases} -P_{1,\infty} + (A_1 + B_1 K_2)^T P_{1,\infty}(A_1 + B_1 K_2) - C^T C - K_2^T K_2 \\ \quad -K_3 H_1(P_{1,\infty})^{-1}K_3^T = 0, \\ H_1(P_{1,\infty}) > 0, \end{cases} \quad (3.110)$$

$$K_1 = -H_1(P_{1,\infty})^{-1}K_3^T, \quad (3.111)$$

$$-P_{2,\infty} + (A_1 + C_1 K_1)^T P_{2,\infty}(A_1 + C_1 K_1) + C^T C - K_4 H_2(P_{2,\infty})^{-1}K_4^T = 0, \quad (3.112)$$

and

$$K_2 = -H_2(P_{2,\infty})^{-1}K_4^T \quad (3.113)$$

admit a solution $(P_{1,\infty} \le 0, K_1; P_{2,\infty} \ge 0, K_2)$ with

$$K_3 = (A_1 + B_1 K_2)^T P_{1,\infty} C_1, \quad H_1(P_{1,\infty}) = \gamma^2 I + C_1^T P_{1,\infty} C_1,$$

$$K_4 = (A_1 + C_1 K_1)^T P_{2,\infty} B_1, \quad H_2(P_{2,\infty}) = I + B_1^T P_{2,\infty} B_1,$$

then the discrete-time mixed H_2/H_∞ control has a solution (u_∞^*, v_∞^*) with

$$u_{\infty,k}^* = K_2 x_k, \quad v_{\infty,k}^* = K_1 x_k.$$

REMARK 3.14 Corollary 3.5 appears to be new, which can be viewed as a discrete-time version of [113]. It is, however, worth noting that, different from the continuous-time case [113] where the solution involves two coupled matrix differential equations, the deterministic discrete-time H_2/H_∞ control in Corollary 3.5 is still associated with four coupled matrix-valued equations.

A converse result of Theorem 3.8 is given in the following.

THEOREM 3.9
Assume that the stochastic H_2/H_∞ control problem admits a solution (u_∞^*, v_∞^*) with

$$u_{\infty,k}^* = K_2 x_k, \quad v_{\infty,k}^* = K_1 x_k,$$

where K_2 and K_1 are constant matrices, and that $(A_1 + C_1 K_1, A_2 + C_2 K_1 | C)$ is exactly detectable. Then the four coupled matrix-valued equations (3.90)–(3.93) have a unique quaternion solution $(P_{1,\infty} \leq 0, K_1; P_{2,\infty} \geq 0, K_2)$.

Proof. Implementing $u_k = u_{\infty,k}^* = K_2 x_k$ in (3.54) yields the closed-loop system (3.98). Hence, $(A_1 + B_1 K_2, A_2)$ is stable and $\| \mathcal{L}_\infty \| < \gamma$. From Lemma 3.7, GARE (3.90) has a unique solution $P_{1,\infty} \leq 0$. Moreover, $(A_1 + B_1 K_2 + C_1 K_1, A_2 + C_2 K_1)$ is Schur stable.

In system (3.98), for any $v \in l_w^2(\mathcal{N}, \mathcal{R}^{n_v})$, we have $x \in l_w^2(\mathcal{N}, \mathcal{R}^n)$ as a result of the stability of $(A_1 + B_1 K_2, A_2)$ from Remark 2.9 [65]. Considering (3.90) and (3.98) together with Lemma 3.1, we have

$$J_{1,\infty}(u_\infty^*, v) = \sum_{k=0}^{\infty} \mathcal{E}(\gamma^2 \| v_k \|^2 - \| z_k \|^2)$$

$$= \sum_{k=0}^{\infty} \mathcal{E}[v_k + H_1(P_{1,\infty})^{-1} K_3^T x_k]^T H_1(P_{1,\infty})[v_k + H_1(P_{1,\infty})^{-1} K_3^T x_k]$$

$$+ x_0^T P_{1,\infty} x_0$$

$$\geq J_{1,\infty}(u_\infty^*, v_\infty^*) = x_0^T P_{1,\infty} x_0. \tag{3.114}$$

Hence, $K_1 = -H_1(P_{1,\infty})^{-1} K_3^T$. Substituting $v_{\infty,k}^*$ into (3.54) gives (3.99), subject to which, the optimization problem (3.100) is a standard LQ optimal control problem. The stability of $(A_1 + C_1 K_1 + B_1 K_2, A_2 + C_2 K_1)$ implies the stabilizability of $(A_1 + C_1 K_1, B_1; A_2 + C_2 K_1)$. In addition, by assumption, $(A_1 + C_1 K_1, A_2 + C_2 K_1 | C)$ is exactly detectable. So by Lemma 3.6, (3.92) has a unique solution $P_{2,\infty} \geq 0$. Furthermore,

$$\min_{u \in \mathcal{U}_{ad}} J_{2,\infty}(u, v_\infty^*) = J_{2,\infty}(u_\infty^*, v_\infty^*) = x_0^T P_{2,\infty} x_0,$$

where $u_{\infty,k}^* = -H_2(P_{2,\infty})^{-1} K_4^T x_k$, i.e., $K_2 = -H_2(P_{2,\infty})^{-1} K_4^T$. The proof is completed. $\square$

Theorems 3.8–3.9 tell us that, to some extent, e.g., $C^T C > 0$, the existence of a stochastic H_2/H_∞ static state feedback controller is equivalent to the solvability of coupled equations (3.90)–(3.93). Under a stronger condition of exact observability, Theorems 3.8–3.9 hold with $P_{1,\infty} < 0$ and $P_{2,\infty} > 0$, which was discussed in [207].

In the forthcoming section, we will present a convex optimization algorithm for the design of a suboptimal H_2/H_∞ controller in the case of $C_1 = 0$ and an iterative algorithm for solving the four coupled matrix-valued equations. At the end of this section, we give the following comments to strengthen some important facts.

REMARK 3.15   Reference [78] also studied a class of stochastic H_2/H_∞ control, where, different from our definition of the H_2/H_∞ control, the H_2 performance $J_{2,\infty}(u,v)$ is minimized under the assumption that v is a white noise. Following the line of development in [78], it is easy to present a suboptimal H_2/H_∞ controller design by solving some convex optimization problem if we consider the H_2/H_∞ control problem as in [78].   ∎

REMARK 3.16   As discussed in the finite horizon case, the pure H_2 control [97] or H_∞ control [65, 78] may be obtained as special cases of the mixed H_2/H_∞ control.   ∎

REMARK 3.17   Compared with the finite horizon stochastic H_2/H_∞ control, the infinite horizon stochastic H_2/H_∞ control is much more challenging due to the requirement of internal stability which is not easy to establish. In particular, in order to obtain a stabilizing stochastic H_2/H_∞ state feedback controller, some new concepts such as exact observability, exact detectability and mean square stability have been introduced so that the stochastic LQ optimal control and SBRL can be studied.   ∎

3.3.5   Numerical algorithms

As noted, it is generally difficult to solve the four coupled matrix-valued equations (3.90)–(3.93) analytically, so in this section, we seek some numerical algorithms to solve the two coupled matrix-valued equations (3.104)–(3.105) and four coupled equations (3.90)–(3.93).

3.3.5.1   Iterative algorithms

We consider the case of $C_1 = 0$. From the proof of Theorem 3.8, it is easy to see that if we replace (3.104) and (3.105) respectively by the following inequalities

$$\begin{cases} -P_{1,\infty} + (A_1 + B_1 K_2)^T P_{1,\infty}(A_1 + B_1 K_2) + A_2^T P_{1,\infty} A_2 - C^T C - K_2^T K_2 \\ \quad - K_3 H_1 (P_{1,\infty})^{-1} K_3^T > 0, \\ \gamma^2 I + C_2^T P_{1,\infty} C_2 > 0 \end{cases}$$

$$(3.115)$$

and

$$-P_{2,\infty} + A_1^T P_{2,\infty} A_1 + (A_2 + C_2 K_1)^T P_{2,\infty}(A_2 + C_2 K_1) + C^T C$$
$$-K_4 H_2(P_{2,\infty})^{-1} K_4^T < 0, \qquad (3.116)$$

then a suboptimal stochastic H_2/H_∞ control is obtained, which satisfies the requirements (i) and (ii) of Definition 3.11 but (iii) with a suboptimal H_2 performance, namely, $\min_{u \in \mathcal{U}_{ad}} J_{2,\infty}(u, v_\infty^*) \le x_0^T P_{2,\infty} x_0$. In order to present a practical algorithm to design a suboptimal H_2/H_∞ controller, we further take $P_{1,\infty} = -P_{2,\infty}$ in (3.115) and (3.116). By some manipulations, (3.115) and (3.116) are respectively simplified as

$$\begin{cases} P_{2,\infty} - A_1^T P_{2,\infty} A_1 - A_2^T P_{2,\infty} A_2 + A_1^T P_{2,\infty} B_1 (I + B_1^T P_{2,\infty} B_1)^{-1} B_1^T P_{2,\infty} A_1 \\ \quad -A_2^T P_{2,\infty} C_2 (\gamma^2 I - C_2^T P_{2,\infty} C_2)^{-1} C_2^T P_{2,\infty} A_2 - C^T C > 0, \\ \gamma^2 I - C_2^T P_{2,\infty} C_2 > 0 \end{cases}$$

$$(3.117)$$

and

$$A_1^T P_{2,\infty} A_1 + A_2^T P_{2,\infty} A_2 - A_1^T P_{2,\infty} B_1 (I + B_1^T P_{2,\infty} B_1)^{-1} B_1^T P_{2,\infty} A_1$$
$$-P_{2,\infty} + C^T C + A_2^T P_{2,\infty} C_2 (\gamma^2 I - C_2^T P_{2,\infty} C_2)^{-1}(2\gamma^2 I - C_2^T P_{2,\infty} C_2)$$
$$\cdot(\gamma^2 I - C_2^T P_{2,\infty} C_2)^{-1} C_2^T P_{2,\infty} A_2 < 0. \qquad (3.118)$$

Using the matrix inequality

$$L^T \begin{bmatrix} S^{-1} & 0 \\ 0 & 0 \end{bmatrix} L \ge \begin{bmatrix} I & 0 \\ 0 & 0 \end{bmatrix} L + L^T \begin{bmatrix} I & 0 \\ 0 & 0 \end{bmatrix} - \begin{bmatrix} S & 0 \\ 0 & 0 \end{bmatrix},$$

(3.117) holds if

$$\begin{cases} P_{2,\infty} - A_1^T P_{2,\infty} A_1 - A_2^T P_{2,\infty} A_2 + A_1^T P_{2,\infty} \bar{B}_2 + \bar{B}_2^T P_{2,\infty} A_1 - \Pi(P_{2,\infty}) \\ \quad -A_2^T P_{2,\infty} C_2 (\gamma^2 I - C_2^T P_{2,\infty} C_2)^{-1} C_2^T P_{2,\infty} A_2 - C^T C > 0, \\ \gamma^2 I - C_2^T P_{2,\infty} C_2 > 0, \end{cases}$$

$$(3.119)$$

where $\bar{B}_2 = [B_1 \ 0]$ and $\Pi(P_{2,\infty}) = \begin{bmatrix} I + B_1^T P_{2,\infty} B_1 & 0 \\ 0 & 0 \end{bmatrix}$. By Schur's complement, (3.119) can be transformed into the following **LMI**:

$$\begin{bmatrix} \{P_{2,\infty} - A_1^T P_{2,\infty} A_1 - A_2^T P_{2,\infty} A_2 + A_1^T P_{2,\infty} \bar{B}_2 & A_2^T P_{2,\infty} C_2 \\ \quad +\bar{B}_2^T P_{2,\infty} A_1 - \Pi(P_{2,\infty}) - C^T C\} & \\ \quad C_2^T P_{2,\infty} A_2 & \gamma^2 I - C_2^T P_{2,\infty} C_2 \end{bmatrix} > 0. \quad (3.120)$$

On the other hand, (3.118) holds if

$$-P_{2,\infty} + A_2^T P_{2,\infty} A_2 - A_1^T P_{2,\infty} B_1 (I + B_1^T P_{2,\infty} B_1)^{-1} B_1^T P_{2,\infty} A_1$$
$$+A_1^T P_{2,\infty} A_1 + +C^T C + A_2^T P_{2,\infty} C_2 M^{-1} C_2^T P_{2,\infty} A_2 < 0, \qquad (3.121)$$

where

$$M = (\gamma^2 I - C_2^T P_{2,\infty} C_2)(2\gamma^2 I - C_2^T P_{2,\infty} C_2)^{-1}(\gamma^2 I - C_2^T P_{2,\infty} C_2).$$

(3.121) is further implied by

$$-P_{2,\infty} + A_1^T P_{2,\infty} A_1 + A_2^T P_{2,\infty} A_2 - A_1^T P_{2,\infty} \bar{B}_2 - \bar{B}_2^T P_{2,\infty} A_1 + \Pi(P_{2,\infty})$$
$$+ C^T C + A_2^T P_{2,\infty} C_2 M^{-1} C_2^T P_{2,\infty} A_2 < 0. \quad (3.122)$$

We note that

$$A_2^T P_{2,\infty} C_2 M^{-1} C_2^T P_{2,\infty} A_2 = A_2^T P_{2,\infty} C_2 (\gamma^2 I - C_2^T P_{2,\infty} C_2)^{-1} C_2^T P_{2,\infty} A_2$$
$$+ \gamma^2 A_2^T P_{2,\infty} C_2 (\gamma^2 I - C_2^T P_{2,\infty} C_2)^{-2} C_2^T P_{2,\infty} A_2$$
$$\leq A_2^T P_{2,\infty} C_2 (\gamma^2 I - C_2^T P_{2,\infty} C_2)^{-1} C_2^T P_{2,\infty} A_2$$
$$+ A_2^T P_{2,\infty} C_2 (\gamma^2 I - 2 C_2^T P_{2,\infty} C_2)^{-1} C_2^T P_{2,\infty} A_2,$$

provided that $(\gamma^2 I - 2 C_2^T P_{2,\infty} C_2)$ is nonsingular. By Schur's complement, (3.122) is valid if the following LMI

$$\begin{bmatrix} M_{11} & A_2^T P_{2,\infty} C_2 & A_2^T P_{2,\infty} C_2 \\ C_2^T P_{2,\infty} A_2 & C_2^T P_{2,\infty} C_2 - \gamma^2 I & 0 \\ C_2^T P_{2,\infty} A_2 & 0 & 2 C_2^T P_{2,\infty} C_2 - \gamma^2 I \end{bmatrix} < 0 \quad (3.123)$$

has a solution $P_{2,\infty} > 0$, where

$$M_{11} = -P_{2,\infty} + A_1^T P_{2,\infty} A_1$$
$$+ A_2^T P_{2,\infty} A_2 - A_1^T P_{2,\infty} \bar{B}_2 - \bar{B}_2^T P_{2,\infty} A_1 + \Pi(P_{2,\infty}) + C^T C.$$

Summarizing the above discussion, we obtain the following theorem.

THEOREM 3.10
For the case when $C_1 = 0$ in (3.54), a suboptimal stochastic H_2/H_∞ controller can be obtained by solving the following convex optimization problem:

$$\min \quad \mathrm{Trace}(P_{2,\infty})$$
$$subject \ to \ (3.120), \ (3.123) \ and \ P_{2,\infty} > 0$$

with

$$u_{\infty,k}^* = -(I + B_1^T P_{2,\infty} B_1)^{-1} B_1^T P_{2,\infty} A_1 x_k,$$
$$v_{\infty,k}^* = (\gamma^2 I - C_2^T P_{2,\infty} C_2)^{-1} C_2^T P_{2,\infty} A_2 x_k.$$

3.3.5.2 Iterative algorithms

We first present an iterative algorithm to compute the feedback stabilizing solution of the GARE

$$\begin{cases} A_1^T P A_1 + A_2^T P A_2 + C^T C - A_1^T P B_1 (I + B_1^T P B_1)^{-1} B_1^T P A_1 = P \\ I + B_1^T P B_1 > 0 \end{cases}$$

$$(3.124)$$

arising from the stochastic LQ optimal control for systems with state-dependent noise. We make the following assumptions:

(i) $(A_1, B_1; A_2)$ is stabilizable.

(ii) $(A_1, A_2|C)$ is exactly detectable.

Under assumptions (i)–(ii), GARE (3.124) admits a unique feedback stabilizing solution $\bar{P} \geq 0$ by Lemma 3.6. We construct a numerical algorithm to obtain $\bar{P}$.

Rewrite (3.124) as

$$-P + (A_1 + B_1 K)^T P(A_1 + B_1 K) + A_2^T P A_2 = -C^T C - K^T K \quad (3.125)$$

where

$$K = -(I + B_1^T P B_1)^{-1} B_1^T P A_1.$$

We construct the following iteration formula

$$-P_{i+1} + (A_1 + B_1 K_i)^T P_{i+1}(A_1 + B_1 K_i) + A_2^T P_{i+1} A_2 = -C^T C - K_i^T K_i, \quad (3.126)$$

$$K_i = -(I + B_1^T P_i B_1)^{-1} B_1^T P_i A_1, \ i \in \mathcal{N}. \quad (3.127)$$

It is easy to verify that P_i and K_i satisfy

$$-P_i + (A_1 + B_1 K_i)^T P_i(A_1 + B_1 K_i) + A_2^T P_i A_2$$
$$= -C^T C - K_i^T K_i - (K_i - K_{i-1})^T (I + B_1^T P_i B_1)(K_i - K_{i-1}). \quad (3.128)$$

Subtracting (3.128) from (3.126) yields

$$-(P_{i+1} - P_i) + (A_1 + B_1 K_i)^T (P_{i+1} - P_i)(A_1 + B_1 K_i) + A_2^T (P_{i+1} - P_i) A_2$$
$$= (K_i - K_{i-1})^T (I + B_1^T P_i B_1)(K_i - K_{i-1}). \quad (3.129)$$

Select a K_0 such that $(A_1 + B_1 K_0, A_2)$ is Schur stable. K_0 must exist due to assumption (i). (3.126) has a solution $P_1 \geq 0$ by Lemma 3.5, and accordingly, K_1 is obtained from (3.127). Equation (3.128) guarantees that $(A_1 + B_1 K_1, A_2)$ is Schur stable by our Assumption (ii) and Lemma 3.5. Repeating the above procedure, we can obtain $P_1, P_2, \cdots$. Applying (3.129) and Lemma 3.5, we know that $\{P_i\}_{i \geq 1}$ is monotonically decreasing with $P_i \geq 0$, i.e., $P_1 \geq P_2 \geq \cdots \geq P_n > \cdots \geq 0$, so $\lim_{i \to \infty} P_i = \bar{P} \geq 0$, which is a solution of (3.70).

As shown in Section 3.1, the existence of a finite horizon H_2/H_∞ static feedback controller on $[0, \mathcal{T}]$ is equivalent to the solvability of the following four finite horizon coupled difference equations

$$\begin{cases} P_{1,\mathcal{T}}^k = (A_1 + B_1 K_{2,\mathcal{T}}^k)^T P_{1,\mathcal{T}}^{k+1}(A_1 + B_1 K_{2,\mathcal{T}}^k) + A_2^T P_{1,\mathcal{T}}^{k+1} A_2 \\ \qquad\quad -C^T C - (K_{2,\mathcal{T}}^k)^T K_{2,\mathcal{T}}^k - K_{3,\mathcal{T}}^k H_1(P_{1,\mathcal{T}}^{k+1})^{-1}(K_{3,\mathcal{T}}^k)^T, \\ P_{1,\mathcal{T}}^{\mathcal{T}+1} = 0, \\ H_1(P_{1,\mathcal{T}}^{k+1}) > 0, \end{cases} \quad (3.130)$$

$$K_{1,\mathcal{T}}^k = -H_1(P_{1,\mathcal{T}}^{k+1})^{-1}(K_{3,\mathcal{T}}^k)^T, \quad (3.131)$$

$$\begin{cases} P_{2,\mathcal{T}}^k = (A_1 + C_1 K_{1,\mathcal{T}}^k)^T P_{2,\mathcal{T}}^{k+1}(A_1 + C_1 K_{1,\mathcal{T}}^k) \\ \qquad + (A_2 + C_2 K_{1,\mathcal{T}}^k)^T P_{2,\mathcal{T}}^{k+1}(A_2 + C_2 K_{1,\mathcal{T}}^k) \\ \qquad + C^T C - K_{4,\mathcal{T}}^k H_2(P_{2,\mathcal{T}}^{k+1})^{-1}(K_{4,\mathcal{T}}^k)^T, \\ P_{2,\mathcal{T}}^{\mathcal{T}+1} = 0, \end{cases} \quad (3.132)$$

and

$$K_{2,\mathcal{T}}^k = -H_2(P_{2,\mathcal{T}}^{k+1})^{-1}(K_{4,\mathcal{T}}^k)^T, \quad (3.133)$$

where

$$K_{3,\mathcal{T}}^k = (A_1 + B_1 K_{2,\mathcal{T}}^{k+1})^T P_{1,\mathcal{T}}^{k+1} C_1 + A_2^T P_{1,\mathcal{T}}^{k+1} C_2, \quad (3.134)$$

$$K_{4,\mathcal{T}}^k = (A_1 + C_1 K_{1,\mathcal{T}}^{k+1})^T P_{2,\mathcal{T}}^{k+1} B_1, \quad (3.135)$$

and

$$H_2(P_{2,\mathcal{T}}^{k+1}) = I + B_1^T P_{2,\mathcal{T}}^{k+1} B_1, \quad (3.136)$$

$$H_1(P_{1,\mathcal{T}}^{k+1}) = \gamma^2 I + C_1^T P_{1,\mathcal{T}}^{k+1} C_1 + C_2^T P_{2,\mathcal{T}}^{k+1} C_2. \quad (3.137)$$

An iterative procedure for solving (3.90)–(3.93) based on the above recursions is proposed as follows:

1. Given T and the initial conditions $P_{1,\mathcal{T}}^{\mathcal{T}+1} = 0$, $P_{2,\mathcal{T}}^{\mathcal{T}+1} = 0$, we have

$$H_1(P_{1,\mathcal{T}}^{\mathcal{T}+1}) = \gamma^2 I, \ H_2(P_{2,\mathcal{T}}^{\mathcal{T}+1}) = I$$

   by (3.136)–(3.137), and

$$K_{3,\mathcal{T}}^{\mathcal{T}} = 0, \ K_{4,\mathcal{T}}^{\mathcal{T}} = 0$$

   by (3.134)–(3.135).

2. Given $K_{3,\mathcal{T}}^k$, $K_{4,\mathcal{T}}^k$, $H_1(P_{1,\mathcal{T}}^{k+1})$ and $H_2(P_{2,\mathcal{T}}^{k+1})$, then $K_{1,\mathcal{T}}^k$ and $K_{2,\mathcal{T}}^k$ can be computed from (3.131) and (3.133), and hence $P_{1,\mathcal{T}}^k$ and $P_{2,\mathcal{T}}^k$ from (3.130) and (3.132), respectively.

3. With the given $P_{1,\mathcal{T}}^k$ and $P_{2,\mathcal{T}}^k$, repeat steps 1–2 for $k := k - 1$ until convergence is found.

From the proof of Theorem 3.1, if (3.130)–(3.133) have a solution $(P_{1,\mathcal{T}}^k \leq 0, K_{1,\mathcal{T}}^k;$ $P_{2,\mathcal{T}}^k \geq 0, K_{2,\mathcal{T}}^k)$ for any $\mathcal{T} \in \mathcal{N}$, then

$$J_{1,\mathcal{T}}(u_{\mathcal{T}}^*, v) = \sum_{k=0}^{\mathcal{T}} \mathcal{E}\left(\gamma^2 \|v_k\|^2 - \|z_k\|^2\right) = x_0^T P_{1,\mathcal{T}}^0 x_0,$$

$$J_{2,T}(u, v_T^*) = \sum_{k=0}^{T} \mathcal{E}\|z_k\|^2 = x_0^T P_{2,T}^0 x_0,$$

$$u_{T,k}^* = K_{2,T}^0 x_k, \quad v_{T,k}^* = K_{1,T}^0 x_k.$$

As in [5] where the asymptotic analysis of the GARE arising from the continuous-time stochastic LQ optimal control is discussed, under the assumptions of stabilizability and exact detectability, we have for any $x_0 \in \mathcal{R}^n$,

$$\lim_{T\to\infty} \min_{v\in l_w^2(N_T, \mathcal{R}^{n_v})} J_{1,T}(u_T^*, v) = \lim_{T\to\infty} x_0^T P_{1,T}^0 x_0$$

$$= \min_{v\in l_w^2(N, \mathcal{R}^{n_v})} J_{1,\infty}(u_\infty^*, v) = x_0^T P_{1,\infty} x_0,$$

$$\lim_{T\to\infty} K_{1,T}^0 = K_1,$$

$$\lim_{T\to\infty} \min_{u\in l_w^2(N_T, \mathcal{R}^{n_u})} J_{2,T}(u, v_T^*) = \lim_{T\to\infty} x_0^T P_{2,T}^0 x_0$$

$$= \min_{u\in l_w^2(N, \mathcal{R}^{n_u})} J_2^\infty(u, v_T^*) = x_0^T P_{2,\infty} x_0,$$

$$\lim_{T\to\infty} K_{2,T}^0 = K_2.$$

Therefore, we have

$$\lim_{T\to\infty} (P_{1,T}^0, K_{1,T}^0; P_{2,T}^0, K_{2,T}^0) = (P_{1,\infty}, K_1; P_{2,\infty}, K_2),$$

where $(P_{1,\infty}, K_1; P_{2,\infty}, K_2)$ is a solution of (3.90)–(3.93).

REMARK 3.18   Several approaches have appeared in dealing with deterministic and stochastic discrete-time $H2/H_\infty$ control, for example, [44] and [46] applied an LMI optimization approach to the design of a H_2/H_∞ controller while an exact solution to the suboptimal deterministic H_2/H_∞ control problem was studied via convex optimization in [159]. In this section, we have proposed two numerical algorithms to solve the stochastic H_2/H_∞ control which are different from those of deterministic H_2/H_∞ control in [46, 159].
∎

Example 3.2
Consider the following second-order discrete-time stochastic system

$$\begin{cases} x_{k+1} = A_1 x_k + B_1 u_k + C_1 v_k + (A_2 x_k + C_2 v_k) w_k, \\ z_k = \begin{bmatrix} C x_k \\ u_k \end{bmatrix}, \quad x_0 \in \mathcal{R}^n, \ k \in \mathcal{N}, \end{cases} \tag{3.138}$$

where

$$A_1 = \begin{bmatrix} 0.8 & 0 \\ 0 & 0.75 \end{bmatrix}, \quad A_2 = \begin{bmatrix} 0.35 & 0 \\ 0 & 0.4 \end{bmatrix}, \quad C_1 = \begin{bmatrix} 0.6 \\ 0.4 \end{bmatrix},$$

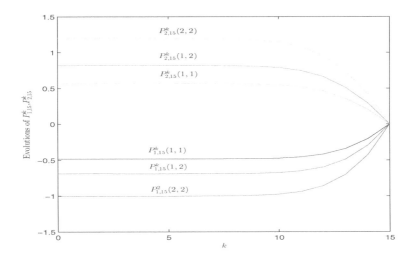

FIGURE 3.1

Convergence of $P_{1,15}^k$ and $P_{2,15}^k$.

$$C_2 = \begin{bmatrix} 0.4 \\ 0.55 \end{bmatrix}, \quad B_1 = \begin{bmatrix} 0.6 \\ 0.45 \end{bmatrix}, \quad C = \begin{bmatrix} 0.45 & 0.65 \end{bmatrix}.$$

Set $\gamma = 2.3$, $\mathcal{T} = 15$. By applying the above iterative algorithm, the evolution of $(P_{1,15}^k, K_{1,15}^k; P_{2,15}^k, K_{2,15}^k)$ is given in Figures 3.1–3.2 which clearly show the convergence of the backward iterations. The solution to the four coupled matrix-valued equations (3.90)–(3.93) obtained from the iterations is given by

$$P_{1,\infty} := \begin{bmatrix} P_{1,\infty}(1,1) & P_{1,\infty}(1,2) \\ P_{1,\infty}(1,2) & P_{1,\infty}(2,2) \end{bmatrix} = \begin{bmatrix} -0.4834 & -0.6916 \\ -0.6916 & -1.0045 \end{bmatrix},$$

$$P_{2,\infty} := \begin{bmatrix} P_{2,\infty}(1,1) & P_{2,\infty}(1,2) \\ P_{2,\infty}(1,2) & P_{2,\infty}(2,2) \end{bmatrix} = \begin{bmatrix} 0.5680 & 0.8203 \\ 0.8203 & 1.2024 \end{bmatrix},$$

$$K_1 = \begin{bmatrix} K_1(1,1) & K_1(1,2) \end{bmatrix} = \begin{bmatrix} 0.0894 & 0.1325 \end{bmatrix},$$

$$K_2 = \begin{bmatrix} K_2(1,1) & K_2(1,2) \end{bmatrix} = \begin{bmatrix} -0.3400 & -0.4686 \end{bmatrix}.$$

□

3.3.6 H_2/H_∞ control with (x, u)- and (x, u, v)-dependent noise

Because it is very similar to Theorems 3.8–3.9, we only state the following facts without proof.

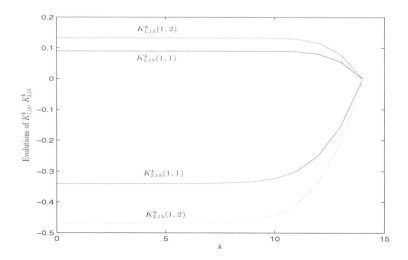

FIGURE 3.2

Convergence of $K_{1,15}^k$ and $K_{2,15}^k$.

(i) Theorems 3.8–3.9 still hold for the stochastic H_2/H_∞ control of systems with (x, u)-dependent noise if we replace the coupled equations (3.90)–(3.93) with the following four coupled ones:

$$\begin{cases} -P_{1,\infty} + (A_1 + B_1 K_2)^T P_{1,\infty}(A_1 + B_1 K_2) - C^T C - K_2^T K_2 \\ \quad +(A_2 + B_2 K_2)^T P_{1,\infty}(A_2 + B_2 K_2) - K_3 H_1(P_{1,\infty})^{-1} K_3^T = 0, \\ H_1(P_{1,\infty}) > 0, \end{cases}$$
$$\tag{3.139}$$

$$K_1 = -H_1(P_{1,\infty})^{-1} K_3^T, \tag{3.140}$$

$$\begin{cases} -P_{2,\infty} + (A_1 + C_1 K_1)^T P_{2,\infty}(A_1 + C_1 K_1) + A_2^T P_{2,\infty} A_2 \\ \quad +C^T C - K_4 H_2(P_{2,\infty})^{-1} K_4^T = 0, \\ H_2(P_{2,\infty}) > 0, \end{cases} \tag{3.141}$$

and

$$K_2 = -H_2(P_{2,\infty})^{-1} K_4^T, \tag{3.142}$$

where in (3.139)–(3.142),

$$K_3 = (A_1 + B_1 K_2)^T P_{1,\infty} C_1,$$

$$H_1(P_{1,\infty}) = \gamma^2 I + C_1^T P_{1,\infty} C_1,$$

$$K_4 = (A_1 + C_1 K_1)^T P_{2,\infty} B_1 + A_2^T P_{2,\infty} B_2,$$

$$H_2(P_{2,\infty}) = I + B_1^T P_{2,\infty} B_1 + B_2^T P_{2,\infty} B_2.$$

(ii) Theorems 3.8–3.9 still hold for the stochastic H_2/H_∞ control for systems with (x, u, v)-dependent noise if we replace the coupled equations (3.90)–(3.93) with the following four coupled ones:

$$\begin{cases} -P_{1,\infty} + (A_1 + B_1 K_2)^T P_{1,\infty} (A_1 + B_1 K_2) - C^T C - K_2^T K_2 \\ \quad +(A_2^T + B_2 K_2)^T P_{1,\infty} (A_2 + B_2 K_2) - K_3 H_1(P_{1,\infty})^{-1} K_3^T = 0, \\ H_1(P_{1,\infty}) > 0, \end{cases}$$

$$(3.143)$$

$$K_1 = -H_1(P_{1,\infty})^{-1} K_3^T, \tag{3.144}$$

$$\begin{cases} -P_{2,\infty} + (A_1 + C_1 K_1)^T P_{2,\infty} (A_1 + C_1 K_1) + C^T C \\ \quad +(A_2 + C_2 K_1)^T P_{2,\infty} (A_2 + C_2 K_1) - K_4 H_2(P_{2,\infty})^{-1} K_4^T = 0, \quad (3.145) \\ H_2(P_{2,\infty}) > 0, \end{cases}$$

and

$$K_2 = -H_2(P_{2,\infty})^{-1} K_4^T, \tag{3.146}$$

where in (3.143)–(3.146),

$$\begin{aligned} K_3 &= (A_1 + B_1 K_2)^T P_{1,\infty} C_1 + (A_2 + B_2 K_2)^T P_{1,\infty} C_2, \\ K_4 &= (A_1 + C_1 K_1)^T P_{2,\infty} B_1 + (A_2 + C_2 K_1)^T P_{2,\infty} B_2, \\ H_1(P_{1,\infty}) &= \gamma^2 I + C_1^T P_{1,\infty} C_1 + C_2^T P_{1,\infty} C_2, \\ H_2(P_{2,\infty}) &= I + B_1^T P_{2,\infty} B_1 + B_2^T P_{2,\infty} B_2. \end{aligned}$$

3.4 Infinite Horizon Indefinite LQ Control

To study the relationship between the infinite horizon stochastic H_2/H_∞ control and the existence of Nash equilibrium point (u_∞^*, v_∞^*) of the following two-person non-zero sum Nash game

$$J_{1,\infty}(u_\infty^*, v_\infty^*) \le J_{1,\infty}(u_\infty^*, v), \quad J_{2,\infty}(u_\infty^*, v_\infty^*) \le J_{2,\infty}(u, v_\infty^*) \tag{3.147}$$

for $(u_\infty^*, v_\infty^*) \in l_w^2(\mathcal{N}, \mathcal{R}^{n_u}) \times l_w^2(\mathcal{N}, \mathcal{R}^{n_v})$, we need to develop the theory of indefinite stochastic LQ control. That is, under the constraint of the time-invariant control system

$$\begin{cases} x_{k+1} = A_1 x_k + B_1 u_k + (A_2 x_k + B_2 u_k) w_k, \\ x_0 \in \mathcal{R}^n, \ k \in \mathcal{N}, \end{cases} \tag{3.148}$$

we consider the optimization problem of

$$J^\infty(u_s^*) = \Theta(x_0) := \inf_{u \in \mathcal{U}_{ad}} J^\infty(u) \tag{3.149}$$

with

$$J^\infty(u) := \sum_{k=0}^\infty \mathcal{E}(x_k^T Q x_k + u_k^T R x_k), \tag{3.150}$$

where in (3.150), Q and R are indefinite symmetric matrices. The indefinite LQ control problem (3.148)–(3.149) is associated with the GARE

$$\begin{cases} F(P) - H^T(P)G^+(P)H(P) = 0, \\ G(P)G^+(P)H(P) - H(P) = 0, \ G(P) \ge 0, \end{cases} \qquad (3.151)$$

where

$$\begin{cases} F(P) = A_1^T P A_1 + A_2^T P A_2 - P + Q, \\ H(P) = B_1^T P A_1 + B_2^T P A_2, \\ G(P) = R + B_1^T P B_1 + B_2^T P B_2. \end{cases}$$

Throughout this section, we assume that $(A_1, B_1; A_2, B_2)$ is stabilizable.

THEOREM 3.11
If the GARE (3.151) admits a solution P and there exist $Y \in l_w^2(\mathcal{N}, \mathcal{R}^{n_u \times n})$ and $Z \in l_w^2(\mathcal{N}, \mathcal{R}^{n_u})$ such that

$$u_{s,k}^* = -[G^+(P)H(P) - Y_k + G^+(P)G(P)Y_k]x_k$$
$$-G^+(P)G(P)Z_k + Z_k \qquad (3.152)$$

is a stabilizing control, then the LQ control problem (3.148)–(3.149) is attainable. Furthermore, $u_{s,k}^*$ is the optimal control with the optimal cost value uniquely determined by

$$\Theta(x_0) = x_0^T P x_0.$$

Proof. We note that

$$J_\nu(0, x_0; u_0, \cdots, u_\nu) = \sum_{k=0}^{\nu} \mathcal{E}(x_k^T Q_k x_k + u_k^T R_k u_k)$$

$$= \sum_{k=0}^{\nu} \mathcal{E}(x_k^T Q_k x_k + u_k^T R_k u_k + x_{k+1}^T P x_{k+1} - x_k^T P x_k)$$

$$+ x_0^T P x_0 - \mathcal{E}(x_{k+1}^T P x_{k+1})$$

$$= \sum_{k=0}^{\nu} \mathcal{E}[x_k^T (Q + A_1^T P A_1 + A_2^T P A_2 - P)x_k$$

$$+ 2x_k^T (A_1^T P B_1 + A_2^T P B_2)u_k + u_k^T (R + B_1^T P B_1 + B_2^T P B_2)u_k]$$

$$+ x_0^T P x_0 - \mathcal{E}(x_{k+1}^T P x_{k+1})$$

$$= \sum_{k=0}^{\nu} \mathcal{E}[x_k^T F(P)x_k + 2x_k^T H^T(P)u_k + u_k^T G(P)u_k]$$

$$+ x_0^T P x_0 - \mathcal{E}(x_{k+1}^T P x_{k+1}).$$

Set

$$M_k^1 = G^+(P)G(P)Y_k - Y_k, \ M_k^2 = G^+(P)G(P)Z_k - Z_k.$$

Then, by the properties of Pseudo inverse, it can be seen that

$$G(P)M_k^i = 0, \ i = 1, 2, \ k \in \mathcal{N}.$$

Therefore,

$$\begin{aligned}
J_\nu(0, x_0; u_0, \cdots, u_\nu) &= x_0^T P x_0 - \mathcal{E}(x_{\nu+1}^T P x_{\nu+1}) \\
&+ \sum_{k=0}^{\nu} \mathcal{E}\left\{ [u_k + (G^+(P)H(P) + M_k^1)x_k + M_k^2]^T G(P) \right. \\
&\quad \left. \cdot [u_k + (G^+(P)H(P) + M_k^1)x_k + M_k^2] \right\} \\
&+ \sum_{k=0}^{\nu} \mathcal{E}[x_k^T (F(P) - H^T(P)G^+(P)H(P))x_k] \\
&= \sum_{k=0}^{\nu} \mathcal{E}\left\{ [u_k + (G^+(P)H(P) + M_k^1)x_k + M_k^2]^T G(P) \right. \\
&\quad \left. \cdot [u_k + (G^+(P)H(P) + M_k^1)x_k + M_k^2] \right\} \\
&+ x_0^T P x_0 - \mathcal{E}(x_{\nu+1}^T P x_{\nu+1}).
\end{aligned} \tag{3.153}$$

So for any $u \in \mathcal{U}_{ad}$,

$$J_\nu(0, x_0; u_0, \cdots, u_\nu) \geq x_0^T P x_0 - \mathcal{E}(x_{\nu+1}^T P x_{\nu+1}), \tag{3.154}$$

while

$$J_\nu(0, x_0; u_{s,0}^*, \cdots, u_{s,\nu}^*) = x_0^T P x_0 - \mathcal{E}(x_{\nu+1}^T P x_{\nu+1}). \tag{3.155}$$

Taking $\nu \to \infty$ in (3.154) and (3.155) yields

$$J^\infty(u) := J_\infty(0, x_0; u_0, u_1, \cdots) \geq x_0^T P x_0 = \Theta(x_0) = J^\infty(u_s^*).$$

This completes the proof. □

The following result can be proved by following the line of [11].

LEMMA 3.9
The LQ control problem (3.148)–(3.149) is well posed iff there exists a constant symmetric matrix P such that

$$\Theta(x_0) = x_0^T P x_0, \quad \forall x_0 \in \mathcal{R}^n. \tag{3.156}$$

We introduce the following convex set $\mathcal{P}$ in $\mathcal{S}^n$:

$$\mathcal{P} = \left\{ P \in \mathcal{S}_n \mid \begin{bmatrix} F(P) & H^T(P) \\ H(P) & G(P) \end{bmatrix} \geq 0 \right\}. \tag{3.157}$$

LEMMA 3.10 (Extended Schur's lemma)[3]
Let the matrices $M = M^T$, N, $R = R^T$ be given with appropriate dimensions. Then, the following conditions are equivalent:

(i) $M - NR^+N^T \geq 0, R \geq 0,$ and $N(I - RR^+) = 0.$

(ii) $\begin{bmatrix} M & N \\ N^T & R \end{bmatrix} \geq 0.$

(iii) $\begin{bmatrix} R & N^T \\ N & M \end{bmatrix} \geq 0.$

LEMMA 3.11 [7]

Let the matrices L, M and N be given. Then, the matrix equation $LXM = N$ has a solution X iff $LL^+NMM^+ = N$. In this situation, X is given by $X = L^+NM^+ + Z - L^+LYMM^+$, where Z is a matrix with an appropriate dimension.

THEOREM 3.12

The indefinite LQ control problem (3.148)–(3.149) is well posed iff the set $\mathcal{P}$ is nonempty.

Proof. For any $\tilde{P} \in \mathcal{P}$, and $u \in \mathcal{U}_{ad}$, $v \in \mathcal{N}$, the sufficiency can be shown by noticing that

$$J_v(0, x_0; u_0, \cdots, u_v) = \sum_{k=0}^{v} \mathcal{E} \begin{bmatrix} x_k \\ u_k \end{bmatrix}^T \begin{bmatrix} F(\tilde{P}) & H^T(\tilde{P}) \\ H(\tilde{P}) & G(\tilde{P}) \end{bmatrix} \begin{bmatrix} x_k \\ u_k \end{bmatrix}$$
$$+ x_0^T \tilde{P} x_0 - \mathcal{E}(x_{v+1}^T \tilde{P} x_{v+1})$$
$$\geq x_0^T \tilde{P} x_0 - \mathcal{E}(x_{v+1}^T \tilde{P} x_{v+1}). \qquad (3.158)$$

Taking $v \to \infty$ on the above yields $J^\infty(u) \geq x_0^T \tilde{P} x_0$, so

$$\Theta(x_0) \geq x_0^T \tilde{P} x_0 > -\infty. \qquad (3.159)$$

$\Theta(x_0) < +\infty$ is due to the stabilization of $(A_1, B_1; A_2, B_2)$. The sufficiency is proved.

Necessity: By Lemma 3.9, there exists a symmetric matrix P such that $\Theta(x_0) = x_0^T P x_0$. By the dynamic programming principle, for $\forall u \in \mathcal{U}_{ad}$, we obtain

$$x_0^T P x_0 \leq \sum_{k=0}^{h} \mathcal{E}(x_k^T Q x_k + u_k^T R u_k) + \mathcal{E}(x_{h+1}^T P x_{h+1}), \ \forall h \in \mathcal{N}.$$

Hence,

$$\mathcal{E}(x_{h+1}^T P x_{h+1}) - x_0^T P x_0 + \sum_{k=0}^{h} \mathcal{E}(x_k^T Q x_k + u_k^T R u_k)$$
$$= \sum_{k=0}^{h} \mathcal{E}(x_{k+1}^T P x_{k+1} - x_k^T P x_k) + \sum_{k=0}^{h} \mathcal{E}(x_k^T Q x_k + u_k^T R u_k)$$

$$= \sum_{k=0}^{h} \mathcal{E} \begin{bmatrix} x_k \\ u_k \end{bmatrix}^T \begin{bmatrix} F(P) & H^T(P) \\ H(P) & G(P) \end{bmatrix} \begin{bmatrix} x_k \\ u_k \end{bmatrix} \geq 0,$$

which implies that for any deterministic vectors $x_0 \in \mathcal{R}^n$ and $u_0 \in \mathcal{R}^{n_u}$,

$$\begin{bmatrix} x_0 \\ u_0 \end{bmatrix}^T \begin{bmatrix} F(P) & H^T(P) \\ H(P) & G(P) \end{bmatrix} \begin{bmatrix} x_0 \\ u_0 \end{bmatrix} \geq 0,$$

or equivalently,

$$\begin{bmatrix} F(P) & H^T(P) \\ H(P) & G(P) \end{bmatrix} \geq 0, \tag{3.160}$$

i.e., $\mathcal{P}$ is nonempty. $\square$

THEOREM 3.13

If the indefinite LQ control problem (3.148)–(3.149) is attainable for any x_0, then the GARE (3.151) has a stabilizing solution. Moreover, any optimal control is given by (3.152) with $Y_k \equiv 0$ for $k \in \mathcal{N}$.

Proof. Because the indefinite LQ control problem (3.148)–(3.149) is attainable, it is well posed. By Theorem 3.12 and Lemma 3.9, there is $P \in \mathcal{P}$ such that (3.160) holds and $\Theta(x_0) = x_0^T P x_0$. Applying Lemma 3.10, we have

$$\begin{aligned} F(P) - H^T(P)G^+(P)H(P) &\geq 0, \\ H^T(P)(I - G(P)G^+(P)) &= 0, \\ G(P) &\geq 0. \end{aligned} \tag{3.161}$$

By (3.153), if $u_s^* \in \mathcal{U}_{ad}$ is an optimal control, then

$$\Theta(x_0) = J^\infty(u_s^*) = x_0^T P x_0$$

$$= x_0^T P x_0 + \sum_{k=0}^{\infty} \mathcal{E}[x_k^T (F(P) - H^T(P)G^+(P)H(P))x_k]$$

$$+ \sum_{k=0}^{\infty} \mathcal{E}\{[u_{s,k}^* + G^+(P)H(P)x_k]^T G(P)[u_{s,k}^* + G^+(P)H(P)x_k]\}. \tag{3.162}$$

From (3.162), the following are derived:

$$\sum_{k=0}^{\infty} \mathcal{E}\{x_k^T[F(P) - H^T(P)G^+(P)H(P)]x_k\} = 0, \tag{3.163}$$

$$\sum_{k=0}^{\infty} \mathcal{E}\{[u_{s,k}^* + G^+(P)H(P)x_k]^T G(P)[u_{s,k}^* + G^+(P)H(P)x_k]\} = 0. \tag{3.164}$$

(3.163) implies that

$$x_0^T[F(P) - H^T(P)G^+(P)H(P)]x_0 = 0, \ \forall x_0 \in \mathcal{R}^n$$

which leads to

$$F(P) - H^T(P)G^+(P)H(P) = 0. \tag{3.165}$$

The second equality of (3.161) is equivalent to

$$[I - G(P)G^+(P)]H(P) = 0 \tag{3.166}$$

due to $[GG^+]^T = GG^+$ [149]. Considering (3.165), (3.166) and the third inequality of (3.161), it is shown that P is a solution to GARE (3.151).

Due to $G(P) \geq 0$, (3.164) gives

$$G(P)[u_{s,k}^* + G^+(P)H(P)x_k] = 0, \forall k \in \mathcal{N}.$$

In Lemma 3.11, taking $L = G(P)$, $M = I$, $N = -G(P)G^+(P)H(P)x_k$, we have

$$u_{s,k}^* = -G^+(P)H(P)x_k + Z_k - G^+(P)G(P)Z_k,$$

which is a special case of (3.152) with $Y_k \equiv 0$. The theorem is shown. $\square$

The indefinite LQ control problem (3.148)–(3.149) is associated with the following semidefinite programming (SDP):

$$\begin{aligned} \text{minimize} \quad & -\operatorname{Trace}(P) \\ \text{subject to} \quad & \begin{bmatrix} F(P) & H^T(P) \\ H(P) & G(P) \end{bmatrix} \geq 0. \end{aligned} \tag{3.167}$$

Similar to Theorem 5.1 of [6], it is easy to obtain [225]:

THEOREM 3.14

For the LQ control problem (3.148)–(3.149) and the SDP (3.167), we have the following results:

(i) *The SDP (3.167) is feasible iff the LQ control problem (3.148)–(3.149) is well posed.*

(ii) *If P^* is the unique optimal solution to SDP (3.167), i.e., SDP (3.167) is feasible, then $\Theta(x_0) = x_0^T P^* x_0$.*

(iii) *If the LQ (3.148)–(3.149) is attainable, then the unique solution of (3.167) is a feedback stabilizing solution to GARE (3.151).*

3.5 Comments on Stochastic H_2/H_∞ and Nash Game

Similar to Theorem 2.21, based on the results of Section 3.4, for the infinite horizon two-person non-zero sum Nash game (3.147), we have the following result. The proof is omitted.

THEOREM 3.15
Consider system (3.52). There exists a Nash equilibrium strategy $(u^*_{\infty,k} = K_2 x_k, v^*_{\infty,k} = K_1 x_k)_{k \in \mathcal{N}} \in \mathcal{U}_{ad}(v^*_\infty) \times \mathcal{V}_{ad}(u^*_\infty)$ iff the following four coupled matrix-valued equations

$$\begin{cases} -P_{1,\infty} + (A_1 + B_1 K_2)^T P_{1,\infty}(A_1 + B_1 K_2) - C^T C \\ \quad +(A_2 + B_2 K_2)^T P_{1,\infty}(A_2 + B_2 K_2) - K_2^T K_2 - K_3 H_1(P_{1,\infty})^+ K_3^T = 0, \\ H_1(P_{1,\infty})H_1(P_{1,\infty})^+ K_3^T - K_3^T = 0, \\ H_1(P_{1,\infty}) \geq 0, \end{cases} \quad (3.168)$$

$$K_1 = -H_1(P_{1,\infty})^+ K_3^T, \quad (3.169)$$

$$\begin{cases} -P_{2,\infty} + (A_1 + C_1 K_1)^T P_{2,\infty}(A_1 + C_1 K_1) \\ \quad +(A_2 + C_2 K_1)^T P_{2,\infty}(A_2 + C_2 K_1) + C^T C - K_4 H_2(P_{2,\infty})^{-1} K_4^T = 0, \quad (3.170) \\ H_2(P_{2,\infty}) > 0, \end{cases}$$

and

$$K_2 = -H_2(P_{2,\infty})^{-1} K_4^T \quad (3.171)$$

have a unique quaternion solution $(P_{1,\infty} \leq 0, K_1; P_{2,\infty} \geq 0, K_2)$, where in (3.168)–(3.171), K_3, K_4, $H_1(P_{1,\infty})$ and $H_2(P_{2,\infty})$ are the same as in (3.143)–(3.146).

Obviously, the infinite horizon stochastic H_2/H_∞ control is equivalent to the Nash game problem (3.147) iff $H_1(P_{1,\infty}) > 0$; such a fact is first revealed in this book, which is mistaken to be unconditionally equivalent in [207, 208].

3.6 Notes and References

By around year 2000, the deterministic H_2/H_∞ control had been extensively studied and gradually reached its maturity; see [19, 31, 188, 189, 227]. On the other hand, since the work of [65], there have been a lot of studies on the discrete-time stochastic H_∞ control and filtering for systems with multiplicative noises; see [16, 78, 79, 81, 162]. Also, the H_2 (LQ) optimal control has been studied in [7, 18, 97]. Though the H_2/H_∞ control is closely related to game theory, it is worth noting that the discrete-time H_2/H_∞ control is not completely equivalent to the existence of a two-person zero sun Nash equilibrium point as commented in Remark 3.6. We note that the study of this chapter can be extended to the $\mathcal{H}_-$ control [125] for stochastic systems. Further research topics include the H_2/H_∞ control of stochastic systems via output feedback. It is worth mentioning that discrete-time and continuous-time stochastic systems have applications in networked control [39, 187, 199, 201] and synthesis gene networks [38, 184, 181].

The materials of this chapter mainly come from [207, 208].

4

H_2/H_∞ *Control for Linear Discrete Time-Varying Stochastic Systems*

In Section 3.3, we were concerned with linear discrete time-invariant stochastic systems. In this chapter, we shall investigate the infinite horizon H_2/H_∞ control for linear discrete time-varying (LDTV) systems that can be viewed as an extension of Section 3.3. To this end, we have to develop some essential theories on time-varying systems with multiplicative noise, including detectability, observability and properties of GLEs.

4.1 Stability and Uniform Detectability

For notational simplicity, in this chapter, we express an LDTV system of the form

$$\begin{cases} x_{k+1} = F_k x_k + G_k x_k w_k, \\ x(0) = x_0 \in \mathcal{R}^n \\ y_k = H_k x_k, \ k \in \mathcal{N}. \end{cases} \qquad (4.1)$$

In (4.1), x_k and y_k are respectively the system state and the measurement output. $\{w_k\}_{k \geq 0}$ is defined as in Chapter 3. F_k, G_k and H_k are time-varying matrices of suitable dimensions for $k = 0, 1, 2, \cdots$.

DEFINITION 4.1   *System (4.1) is said to be exponentially stable in a mean square (ESMS) sense if there exist $\beta \geq 1$ and $\lambda \in (0,1)$ such that for any $0 \leq k_0 \leq k < +\infty$, there holds*

$$\mathcal{E}\|x_k\|^2 \leq \beta \mathcal{E}\|x_{k_0}\|^2 \lambda^{(k-k_0)}. \qquad (4.2)$$

To define uniform detectability for (4.1), we establish the following lemmas.

LEMMA 4.1
For system (4.1), $\mathcal{E}\|x_l\|^2 = \mathcal{E}\|\phi_{l,k} x_k\|^2$ for $k \leq l$, where it is assumed that

$\phi_{k,k} = I$, *and* $\phi_{l,k}$ *is given by the following iterative relation*

$$\phi_{l,k} = \begin{bmatrix} \phi_{l,k+1}F_k \\ \phi_{l,k+1}G_k \end{bmatrix}, \ l > k. \tag{4.3}$$

Proof. We prove this lemma by induction. For $k = l - 1$, we have

$$\begin{aligned} \mathcal{E}\|x_l\|^2 &= \mathcal{E}[(F_{l-1}x_{l-1} + G_{l-1}x_{l-1}w_{l-1})^T(F_{l-1}x_{l-1} + G_{l-1}x_{l-1}w_{l-1})] \\ &= \mathcal{E}[x_{l-1}^T(F_{l-1}^T F_{l-1} + G_{l-1}^T G_{l-1})x_{l-1}] \\ &= \mathcal{E}\|\phi_{l,l-1}x_{l-1}\|^2. \end{aligned}$$

Hence, (4.3) holds for $k = l - 1$. Assume that for $k = m < l - 1$, $\mathcal{E}\|x_l\|^2 = \mathcal{E}\|\phi_{l,m}x_m\|^2$. Next, we prove $\mathcal{E}\|x_l\|^2 = \mathcal{E}\|\phi_{l,m-1}x_{m-1}\|^2$. It can be seen that

$$\begin{aligned} \mathcal{E}\|x_l\|^2 &= \mathcal{E}[x_m^T \phi_{l,m}^T \phi_{l,m} x_m] \\ &= \mathcal{E}[(F_{m-1}x_{m-1} + G_{m-1}x_{m-1}w_{m-1})^T \phi_{l,m}^T \phi_{l,m} \\ &\quad \cdot (F_{m-1}x_{m-1} + G_{m-1}x_{m-1}w_{m-1})] \\ &= \mathcal{E}[x_{m-1}^T(F_{m-1}^T \phi_{l,m}^T \phi_{l,m}F_{m-1} + G_{m-1}^T \phi_{l,m}^T \phi_{l,m}G_{m-1})x_{m-1}] \\ &= \mathcal{E}\|\phi_{l,m-1}x_{m-1}\|^2. \end{aligned}$$

This completes the proof. $\square$

LEMMA 4.2
For system (4.1), there holds $\sum_{i=k}^{l} \mathcal{E}\|y_i\|^2 = \mathcal{E}\|H_{l,k}x_k\|^2$ for $0 \le k \le l$, where

$$H_{l,k} = \begin{bmatrix} H_k \\ (I_2 \otimes H_{k+1})\phi_{k+1,k} \\ (I_{2^2} \otimes H_{k+2})\phi_{k+2,k} \\ \vdots \\ (I_{2^{l-k}} \otimes H_l)\phi_{l,k} \end{bmatrix} \tag{4.4}$$

with $H_{k,k} = H_k$ and $\phi_{j,k}(j = k + 1, \cdots, l)$ given by (4.3).

Proof. This lemma can also be proved by induction. First, by some straightforward computations, the conclusion holds in the case of $k = l, l-1$. Next, we assume that for $k = m < l - 1$, $\sum_{i=m}^{l} \mathcal{E}\|y_i\|^2 = \mathcal{E}\|H_{l,m}x_m\|^2$ holds, then we only need to prove $\sum_{i=m-1}^{l} \mathcal{E}\|y_i\|^2 = \mathcal{E}\|H_{l,m-1}x_{m-1}\|^2$. It can be verified that

$$\begin{aligned} \sum_{i=m-1}^{l} \mathcal{E}\|y_i\|^2 &= \sum_{i=m}^{l} \mathcal{E}\|y_i\|^2 + \mathcal{E}\|y_{m-1}\|^2 \\ &= \mathcal{E}\|H_{l,m}x_m\|^2 + \mathcal{E}\|y_{m-1}\|^2 \\ &= \mathcal{E}(x_m^T H_{l,m}^T H_{l,m}x_m) + \mathcal{E}(x_{m-1}^T H_{m-1}^T H_{m-1}x_{m-1}) \end{aligned}$$

$$
\begin{aligned}
&= \mathcal{E}[(F_{m-1}x_{m-1} + G_{m-1}x_{m-1}w_{m-1})^T H_{l,m}^T H_{l,m} \\
&\quad \cdot (F_{m-1}x_{m-1} + G_{m-1}x_{m-1}w_{m-1})] \\
&\quad + \mathcal{E}(x_{m-1}^T H_{m-1}^T H_{m-1} x_{m-1}) \\
&= \mathcal{E}\left\{ x_{m-1}^T \begin{bmatrix} H_{m-1} \\ H_{l,m}F_{m-1} \\ H_{l,m}G_{m-1} \end{bmatrix}^T \begin{bmatrix} H_{m-1} \\ H_{l,m}F_{m-1} \\ H_{l,m}G_{m-1} \end{bmatrix} x_{m-1} \right\}.
\end{aligned}
\tag{4.5}
$$

By (4.4), it follows that

$$
\begin{bmatrix} H_{m-1} \\ H_{l,m}F_{m-1} \\ H_{l,m}G_{m-1} \end{bmatrix} = \begin{bmatrix} H_{m-1} \\ H_m F_{m-1} \\ (I_2 \otimes H_{m+1})\phi_{m+1,m}F_{m-1} \\ \vdots \\ (I_{2^{l-m}} \otimes H_l)\phi_{l,m}F_{m-1} \\ H_m G_{m-1} \\ (I_2 \otimes H_{m+1})\phi_{m+1,m}G_{m-1} \\ \vdots \\ (I_{2^{l-m}} \otimes H_l)\phi_{l,m}G_{m-1} \end{bmatrix}.
\tag{4.6}
$$

On the other hand, it can be deduced from (4.3) and (4.4) that

$$
H_{l,m-1} = \begin{bmatrix} H_{m-1} \\ (I_2 \otimes H_m)\begin{bmatrix} F_{m-1} \\ G_{m-1} \end{bmatrix} \\ (I_{2^2} \otimes H_{m+1})\begin{bmatrix} \phi_{m+1,m}F_{m-1} \\ \phi_{m+1,m}G_{m-1} \end{bmatrix} \\ \vdots \\ (I_{2^{l-m+1}} \otimes H_l)\begin{bmatrix} \phi_{l,m}F_{m-1} \\ \phi_{l,m}G_{m-1} \end{bmatrix} \end{bmatrix}.
\tag{4.7}
$$

Combining (4.6) and (4.7) together results in

$$
\begin{bmatrix} H_{m-1} \\ H_{l,m}F_{m-1} \\ H_{l,m}G_{m-1} \end{bmatrix}^T \begin{bmatrix} H_{m-1} \\ H_{l,m}F_{m-1} \\ H_{l,m}G_{m-1} \end{bmatrix} = H_{l,m-1}^T H_{l,m-1}.
\tag{4.8}
$$

Hence, $\sum_{i=m-1}^l \mathcal{E}\|y_i\|^2 = \mathcal{E}\|H_{l,m-1}x_{m-1}\|^2$. This lemma is shown. $\square$

DEFINITION 4.2 *System (4.1) or $(F_k, G_k|H_k)$ is said to be uniformly detectable if there exist integers $s, t \geq 0$, and positive constants d, b with $0 \leq d < 1$ and $0 < b < \infty$ such that whenever*

$$
\mathcal{E}\|x_{k+t}\|^2 = \mathcal{E}\|\phi_{k+t,k}x_k\|^2 \geq d^2 \mathcal{E}\|x_k\|^2,
\tag{4.9}
$$

there holds

$$\sum_{i=k}^{k+s} \mathcal{E}\|y_i\|^2 = \mathcal{E}\|H_{k+s,k}x_k\|^2 \geq b^2 \mathcal{E}\|x_k\|^2, \qquad (4.10)$$

where $k \in \mathcal{N}$, and $\phi_{k+t,k}$ and $H_{k+s,k}$ are the same as defined in Lemma 4.2.

Obviously, without loss of generality, in Definition 4.2 we can assume that $t \leq s$. By Lemmas 4.1–4.2, the uniform detectability of $(F_k, G_k|H_k)$ implies, roughly speaking, that the state trajectory decays faster than the output energy does. In what follows, $\mathcal{O}_{k+s,k} := H_{k+s,k}^T H_{k+s,k}$ is called an observability Gramian matrix, and $\phi_{l,k}$ the state transition matrix of stochastic system (4.1) from x_k to x_l in a mean square sense. So (4.10) can be written as $\mathcal{E}[x_k^T \mathcal{O}_{k+s,k}x_k] \geq b^2 \mathcal{E}\|x_k\|^2$. If $G_k \equiv 0$ for $k \geq 0$, then system (4.1) reduces to the following deterministic system

$$\begin{cases} x_{k+1} = F_k x_k, \ x_0 \in \mathcal{R}^n, \\ y_k = H_k x_k, \end{cases} \qquad (4.11)$$

which was discussed in [10, 152]. When $\{F_k\}_{k\geq 0}$ and $\{G_k\}_{k\geq 0}$ in (4.11) are sequences of independent random matrices with constant statistics, the corresponding mean square (ms)-observability and ms-detectability were defined in [63, 64].

DEFINITION 4.3 [52] $(F(t), G(t)|H(t))$ *is said to be stochastically detectable, if there exists a matrix sequence $\{K(t)\}$ such that $(F(t) + K(t)H(t), G(t))$ is ESMS.*

Similarly, uniform observability can be defined as follows:

DEFINITION 4.4   *System (4.1) or $(F_k, G_k|H_k)$ is said to be uniformly observable if there exist an integer $s \geq 0$ and a positive constant $b > 0$ such that*

$$\mathcal{E}\|H_{k+s,k}x_k\|^2 \geq b^2 \mathcal{E}\|x_k\|^2 \qquad (4.12)$$

holds for each initial condition $x_k \in l^2_{\mathcal{F}_{k-1}}$, $k \in \mathcal{N}$.

REMARK 4.1   Different from the uniform detectability, uniform observability requires that any model (unstable or stable) should be reflected by the output. Uniform observability is also an important concept, which needs further study. ∎

PROPOSITION 4.1
If system (4.1) is ESMS, then for any bounded matrix sequence $\{H_k\}_{k\geq 0}$, system (4.1) is uniformly detectable.

Proof. By Definition 4.1, for any $k, t \geq 0$, we always have

$$\mathcal{E}\|x_{k+t}\|^2 = \mathcal{E}\|\phi_{k+t,k}x_k\|^2 \leq \beta \mathcal{E}\|x_k\|^2 \lambda^t, \ \beta > 1, \ 0 < \lambda < 1. \qquad (4.13)$$

By (4.13), $\beta\lambda^t \to 0$ as $t \to \infty$. Set a large $t_0 > 0$ such that $0 \le d^2 := \beta\lambda^{t_0} < 1$. Then, for any fixed $t > t_0$, (4.9) holds only for $x_k = 0$, which makes (4.10) valid for any $s \ge t > t_0$ and $b > 0$. So system (4.1) is uniformly detectable. $\square$

LEMMA 4.3 [77]
For a nonnegative real sequence $\{s_k\}_{k \ge k_0}$, if there exist constants $M_0 \ge 1$, $\delta_0 \in (0,1)$, and an integer $h_0 > 0$ such that $s_{k+1} \le M_0 s_k$ and $\min_{k+1 \le i \le k+h_0} s_i \le \delta_0 s_k$, then

$$s_k \le (M_0^{h_0}\delta_0^{-1})(\delta_0^{h_0})^{k-k_0} s_{k_0}, \quad \forall k \ge k_0.$$

The following proposition extends Lemma 2.2 in [10].

PROPOSITION 4.2
Suppose that $(F_k, G_k|H_k)$ is uniformly detectable, and F_k and G_k are uniformly bounded, i.e., $\|F_k\| \le M, \|G_k\| \le M, M > 0$. Then $\lim_{k \to \infty} \mathcal{E}\|y_k\|^2 = 0$ implies $\lim_{k \to \infty} \mathcal{E}\|x_k\|^2 = 0$.

Proof. If there exists some integer k_0 such that for all $k \ge k_0$, $\mathcal{E}\|x_{k+t}\|^2 = \mathcal{E}\|\phi_{k+t,k}x_k\|^2 < d^2\mathcal{E}\|x_k\|^2$, then $\min_{k+1 \le i \le k+t} \mathcal{E}\|x_i\|^2 < d^2\mathcal{E}\|x_k\|^2$. Moreover, $\mathcal{E}\|x_{i+1}\|^2 = \mathcal{E}\|\phi_{i+1,i}x_i\|^2 = \mathcal{E}[x_i^T(F_i^T F_i + G_i^T G_i)x_i] \le 2M^2\mathcal{E}\|x_i\|^2 \le M_0\mathcal{E}\|x_i\|^2$, where $M_0 = \max\{2M^2, 1\} \ge 1$. By Lemma 4.3, not only does $\lim_{k \to \infty} \mathcal{E}\|x_k\|^2 = 0$, but also system (4.1) is ESMS. Otherwise, there exists a subsequence $\{k_i\}_{i \ge 0}$ such that $\mathcal{E}\|\phi_{k_i+t,k_i}x_{k_i}\|^2 \ge d^2\mathcal{E}\|x_{k_i}\|^2$. Now, for $k \in (k_i, k_{i+1})$, we write $k = k_i + 1 + t\alpha + \beta$ with $\beta < t$, then

$$\mathcal{E}\|x_{k_i+1+\alpha t}\|^2 \le d^\alpha\mathcal{E}\|x_{k_i+1}\|^2,$$

$$\mathcal{E}\|x_{k_i+1+\alpha t+\beta}\|^2 \le (2M^2)^\beta\mathcal{E}\|x_{k_i+1+\alpha t}\|^2,$$

$$\mathcal{E}\|x_{k_i+1}\|^2 \le 2M^2\mathcal{E}\|x_{k_i}\|^2.$$

Therefore, we have

$$\begin{aligned}
\mathcal{E}\|x_k\|^2 = \mathcal{E}\|x_{k_i+1+\alpha t+\beta}\|^2 &\le (2M^2)^\beta d^\alpha\mathcal{E}\|x_{k_i+1}\|^2 \\
&\le (2M^2)^{\beta+1}d^\alpha\mathcal{E}\|x_{k_i}\|^2.
\end{aligned} \tag{4.14}$$

Obviously, in order to show $\lim_{k \to \infty} \mathcal{E}\|x_k\|^2 = 0$, we only need to show

$$\lim_{k_i \to \infty} \mathcal{E}\|x_{k_i}\|^2 = 0.$$

If it is not so, then there exist a subsequence $\{n_i\}_{i \ge 0}$ of $\{k_i\}_{i \ge 0}$ and $\varsigma > 0$, such that $\mathcal{E}\|x_{n_i}\|^2 > \varsigma, \mathcal{E}\|\phi_{n_i+t,n_i}x_{n_i}\|^2 \ge d^2\mathcal{E}\|x_{n_i}\|^2$. By Definition 4.2,

$$\sum_{i=n_i}^{n_i+s} \mathcal{E}\|y_i\|^2 = \mathcal{E}(x_{n_i}^T \mathcal{O}_{n_i+s,n_i}x_{n_i}) \ge b^2\mathcal{E}\|x_{n_i}\|^2 > b^2\varsigma. \tag{4.15}$$

Taking $n_i \to \infty$ in (4.15), we have $0 > b^2 \varsigma > 0$, which is a contradiction. Hence, the proof is complete. $\square$

A static output feedback $u_k = K_k y_k$ does not change uniform detectability of the original system, i.e., if $(F_k, G_k|H_k)$ is uniformly detectable, then if we let $u_k = K_k y_k$ in the control system

$$\begin{cases} x_{k+1} = (F_k x_k + M_k u_k) + (G_k x_k + N_k u_k)w_k, \\ y_k = H_k x_k, \ k \in \mathcal{N}, \end{cases} \tag{4.16}$$

the resulting closed-loop system

$$\begin{cases} x_{k+1} = (F_k + M_k K_k H_k)x_k + (G_k + N_k K_k H_k)x_k w_k, \\ y_k = H_k x_k, \ k \in \mathcal{N} \end{cases} \tag{4.17}$$

is still uniformly detectable, which is given in the following proposition.

THEOREM 4.1
If $(F_k, G_k|H_k)$ is uniformly detectable, so is

$$(F_k + M_k K_k H_k, G_k + N_k K_k H_k|H_k).$$

Proof. By Lemma 4.2, the observability Gramian for system (4.17) is

$$\bar{\mathcal{O}}_{k+s,k} = \bar{H}_{k+s,k}^T \bar{H}_{k+s,k},$$

where

$$\bar{H}_{k+s,k} = \begin{bmatrix} H_k \\ (I_2 \otimes H_{k+1})\bar{\phi}_{k+1,k} \\ (I_{2^2} \otimes H_{k+2})\bar{\phi}_{k+2,k} \\ \vdots \\ (I_{2^s} \otimes H_{k+s})\bar{\phi}_{k+s,k} \end{bmatrix}, \ \bar{\phi}_{k+i,k} = \begin{bmatrix} \bar{\phi}_{k+i,k+1}\bar{F}_k \\ \bar{\phi}_{k+i,k+1}\bar{G}_k \end{bmatrix}, \ i = 1, \cdots, s.$$

$$\bar{F}_j = F_j + M_j K_j H_j, \ \bar{G}_j = G_j + N_j K_j H_j, \ j = k, k+1, \cdots, k+s.$$

To prove that $(\bar{F}_k, \bar{G}_k|H_k)$ is uniformly detectable, it suffices to show that there are constants $\bar{b} > 0, \ 0 < \bar{d} < 1, \ s, t \geq 0$ such that for $x_k \in l_{\mathcal{F}_{k-1}}^2, \ k \in \mathcal{N}$, whenever

$$\mathcal{E}(x_k^T \bar{\mathcal{O}}_{k+s,k} x_k) < \bar{b}^2 \mathcal{E}\|x_k\|^2, \tag{4.18}$$

we have

$$\mathcal{E}\|\bar{\phi}_{k+t,k} x_k\|^2 < \bar{d}^2 \mathcal{E}\|x_k\|^2. \tag{4.19}$$

It is easy to show

$$\bar{H}_{k+s,k} = Q_{k+s,k} H_{k+s,k}, \ Q_{k+s,k} = \begin{bmatrix} I & 0 & \cdots & 0 \\ * & I & \cdots & 0 \\ \vdots & \vdots & & \vdots \\ * & * & \cdots & I \end{bmatrix},$$

where * denotes an element involving H_i, M_i, K_i and N_i, $i = k, k+1, \cdots, k+s$. Hence, for any $x_k \in l^2_{\mathcal{F}_{k-1}}$,

$$\rho \mathcal{E}(x_k^T \mathcal{O}_{k+s,k} x_k) \leq \mathcal{E}(x_k^T \bar{\mathcal{O}}_{k+s,k} x_k) \leq \varrho \mathcal{E}(x_k^T \mathcal{O}_{k+s,k} x_k), \qquad (4.20)$$

where $\rho = \lambda_{\min}(Q_{k+s,k}^T Q_{k+s,k}) > 0$, $\varrho = \lambda_{\max}(Q_{k+s,k}^T Q_{k+s,k}) > 0$. In addition, by observation, for any $l > k \geq 0$,

$$\bar{\phi}_{l,k} = \phi_{l,k} + R_{l,k} H_{l,k},$$

where $R_{l,k}$ is a matrix involving H_i, M_i, K_i and N_i, $i = k, k+1, \cdots, l-1$. If we take $0 < \bar{b} \leq \sqrt{\rho} b$, then it follows from (4.20) that $\mathcal{E}(x_k^T \bar{\mathcal{O}}_{k+s,k} x_k) < \frac{1}{\rho} \mathcal{E}(x_k^T \bar{\mathcal{O}}_{k+s,k} x_k) \leq \frac{\bar{b}^2}{\rho} \mathcal{E}\|x_k\|^2 \leq b^2 \mathcal{E}\|x_k\|^2$. By the uniform observability of $(F_k, G_k | H_k)$, it follows that

$$\begin{aligned}
\mathcal{E}\|\bar{\phi}_{k+t,k} x_k\|^2 &= \mathcal{E}\|\phi_{k+t,k} x_k + R_{k+t,k} H_{k+t,k} x_k\|^2 \\
&\leq 2\mathcal{E}\|\phi_{k+t,k} x_k\|^2 + 2\mu^2 \mathcal{E}\|H_{k+t,k} x_k\|^2 \\
&\leq 2d^2 \mathcal{E}\|x_k\|^2 + 2\mu^2 \mathcal{E}(x_k^T \mathcal{O}_{k+s,k} x_k) \\
&\leq \left(2d^2 + 2\mu^2 \frac{\bar{b}^2}{\rho}\right) \mathcal{E}\|x_k\|^2 \\
&= \bar{d} \mathcal{E}\|x_k\|^2,
\end{aligned}$$

where $\mu = \sup_k \|R_{k+t,k}\|$, $\bar{d} = 2d^2 + 2\mu^2 \frac{\bar{b}^2}{\rho}$. If we take $\bar{b}$ to be sufficiently small, then $\bar{d} < 1$, which yields the uniform detectability of $(\bar{F}_k, \bar{G}_k | H_k)$. Hence, the proof of this theorem is complete. $\square$

Example 4.1

For simplicity, we set $s = 1$. Then it can be computed that

$$\bar{H}_{k+1,k} = \begin{bmatrix} H_k \\ (I_2 \otimes H_{k+1})\bar{\phi}_{k+1,k} \end{bmatrix} = \begin{bmatrix} H_k \\ H_{k+1}(F_k + M_k K_k H_k) \\ H_{k+1}(G_k + N_k K_k H_k) \end{bmatrix},$$

$$H_{k+1,k} = \begin{bmatrix} H_k \\ (I_2 \otimes H_{k+1})\phi_{k+1,k} \end{bmatrix} = \begin{bmatrix} H_k \\ H_{k+1}F_k \\ H_{k+1}G_k \end{bmatrix}.$$

Obviously,

$$Q_{k+1,k} = \begin{bmatrix} I & 0 & 0 \\ H_{k+1}M_k K_k & I & 0 \\ H_{k+1}N_k K_k & 0 & I \end{bmatrix}.$$

〇

Example 4.2
By our definition, we have

$$\bar{\phi}_{k+1,k} = \begin{bmatrix} F_k + M_k K_k H_k \\ G_k + N_k K_k H_k \end{bmatrix}, \quad \phi_{k+1,k} = \begin{bmatrix} F_k \\ G_k \end{bmatrix}.$$

Hence, $\bar{\phi}_{k+1,k} = \phi_{k+1,k} + R_{k+1,k} H_{k+1,k}$ with $R_{k+1,k} = \begin{bmatrix} M_k K_k & 0 & 0 \\ N_k K_k & 0 & 0 \end{bmatrix}$. $\quad\Box$

Another detectability called W-detectability was introduced in [42] for the discrete time-invariant Markov jump system

$$\begin{cases} x_{k+1} = F(\theta_k)x_k, \ x_0 = x_0, \\ y_k = H(\theta_k)x_k, \ \theta_0 \sim \mu_0, \end{cases}$$

where θ_k is a homogeneous Markov chain $\Theta = \{\theta_k, k \in \mathcal{N}\}$ with the state space $\mathcal{S} = \{1, 2, \cdots, S\}$ and the state transition matrix $\mathcal{P}_{S \times S} = (p_{ij})_{S \times S}, i, j = 1, 2, \cdots, S$. Definition 5 of [42] can be extended to the linear time-varying Markov jump system

$$\begin{cases} x_{k+1} = F(k, \theta_k)x_k, \\ y_k = H(k, \theta_k)x_k, \ \theta_0 \sim \mu_0, \end{cases}$$

where it only needs to replace

$$W^N(X) = \mathcal{E}_{x_0,\mu_0} \left\{ \sum_{i=0}^{N-1} \|y_i\|^2 \right\}$$

therein with

$$W^{k+j,k}(X) = \mathcal{E}_{x_0,\mu_0} \left\{ \sum_{i=k}^{k+j-1} \|y_i\|^2 \right\}.$$

It is easy to show that the uniform detectability of $(F_k, G_k | H_k)$ is equivalent to the W-detectability of

$$v_{k+1} = A(k, r_k)v_k, \ v_0 = x_0, \tag{4.21}$$

with

$$\mathcal{P}_{2 \times 2} = \begin{bmatrix} \frac{1}{2} & \frac{1}{2} \\ \frac{1}{2} & \frac{1}{2} \end{bmatrix}, \ A(k,1) = \sqrt{2}F_k, \ A(k,2) = \sqrt{2}G_k.$$

The above fact reveals that discrete time-varying systems with state-dependent noise are closely related to time-varying Markov systems without state-dependent noise.

4.2 Lyapunov-Type Theorem under Uniform Detectability

In the following, we will further study the time-varying GLE

$$-P_k + F_k^T P_{k+1} F_k + G_k^T P_{k+1} G_k + H_k^T H_k = 0, \ k \in \mathcal{N} \tag{4.22}$$

under uniform detectability. The aim is to extend the classical Lyapunov theorem to LDTV stochastic systems. We first introduce the following finite time backward difference equation

$$\begin{cases} -P_{k,\mathcal{T}} + F_k^T P_{k+1,\mathcal{T}} F_k + G_k^T P_{k+1,\mathcal{T}} G_k + H_k^T H_k = 0, \\ P_{\mathcal{T},\mathcal{T}} = 0, \ k = 0, 1, \cdots, \mathcal{T} - 1; \ \mathcal{T} \in \mathbb{Z}_{1+} := \{1, 2, \cdots, \}. \end{cases} \quad (4.23)$$

Obviously, the equation (4.23) has nonnegative definite solutions $P_{k,\mathcal{T}} \geq 0$.

PROPOSITION 4.3
$P_{k,\mathcal{T}}$ is monotonically increasing with respect to $\mathcal{T}$, i.e., for any $k_0 \leq \mathcal{T}_1 \leq \mathcal{T}_2 < +\infty$,

$$P_{k_0,\mathcal{T}_1} \leq P_{k_0,\mathcal{T}_2}, \ k_0 \in N_{\mathcal{T}_1} := \{0, 1, \cdots, \mathcal{T}_1\}.$$

Proof. Obviously, $P_{k,\mathcal{T}_1}$ and $P_{k,\mathcal{T}_2}$ solve

$$\begin{cases} -P_{k,\mathcal{T}_1} + F_k^T P_{k+1,\mathcal{T}_1} F_k + G_k^T P_{k+1,\mathcal{T}_1} G_k + H_k^T H_k = 0, \\ P_{\mathcal{T}_1,\mathcal{T}_1} = 0, \ k \in N_{\mathcal{T}_1-1}, \end{cases} \quad (4.24)$$

and

$$\begin{cases} -P_{k,\mathcal{T}_2} + F_k^T P_{k+1,\mathcal{T}_2} F_k + G_k^T P_{k+1,\mathcal{T}_2} G_k + H_k^T H_k = 0, \\ P_{\mathcal{T}_2,\mathcal{T}_2} = 0, \ k \in N_{\mathcal{T}_2-1}, \end{cases} \quad (4.25)$$

respectively. Consider the following LDTV stochastic system with a deterministic initial state x_{k_0}:

$$\begin{cases} x_{k+1} = F_k x_k + G_k x_k w_k, \\ x_{k_0} \in \mathcal{R}^n, \ k \in \mathcal{N}_{k_0}^\infty := \{k_0, k_0 + 1, \cdots, \}. \end{cases} \quad (4.26)$$

Associated with (4.26), in view of (4.24), we have

$$\sum_{k=k_0}^{\mathcal{T}_1-1} \mathcal{E}(x_k^T H_k^T H_k x_k)$$

$$= \sum_{k=k_0}^{\mathcal{T}_1-1} \mathcal{E}(x_k^T H_k^T H_k x_k + x_{k+1}^T P_{k+1,\mathcal{T}_1} x_{k+1} - x_k^T P_{k,\mathcal{T}_1} x_k)$$

$$+ x_{k_0}^T P_{k_0,\mathcal{T}_1} x_{k_0} - \mathcal{E}(x_{\mathcal{T}_1}^T P_{\mathcal{T}_1,\mathcal{T}_1} x_{\mathcal{T}_1})$$

$$= \sum_{k=k_0}^{\mathcal{T}_1-1} \mathcal{E}\left[x_k^T (-P_{k,\mathcal{T}_1} + F_k^T P_{k+1,\mathcal{T}_1} F_k + G_k^T P_{k+1,\mathcal{T}_1} G_k + H_k^T H_k) x_k \right]$$

$$+ x_{k_0}^T P_{k_0,\mathcal{T}_1} x_{k_0}$$

$$= x_{k_0}^T P_{k_0,\mathcal{T}_1} x_{k_0}. \quad (4.27)$$

Similarly,

$$\sum_{k=k_0}^{\mathcal{T}_2-1} \mathcal{E}(x_k^T H_k^T H_k x_k) = x_{k_0}^T P_{k_0,\mathcal{T}_2} x_{k_0}. \quad (4.28)$$

From (4.27)–(4.28), it follows that

$$0 \leq \sum_{k=k_0}^{\mathcal{T}_1-1} \mathcal{E}(x_k^T H_k^T H_k x_k) = x_{k_0}^T P_{k_0,\mathcal{T}_1} x_{k_0} \leq \sum_{k=k_0}^{\mathcal{T}_2-1} \mathcal{E}(x_k^T H_k^T H_k x_k)$$

$$= x_{k_0}^T P_{k_0,\mathcal{T}_2} x_{k_0}. \tag{4.29}$$

The above expression holds for any $x_{k_0} \in \mathcal{R}^n$, which yields $P_{k_0,\mathcal{T}_1} \leq P_{k_0,\mathcal{T}_2}$. Thus, the proof is complete. $\square$

Another simple approach to proving Proposition 4.3 is to construct the following difference equation: write $\Delta_k = P_{k,\mathcal{T}_2} - P_{k,\mathcal{T}_1}$, then subtracting (4.24) from (4.25) yields

$$\Delta_k = F_k^T \Delta_{k+1} F_k + G_k^T \Delta_{k+1} G_k, \ \Delta_{\mathcal{T}_1} = P_{\mathcal{T}_1,\mathcal{T}_2} \geq 0, \ k \in N_{\mathcal{T}_1-1},$$

which allows us to obtain recursively that $\Delta_k \geq 0$ for $k \in N_{\mathcal{T}_1-1}$.

PROPOSITION 4.4
If system (4.1) is ESMS, and H_k is uniformly bounded (i.e., there exists $M > 0$ such that $\|H_k\| \leq M, \ \forall k \in \mathcal{N}$), then the solution $P_{k,\mathcal{T}}$ of (4.23) is uniformly bounded for any $\mathcal{T} \in \mathbb{Z}_{1+}$.

Proof. By (4.27), for any deterministic $x_k \in \mathcal{R}^n$, we have

$$x_k^T P_{k,\mathcal{T}} x_k = \sum_{i=k}^{\mathcal{T}-1} \mathcal{E}(x_i^T H_i^T H_i x_i) \leq \sum_{i=k}^{\infty} \mathcal{E}(x_i^T H_i^T H_i x_i)$$

$$\leq M^2 \|x_k\|^2 \beta \sum_{i=k}^{\infty} \lambda^{(i-k)} = M^2 \|x_k\|^2 \beta \frac{1}{1-\lambda},$$

which leads to that $0 \leq P_{k,\mathcal{T}} \leq \frac{\beta M^2}{1-\lambda} I$ since x_k is arbitrary. Hence, the proof is complete. $\square$

Combining Proposition 4.3 with Proposition 4.4 yields that $P_k := \lim_{\mathcal{T}\to\infty} P_{k,\mathcal{T}}$ exists, which is a solution of (4.22). Hence, we obtain the following Lyapunov-type theorem.

THEOREM 4.2
If system (4.1) is ESMS and $\{H_k\}_{k\in\mathcal{N}}$ is uniformly bounded, then (4.22) admits a unique nonnegative definite solution $\{P_k\}_{k\in\mathcal{N}}$.

The converse of Theorem 4.2 still holds, which is given by the following converse Lyapunov-type theorem.

THEOREM 4.3
Suppose that $(F_k, G_k|H_k)$ is uniformly detectable and F_k and G_k are uniformly bounded with an upper bound $M > 0$. If there is a bounded positive

definite symmetric matrix sequence $\{P_k\}_{k\geq 0}$ *with* $\inf_k \lambda_{min}(P_k) > 0$ *solving GLE (4.22), then system (4.1) is ESMS.*

Proof. For system (4.1), we take a Lyapunov function candidate as

$$V_k(x) = x^T(P_k + \varepsilon I)x,$$

where $\varepsilon > 0$ is to be determined. For simplicity, in the sequel, we let $V_k := V_k(x_k)$. It is easy to compute

$$\begin{aligned}
\mathcal{E}V_k - \mathcal{E}V_{k+1} &= \mathcal{E}[x_k^T(P_k + \varepsilon I)x_k] - \mathcal{E}[x_{k+1}^T(P_{k+1} + \varepsilon I)x_{k+1}] \\
&= \mathcal{E}[x_k^T(P_k + \varepsilon I)x_k] - \mathcal{E}[(F_k x_k + G_k x_k w_k)^T(P_{k+1} + \varepsilon I)(F_k x_k + G_k x_k w_k)] \\
&= \mathcal{E}[x_k^T(P_k - F_k^T P_{k+1} F_k - G_k^T P_{k+1} G_k)x_k] + \varepsilon \mathcal{E}[x_k^T(I - F_k^T F_k - G_k^T G_k)x_k] \\
&= \mathcal{E}\|y_k\|^2 + \varepsilon \mathcal{E}[x_k^T(I - F_k^T F_k - G_k^T G_k)x_k] \\
&= \mathcal{E}\|y_k\|^2 + \varepsilon \mathcal{E}\|x_k\|^2 - \varepsilon \mathcal{E}\|x_{k+1}\|^2.
\end{aligned} \tag{4.30}$$

Identity (4.30) yields

$$\begin{aligned}
\mathcal{E}V_k - \mathcal{E}V_{k+s+1} &= (\mathcal{E}V_k - \mathcal{E}V_{k+1}) + (\mathcal{E}V_{k+1} - \mathcal{E}V_{k+2}) \\
&\quad + \cdots + (\mathcal{E}V_{k+s} - \mathcal{E}V_{k+s+1}) \\
&= \sum_{i=k}^{k+s} \mathcal{E}\|y_i\|^2 + \varepsilon \mathcal{E}\|x_k\|^2 - \varepsilon \mathcal{E}\|x_{k+s+1}\|^2.
\end{aligned} \tag{4.31}$$

When $\sum_{i=k}^{k+s} \mathcal{E}\|y_i\|^2 \geq b^2 \mathcal{E}\|x_k\|^2$, we first note that

$$\begin{aligned}
\mathcal{E}\|x_{k+s+1}\|^2 &= \mathcal{E}\{x_{k+s}^T(F_{k+s}^T F_{k+s} + G_{k+s}^T G_{k+s})x_{k+s}\} \\
&\leq 2M^2 \mathcal{E}\|x_{k+s}\|^2 \leq (2M^2)^2 \mathcal{E}\|x_{k+s-1}\|^2 \leq \cdots \\
&\leq (2M^2)^{s+1} \mathcal{E}\|x_k\|^2.
\end{aligned} \tag{4.32}$$

Then, by (4.31), we still have

$$\begin{aligned}
\mathcal{E}V_k - \mathcal{E}V_{k+s+1} &\geq b^2 \mathcal{E}\|x_k\|^2 + \varepsilon \mathcal{E}\|x_k\|^2 - \varepsilon(2M^2)^{s+1}\mathcal{E}\|x_k\|^2 \\
&= [b^2 + \varepsilon - \varepsilon(2M^2)^{s+1}]\mathcal{E}\|x_k\|^2.
\end{aligned} \tag{4.33}$$

From (4.33), it readily follows that

$$\begin{aligned}
\mathcal{E}V_{k+s+1} &\leq \mathcal{E}V_k - \{b^2 + \varepsilon[1 - (2M^2)^{s+1}]\}\mathcal{E}\|x_k\|^2 \\
&\leq \left\{1 - \frac{b^2 + \varepsilon[1 - (2M^2)^{s+1}]}{\lambda_{max}(P_k + \varepsilon I)}\right\}\mathcal{E}V_k.
\end{aligned} \tag{4.34}$$

Considering that $\{P_k \geq 0\}_{k \in \mathcal{N}}$ is uniformly bounded, so there exists a sufficiently large $\tilde{M} > 0$ such that $P_k \leq \tilde{M}I_{n \times n}$. Hence, (4.34) leads to

$$\mathcal{E}V_{k+s+1} \leq \left\{1 - \frac{b^2 + \varepsilon[1 - (2M^2)^{s+1}]}{\tilde{M} + \varepsilon}\right\}\mathcal{E}V_k, \; \forall s \in \mathcal{N}.$$

Because $V_k(x) > 0$ for $x \neq 0$, we must have

$$0 < \delta := 1 - \frac{b^2 + \varepsilon[1 - (2M^2)^{s+1}]}{\tilde{M} + \varepsilon I} < 1.$$

for sufficiently large $\tilde{M} > 0$ and sufficiently small $\varepsilon > 0$. So

$$\mathcal{E}V_{k+s+1} \leq \delta \mathcal{E}V_k. \tag{4.35}$$

When $\sum_{i=k}^{k+s} \mathcal{E}\|y_i\|^2 \leq b^2 \mathcal{E}\|x_k\|^2$, by uniform detectability we have $\mathcal{E}\|x_{k+t}\|^2 \leq d^2 \mathcal{E}\|x_k\|^2$. From (4.31), it follows that

$$\begin{aligned}
\mathcal{E}V_k - \mathcal{E}V_{k+t} &\geq \varepsilon \mathcal{E}\|x_k\|^2 - \varepsilon d^2 \mathcal{E}\|x_k\|^2 \\
&= \varepsilon(1 - d^2)\mathcal{E}\|x_k\|^2.
\end{aligned} \tag{4.36}$$

Similarly, we can show that there exists a constant $\delta_1 \in (0, 1)$ such that

$$\mathcal{E}V_{k+t} \leq \delta_1 \mathcal{E}V_k. \tag{4.37}$$

Set $\delta_0 := \max\{\delta, \delta_1\}$, in view of (4.35) and (4.37), we have

$$\min_{k+1 \leq i \leq k+s+1} \mathcal{E}V_i \leq \delta_0 \mathcal{E}V_k, \ \forall k \geq 0. \tag{4.38}$$

From (4.30), we know

$$\mathcal{E}V_{k+1} \leq \mathcal{E}V_k + \varepsilon \mathcal{E}\|x_{k+1}\|^2 \leq \mathcal{E}V_k + \frac{\varepsilon}{\varrho + \varepsilon}\mathcal{E}V_{k+1}, \tag{4.39}$$

where $\varrho = \inf_k \lambda_{min}(P_k) > 0$. Therefore, there exists a positive constant

$$M_0 := \frac{1}{1 - \frac{\varepsilon}{\varrho + \varepsilon}} > 1$$

satisfying

$$\mathcal{E}V_{k+1} \leq M_0 \mathcal{E}V_k, \ \forall k \geq 0. \tag{4.40}$$

Applying Lemma 4.3 with $s_k = \mathcal{E}V_k$, $h_0 = s + 1$, $\beta = M_0^{h_0} \delta_0^{-1}$, $\lambda = \delta_0^{h_0}$, it follows that

$$\mathcal{E}V_k \leq \beta \lambda^{(k-k_0)} \mathcal{E}V_{k_0} \leq \lambda_{\max}(P_{k_0} + \varepsilon I)\beta \lambda^{(k-k_0)}\mathcal{E}\|x_{k_0}\|^2,$$

which implies that system (4.1) is ESMS due to the fact that $\{P_k\}_{k \geq 0}$ is uniformly bounded. $\square$

4.3 Exact Detectability

In this section, we will study exact detectability of the stochastic system (4.1), from which it can be found that there are some essential differences between time-varying and time-invariant systems. In addition, Lyapunov-type theorems are also presented.

We first give several definitions.

DEFINITION 4.5 *For system (4.1), $x_{k_0} \in l^2_{\mathcal{F}_{k_0-1}}$ is called a k_0^∞-unobservable state if $y_k \equiv 0$ a.s. for $k \in \{k_0, k_0+1, \cdots, \infty\}$, and $x_{k_0} \in l^2_{\mathcal{F}_{k_0-1}}$ is called a $k_0^{s_0}$-unobservable state if $y_k \equiv 0$ a.s. for $k \in \{k_0, k_0+1, \cdots, k_0+s_0\}$.*

REMARK 4.2 From Definition 4.5, we point out the following obvious facts: (i) If x_{k_0} is a k_0^∞-unobservable state, then for any $s_0 \geq 0$, it must be a $k_0^{s_0}$-unobservable state. (ii) If x_{k_0} is a $k_0^{s_1}$-unobservable state, then for any $0 \leq s_0 \leq s_1$, it must be a $k_0^{s_0}$-unobservable state. ∎

Example 4.3
In system (4.1), if we take $H_k \equiv 0$ for $k \geq k_0$, then any state $x_{k_0} \in l^2_{\mathcal{F}_{k_0-1}}$ is a k_0^∞-unobservable state. For any $k_0 \geq 0$, $x_{k_0} = 0$ is a trivial k_0^∞-unobservable state.
∎

Different from the linear time-invariant system

$$\begin{cases} x_{k+1} = Fx_k + Gx_kw_k, \ x_0 \in \mathcal{R}^n \\ y_k = Hx_k, \ k \in \mathcal{N}, \end{cases} \tag{4.41}$$

even if $x_{k_0} = \zeta$ is a k_0^∞-unobservable state, $x_{k_1} = \zeta$ may not be a $k_1^{s_1}$-unobservable state for any $s_1 \geq 0$, which is seen from the next example.

Example 4.4
Consider the deterministic linear time-varying system with $G_k = 0$ and

$$H_k = F_k = \begin{cases} \begin{bmatrix} 1 & 0 \\ 0 & 0 \end{bmatrix}, \text{ if } k \text{ is even}, \\ \begin{bmatrix} 0 & 0 \\ 0 & 1 \end{bmatrix}, \text{ if } k \text{ is odd}. \end{cases}$$

Obviously, $x_0 = \begin{bmatrix} 0 \\ 1 \end{bmatrix}$ is a 0^∞-unobservable state due to $y_k = 0$ for $k \geq 0$, but $x_1 = \begin{bmatrix} 0 \\ 1 \end{bmatrix}$ is not a 1^{s_1}-unobservable state for any $s_1 \geq 0$ due to $y_1 = H_1 \begin{bmatrix} 0 \\ 1 \end{bmatrix} \neq 0$, let alone 1^∞-unobservable state. ∎

DEFINITION 4.6   *System (4.1) is called k_0^∞-exactly detectable if any k_0^∞-unobservable initial state ξ leads to an exponentially stable trajectory, i.e., there are constants $\beta \geq 1$, $0 < \lambda < 1$ such that*

$$\mathcal{E}\|x_k\|^2 \leq \beta \mathcal{E}\|\xi\|^2 \lambda^{(k-k_0)}, \quad \forall k \geq k_0. \tag{4.42}$$

Similarly, system (4.1) is called $k_0^{s_0}$-exactly detectable if (4.42) holds for each $k_0^{s_0}$-unobservable initial state ξ.

DEFINITION 4.7   *System (4.1) (or $(F_k, G_k | H_k)$) is said to be $\mathcal{K}^\infty$-exactly detectable if it is k^∞-exactly detectable for any $k \geq 0$. If there exists a nonnegative integer sequence $\{s_k\}_{k \geq 0}$ with the upper limit $\overline{\lim}_{k \to \infty} s_k = +\infty$ such that system (4.1) is k^{s_k}-exactly detectable, i.e., for any k^{s_k}-unobservable initial state ξ_k,*

$$\mathcal{E}\|x_t\|^2 \leq \beta \mathcal{E}\|\xi_k\|^2 \lambda^{(t-k)}, \ \beta \geq 1, \ 0 < \lambda < 1, \ t \geq k,$$

then system (4.1) is said to be weakly finite time or $\mathcal{K}^{WFT}$-exactly detectable. If $\overline{\lim}_{k \to \infty} s_k < +\infty$, then system (4.1) is said to be finite time or $\mathcal{K}^{FT}$-exactly detectable.

A special case of $\mathcal{K}^{FT}$-exact detectability is the so-called $\mathcal{K}^N$-exact detectability, which will be used to study GLEs.

DEFINITION 4.8   *If there exists an integer $N \geq 0$ such that for any time $k_0 \in [0, \infty)$, system (4.1) (or $(F_k, G_k | H_k)$) is k_0^N-exactly detectable, then system (4.1) (or $(F_k, G_k | H_k)$) is said to be $\mathcal{K}^N$-exactly detectable.*

From Definitions 4.7–4.8, we have the following inclusion relation

$$\mathcal{K}^N\text{-exact detectability} \implies \mathcal{K}^{FT}\text{-exact detectability}$$
$$\implies \mathcal{K}^{WFT}\text{-exact detectability} \implies \mathcal{K}^\infty\text{-exact detectability}.$$

Obviously, $\mathcal{K}^N$-exact detectability implies $\mathcal{K}^\infty$-exact detectability, but the converse is not true. We present the following examples to illustrate various relations among several definitions on detectability. For illustration simplicity, we consider deterministic systems only.

Example 4.5
In system (4.11), we take $F_k = 1$ for $k \geq 0$, and

$$H_k = \begin{cases} 1, & \text{for } k = n^2, \ n = 1, 2, \cdots, \\ 0, & \text{otherwise.} \end{cases}$$

In this case, system (4.11) (or $(F_k | H_k)$) is $\mathcal{K}^\infty$-exactly detectable, and the zero is the unique k^∞-unobservable state. $(F_k | H_k)$ is also $\mathcal{K}^{WFT}$-exactly detectable, where

$s_k = k^2 - k \to \infty$. However, $(F_k|H_k)$ is not $\mathcal{K}^{FT}$-exactly detectable, and, accordingly, is not $\mathcal{K}^N$-exactly detectable for any $N \geq 0$. ☐

Example 4.6
In system (4.11), if we take $F_k = 1$ and $H_k = \frac{1}{k}$ for $k \geq 0$, then $(F_k|H_k)$ is $\mathcal{K}^N$-exactly detectable for any $N \geq 0$, but $(F_k|H_k)$ is not uniformly detectable. This is because for any $t \geq 0$, $0 \leq d < 1$ and $\xi \in \mathcal{R}^n$, we always have $\|\phi_{k+t,k}\xi\|^2 = \|\xi\|^2 \geq d^2\|\xi\|^2$. But there do not exist $b > 0$ and $s \geq 0$ satisfying (4.10), because $\xi^T \mathcal{O}_{k+s,k}\xi = \|\xi\|^2 \sum_{i=k}^{k+s} \frac{1}{i^2}$ while $\lim_{k \to \infty} \sum_{i=k}^{k+s} \frac{1}{i^2} = 0$. ☐

Example 4.7
In system (4.11), if we take $F_k = 1$ for $k \geq 0$, and $H_{2n} = 1$ and $H_{2n+1} = 0$ for $n \in \mathcal{N}$, then $(F_k|H_k)$ is uniformly detectable and $\mathcal{K}^1$-exactly detectable, but it is not $\mathcal{K}^0$-exactly detectable. ☐

REMARK 4.3 Examples 4.6–4.7 show that there is no inclusion relation between uniform detectability and $\mathcal{K}^N$-exact detectability for some $N > 0$. However, they are consistent for time-invariant systems. ∎

The following lemma is obvious.

LEMMA 4.4
At any time k_0, $x_{k_0} = 0$ is not only a k_0^∞-but also a $k_0^{s_0}$-unobservable state for any $s_0 \geq 0$.

By Lemma 4.4, if we let $\Theta_{k_0}^\infty$ denote the set of all the k_0^∞-unobservable states of system (4.1) at time k_0, then $\Theta_{k_0}^\infty$ is not empty. Furthermore, it is easy to show that $\Theta_{k_0}^\infty$ is a linear vector space.

LEMMA 4.5
For $k_0 \in \mathcal{N}$, if there does not exist a nonzero $\zeta \in \mathcal{R}^n$ such that $H_{k_0}\zeta = 0$, $(I_{2^{l-k_0}} \otimes H_l)\phi_{l,k_0}\zeta = 0$, $l \in \mathcal{N}_{k_0+1} := \{k_0 + 1, k_0 + 2, \cdots, \}$, then $y_k \equiv 0$ a.s. with $k \geq k_0$ implies $x_{k_0} = 0$ a.s..

Proof. From $y_{k_0} \equiv 0$ a.s., it follows that

$$\mathcal{E}(x_{k_0}^T H_{k_0}^T H_{k_0} x_{k_0}) = 0. \tag{4.43}$$

From $y_l \equiv 0$ a.s., $l = k_0 + 1, \cdots$, it follows from Lemma 4.2 that

$$\mathcal{E}\left[x_{k_0}^T \phi_{l,k_0}^T (I_{2^{l-k_0}} \otimes H_l^T)(I_{2^{l-k_0}} \otimes H_l)\phi_{l,k_0} x_{k_0}\right] = 0. \tag{4.44}$$

Let $R_{k_0} = \mathcal{E}(x_{k_0} x_{k_0}^T)$, $\text{rank}(R_{k_0}) = r$. When $r = 0$, this implies $x_{k_0} = 0$ a.s., and this lemma is proved. For $1 \leq r \leq n$, by the result of [157], there are real nonzero vectors $z_1, z_2, \cdots, z_r$ such that $R_{k_0} = \sum_{i=1}^{r} z_i z_i^T$. By (4.43), we have

$$
\begin{aligned}
\mathcal{E}(x_{k_0}^T H_{k_0}^T H_{k_0} x_{k_0}) &= \text{trace}\mathcal{E}(H_{k_0}^T H_{k_0} x_{k_0} x_{k_0}^T) \\
&= \text{trace}\{H_{k_0}^T H_{k_0} \mathcal{E}(x_{k_0} x_{k_0}^T)\} \\
&= \text{trace}\{H_{k_0}^T H_{k_0} \sum_{i=1}^{r} z_i z_i^T\} \\
&= \sum_{i=1}^{r} (z_i^T H_{k_0}^T H_{k_0} z_i) = 0,
\end{aligned}
$$

which gives $H_{k_0} z_i = 0$ for $i = 1, 2, \cdots, r$. Similarly, (4.44) yields

$$
(I_{2^{l-k_0}} \otimes H_l)\phi_{l,k_0} z_i = 0, \ i = 1, 2, \cdots, r.
$$

According to the given assumptions, we must have $z_i = 0$, $i = 1, 2, \cdots, r$, which again implies $x_{k_0} = 0$ a.s.. $\square$

By Lemma 4.5, it is known that under the conditions of Lemma 4.5, $x_{k_0} = 0$ is the unique k_0^∞-unobservable state, i.e., $\Theta_{k_0}^\infty = \{0\}$.

LEMMA 4.6

Uniform detectability implies $\mathcal{K}^\infty$-exact detectability.

Proof. For any k_0^∞-unobservable state $x_{k_0} = \xi$, by Definition 4.2 and Definition 4.7, we must have $\mathcal{E}\|\phi_{k+t,k} x_k\|^2 < d^2 \mathcal{E}\|x_k\|^2$ or $x_k \equiv 0$ for $k \geq k_0$; otherwise, it will lead to a contradiction since

$$
0 = \sum_{i=k}^{k+s} \mathcal{E}\|y_i\|^2 \geq b\mathcal{E}\|x_k\|^2 > 0.
$$

In any case, the following system

$$
\begin{cases}
x_{k+1} = F_k x_k + G_k x_k w_k, \\
x_{k_0} = \xi \in \Theta_{k_0}^\infty, \\
y_k = H_k x_k, \ k \in \mathcal{N}
\end{cases} \tag{4.45}
$$

is ESMS, so $(F_k, G_k | H_k)$ is $\mathcal{K}^\infty$-exactly detectable. $\square$

Corresponding to Theorem 4.1, we also have the following theorem for exact detectability, whose proof is very simple.

THEOREM 4.4

If $(F_k, G_k | H_k)$ is $\mathcal{K}^\infty$-exactly detectable, so is $(F_k + M_k K_k H_k, G_k + N_k K_k H_k | H_k)$ for any output feedback $u_k = K_k y_k$.

Proof. We prove this theorem by contradiction. Assume that $(F_k + M_k K_k H_k, G_k + N_k K_k H_k | H_k)$ is not $\mathcal{K}^\infty$-exactly detectable. By Definition 4.7, for system (4.17), although the measurement equation becomes $y_k = H_k x_k \equiv 0$ for $k \in \mathcal{N}$, the state equation

$$x_{k+1} = (F_k + M_k K_k H_k)x_k + (G_k + N_k K_k H_k)x_k w_k \tag{4.46}$$

is not ESMS. In view of $y_k = H_k x_k \equiv 0$, (4.46) is equivalent to

$$x_{k+1} = F_k x_k + G_k x_k w_k. \tag{4.47}$$

Hence, under the condition of $y_k = H_k x_k \equiv 0$ for $k \in \mathcal{N}$, if (4.46) is not ESMS, so is (4.47), which contradicts the $\mathcal{K}^\infty$-exact detectability of $(F_k, G_k|H_k)$. $\square$

It should be pointed out that Theorem 4.4 does not hold for $\mathcal{K}^N$-exact detectability. That is, even if $(F_k, G_k|H_k)$ is $\mathcal{K}^N$-exactly detectable for $N \geq 0$, $(F_k + M_k K_k H_k, G_k + N_k K_k H_k|H_k)$ may not be so, and such a counterexample can be easily constructed.

PROPOSITION 4.5

If there exists a matrix sequence $\{K_k\}_{k\in\mathcal{N}}$ such that

$$x_{k+1} = (F_k + K_k H_k)x_k + G_k x_k w_k \tag{4.48}$$

is ESMS, then $(F_k, G_k|H_k)$ is $\mathcal{K}^\infty$-exactly detectable.

Proof. Because (4.48) is ESMS, by Proposition 4.1 and Lemma 4.6, $(F_k + K_k H_k, G_k |H_k)$ is $\mathcal{K}^\infty$-exactly detectable. By Theorem 4.4, for any matrix sequence $\{L_k\}_{k\in\mathcal{N}}$, $(F_k + K_k H_k + L_k H_k, G_k|H_k)$ is also $\mathcal{K}^\infty$-exactly detectable. Taking $L_k = -K_k$, we obtain that $(F_k, G_k|H_k)$ is $\mathcal{K}^\infty$-exactly detectable. Thus, this proposition is proved. $\square$

REMARK 4.4   In some previous references such as [52, 170], if system (4.48) is ESMS for some matrix sequence $\{K_k\}_{k\in\mathcal{N}}$, then $(F_k, G_k|H_k)$ is called stochastically detectable or detectable in conditional mean [170]. In view of Proposition 4.1 and Theorem 4.1, stochastic detectability implies uniform detectability, but the converse is not true. Proposition 4.5 tells us that stochastic detectability implies $\mathcal{K}^\infty$-exact detectability, but the converse is not true. Such a counterexample can be easily constructed; see Example 4.8 below. The $\mathcal{K}^\infty$-exact detectability implies that any k_0^∞-unobservable initial state ξ leads to an exponentially stable trajectory for any $k_0 \geq 0$. However, in the time-invariant system (4.41), the stochastic detectability of (4.41) (or $(F, G|H)$ for short) is equivalent to there being a constant output feedback gain matrix K, rather than necessarily a time-varying feedback gain matrix sequence $\{K_k\}_{k\in\mathcal{N}_0}$, such that

$$x_{k+1} = (F + KH)x_k + Gx_k w_k \tag{4.49}$$

is ESMS; see [52].   ∎

Example 4.8

Let $G_k = 3$ for $k \geq 0$, and

$$F_k = H_k = \begin{cases} 1, & \text{for } k = 3n, \ n \in \mathbb{Z}_{1+}, \\ 0, & \text{otherwise.} \end{cases}$$

By Lemma 4.1, it can be shown that, for any output feedback $u_k = K_k y_k$, the closed-loop state trajectory of

$$x_{k+1} = (F_k + K_k H_k)x_k + 3x_k w_k,$$

is not ESMS. So $(F_k, G_k | H_k)$ is not stochastically detectable. However, $(F_k, G_k | H_k)$ is not only $\mathcal{K}^\infty$- but also $\mathcal{K}^3$-exactly detectable, and 0 is the unique k^3-unobservable state.   □

REMARK 4.5   According to the linear system theory, for the deterministic linear time-invariant system

$$x_{k+1} = Fx_k, \quad x_0 \in \mathcal{R}^n, \quad y_k = Hx_k, \quad k \in \mathcal{N}, \qquad (4.50)$$

the $\mathcal{K}^\infty$- and $\mathcal{K}^{n-1}$-exact detectability are equivalent. By the $\mathcal{H}$-representation theory [212], for (4.41), the $\mathcal{K}^\infty$- and $\mathcal{K}^{[\frac{n(n+1)}{2}-1]}$-exact detectability are also equivalent. So, in what follows, system (4.41) is simply called exactly detectable.   ∎

REMARK 4.6   In Example 4.5, $(F_k | H_k)$ is stochastically detectable, but it is not $\mathcal{K}^N$-exactly detectable for any $N \geq 0$. In Example 4.8, $(F_k | H_k)$ is not stochastically detectable, but it is $\mathcal{K}^N$-exactly detectable for $N \geq 3$. Hence, it seems that there is no inclusion relation between stochastic detectability and $\mathcal{K}^N$-exact detectability.
∎

4.4   Lyapunov-Type Theorems for Periodic Systems under Exact Detectability

At present, we do not know whether Theorem 4.3 holds under exact detectability, but we are able to prove a similar result to Theorem 4.3 for a periodic system, namely, in (4.1), $F_{k+\tau} = F_k$, $G_{k+\tau} = G_k$, $H_{k+\tau} = H_k$. Periodic systems are a class of very important time-varying systems, which have been studied by many researchers; see [32].

THEOREM 4.5

Assume that system (4.1) is a periodic system with the period $\tau > 0$. If system (4.1) is $\mathcal{K}^N$-exactly detectable for any fixed $N \geq 0$ and $\{P_k > 0\}_{k \geq 0}$ is *a positive definite matrix sequence which solves GLE (4.22), then the periodic system (4.1) is ESMS.*

Proof. By periodicity, $P_k = P_{k+\tau}$. Select an integer $\bar{\kappa} > 0$ satisfying $\bar{\kappa}\tau - 1 \geq N$. For $\kappa \geq \bar{\kappa}$, we introduce the following backward difference equation

$$\begin{cases} -P_0^{\kappa\tau-1}(k) + F_k^T P_0^{\kappa\tau-1}(k+1)F_k + G_k^T P_0^{\kappa\tau-1}(k+1)G_k + H_k^T H_k = 0, \\ P_0^{\kappa\tau-1}(\kappa\tau) = 0, \ k = 0, 1, \cdots, \kappa\tau - 1. \end{cases}$$

$$(4.51)$$

Set $V_k = x_k^T P_k x_k$, then associated with (4.51), we have

$$\begin{aligned} \mathcal{E}V_0 - \mathcal{E}V_{\kappa\tau} &= x_0^T P_0 x_0 - \mathcal{E}(x_{\kappa\tau}^T P_{\kappa\tau} x_{\kappa\tau}) = x_0^T P_0 x_0 - \mathcal{E}(x_{\kappa\tau}^T P_0 x_{\kappa\tau}) \\ &= \sum_{i=0}^{\kappa\tau-1} \mathcal{E}\|y_i\|^2 = x_0^T P_0^{\kappa\tau-1}(0)x_0, \end{aligned}$$

$$(4.52)$$

where the last equality is derived by using the completing squares technique. We assert that $P_0^{\kappa\tau-1}(0) > 0$. Otherwise, there exists a nonzero x_0 satisfying $x_0^T P_0^{\kappa\tau-1}(0)$ $\cdot x_0 = 0$ due to $P_0^{\kappa\tau-1}(0) \geq 0$. Hence, by $\mathcal{K}^N$-exact detectability, (4.52) leads to

$$\begin{aligned} 0 = \sum_{i=0}^{\kappa\tau-1} \mathcal{E}\|y_i\|^2 &\geq \lambda_{\min}(P_0)\|x_0\|^2 - \lambda_{\max}(P_0)\beta\lambda^{\kappa\tau}\|x_0\|^2 \\ &= (\lambda_{\min}(P_0) - \lambda_{\max}(P_0)\beta\lambda^{\kappa\tau})\|x_0\|^2, \end{aligned}$$

$$(4.53)$$

where $\beta \geq 1$ and $0 < \lambda < 1$ are defined in (4.2). If κ is taken sufficiently large such that $\kappa \geq \kappa_0 > 0$ with $\kappa_0 > 0$ being a minimal integer satisfying $\lambda_{\min}(P_0) - \lambda_{\max}(P_0)\beta\lambda^{\kappa_0\tau} > 0$, then (4.53) yields $x_0 = 0$, which renders a contradiction.

If we let $P_{(n-1)\kappa\tau}^{n\kappa\tau-1}((n-1)\kappa\tau + k)$ denote the solution of

$$\begin{cases} -P_{(n-1)\kappa\tau}^{n\kappa\tau-1}((n-1)\kappa\tau + k) + F_{(n-1)\kappa\tau+k}^T P_{(n-1)\kappa\tau}^{n\kappa\tau-1}((n-1)\kappa\tau + k + 1)F_{(n-1)\kappa\tau+k} \\ \quad + G_{(n-1)\kappa\tau+k}^T P_{(n-1)\kappa\tau}^{n\kappa\tau-1}((n-1)\kappa\tau + k + 1)G_{(n-1)\kappa\tau+k} \\ \quad + H_{(n-1)\kappa\tau+k}^T H_{(n-1)\kappa\tau+k} = 0, \\ P_{(n-1)\kappa\tau}^{n\kappa\tau-1}(n\kappa\tau) = 0, \ k = 0, 1, \cdots, \kappa\tau - 1; \ n \in \mathbb{Z}_{1+}, \end{cases}$$

then by periodicity, $P_0^{\kappa\tau-1}(0) = P_{(n-1)\kappa\tau}^{n\kappa\tau-1}((n-1)\kappa\tau) > 0$, and

$$\begin{aligned} \mathcal{E}V_{(n-1)\kappa\tau} - \mathcal{E}V_{n\kappa\tau} &= \sum_{i=(n-1)\kappa\tau}^{n\kappa\tau-1} \mathcal{E}\|y_i\|^2 \\ &= \mathcal{E}[x_{(n-1)\kappa\tau}^T P_{(n-1)\kappa\tau}^{n\kappa\tau-1}((n-1)\kappa\tau)x_{(n-1)\kappa\tau}] \\ &= \mathcal{E}[x_{(n-1)\kappa\tau}^T P_0^{\kappa\tau-1}(0)x_{(n-1)\kappa\tau}] \geq \varrho_0\mathcal{E}\|x_{(n-1)\kappa\tau}\|^2, \end{aligned}$$

where $\varrho_0 = \lambda_{\min}(P_0^{\kappa\tau-1}(0)) > 0$. Generally, for $0 \leq s \leq \kappa\tau - 1$, we define

$P_{(n-1)\kappa\tau+s}^{n\kappa\tau+s-1}((n-1)\kappa\tau+s+k)$ as the solution to

$$
\begin{cases}
-P_{(n-1)\kappa\tau+s}^{n\kappa\tau+s-1}((n-1)\kappa\tau+s+k) \\
+F_{(n-1)\kappa\tau+s+k}^{T} P_{(n-1)\kappa\tau+s}^{n\kappa\tau+s-1}((n-1)\kappa\tau+s+k+1)F_{(n-1)\kappa\tau+s+k} \\
+G_{(n-1)\kappa\tau+s+k}^{T} P_{(n-1)\kappa\tau+s}^{n\kappa\tau+s-1}((n-1)\kappa\tau+s+k+1)G_{(n-1)\kappa\tau+s+k} \\
+H_{(n-1)\kappa\tau+s+k}^{T} H_{(n-1)\kappa\tau+s+k} = 0, \\
P_{(n-1)\kappa\tau+s}^{n\kappa\tau+s-1}(n\kappa\tau+s) = 0, \ k=0,1,\cdots,\kappa\tau-1; \ n \in \mathbb{Z}_{1+}.
\end{cases}
$$

It can be shown that $P_{(n-1)\kappa\tau+s}^{n\kappa\tau+s-1}((n-1)\kappa\tau+s) = P_s^{\kappa\tau+s-1}(s) > 0$ and

$$
\sum_{i=(n-1)\kappa\tau+s}^{n\kappa\tau+s-1} \mathcal{E}\|y_i\|^2 = \mathcal{E}[x_{(n-1)\kappa\tau+s}^{T} P_s^{\kappa\tau+s-1}(s)x_{(n-1)\kappa\tau+s}],
$$

provided that we take $\kappa \geq \max_{0 \leq s \leq \kappa\tau-1} \kappa_s$, where $\kappa_s > 0$ is the minimal integer satisfying $\lambda_{\min}(P_s^{\kappa\tau+s-1}(s)) - \lambda_{\max}(P_s^{\kappa\tau+s-1}(s))\beta\lambda^{\kappa_s\tau} > 0$.
Summarizing the above discussions, for any $k \geq 0$ and

$$
\hat{\kappa} > \max\{\bar{\kappa}, \max_{0 \leq s \leq \kappa\tau-1} \kappa_s\},
$$

we have

$$
\mathcal{E}V_k - \mathcal{E}V_{k+\hat{\kappa}\tau} = \sum_{i=k}^{k+\hat{\kappa}\tau-1} \mathcal{E}\|y_i\|^2 \geq \rho\mathcal{E}\|x_k\|^2,
$$

where $\rho = \min_{0 \leq s \leq \hat{\kappa}\tau-1} \rho_s > 0$ with $\rho_s = \lambda_{\min}[P_s^{\hat{\kappa}\tau+s-1}(s)]$. The rest is similar to the proof of Theorem 4.3 and thus is omitted. $\square$

4.5   Further Remarks on LDTV Systems

It should be noted that for the LDTV system (4.1), Theorem 3.6 does not hold. Specially, we cannot in general use the spectrum of $\mathcal{D}_{F_k,G_k}$ to describe the Schur stability of (4.1) even for the case of $G_k \equiv 0$.

Example 4.9
Let

$$
F_k = \frac{1}{8} \begin{bmatrix} 0 & 9-(-1)^k 7 \\ 9+(-1)^k 7 & 0 \end{bmatrix}, \ k \in \mathcal{N}.
$$

It is easy to compute that the eigenvalues of $\mathcal{D}_{F_k}X := F_k X F_k^T$ are $\sigma(\mathcal{D}_{F_k}) = \{\frac{1}{2}, \frac{1}{2}, -\frac{1}{2}\} \subset \mathcal{D}(0,1)$. However, $x_{k+1} = F_k x_k$ is not stable because the state transition matrix is given by

$$\phi_{k,0} = \begin{cases} \begin{bmatrix} 2^{-2k} & 0 \\ 0 & 2^k \end{bmatrix}, & \text{if } k \text{ is even} \\[2em] \begin{bmatrix} 0 & 2^k \\ 2^{-2k} & 0 \end{bmatrix}, & \text{if } k \text{ is odd.} \end{cases}$$

◻

Example 4.10
We set

$$F_k = \begin{cases} \begin{bmatrix} 1 & 0 \\ 0 & 0 \end{bmatrix}, & \text{if } k \text{ is even} \\[2em] \begin{bmatrix} 0 & 0 \\ 0 & 1 \end{bmatrix}, & \text{if } k \text{ is odd.} \end{cases}$$

It can be computed that $\sigma(\mathcal{D}_{F_k}) = \{1, 0, 0\} \subset \bar{\mathcal{D}}$, but $x_{k+1} = F_k x_k$ is exponentially stable due to the fact that $\phi_{k,0} = 0$ for $k \geq 2$. ◻

Applying the infinite-dimensional operator theory [152], a spectral criterion for stability of system (4.1) is hopeful to be obtained.

4.6 Infinite Horizon Time-Varying H_2/H_∞ Control

For simplicity, we consider the H_2/H_∞ control of the following time-varying system with only (x, v)-dependent noise:

$$\begin{cases} x_{k+1} = A_1^k x_k + B_1^k u_k + C_1^k v_k + (A_2^k x_k + C_2^k v_k) w_k, \\ z_k = \begin{bmatrix} C_k x_k \\ D_k u_k \end{bmatrix}, \quad D_k^T D_k = I, \quad k \in \mathcal{N}, \quad x_0 \in \mathcal{R}^n. \end{cases} \tag{4.54}$$

Similar to Definition 3.11, we can define the H_2/H_∞ control of (4.54). The only difference is that we replace (i) of Definition 3.11 by

(i) $u_\infty^* = \{u_k = u_{\infty,k}^*\}_{k \in \mathcal{N}}$ stabilizes the state equation of system (4.54) internally, i.e.,

$$x_{k+1} = A_1^k x_k + B_1^k u_{\infty,k}^* + A_2^k x_k w_k$$

is ESMS.

To solve the infinite horizon H_2/H_∞ control problem for the time-varying system (4.54), we should first generalize the SBRL (Lemma 3.7) to the time-varying system case.

LEMMA 4.7 (Infinite horizon time-varying SBRL)
For the stochastic system

$$\begin{cases} x_{k+1} = (A_{11}^k x_k + B_{11}^k v_k) + (A_{12}^k x_k + B_{12}^k v_k)w_k, \\ z_1^k = C_{11}^k x_k, \ k \in \mathcal{N}, \end{cases} \tag{4.55}$$

$\| \tilde{\mathcal{L}}_\infty \| < \gamma$ *for some $\gamma > 0$ iff the following constrained backward difference equation*

$$\begin{cases} P_k = L(P_{k+1}) - K(P_{k+1})H(P_{k+1})^{-1}K(P_{k+1})^T, \\ H(P_{k+1}) \geq \epsilon I, \ k \in \mathcal{N} \end{cases} \tag{4.56}$$

has a unique bounded stabilizing solution $P_\infty^k \leq 0, k \in \mathcal{N}$. Here, $\tilde{\mathcal{L}}_\infty$ is defined as in Definition 3.9, i.e.,

$$\tilde{\mathcal{L}}_\infty v_k := z_1^k|_{x_0=0} = C_{11}^k x_k|_{x_0=0}.$$

$L(P_{k+1})$, $K(P_{k+1})$ *and* $H(P_{k+1})$ *are the same as in Lemma 3.3.*

This lemma is a special case of Theorem 5.4 in Chapter 5.

THEOREM 4.6 (Infinite horizon time-varying H_2/H_∞ control)
For system (4.54), assume that $(A_1^k; A_2^k|C_k)$ and $(A_1^k + C_1^k K_1^k; A_2^k + C_2^k K_1^k|C_k)$ are uniformly detectable. Then the infinite horizon H_2/H_∞ control problem of system (4.54) admits a solution $(u_{\infty,k}^*, v_{\infty,k}^*)$ with $u_{\infty,k}^* = K_2^k x_k$, $v_{\infty,k}^* = K_1^k x_k$, if and only if the following four coupled time-varying matrix-valued equations admit a bounded quadruple solution $(P_{1,\infty}^k \leq 0, K_1^k; P_{2,\infty}^k \geq 0, K_2^k)_{k \in \mathcal{N}}$:

$$\begin{cases} P_{1,\infty}^k = (A_1^k + B_1^k K_2^k)^T P_{1,\infty}^{k+1}(A_1^k + B_1^k K_2^k) + (A_2^k)^T P_{1,\infty}^{k+1} A_2^k - C_k^T C_k \\ \quad -(K_2^k)^T K_2^k - K_3^k H_1(P_{1,\infty}^k)^{-1}(K_3^k)^T, \\ H_1(P_{1,\infty}^k) > \epsilon_0 I, \epsilon_0 > 0, \end{cases} \tag{4.57}$$

$$K_1^k = -H_1(P_{1,\infty}^{k+1})^{-1}(K_3^k)^T, \tag{4.58}$$

$$P_{2,\infty}^k = \left(A_1^k + C_1^k K_1^k\right)^T P_{2,\infty}^{k+1}\left(A_1^k + C_1^k K_1^k\right) + (A_2^k + C_2^k K_1^k)^T P_{2,\infty}^{k+1}$$
$$\cdot(A_2^k + C_2^k K_1^k) + C_k^T C_k - K_4^k H_2(P_{2,\infty}^{k+1})^{-1}(K_4^k)^T, \tag{4.59}$$

and

$$K_2^k = -H_2(P_{2,\infty}^{k+1})^{-1}(K_4^k)^T, \tag{4.60}$$

where in (4.57)–(4.60),

$$K_3^k = (A_1^k + B_1^k K_2^k)^T P_{1,\infty}^{k+1} C_1^k + (A_2^k)^T P_{1,\infty}^{k+1} C_2^k,$$
$$K_4^k = (A_1^k + C_1^k K_1^k)^T P_{2,\infty}^{k+1} B_1^k,$$
$$H_1(P_{1,\infty}^{k+1}) = \gamma^2 I + (C_1^k)^T P_{1,\infty}^{k+1} C_1^k + (C_2^k)^T P_{1,\infty}^{k+1} C_2^k,$$
$$H_2(P_{2,\infty}^{k+1}) = I + (B_1^k)^T P_{2,\infty}^{k+1} B_1^k.$$

Proof. This theorem can be proved along the same line of Theorem 5.5 of Chapter 5, where stochastic detectability, a stronger assumption than uniform detectability by Remark 4.4, is made. □

Under some conditions, $(P_{1,\infty}^k \le 0, K_1^k; P_{2,\infty}^k \ge 0, K_2^k)$ can be obtained as the limit of the solution $(P_{1,\mathcal{T}}^k \le 0, K_{1,\mathcal{T}}^k; P_{2,\mathcal{T}}^k \ge 0, K_{2,\mathcal{T}}^k)$ of coupled equations

$$\begin{cases} P_{1,\mathcal{T}}^k = (A_1^k + B_1^k K_{2,\mathcal{T}}^k)^T P_{1,\mathcal{T}}^{k+1}(A_1^k + B_1^k K_{2,\mathcal{T}}^k) + (A_2^k)^T P_{1,\mathcal{T}}^{k+1} A_2^k - C_k^T C_k \\ \quad - (K_{2,\mathcal{T}}^k)^T K_{2,\mathcal{T}}^k - K_{3,\mathcal{T}}^k H_1(P_{1,\mathcal{T}}^{k+1})^{-1}(K_{3,\mathcal{T}}^k)^T, \\ P_{1,\mathcal{T}}^{\mathcal{T}+1} = 0, \\ H_1(P_{1,\mathcal{T}}^{k+1}) > \epsilon_0 I, \epsilon_0 > 0, \end{cases}$$

$$\tag{4.61}$$

$$K_{1,\mathcal{T}}^k = -H_1(P_{1,\mathcal{T}}^{k+1})^{-1}(K_{3,\mathcal{T}}^k)^T, \tag{4.62}$$

$$\begin{cases} P_{2,\mathcal{T}}^k = (A_1^k + C_1^k K_{1,\mathcal{T}}^k)^T P_{2,\mathcal{T}}^{k+1}(A_1^k + C_1^k K_{1,\mathcal{T}}^k) + (A_2^k + C_2^k K_{1,\mathcal{T}}^k)^T P_{2,\mathcal{T}}^{k+1} \\ \quad \cdot (A_2^k + C_2^k K_{1,\mathcal{T}}^k) + C_k^T C_k - K_{4,\mathcal{T}}^k H_2(P_{2,\mathcal{T}}^{k+1})^{-1}(K_{4,\mathcal{T}}^k)^T, \\ P_{2,\mathcal{T}}^{\mathcal{T}+1} = 0, \end{cases}$$

$$\tag{4.63}$$

and

$$K_{2,\mathcal{T}}^k = -H_2(P_{2,\mathcal{T}}^{k+1})^{-1}(K_{4,\mathcal{T}}^k)^T. \tag{4.64}$$

That is,

$$\lim_{\mathcal{T} \to \infty} (P_{1,\mathcal{T}}^k, K_{1,\mathcal{T}}^k; P_{2,\mathcal{T}}^k, K_{2,\mathcal{T}}^k) = (P_{1,\infty}^k, K_1^k; P_{2,\infty}^k, K_2^k)$$

if the limit exists and $H_1(P_{1,\infty}^k) > \epsilon_0 I$ for $k \in \mathcal{N}$ and some sufficiently small $\epsilon_0 > 0$.

REMARK 4.7 Stochastic detectability is not only stronger than uniform detectability, but also does not possess the output feedback invariance. Hence, we are not able to study the H_2/H_∞ control for systems with (x, u, v)- or (x, u)- or (x, v)-dependent noise under stochastic detectability. ∎

REMARK 4.8 If system (4.54) is a periodic system, then a similar result to Theorem 4.6 can be given under exact detectability. ∎

REMARK 4.9   When all the coefficient matrices in (4.1) are random with time-varying first and second moments as assumed in [174, 175], how to generalize various definitions for detectability and observability deserves further study.   ∎

4.7   Notes and References

It can be found that for LDTV stochastic systems, there are many different properties from linear time-invariant systems, especially in detectability and spectral characterization for stability. Up to now, there is almost no similar result to that of [152] on the study of stability, observability and detectability of LDTV stochastic systems based on the spectrum technique of infinite dimensional operators. There exist many issues that require further research.

The materials in this chapter mainly comes from [218].

5

Linear Markovian Jump Systems with Multiplicative Noise

This chapter extends the H_2/H_∞ control theory developed in Chapters 2–3 to linear Markov jump systems. The linear discrete-time H_2/H_∞ control with Markov switching and multiplicative noise is studied for finite and infinite horizon cases. We also consider the H_2/H_∞ control of linear Itô-type differential systems with Markov jump. The relationship between the solvability of H_2/H_∞ control and the existence of a Nash equilibrium strategy is discussed.

5.1   Introduction

As one of the most basic dynamics models, Markov jump linear systems can be used to represent random failure processes in the manufacturing industry [25] and some investment portfolio models [43, 49, 198], and have been researched extensively in the monographs [52, 53, 131]. Stability analysis of Markovian jumping systems was dealt with in [74, 72, 121, 131, 170, 171], and robust H_∞ control can be found in [52, 53] for systems with multiplicative noise. In [44, 138] and [137], the mixed H_2/H_∞ control of linear Markov jump systems with additive noise and nonlinear Markov jump systems with probabilistic sensor failures were studied, respectively. The references [43] and [49] discussed the LQ optimal control problem and its applications to investment portfolio optimization for discrete-time Markov jump linear systems with multiplicative noise. In addition, the observability and detectability of Markov jump systems have been studied in, for example, [41, 50, 51, 91, 161, 172, 217].

   The objective of this chapter is to extend the results of Chapters 2–3 on the H_2/H_∞ control to Markov jump linear systems with multiplicative noise, which have many practical applications. It can be found that when the Markovian jump process is a finite-state homogeneous Markov chain, most results are trivial generalizations of those of Chapters 2–3. Otherwise, the related problems remain unsolved.

5.2 Finite Horizon H_2/H_∞ Control of Discrete-Time Markov Jump Systems

Consider the following discrete-time Markov jump linear system with state-, disturbance- and control-dependent noise:

$$\begin{cases} x_{k+1} = A_1^{k,\theta_k} x_k + B_1^{k,\theta_k} u_k + C_1^{k,\theta_k} v_k + (A_2^{k,\theta_k} x_k + B_2^{k,\theta_k} u_k + C_2^{k,\theta_k} v_k) w_k, \\ z_k = \begin{bmatrix} C_{k,\theta_k} x_k \\ D_{k,\theta_k} u_k \end{bmatrix}, \quad D_{k,\theta_k}^T D_{k,\theta_k} = I_{n_u}, \\ x_0 \in \mathcal{R}^n, \ k \in N_{\mathcal{T}}, \ \theta_0 \in \overline{N} = \{1, 2, \cdots, N\}, \end{cases}$$

$$(5.1)$$

where $x_k \in \mathcal{R}^n$, $u_k \in \mathcal{R}^{n_u}$, $v_k \in \mathcal{R}^{n_v}$ and $z_k \in \mathcal{R}^{n_z}$ represent, as in previous chapters, the state, control input, disturbance signal and controlled output of system (5.1), respectively. θ_k is a nonhomogeneous Markov chain taking values in $\overline{N}$ with the transition probability matrix $\mathcal{P}_{N \times N}^k = (p_{ij}^k)_{N \times N}$, $p_{ij}^k = \mathcal{P}(\theta_{k+1} = j|\theta_k = i)$. A_1^{k,θ_k}, A_2^{k,θ_k}, B_1^{k,θ_k}, B_2^{k,θ_k}, C_1^{k,θ_k}, C_2^{k,θ_k}, C_{k,θ_k} and D_{k,θ_k} are matrix-valued functions of suitable dimensions. w_k, $k \in N_{\mathcal{T}}$, are real random variables independent of each other with $\mathcal{E} w_k = 0$ and $\mathcal{E}(w_k w_s) = \delta_{ks}$. The random variables $\{w_k, k \in N_{\mathcal{T}}\}$ are independent of the Markov chain $\{\theta_k\}_{k \in N_{\mathcal{T}}}$. Denote $\mathcal{F}_k$ as the σ-field generated by $\{(\theta_s, w_j) : 0 \le s \le k+1; 0 \le j \le k\}$, in particular, $\mathcal{F}_0 := \sigma\{\theta_0, \theta_1, w_0\}$, $\mathcal{F}_{-1} := \sigma\{\theta_0\}$. Correspondingly, $\mathcal{L}_{\mathcal{F}_{i-1}}^2(\Omega, \mathcal{R}^k)$ ($i \in N_{\mathcal{T}}$), $l_w^2(N_{\mathcal{T}}, \mathcal{R}^k)$ and $\|\cdot\|_{l_w^2(N_{\mathcal{T}}, \mathcal{R}^k)}$ can be defined as in Chapter 3. We assume $(x, u, v) \in l_w^2(N_{\mathcal{T}}, \mathcal{R}^n) \times l_w^2(N_{\mathcal{T}}, \mathcal{R}^{n_u}) \times l_w^2(N_{\mathcal{T}}, \mathcal{R}^{n_v})$. The finite horizon stochastic H_2/H_∞ control of system (5.1) is stated as follows:

DEFINITION 5.1

Given $\mathcal{T} \in \mathcal{N}$ and a disturbance attenuation level $\gamma > 0$, the mixed H_2/H_∞ control problem is concerned with the design of a state feedback controller $u = u_{\mathcal{T}}^* \in l_w^2(N_{\mathcal{T}}, \mathcal{R}^{n_u})$ such that the following hold:

(i) When $u = u_{\mathcal{T}}^*$ is performed in (5.1), $\|\mathcal{L}_{\mathcal{T}}\| < \gamma$, where

$$\|\mathcal{L}_{\mathcal{T}}\| := \sup_{v \in l_w^2(N_{\mathcal{T}}, \mathcal{R}^{n_v}), v \neq 0, \theta_0 \in \overline{N}, u = u_{\mathcal{T}}^*, x_0 = 0} \frac{\|z\|_{l_w^2(N_{\mathcal{T}}, \mathcal{R}^{n_z})}}{\|v\|_{l_w^2(N_{\mathcal{T}}, \mathcal{R}^{n_v})}}.$$

(ii) When the worst-case disturbance $v_{\mathcal{T}}^*(\cdot) \in l_w^2(N_{\mathcal{T}}, \mathcal{R}^{n_v})$ is imposed on (5.1), $u_{\mathcal{T}}^*$ solves

$$\min_{u \in l_w^2(N_{\mathcal{T}}, \mathcal{R}^{n_u})} \{J_{2,\mathcal{T}}(u, v_{\mathcal{T}}^*) = \sum_{k=0}^{\mathcal{T}} \mathcal{E}\|z_k\|^2\}$$

for any $x_0 \in \mathcal{R}^n$ and $\theta_0 \in \overline{N}$. If the above $(u_{\mathcal{T}}^*, v_{\mathcal{T}}^*)$ exist, the mixed H_2/H_∞ control problem is called solvable.

5.2.1 An SBRL

Consider the following Markov jump linear system with multiplicative noise:

$$\begin{cases} x_{k+1} = A_{11}^{k,\theta_k} x_k + B_{11}^{k,\theta_k} v_k + (A_{12}^{k,\theta_k} x_k + B_{12}^{k,\theta_k} v_k) w_k, \\ z_1^k = C_{11}^{k,\theta_k} x_k, \\ x_0 \in \mathcal{R}^n, \ k \in N_T, \theta_0 \in \overline{N}. \end{cases} \quad (5.2)$$

For system (5.2), the perturbed operator $\tilde{\mathcal{L}}_T$ can be defined as follows:

DEFINITION 5.2
The perturbed operator of system (5.2) is defined by

$$\tilde{\mathcal{L}}_T : l_w^2(N_T, \mathcal{R}^{n_v}) \mapsto l_w^2(N_T, \mathcal{R}^{n_{z_1}}),$$

$$\tilde{\mathcal{L}}_T(v_k) = z_1^k|_{x_0=0,\theta_0\in\overline{N}} = C_{11}^{k,\theta_k} x_k|_{x_0=0,\theta_0\in\overline{N}}, \ \forall v_k \in l_w^2(N_T, \mathcal{R}^{n_v})$$

with its norm given by

$$\|\tilde{\mathcal{L}}_T\| := \sup_{v\in l_w^2(N_T,\mathcal{R}^{n_v}),v\neq 0,x_0=0,\theta_0\in\overline{N}} \frac{\|z_1\|_{l_w^2(N_T,\mathcal{R}^{n_{z_1}})}}{\|v\|_{l_w^2(N_T,\mathcal{R}^{n_v})}}.$$

For the simplicity of presentation, we introduce some notations that will be used later. Let $\mathbb{P} = \{P^{k,\theta_k} \in \mathcal{S}_n : k \in N_{T+1}, \theta_k \in \overline{N}\}$, $P^k = [P^{k,1} \ P^{k,2} \ \cdots \ P^{k,N}]$ for $P^{k,i} \in \mathcal{S}_n, k \in N_T, i \in \overline{N}$. In particular, we write $P^k \geq 0(> 0)$ if $P^{k,i} \geq 0(> 0)$. For $i \in \overline{N}$, set

$$\Psi_i(P^{k+1}) = \sum_{j=1}^N p_{ij}^k P^{k+1,j},$$

$$R_i(P^{k+1}) := (A_{11}^{k,i})^T \Psi_i(P^{k+1}) A_{11}^{k,i} + (A_{12}^{k,i})^T \Psi_i(P^{k+1}) A_{12}^{k,i} - P^{k,i},$$

$$K_i(P^{k+1}) = (A_{11}^{k,i})^T \Psi_i(P^{k+1}) B_{11}^{k,i} + (A_{12}^{k,i})^T \Psi_i(P^{k+1}) B_{12}^{k,i},$$

$$T_i(P^{k+1}) = (B_{11}^{k,i})^T \Psi_i(P^{k+1}) B_{11}^{k,i} + (B_{12}^{k,i})^T \Psi_i(P^{k+1}) B_{12}^{k,i},$$

$$L_i(P^{k+1}) = (A_{11}^{k,i})^T \Psi_i(P^{k+1}) A_{11}^{k,i} + (A_{12}^{k,i})^T \Psi_i(P^{k+1}) A_{12}^{k,i} - (C_{11}^{k,i})^T C_{11}^{k,i},$$

$$H_i(P^{k+1}) = \gamma^2 I + T_i(P^{k+1}). \quad (5.3)$$

In addition, associated with the following system

$$\begin{cases} x_{k+1} = A_{11}^{k,\theta_k} x_k + B_{11}^{k,\theta_k} v_k + (A_{12}^{k,\theta_k} x_k + B_{12}^{k,\theta_k} v_k) w_k, \\ z_1^k = C_{11}^{k,\theta_k} x_k, \\ x_{k_0} \in \mathcal{R}^n, \ \theta_{k_0} \in \overline{N}, k \in \{k_0, k_0+1, \cdots, T\}, \end{cases}$$

we denote

$$J_1^T(x, v; x_{k_0}, \theta_{k_0}, k_0) := \sum_{k=k_0}^T \mathcal{E}(\gamma^2 \|v_k\|^2 - \|z_1^k\|^2).$$

LEMMA 5.1
Given $v \in l_w^2(N_T, \mathcal{R}^{n_v})$, $x_0 \in \mathcal{R}^n$ and $\theta_0 \in \overline{N}$, $P = \{P^{k,\theta_k} \in S_n : k \in N_{T+1}, \theta_k \in \overline{N}\}$. Then, for any fixed $T \in \mathcal{N}$,

$$J_1^T(x, v; x_0, \theta_0, 0) = \sum_{k=0}^{T} \mathcal{E}(\gamma^2 \|v_k\|^2 - \|z_1^k\|^2)$$

$$= \sum_{k=0}^{T} \mathcal{E} \begin{bmatrix} x_k \\ v_k \end{bmatrix}^T M_{\theta_k}(P^k) \begin{bmatrix} x_k \\ v_k \end{bmatrix} + \mathcal{E}(x_0^T P^{0,\theta_0} x_0)$$

$$- \mathcal{E}(x_{T+1}^T P^{T+1,\theta_{T+1}} x_{T+1}),$$

where $M_{\theta_k}(P^k) = M_i(P^k)$ for $\theta_k = i$ and

$$M_i(P^k) = \begin{bmatrix} L_i(P^{k+1}) - P^{k,i} & K_i(P^{k+1}) \\ K_i(P^{k+1})^T & H_i(P^{k+1}) \end{bmatrix}. \qquad (5.4)$$

Proof. By the independence assumption on $\{w_k\}$ and $\{\theta_k\}$, we have

$$\mathcal{E}\{(A_{11}^{k,\theta_k} x_k + B_{11}^{k,\theta_k} v_k)^T P^{k+1,\theta_{k+1}}(A_{12}^{k,\theta_k} x_k + B_{12}^{k,\theta_k} v_k) w_k | \mathcal{F}_{k-1}\}$$
$$= (A_{11}^{k,\theta_k} x_k + B_{11}^{k,\theta_k} v_k)^T \mathcal{E} P^{k+1,\theta_{k+1}}(A_{12}^{k,\theta_k} x_k + B_{12}^{k,\theta_k} v_k) \mathcal{E} w_k$$
$$= 0 \quad a.s..$$

For $\theta_k = i$, it gives

$$\mathcal{E}\left(x_{k+1}^T P^{k+1,\theta_{k+1}} x_{k+1} - x_k^T P^{k,\theta_k} x_k \mid \mathcal{F}_{k-1}, \theta_k = i\right)$$

$$= \begin{bmatrix} x_k \\ v_k \end{bmatrix}^T \begin{bmatrix} R_i(P^{k+1}) & K_i(P^{k+1}) \\ K_i(P^{k+1})^T & T_i(P^{k+1}) \end{bmatrix} \begin{bmatrix} x_k \\ v_k \end{bmatrix} \quad a.s..$$

Hence,

$$\mathcal{E}\left(x_{k+1}^T P^{k+1,\theta_{k+1}} x_{k+1} - x_k^T P^{k,\theta_k} x_k \mid \mathcal{F}_{k-1}, \theta_k\right)$$

$$= \begin{bmatrix} x_k \\ v_k \end{bmatrix}^T \begin{bmatrix} R_{\theta_k}(P^{k+1}) & K_{\theta_k}(P^{k+1}) \\ K_{\theta_k}(P^{k+1})^T & T_{\theta_k}(P^{k+1}) \end{bmatrix} \begin{bmatrix} x_k \\ v_k \end{bmatrix}, \quad a.s.. \qquad (5.5)$$

By taking the mathematical expectation and summation from $k = 0$ to T on both sides of (5.5), it follows that

$$\mathcal{E}(x_{T+1}^T P^{T+1,\theta_{T+1}} x_{T+1} - x_0^T P^{0,\theta_0} x_0)$$

$$= \sum_{k=0}^{T} \left\{ \mathcal{E} \begin{bmatrix} x_k \\ v_k \end{bmatrix}^T \begin{bmatrix} R_{\theta_k}(P^{k+1}) & K_{\theta_k}(P^{k+1}) \\ K_{\theta_k}(P^{k+1})^T & T_{\theta_k}(P^{k+1}) \end{bmatrix} \begin{bmatrix} x_k \\ v_k \end{bmatrix} \right\},$$

which leads to our desired result

$$J_1^T(x, v; x_0, \theta_0, 0) = \sum_{k=0}^{T} \mathcal{E}[\gamma^2 v_k^T v_k - x_k^T (C_{11}^{k,\theta_k})^T C_{11}^{k,\theta_k} x_k]$$

$$+\sum_{k=0}^{\mathcal{T}} \mathcal{E} \begin{bmatrix} x_k \\ v_k \end{bmatrix}^T \begin{bmatrix} R_{\theta_k}(P^{k+1}) & K_{\theta_k}(P^{k+1}) \\ K_{\theta_k}(P^{k+1})^T & T_{\theta_k}(P^{k+1}) \end{bmatrix} \begin{bmatrix} x_k \\ v_k \end{bmatrix}$$

$$+\mathcal{E}(x_0^T P^{0,\theta_0} x_0 - x_{\mathcal{T}+1}^T P^{\mathcal{T}+1,\theta_{\mathcal{T}+1}} x_{\mathcal{T}+1})$$

$$=\sum_{k=0}^{\mathcal{T}} \mathcal{E} \begin{bmatrix} x_k \\ v_k \end{bmatrix}^T M_{\theta_k}(P^k) \begin{bmatrix} x_k \\ v_k \end{bmatrix} + \mathcal{E}(x_0^T P^{0,\theta_0} x_0)$$

$$-\mathcal{E}(x_{\mathcal{T}+1}^T P^{\mathcal{T}+1,\theta_{\mathcal{T}+1}} x_{\mathcal{T}+1}).$$

□

The following is an SBRL associated with system (5.2).

THEOREM 5.1
For system (5.2) and a given disturbance attenuation level $\gamma > 0$, the following statements are equivalent:
(i) $\|\tilde{\mathcal{L}}_{\mathcal{T}}\| < \gamma$.
(ii) For any $(i,k) \in \overline{N} \times N_{\mathcal{T}}$, the following recursion is solvable.

$$\begin{cases} P^{k,i} = L_i(P^{k+1}) - K_i(P^{k+1})H_i(P^{k+1})^{-1}K_i(P^{k+1})^T, \\ P^{k,i} \le 0, \ P^{\mathcal{T}+1,i} = 0, \\ H_i(P^{k+1}) > 0, \end{cases} \tag{5.6}$$

where $L_i(P^{k+1})$, $K_i(P^{k+1})$ and $H_i(P^{k+1})$ are as in (5.3).

Proof. (ii) $\Rightarrow$ (i). Suppose that (ii) holds for $k = \mathcal{T}, \cdots, 0$. Fix $\theta_k = i \in \overline{N}$, then for any non-zero $v_k \in l_w^2(N_{\mathcal{T}}, \mathcal{R}^{n_v})$, $\theta_0 \in \overline{N}$ and $x_0 = 0$, it yields from Lemma 5.1 that

$$J_1^{\mathcal{T}}(x, v; 0, \theta_0, 0) = \sum_{k=0}^{\mathcal{T}} \mathcal{E} \left\{ \begin{bmatrix} x_k \\ v_k \end{bmatrix}^T M_{\theta_k}(P^k) \begin{bmatrix} x_k \\ v_k \end{bmatrix} \right\}$$

$$= \sum_{k=0}^{\mathcal{T}} \sum_{i=0}^{N} \mathcal{E} \left\{ \begin{bmatrix} x_k \\ v_k \end{bmatrix}^T M_i(P^k) \begin{bmatrix} x_k \\ v_k \end{bmatrix} I_{\{\theta_k = i\}} \right\}.$$

By means of the first equality in (5.6) and the completing squares technique, it concludes that

$$J_1^{\mathcal{T}}(x, v; 0, \theta_0, 0)$$

$$= \sum_{k=0}^{\mathcal{T}} \sum_{i=0}^{N} \mathcal{E} \left[(v_k - v_k^*)^T H_i(P^{k+1})(v_k - v_k^*) I_{\{\theta_k = i\}} \right]$$

$$\ge 0 \tag{5.7}$$

due to $H_i(P^{k+1}) > 0$, where $v_k^* = -H_i(P^{k+1})^{-1}K_i(P^{k+1})^T x_k$. Equation (5.7) implies $\|\tilde{\mathcal{L}}_{\mathcal{T}}\| \le \gamma$. Taking a similar procedure as in Lemma 3.3, we can further show $\|\tilde{\mathcal{L}}_{\mathcal{T}}\| < \gamma$, so (i) is derived.

(i)$\Rightarrow$(ii). Note that $H_i(P^{\mathcal{T}+1}) = \gamma^2 I > 0$ for any $i \in \overline{N}$, we can solve $P^{\mathcal{T},i}$ from the recursion (5.6), which is given by

$$P^{\mathcal{T},i} = L_i(P^{\mathcal{T}+1}) - \gamma^{-2}K_i(P^{\mathcal{T}+1})K_i(P^{\mathcal{T}+1})^T, \quad i \in \overline{N}.$$

The recursion (5.6) can be solved backward iff $H_i(P^{k+1}) > 0$ for all $k \in N_{\mathcal{T}}$ and $i \in \overline{N}$.

If (5.6) fails for some $k = \mathcal{T}_0 \in N_{\mathcal{T}-1}$, then there must exist at least one $j \in \overline{N}$ such that $H_j(P^{\mathcal{T}_0+1})$ has at least one zero or negative eigenvalue. Below, we show this is impossible by induction. Introduce the following difference equation:

$$\begin{cases} F_i(P^{k+1}) = -H_i(P^{k+1})^{-1}K_i(P^{k+1})^T, \quad \mathcal{T}_0+1 \le k \le \mathcal{T}, \\ F_i(P^{k+1}) = 0, \ 0 \le k \le \mathcal{T}_0, \ i \in \overline{N}. \end{cases}$$

Consider the following backward matrix recursion

$$\begin{cases} \overline{P}^{k,i} = L_i(\overline{P}^{k+1}) + K_i(\overline{P}^{k+1})F_i(\overline{P}^{k+1}) + F_i(\overline{P}^{k+1})^T K_i(\overline{P}^{k+1})^T \\ \quad + F_i(\overline{P}^{k+1})^T H_i(\overline{P}^{k+1})F_i(\overline{P}^{k+1}), \\ \overline{P}^{k,i} \le 0, \ \overline{P}(\mathcal{T}+1) = 0, \\ H_i(\overline{P}^{k+1}) > 0, \ i \in \overline{N}, \end{cases} \tag{5.8}$$

which admits solutions $\overline{P}^{k,i}$ on $(i,k) \in \overline{N} \times N_{\mathcal{T}}$. Moreover, $\overline{P}^{k,i} = P^{k,i}$ for $k = \mathcal{T}_0 + 1, \cdots, \mathcal{T}$. Now we will show that $H_j(\overline{P}^{\mathcal{T}_0+1}) > 0$ for all $j \in \overline{N}$. Otherwise, there must exist $v_0 \in \mathcal{R}^{n_v}$ with $\|v_0\| = 1$ such that $\mathcal{E}[v_0^T H_j(\overline{P}^{\mathcal{T}_0+1})v_0] \le 0$ (i.e., v_0 is a unit eigenvector corresponding to an eigenvalue $\lambda \le 0$ of $H_j(\overline{P}^{\mathcal{T}_0+1})$). Let $v_k = v_0$ for $k = \mathcal{T}_0$ and $v_k = 0$ for $k \ne \mathcal{T}_0$. Denote $\overline{v}_k = v_k + F_{\theta_k}(\overline{P}^{k+1})\bar{x}_k$, where $\bar{x}_k$ is the state of system (5.2) corresponding to $\overline{v}_k$, that is,

$$\begin{cases} \bar{x}_{k+1} = A_{11}^{k,\theta_k}\bar{x}_k + B_{11}^{k,\theta_k}\overline{v}_k + (A_{12}^{k,\theta_k}\bar{x}_k + B_{12}^{k,\theta_k}\overline{v}_k)w_k, \\ \theta_0 \in \overline{N}, \ \bar{x}_0 = x_0 \in \mathcal{R}^n, \ k \in N_{\mathcal{T}}. \end{cases} \tag{5.9}$$

Combining Lemma 5.1 with the definition of $\overline{v}_k$, it follows that

$$J_1^{\mathcal{T}}(\bar{x},\overline{v};0,\theta_0,0) = \sum_{k=0}^{\mathcal{T}} \mathcal{E}\left\{ \begin{bmatrix} \bar{x}_k \\ \overline{v}_k \end{bmatrix}^T M_{\theta_k}(\overline{P}^{k+1}) \begin{bmatrix} \bar{x}_k \\ \overline{v}_k \end{bmatrix} \right\}$$

$$= \sum_{k=0}^{\mathcal{T}} \sum_{i=0}^{N} \mathcal{E}\left\{ \bar{x}_k^T \begin{bmatrix} I \\ F_i(\overline{P}^{k+1}) \end{bmatrix}^T M_i(\overline{P}^{k+1}) \begin{bmatrix} I \\ F_i(\overline{P}^{k+1}) \end{bmatrix} \bar{x}_k I_{\{\theta_k=i\}} \right.$$

$$+ v_k^T N_i(\overline{P}^{k+1})\bar{x}_k I_{\{\theta_k=i\}} + \bar{x}_k^T N_i(\overline{P}^{k+1})^T v_k I_{\{\theta_k=i\}}$$

$$\left. + v_k^T H_i(\overline{P}^{k+1})v_k I_{\{\theta_k=i\}} \right\}, \tag{5.10}$$

where we adopt $N_i(\overline{P}^{k+1}) = K_i(\overline{P}^{k+1})^T + H_i(\overline{P}^{k+1})F_i(\overline{P}^{k+1})$, $i \in \overline{N}$. The first term in (5.10) becomes zero due to (5.8). For the other terms in (5.10), keep the definition of v_k in mind and note the linearity of (5.9), which results in $\bar{x}_k = 0$ for $k \leq \mathcal{T}_0$, while $v_k = 0$ for $k > \mathcal{T}_0$. So (5.10) reduces to

$$J_1^{\mathcal{T}}(\bar{x}, \overline{v}; 0, \theta_0, 0) = \mathcal{E}[v_0^T H_{\theta_{\mathcal{T}_0}}(\overline{P}^{\mathcal{T}_0+1})v_0]. \tag{5.11}$$

Recall the preceding definition of v_0, in the case of $\theta_{\mathcal{T}_0} = j$, (5.11) immediately implies $J_1^{\mathcal{T}}(\bar{x}, \overline{v}; 0, \theta_0, 0) = \lambda \leq 0$, which contradicts the condition (i). Thus, there must be $H_j(\overline{P}^{\mathcal{T}_0+1}) > 0$ for all $j \in \overline{N}$. Note that $\overline{P}^{\mathcal{T}_0+1} = P^{\mathcal{T}_0+1}$; we conclude that $H_j(P^{\mathcal{T}_0+1}) > 0$ for all $j \in \overline{N}$ and $\mathcal{T}_0 \in N_{\mathcal{T}}$. So the recursive procedure can proceed for $k = \mathcal{T}_0, \cdots, 0$ and (5.6) admits solutions $\{P^{k,i} : k \in N_{\mathcal{T}}, i \in \overline{N}\}$.

Now we examine the non-positivity of $P^{i,k}$ for any $(i, k) \in \overline{N} \times N_{\mathcal{T}}$. Given any $k \in N_{\mathcal{T}}, \theta_k = i \in \overline{N}$ and $x_k = \hat{x} \in \mathcal{R}^n$, from Lemma 5.1 it follows that

$$J_1^{\mathcal{T}}(x, v; \hat{x}, i, k) = \sum_{s=k}^{\mathcal{T}} \mathcal{E}(\gamma^2 \|v_s\|^2 - \|z_1^s\|^2)$$

$$= \sum_{s=k}^{\mathcal{T}} \mathcal{E}\{(v_s - v_s^*)^T H_{\theta_s}(P^{s+1})(v_s - v_s^*)\} + \hat{x}^T P^{k,i}\hat{x}.$$

Since $H_{\theta_s}(P^{s+1}) > 0$, we have

$$\min_{v \in l_w^2(N_{\mathcal{T}}, \mathcal{R}^l)} J_1^{\mathcal{T}}(x, v; \hat{x}, i, k) = J_1^{\mathcal{T}}(x, v^*; \hat{x}, i, k) = \hat{x}^T P^{k,i}\hat{x}$$

$$\leq J_1^{\mathcal{T}}(x, 0; \hat{x}, i, k) = -\sum_{s=k}^{\mathcal{T}} \mathcal{E}\|z_1^s\|^2. \tag{5.12}$$

Due to the arbitrariness of $\hat{x}$, (5.12) yields $P^{k,i} \leq 0$ for all $i \in \overline{N}$ and $k \in N_{\mathcal{T}}$. $\square$

REMARK 5.1   If the state space of the Markov chain θ_k consists of only one value, or equivalently, there is no jump, Theorem 5.1 reduces to Lemma 3.3.   ∎

5.2.2   Results on the H_2/H_∞ control

Before presenting the main result of this section, we give the following four coupled difference matrix-valued recursion equations (CDMREs) on $(i, k) \in \overline{N} \times N_{\mathcal{T}}$:

$$\begin{cases} P_{1,\mathcal{T}}^{k,i} = (A_1^{k,i} + B_1^{k,i}K_{2,\mathcal{T}}^{k,i})^T \Psi_i(P_{1,\mathcal{T}}^{k+1})(A_1^{k,i} + B_1^{k,i}K_{2,\mathcal{T}}^{k,i}) \\ \quad + (A_2^{k,i} + B_2^{k,i}K_{2,\mathcal{T}}^{k,i})^T \Psi_i(P_{1,\mathcal{T}}^{k+1})(A_2^{k,i} + B_2^{k,i}K_{2,\mathcal{T}}^{k,i}) - C_{k,i}^T C_{k,i} \\ \quad - K_{3,\mathcal{T}}^{k,i} H_i^1(P_{1,\mathcal{T}}^{k+1})^{-1}(K_{3,\mathcal{T}}^{k,i})^T - (K_{2,\mathcal{T}}^{k,i})^T K_{2,\mathcal{T}}^{k,i}, \\ P_{1,\mathcal{T}}^{\mathcal{T}+1} = 0, \ H_i^1(P_{1,\mathcal{T}}^{k+1}) > 0, \end{cases} \tag{5.13}$$

$$K_{1,\mathcal{T}}^{k,i} = -H_i^1(P_{1,\mathcal{T}}^{k+1})^{-1}(K_{3,\mathcal{T}}^{k,i})^T, \tag{5.14}$$

$$\begin{cases} P_{2,\mathcal{T}}^{k,i} = (A_1^{k,i} + C_1^{k,i}K_{1,\mathcal{T}}^{k,i})^T \Psi_i(P_{2,\mathcal{T}}^{k+1})(A_1^{k,i} + C_1^{k,i}K_{1,\mathcal{T}}^{k,i}) \\ \quad + (A_2^{k,i} + C_2^{k,i}K_{1,\mathcal{T}}^{k,i})^T \Psi_i(P_{2,\mathcal{T}}^{k+1})(A_2^{k,i} + C_2^{k,i}K_{1,\mathcal{T}}^{k,i}) \\ \quad + C_{k,i}^T C_{k,i} - K_{4,\mathcal{T}}^{k,i} H_i^2(P_{2,\mathcal{T}}^{k+1})^{-1}(K_{4,\mathcal{T}}^{k,i})^T, \\ P_{2,\mathcal{T}}^{\mathcal{T}+1} = 0, \ H_i^2(P_{2,\mathcal{T}}^{k+1}) > 0, \end{cases} \tag{5.15}$$

$$K_{2,\mathcal{T}}^{k,i} = -H_i^2(P_{2,\mathcal{T}}^{k+1})^{-1}(K_{4,\mathcal{T}}^{k,i})^T, \qquad (5.16)$$

where in (5.13)–(5.16), $\Psi_i(\cdot)$ is defined by (5.3), and

$$
\begin{aligned}
H_i^1(P_{1,\mathcal{T}}^{k+1}) &= \gamma^2 I + (C_1^{k,i})^T \Psi_i(P_{1,\mathcal{T}}^{k+1})C_1^{k,i} + (C_2^{k,i})^T \Psi_i(P_{1,\mathcal{T}}^{k+1})C_2^{k,i}, \\
H_i^2(P_{2,\mathcal{T}}^{k+1}) &= I + (B_1^{k,i})^T \Psi_i(P_{2,\mathcal{T}}^{k+1})B_1^{k,i} + (B_2^{k,i})^T \Psi_i(P_{2,\mathcal{T}}^{k+1})B_2^{k,i}, \qquad (5.17) \\
K_{3,\mathcal{T}}^{k,i} &= (A_1^{k,i} + B_1^{k,i}K_{2,\mathcal{T}}^{k,i})^T \Psi_i(P_{1,\mathcal{T}}^{k+1})C_1^{k,i} + (A_2^{k,i} + B_2^{k,i}K_{2,\mathcal{T}}^{k,i})^T \Psi_i(P_{1,\mathcal{T}}^{k+1})C_2^{k,i}, \\
K_{4,\mathcal{T}}^{k,i} &= (A_1^{k,i} + C_1^{k,i}K_{1,\mathcal{T}}^{k,i})^T \Psi_i(P_{2,\mathcal{T}}^{k+1})B_1^{k,i} + (A_2^{k,i} + C_2^{k,i}K_{1,\mathcal{T}}^{k,i})^T \Psi_i(P_{2,\mathcal{T}}^{k+1})B_2^{k,i}.
\end{aligned}
$$

THEOREM 5.2

For a prescribed disturbance attenuation level $\gamma > 0$, the finite horizon H_2/H_∞ control problem of system (5.1) on $N_{\mathcal{T}}$ is solvable with $u_k^* = u_{\mathcal{T},k}^* = K_{2,\mathcal{T}}^{k,\theta_k} x_k$, $v_k^* = v_{\mathcal{T},k}^* = K_{1,\mathcal{T}}^{k,\theta_k} x_k$, iff the CDMREs (5.13)–(5.16) admit a unique solution $(P_{1,\mathcal{T}}^{k,i} \leq 0, K_{1,\mathcal{T}}^{k,i}; P_{2,\mathcal{T}}^{k,i} \geq 0, K_{2,\mathcal{T}}^{k,i})$ for $(i,k) \in \overline{N} \times N_{\mathcal{T}}$.

The proof is similar to that of Theorem 3.1 and is given in the following.

Proof. Sufficiency: If the CDMREs (5.13)–(5.16) admit a unique solution $(P_{1,\mathcal{T}}^{k,i}, K_{1,\mathcal{T}}^{k,i}; P_{2,\mathcal{T}}^{k,i}, K_{2,\mathcal{T}}^{k,i})$ on $(i,k) \in \overline{N} \times N_{\mathcal{T}}$, set $u_{\mathcal{T},k}^* = K_{2,\mathcal{T}}^{k,\theta_k} x_k$ and substitute $u_{\mathcal{T}}^*$ into system (5.1). By Theorem 5.1 and recursion (5.13), it yields $\|\mathcal{L}_{\mathcal{T}}\| < \gamma$ for all non-zero $v \in l_w^2(N_{\mathcal{T}}, \mathcal{R}^{n_v})$, $\theta_0 \in \overline{N}$ and $x_0 = 0$. Associated with (5.1), define

$$J_{1,\mathcal{T}}(u,v) := \sum_{k=0}^{\mathcal{T}} \mathcal{E}(\gamma^2 \|v_k\|^2 - \|z_k\|^2).$$

Considering (5.13), by Lemma 5.1 and the technique of completing squares, for any $x_0 \neq 0$, we have

$$
\begin{aligned}
J_{1,\mathcal{T}}(u_{\mathcal{T}}^*, v) &= \sum_{k=0}^{\mathcal{T}} \mathcal{E}(v_k - v_{\mathcal{T},k}^*)^T H_{\theta_k}^1(P_{1,\mathcal{T}}^{k+1})(v_k - v_{\mathcal{T},k}^*) + \mathcal{E}[x_0^T P_{1,\mathcal{T}}^{0,\theta_0} x_0] \\
&\geq J_{1,\mathcal{T}}(u_{\mathcal{T}}^*, v_{\mathcal{T}}^*) = \mathcal{E}[x_0^T P_{1,\mathcal{T}}^{0,\theta_0} x_0] \\
&= x_0^T \left[\sum_{i=1}^{N} \pi_0(i) P_{1,\mathcal{T}}^{0,i} \right] x_0, \qquad (5.18)
\end{aligned}
$$

where $\pi_0(i) = \mathcal{P}(\theta_0 = i)$, $v_{\mathcal{T},k}^* = K_{1,\mathcal{T}}^{k,\theta_k} x_k$. Equation (5.18) shows that $J_{1,\mathcal{T}}(u_{\mathcal{T}}^*, v)$ is minimized by $v_k = v_{\mathcal{T},k}^*$ for any $x_0 \in \mathcal{R}^n$, so $v_{\mathcal{T}}^*$ is just the worst-case disturbance. Furthermore, by the recursion (5.15) and the technique of completing squares, we have

$$J_{2,\mathcal{T}}(u, v_{\mathcal{T}}^*) = \sum_{k=0}^{\mathcal{T}} \mathcal{E}\|z_k\|^2$$

$$= \sum_{k=0}^{\mathcal{T}} \mathcal{E}(u_k - u^*_{\mathcal{T},k})^T H^2_{\theta_k}(P^{k+1}_{2,\mathcal{T}})(u_k - u^*_{\mathcal{T},k}) + \mathcal{E}[x_0^T P^{0,\theta_0}_{2,\mathcal{T}} x_0]$$

$$\geq J_{2,\mathcal{T}}(u^*_{\mathcal{T}}, v^*_{\mathcal{T}}) = x_0^T \left[\sum_{i=1}^{N} \pi_0(i) P^{0,i}_{2,\mathcal{T}} \right] x_0, \tag{5.19}$$

where $u^*_{\mathcal{T},k} = K^{k,\theta_k}_{2,\mathcal{T}} x_k$ with $K^{k,i}_{2,\mathcal{T}}$ given by (5.16). (5.19) implies that $u^*_{\mathcal{T}}$ minimizes $J_{2,\mathcal{T}}(u, v^*_{\mathcal{T}})$. Therefore, $(u^*_{\mathcal{T}}, v^*_{\mathcal{T}})$ solves the finite horizon H_2/H_∞ control problem of system (5.1).

Necessity: If the finite horizon H_2/H_∞ control problem for system (5.1) is solved by $u^*_{\mathcal{T},k} = K^{k,\theta_k}_{2,\mathcal{T}} x_k, v^*_{\mathcal{T},k} = K^{k,\theta_k}_{1,\mathcal{T}} x_k$. Substituting $u^*_{\mathcal{T},k}$ into (5.1), we get

$$\begin{cases} x_{k+1} = (A^{k,\theta_k}_1 + B^{k,\theta_k}_1 K^{k,\theta_k}_{2,\mathcal{T}})x_k + C^{k,\theta_k}_1 v_k \\ \quad\quad + (A^{k,\theta_k}_2 + B^{k,\theta_k}_2 K^{k,\theta_k}_{2,\mathcal{T}})x_k w_k + C^{k,\theta_k}_2 v_k w_k, \\ z_k = \begin{bmatrix} C_{k,\theta_k} x_k \\ D_{k,\theta_k} K^{k,\theta_k}_{2,\mathcal{T}} x_k \end{bmatrix}, \quad D^T_{k,\theta_k} D_{k,\theta_k} = I, \\ x_0 \in \mathcal{R}^n, \ k \in N_{\mathcal{T}}, \ \theta_0 \in \overline{N}. \end{cases} \tag{5.20}$$

Applying Theorem 5.1 to (5.20), we can derive that $P^{k,i}_{1,\mathcal{T}}$ satisfies recursion (5.13) on $N_{\mathcal{T}}$ with $P^k_{1,\mathcal{T}} = \begin{bmatrix} P^{k,1}_{1,\mathcal{T}} & P^{k,2}_{1,\mathcal{T}} & \cdots & P^{k,N}_{1,\mathcal{T}} \end{bmatrix} \leq 0$. From the proof of sufficiency, the worst disturbance $v^*_{\mathcal{T}} = K^{k,\theta_k}_{1,\mathcal{T}} x_k$ with $K^{k,i}_{1,\mathcal{T}}$ given by (5.14). Imposing $v^*_{\mathcal{T}}$ on system (5.1), we have

$$\begin{cases} x_{k+1} = (A^{k,\theta_k}_1 + C^{k,\theta_k}_1 K^{k,\theta_k}_{1,\mathcal{T}})x_k + (A^{k,\theta_k}_2 + C^{k,\theta_k}_2 K^{k,\theta_k}_{1,\mathcal{T}})x_k w_k \\ \quad\quad + B^{k,\theta_k}_1 u_k + B^{k,\theta_k}_2 u_k w_k, \\ z_k = \begin{bmatrix} C_{k,\theta_k} x_k \\ D_{k,\theta_k} u_k \end{bmatrix}, \quad D^T_{k,\theta_k} D_{k,\theta_k} = I, \\ x_0 \in \mathcal{R}^n, \ k \in N_{\mathcal{T}}, \ \theta_0 \in \overline{N}. \end{cases} \tag{5.21}$$

By the assumption, $u^*_{\mathcal{T},k}$ is optimal for the following optimization problem:

$$\min_{\substack{\text{subject to } (5.21), \ u \in l^2_w(N_{\mathcal{T}}, \mathcal{R}^{n_u})}} J_{2,\mathcal{T}}(u, v^*_{\mathcal{T}}), \tag{5.22}$$

which is a standard finite horizon LQ control problem for a Markov jump linear system and has been discussed in [43]. By Theorem 1 in [43], it is easy to prove that the recursion (5.15) is solved by $P^k_{2,\mathcal{T}} = \begin{bmatrix} P^{k,1}_{2,\mathcal{T}} & P^{k,2}_{2,\mathcal{T}} & \cdots & P^{k,N}_{2,\mathcal{T}} \end{bmatrix} \geq 0$. The proof of this theorem is complete. $\square$

As said in Remark 3.1, by the same procedure, Theorem 5.2 can be extended to the multiple noises case as discussed in [65] without any essential difficulty.

5.2.3 Algorithm and numerical example

This section provides a recursive algorithm, by which the four CDMREs (5.13)–(5.16) will be solved accurately. The algorithm procedure is as follows:

(i) For $k = \mathcal{T}$, $\theta_{\mathcal{T}} = i \in \overline{N}$, $H_i^1(P_{1,\mathcal{T}}^{\mathcal{T}+1})$ and $H_i^2(P_{2,\mathcal{T}}^{\mathcal{T}+1})$ are available by the terminal conditions $P_{1,\mathcal{T}}^{\mathcal{T}+1,\theta_{\mathcal{T}+1}} = 0$ and $P_{2,\mathcal{T}}^{\mathcal{T}+1,\theta_{\mathcal{T}+1}} = 0$.

(ii) Solve the CDMREs (5.14) and (5.16), then $K_{1,\mathcal{T}}^{\mathcal{T},i}$ and $K_{2,\mathcal{T}}^{\mathcal{T},i}$ are obtained.

(iii) Substitute the obtained $K_{1,\mathcal{T}}^{\mathcal{T},i}$ and $K_{2,\mathcal{T}}^{\mathcal{T},i}$ into the CDMREs (5.13) and (5.15) respectively, then $P_{1,\mathcal{T}}^{\mathcal{T},i} \leq 0$ and $P_{2,\mathcal{T}}^{\mathcal{T},i} \geq 0$ for $i \in \overline{N}$ are available.

(iv) Repeat the above procedure, $P_{1,\mathcal{T}}^{k,i}$, $P_{2,\mathcal{T}}^{k,i}$, $K_{1,\mathcal{T}}^{k,i}$ and $K_{2,\mathcal{T}}^{k,i}$ can be computed for $k = \mathcal{T} - 1, \mathcal{T} - 2, \cdots, 0$ and $i \in \overline{N}$.

In order to guarantee this recursive algorithm proceeds backward, the priori conditions $H_i^1(P_{1,\mathcal{T}}^{k+1}) > 0$ and $H_i^2(P_{2,\mathcal{T}}^{k+1}) > 0$ should be checked first. Otherwise, the recursive procedure has to stop. We should point out that $H_i^1(P_{1,\mathcal{T}}^{k+1})$ and $H_i^2(P_{2,\mathcal{T}}^{k+1})$ can be computed, provided that $P_{1,\mathcal{T}}^{k+1,i}$ and $P_{2,\mathcal{T}}^{k+1,i}$ are known. In this case, (5.14) and (5.16) constitute a group of coupled linear difference equations about $K_{1,\mathcal{T}}^{k,i}$ and $K_{2,\mathcal{T}}^{k,i}$. Similarly, after $K_{1,\mathcal{T}}^{k,i}$ and $K_{2,\mathcal{T}}^{k,i}$ have been obtained, (5.13) and (5.15) become two coupled linear difference equations about $P_{1,\mathcal{T}}^{k,i}$ and $P_{2,\mathcal{T}}^{k,i}$.

Example 5.1
Set $\mathcal{T} = 2$, $\gamma = 0.8$, $\overline{N} = \{1, 2\}$. The transition probability matrices of θ_k are given as follows:

$$\mathcal{P}_{2\times2}^0 = \begin{bmatrix} 0.1 & 0.9 \\ 0.5 & 0.5 \end{bmatrix}, \quad \mathcal{P}_{2\times2}^1 = \begin{bmatrix} 0.2 & 0.8 \\ 0.3 & 0.7 \end{bmatrix}, \quad \mathcal{P}_{2\times2}^2 = \begin{bmatrix} 0.85 & 0.15 \\ 0.6 & 0.4 \end{bmatrix}.$$

Tables 5.1–5.1 present the parameters of system (5.1). Utilizing the above algorithm procedure, we can check the existence of the solutions of (5.13)–(5.16) and then compute them backward. The solutions are given in Tables 5.1–5.3.
□

5.2.4　Unified treatment of H_2, H_∞ and H_2/H_∞ control based on Nash game

Associated with system (5.1), we define

$$J_{1,\mathcal{T}}(u, v) = \sum_{k=0}^{\mathcal{T}} \mathcal{E}(\gamma^2 \|v_k\|^2 - \|z_k\|^2), \tag{5.23}$$

$$J_{2,\mathcal{T}}^\rho(u, v) := \sum_{k=0}^{\mathcal{T}} \mathcal{E}(\|z_k\|^2 - \rho^2 \|v_k\|^2), \tag{5.24}$$

where $\gamma > 0$ and $\rho \geq 0$ are two nonnegative real parameters.

TABLE 5.1

Parameters of system (5.1).

mode	$i = 1$		
time	$k = 0$	$k = 1$	$k = 2$
$A_1^{k,i}$	$\begin{bmatrix} 0.3 & 0 \\ 0.1 & 0.2 \end{bmatrix}$	$\begin{bmatrix} 0.5 & 0.1 \\ 0 & 0.4 \end{bmatrix}$	$\begin{bmatrix} 2.0 & 1.0 \\ 0 & 3.0 \end{bmatrix}$
$A_2^{k,i}$	$\begin{bmatrix} 0.4 & 0.1 \\ 0 & 0.5 \end{bmatrix}$	$\begin{bmatrix} 0.5 & 0.2 \\ 0.1 & 0.6 \end{bmatrix}$	$\begin{bmatrix} 1.0 & 2.0 \\ 1.0 & 1.0 \end{bmatrix}$
$C_1^{k,i}$	$\begin{bmatrix} 0.5 & 0 \\ 0 & 0.4 \end{bmatrix}$	$\begin{bmatrix} 0.8 & 0 \\ 0 & 0.5 \end{bmatrix}$	$\begin{bmatrix} 3.0 & 1.0 \\ 2.0 & 2.0 \end{bmatrix}$
$C_2^{k,i}$	$\begin{bmatrix} 0.4 & 0 \\ 0 & 0.5 \end{bmatrix}$	$\begin{bmatrix} 0.5 & 0 \\ 0 & 0.4 \end{bmatrix}$	$\begin{bmatrix} 2.0 & 0 \\ 0 & 3.0 \end{bmatrix}$
$B_1^{k,i}$	$\begin{bmatrix} 0.6 & 0 \\ 0 & 0.5 \end{bmatrix}$	$\begin{bmatrix} 1.0 & 0 \\ 0 & 0 \end{bmatrix}$	$\begin{bmatrix} 1.0 & 0 \\ 0 & 0 \end{bmatrix}$
$B_2^{k,i}$	$\begin{bmatrix} 0.5 & 0 \\ 0 & 0.2 \end{bmatrix}$	$\begin{bmatrix} 1.0 & 0 \\ 0 & 1.0 \end{bmatrix}$	$\begin{bmatrix} 1.0 & 0 \\ 1.0 & 1.0 \end{bmatrix}$
$C_{k,i}$	$\begin{bmatrix} 0.5 & 0 \\ 0 & 0.5 \end{bmatrix}$	$\begin{bmatrix} 1.0 & 0 \\ 0 & 0.5 \end{bmatrix}$	$\begin{bmatrix} 1.0 & 0 \\ 0 & 2.0 \end{bmatrix}$
$D_{k,i}$	$\begin{bmatrix} 0.6 & -0.8 \\ 0.8 & 0.6 \end{bmatrix}$	$\begin{bmatrix} -0.6 & 0.8 \\ 0.8 & 0.6 \end{bmatrix}$	$\begin{bmatrix} 1.0 & 0 \\ 0 & 1.0 \end{bmatrix}$

Given $x_0 \in \mathcal{R}^n$, $\theta_0 \in \overline{N}$ and $\mathcal{T} \in \mathcal{N}$, our objective is to seek a Nash equilibrium strategy $\left(u_{\mathcal{T}}^*(\rho, \gamma), v_{\mathcal{T}}^*(\rho, \gamma)\right) \in l_w^2(N_{\mathcal{T}}, \mathcal{R}^{n_u}) \times l_w^2(N_{\mathcal{T}}, \mathcal{R}^{n_v})$ such that for arbitrary $(u, v) \in l_w^2(N_{\mathcal{T}}, \mathcal{R}^{n_u}) \times l_w^2(N_{\mathcal{T}}, \mathcal{R}^{n_v})$,

$$J_{1,\mathcal{T}}(u_{\mathcal{T}}^*(\rho, \gamma), v_{\mathcal{T}}^*(\rho, \gamma)) \leq J_{1,\mathcal{T}}(u_{\mathcal{T}}^*(\rho, \gamma), v), \tag{5.25}$$

$$J_{2,\mathcal{T}}^{\rho}(u_{\mathcal{T}}^*(\rho, \gamma), v_{\mathcal{T}}^*(\rho, \gamma)) \leq J_{2,\mathcal{T}}^{\rho}(u, v_{\mathcal{T}}^*(\rho, \gamma)). \tag{5.26}$$

If the following parameterized (on ρ and γ) CDMREs

$$\begin{cases} P_{1,\mathcal{T}}^{k,i} = (A_1^{k,i} + B_1^{k,i} K_{2,\mathcal{T}}^{k,i})^T \Psi_i(P_{1,\mathcal{T}}^{k+1})(A_1^{k,i} + B_1^{k,i} K_{2,\mathcal{T}}^{k,i}) \\ \quad + (A_2^{k,i} + B_2^{k,i} K_{2,\mathcal{T}}^{k,i})^T \Psi_i(P_{1,\mathcal{T}}^{k+1})(A_2^{k,i} + B_2^{k,i} K_{2,\mathcal{T}}^{k,i}) \\ \quad - C_{k,i}^T C_{k,i} - K_{3,\mathcal{T}}^{k,i} H_i^1(P_{1,\mathcal{T}}^{k+1})^+ (K_{3,\mathcal{T}}^{k,i})^T - (K_{2,\mathcal{T}}^{k,i})^T K_{2,\mathcal{T}}^{k,i}, \\ P_{1,\mathcal{T}}^{\mathcal{T}+1} = 0, \; H_i^1(P_{1,\mathcal{T}}^{k+1}) \geq 0, \\ K_{3,\mathcal{T}}^{k,i} = K_{3,\mathcal{T}}^{k,i} H_i^1(P_{1,\mathcal{T}}^{k+1})^+ H_i^1(P_{1,\mathcal{T}}^{k+1}), \end{cases} \tag{5.27}$$

$$K_{1,\mathcal{T}}^{k,i} = -H_i^1(P_{1,\mathcal{T}}^{k+1})^+ (K_{3,\mathcal{T}}^{k,i})^T, \tag{5.28}$$

TABLE 5.2
Parameters of system (5.1).

mode	$i = 2$		
time	$k = 0$	$k = 1$	$k = 2$
$A_1^{k,i}$	$\begin{bmatrix} 0.6 & 0 \\ 0.2 & 0.4 \end{bmatrix}$	$\begin{bmatrix} 0.4 & 0 \\ 0.2 & 0.5 \end{bmatrix}$	$\begin{bmatrix} 2.0 & 1.0 \\ 0 & 1.0 \end{bmatrix}$
$A_2^{k,i}$	$\begin{bmatrix} 0.3 & 0 \\ 0 & 0.4 \end{bmatrix}$	$\begin{bmatrix} 0.4 & 0.1 \\ 0 & 0.8 \end{bmatrix}$	$\begin{bmatrix} 2.0 & 0 \\ 0 & 1.0 \end{bmatrix}$
$C_1^{k,i}$	$\begin{bmatrix} 0.4 & 0 \\ 0 & 0.3 \end{bmatrix}$	$\begin{bmatrix} 0.6 & 0 \\ 0 & 0.8 \end{bmatrix}$	$\begin{bmatrix} 2.0 & 0 \\ 0 & 1.0 \end{bmatrix}$
$C_2^{k,i}$	$\begin{bmatrix} 0.2 & 0 \\ 0 & 0.5 \end{bmatrix}$	$\begin{bmatrix} 0.6 & 0 \\ 0 & 0.5 \end{bmatrix}$	$\begin{bmatrix} 3.0 & 0 \\ 1.0 & 1.0 \end{bmatrix}$
$B_1^{k,i}$	$\begin{bmatrix} 1.0 & 0 \\ 0 & 0 \end{bmatrix}$	$\begin{bmatrix} 0.4 & 0 \\ 0.1 & 0.8 \end{bmatrix}$	$\begin{bmatrix} 0 & 0 \\ 0 & 1.0 \end{bmatrix}$
$B_2^{k,i}$	$\begin{bmatrix} 0.2 & 0 \\ 0 & 0.6 \end{bmatrix}$	$\begin{bmatrix} 0.5 & 0 \\ 0 & 0.8 \end{bmatrix}$	$\begin{bmatrix} 1.0 & 0 \\ 0 & 1.0 \end{bmatrix}$
$C_{k,i}$	$\begin{bmatrix} 0.3 & 0 \\ 0 & 0.7 \end{bmatrix}$	$\begin{bmatrix} 0.8 & 0 \\ 0 & 0.6 \end{bmatrix}$	$\begin{bmatrix} 0.2 & 0.3 \\ 0 & 0.1 \end{bmatrix}$
$D_{k,i}$	$\begin{bmatrix} 0.8 & 0.6 \\ 0.6 & -0.8 \end{bmatrix}$	$\begin{bmatrix} -0.8 & 0.6 \\ 0.6 & 0.8 \end{bmatrix}$	$\begin{bmatrix} 0.6 & -0.8 \\ 0.8 & 0.6 \end{bmatrix}$

$$\begin{cases} P_{2,\mathcal{T}}^{k,i} = (A_1^{k,i} + C_1^{k,i}K_{1,\mathcal{T}}^{k,i})^T \Psi_i(P_{2,\mathcal{T}}^{k+1})(A_1^{k,i} + C_1^{k,i}K_{1,\mathcal{T}}^{k,i}) \\ \quad +(A_2^{k,i} + C_2^{k,i}K_{1,\mathcal{T}}^{k,i})^T \Psi_i(P_{2,\mathcal{T}}^{k+1})(A_2^{k,i} + C_2^{k,i}K_{1,\mathcal{T}}^{k,i}) \\ \quad +C_{k,i}^T C_{k,i} - K_{4,\mathcal{T}}^{k,i}H_i^2(P_{2,\mathcal{T}}^{k+1})^+(K_{4,\mathcal{T}}^{k,i})^T - \rho^2(K_{1,\mathcal{T}}^{k,i})^T K_{1,\mathcal{T}}^{k,i}, \quad (5.29) \\ P_{2,\mathcal{T}}^{\mathcal{T}+1} = 0, \; H_i^2(P_{2,\mathcal{T}}^{k+1}) \geq 0, \\ K_{4,\mathcal{T}}^{k,i} = K_{4,\mathcal{T}}^{k,i}H_i^2(P_{2,\mathcal{T}}^{k+1})^+H_i^2(P_{2,\mathcal{T}}^{k+1}), \end{cases}$$

$$K_{2,\mathcal{T}}^{k,i} = -H_i^2(P_{2,\mathcal{T}}^{k+1})^+(K_{4,\mathcal{T}}^{k,i})^T \qquad (5.30)$$

admit a unique solution on $(i,k) \in \overline{N} \times N_{\mathcal{T}}$, we denote it as

$$(P_{1,\mathcal{T}}^k(\rho,\gamma), K_{1,\mathcal{T}}^k(\rho,\gamma); P_{2,\mathcal{T}}^k(\rho,\gamma), K_{2,\mathcal{T}}^k(\rho,\gamma)).$$

Here, in (5.27)–(5.30), $H_i^1(P_{1,\mathcal{T}}^{k+1})$, $H_i^2(P_{2,\mathcal{T}}^{k+1})$, $K_{3,\mathcal{T}}^{k,i}$ and $K_{4,\mathcal{T}}^{k,i}$ are as defined in (5.17).

THEOREM 5.3
For system (5.1), if the four CDMREs (5.27)–(5.30) admit a solution $(P_{1,\mathcal{T}}^{k,i}(\rho,\gamma),$ $K_{1,\mathcal{T}}^{k,i}(\rho,\gamma); P_{2,\mathcal{T}}^{k,i}(\rho,\gamma), K_{2,\mathcal{T}}^{k,i}(\rho,\gamma))$ on $(i,k) \in \overline{N} \times N_{\mathcal{T}}$, then

$$u_{\mathcal{T},k}^*(\rho,\gamma) = K_{2,\mathcal{T}}^{k,\theta_k}(\rho,\gamma)x_k, \quad v_{\mathcal{T},k}^*(\rho,\gamma) = K_{1,\mathcal{T}}^{k,\theta_k}(\rho,\gamma)x_k$$

TABLE 5.3
Solutions for CDMREs (5.13)-(5.16).

mode	$i = 1$		
time	$k = 0$	$k = 1$	$k = 2$
$-P_{1,2}^{k,i}$	$\begin{bmatrix} 0.51 & 0.17 \\ 0.17 & 0.99 \end{bmatrix}$	$\begin{bmatrix} 1.11 & 0.10 \\ 0.10 & 0.82 \end{bmatrix}$	$\begin{bmatrix} 1 & 0 \\ 0 & 4 \end{bmatrix}$
$P_{2,2}^{k,i}$	$\begin{bmatrix} 0.69 & 0.42 \\ 0.42 & 2.23 \end{bmatrix}$	$\begin{bmatrix} 1.15 & 0.17 \\ 0.17 & 1.29 \end{bmatrix}$	$\begin{bmatrix} 1 & 0 \\ 0 & 4 \end{bmatrix}$
$K_{1,2}^{k,i}$	$\begin{bmatrix} 0.40 & 0.06 \\ 0.15 & 0.95 \end{bmatrix}$	$\begin{bmatrix} 0.21 & 0.12 \\ 0.10 & 0.85 \end{bmatrix}$	$\begin{bmatrix} 0 & 0 \\ 0 & 0 \end{bmatrix}$
$-K_{2,2}^{k,i}$	$\begin{bmatrix} 0.38 & 0.19 \\ 0.17 & 0.59 \end{bmatrix}$	$\begin{bmatrix} 0.21 & 0.11 \\ 0.76 & 0.44 \end{bmatrix}$	$\begin{bmatrix} 0 & 0 \\ 0 & 0 \end{bmatrix}$

TABLE 5.4
Solutions for CDMREs (5.13)-(5.16).

mode	$i = 2$		
time	$k = 0$	$k = 1$	$k = 2$
$-P_{1,2}^{k,i}$	$\begin{bmatrix} 0.57 & 0.08 \\ 0.8 & 0.75 \end{bmatrix}$	$\begin{bmatrix} 0.86 & 0.14 \\ 0.14 & 1.16 \end{bmatrix}$	$\begin{bmatrix} 0.04 & 0.06 \\ 0.06 & 0.10 \end{bmatrix}$
$P_{2,2}^{k,i}$	$\begin{bmatrix} 0.77 & 0.09 \\ 0.09 & 0.87 \end{bmatrix}$	$\begin{bmatrix} 1.02 & 0.41 \\ 0.41 & 2.01 \end{bmatrix}$	$\begin{bmatrix} 0.04 & 0.06 \\ 0.06 & 0.10 \end{bmatrix}$
$K_{1,2}^{k,i}$	$\begin{bmatrix} 0.43 & 0.01 \\ -0.01 & 0.29 \end{bmatrix}$	$\begin{bmatrix} 0.38 & 0.10 \\ 0.34 & 1.15 \end{bmatrix}$	$\begin{bmatrix} 0 & 0 \\ 0 & 0 \end{bmatrix}$
$-K_{2,2}^{k,i}$	$\begin{bmatrix} 0.35 & 0.05 \\ 0.17 & 0.47 \end{bmatrix}$	$\begin{bmatrix} 0.09 & 0.03 \\ 0.26 & 1.04 \end{bmatrix}$	$\begin{bmatrix} 0 & 0 \\ 0 & 0 \end{bmatrix}$

are the required two-person non-zero sum Nash equilibrium strategies of (5.25)–(5.26). Conversely, if the Nash game problem (5.25)–(5.26) is solvable with the linear memoryless state feedback strategies $\left(u_T^*(\rho, \gamma), v_T^*(\rho, \gamma)\right)$, then (5.27)–(5.30) admit a unique solution on $(i, k) \in \overline{N} \times N_T$.

Proof. This theorem can be proved following the line of Lemma 2.4 using indefinite LQ control results given by [43]; a detailed proof can be found in [90]. $\square$

Based on Theorem 5.3, we are able to give a unified treatment for H_2, H_∞ and H_2/H_∞ controls.

(i) H_2 (LQ) optimal control: In Theorem 5.3, if we let $\rho = 0$ and $\gamma \to \infty$, then

$\lim_{\gamma \to \infty} P_{2,T}^{k,i}(0,\gamma) = P_T^{k,i}$, which solves

$$\begin{cases} P_T^{k,i} = \sum_{j=1}^{2} (A_j^{k,i})^T \Psi_i(P_T^{k+1}) A_j^{k,i} + C_{k,i}^T C_{k,i} \\ \qquad - \Upsilon_{k,i}^{(A_1,B_1;A_2,B_2)} H_i^2(P_T^{k+1})^{-1} [\Upsilon_{k,i}^{(A_1,B_1;A_2,B_2)}]^T, \\ P_T^{k,i} \geq 0, \ P_T^{T+1,i} = 0, \\ H_i^2(P_T^{k+1}) > 0, \ (k,i) \in N_T \times \overline{N}, \end{cases} \tag{5.31}$$

where

$$\Upsilon_{k,i}^{(A_1,B_1;A_2,B_2)} = \left(\sum_{j=1}^{2} (A_j^{k,i})^T \Psi_i(P_T^{k+1}) B_j^{k,i} \right).$$

Moreover, subject to

$$\begin{cases} x_{k+1} = A_1^{k,\theta_k} x_k + B_1^{k,\theta_k} u_k + (A_2^{k,\theta_k} x_k + B_2^{k,\theta_k} u_k) w_k, \\ z_k = \begin{bmatrix} C_{k,\theta_k} x_k \\ D_{k,\theta_k} u_k \end{bmatrix}, \ D_{k,\theta_k}^T D_{k,\theta_k} = I, \\ x_0 \in \mathcal{R}^n, \ \theta_0 \in \overline{N}, \ (k,i) \in N_T \times \overline{N}, \end{cases}$$

the quadratic optimal performance

$$\min_{u \in l_w^2(N_T, \mathcal{R}^{n_u})} J_{2,T}(u,0) = J_{2,T}(\bar{u}_T^*, 0) = x_0^T P_T^{0,\theta_0} x_0$$

with the optimal control law

$$\bar{u}_{T,k}^* = -H_{\theta_k}^2(P_T^{k+1})^{-1} [\Upsilon_{k,\theta_k}^{(A_1,B_1;A_2,B_2)}]^T x_k.$$

(ii) H_∞ optimal control: Set $\rho = \gamma$, then

$$J_{1,T}(u,v) + J_{2,T}^\gamma(u,v) = 0,$$

so the two-person non-zero sum Nash game (5.25)–(5.26) reduces to a two-person zero-sum game problem as

$$J_{1,T}(u, v_T^*(\gamma,\gamma)) \leq J_{1,T}(u_T^*(\gamma,\gamma), v_T^*(\gamma,\gamma)) \leq J_{1,T}(u_T^*(\gamma,\gamma), v).$$

It is easy to show that

$$P_{1,T}^k(\gamma,\gamma) + P_{2,T}^k(\gamma,\gamma) = 0, \ k \in N_T.$$

Set $P_{\infty,T}^k := P_{1,T}^k(\gamma,\gamma)$, substitute $P_{2,T}^k(\gamma,\gamma) = -P_{\infty,T}^k$ into (5.27) and note $H_i^1(P_{\infty,T}^{k+1}) > 0$ due to Theorem 5.1, it follows that

$$\begin{cases} P_{\infty,T}^{k,i} = (A_1^{k,i} + B_1^{k,i} \tilde{K}_{2,T}^{k,i})^T \Psi_i(P_{\infty,T}^{k+1})(A_1^{k,i} + B_1^{k,i} \tilde{K}_{2,T}^{k,i}) \\ \qquad + (A_2^{k,i} + B_2^{k,i} \tilde{K}_{2,T}^{k,i})^T \Psi_i(P_{\infty,T}^{k+1})(A_2^{k,i} + B_2^{k,i} \tilde{K}_{2,T}^{k,i}) \\ \qquad - C_{k,i}^T C_{k,i} - \tilde{K}_{3,T}^{k,i} H_i^1(P_{\infty,T}^{k+1})^{-1} (\tilde{K}_{3,T}^{k,i})^T - (\tilde{K}_{2,T}^{k,i})^T \tilde{K}_{2,T}^{k,i}, \\ P_{\infty,T}^{T+1} = 0, \ H_i^1(P_{\infty,T}^{k+1}) > 0, \ (k,i) \in N_T \times \overline{N}, \end{cases} \tag{5.32}$$

where

$$\tilde{K}_{2,\mathcal{T}}^{k,i} = K_{2,\mathcal{T}}^{k,i}\Big|_{P_{2,\mathcal{T}}^{k,i}=-P_{\infty,\mathcal{T}}^{k,i},\,P_{1,\mathcal{T}}^{k,i}=P_{\infty,\mathcal{T}}^{k,i}}$$

and

$$\tilde{K}_{3,\mathcal{T}}^{k,i} = K_{3,\mathcal{T}}^{k,i}\Big|_{P_{2,\mathcal{T}}^{k,i}=-P_{\infty,\mathcal{T}}^{k,i},\,P_{1,\mathcal{T}}^{k,i}=P_{\infty,\mathcal{T}}^{k,i}}.$$

In this case, $\tilde{u}_{\mathcal{T},k}^* = \tilde{K}_{2,\mathcal{T}}^{k,i}x_k$ is the H_∞ optimal control, while $\tilde{v}_{\mathcal{T},k}^* = \tilde{K}_{1,\mathcal{T}}^{k,\theta_k}x_k$ is the corresponding worst-case disturbance.

(iii) Setting $\rho = 0$, the mixed H_2/H_∞ control is retrieved.

5.3 Infinite Horizon Discrete Time-Varying H_2/H_∞ Control

In this section, we deal with the infinite horizon stochastic H_2/H_∞ control problem for the following discrete time-varying Markov jump system with state-dependent multiplicative noise:

$$\begin{cases} x_{k+1} = A_1^{k,\theta_k}x_k + B_1^{k,\theta_k}u_k + C_1^{k,\theta_k}v_k + \sum_{j=1}^{r} \bar{A}_j^{k,\theta_k}x_k w_j^k, \\ z_k = \begin{bmatrix} C_{k,\theta_k}x_k \\ D_{k,\theta_k}u_k \end{bmatrix}, \quad D_{k,\theta_k}^T D_{k,\theta_k} = I, \\ x_0 \in \mathcal{R}^n, \ k \in \mathcal{N}, \ \theta_0 \in \overline{N}, \end{cases} \quad (5.33)$$

where $\{w_k|w_k = (w_1^k,\cdots,w_r^k)^T, k \in \mathcal{N}\}$ is a sequence of independent random vectors, which are independent of $\{\theta_k, k \in \mathcal{N}\}$. $\{w_k\}_{k\in\mathcal{N}}$ and $\{\theta_k\}_{k\in\mathcal{N}}$ are defined on the filtered probability space $(\Omega, \mathcal{F}, \{\mathcal{F}_k\}_{k\in\mathcal{N}}, \mathcal{P})$ with $\mathcal{F}_k := \sigma\{\theta_s, w_j : 0 \le s \le k; 0 \le j \le k-1\}$. Here, we consider system (5.33) with multiple noises to be general for some practical needs. Below, we make the following underlying assumptions:

Assumption:

(A1) $\mathcal{E}(w_k) = 0_{r\times 1}$ and $\mathcal{E}(w_k w_k^T) = I_r$, $k \in \mathcal{N}$.

(A2) The transition probability matrix $\mathcal{P}_{N\times N}^k$ is non-degenerate, i.e., $\inf_{k\in\mathcal{N}} \pi_k(i) > 0$ with $\pi_k(i) = \mathcal{P}(\theta_k = i) > 0$, $i \in \overline{N}$.

(A3) All the coefficient matrices of (5.33) are uniformly bounded.

The infinite horizon H_2/H_∞ control of system (5.33) can be defined similarly to Definition 3.11.

DEFINITION 5.3 *For a prescribed disturbance attenuation level $\gamma > 0$, the infinite horizon H_2/H_∞ control problem is concerned with the design of a linear memoryless state feedback controller $u_\infty^* = \{u_{\infty,k}^* = K_2^{k,\theta_k}\}_{k\in\mathcal{N}} \in l_w^2(\mathcal{N}, \mathcal{R}^{n_u})$ such that*

1. *$\{u_k = u_{\infty,k}^*\}_{k\in\mathcal{N}}$ stabilizes system (5.33) internally, i.e., the unperturbed system*

$$\begin{cases} x_{k+1} = A_1^{k,\theta_k}x_k + B_1^{k,\theta_k}u_{\infty,k}^* + \sum_{j=1}^{r} \bar{A}_j^{k,\theta_k}x_k w_j^k, \\ x_0 \in \mathcal{R}^n, \ k \in \mathcal{N}, \ \theta_0 \in \overline{N} \end{cases} \quad (5.34)$$

is strongly exponentially stable in the mean square(SESMS) sense, that is, there exist $\beta \geq 1, q \in (0,1)$, such that for any $k \geq s \geq 0, i \in \overline{N}, x_0 \in \mathcal{R}^n$,

$$\mathcal{E}[\|x_k\|^2 | \theta_s = i] \leq \beta q^{k-s} \mathcal{E}\|x_s\|^2.$$

2.

$$\|\mathcal{L}_\infty\| = \sup_{v \in l_w^2(\mathcal{N}, \mathcal{R}^{n_v}), v \neq 0, x_0 = 0, \theta_0 \in \overline{N}, u = u_\infty^*} \frac{\left(\sum_{k=0}^{\infty} \mathcal{E}\|z_k\|^2\right)^{\frac{1}{2}}}{\left(\sum_{k=0}^{\infty} \mathcal{E}\|v_k\|^2\right)^{\frac{1}{2}}} < \gamma.$$

3. *When the worst-case disturbance $v_\infty^* = \{v_k = v_{\infty,k}^*\}_{k \in \mathcal{N}}$, if it exists, is imposed on (5.33), u_∞^* minimizes the output energy*

$$J_{2,\infty}(u, v_\infty^*) = \sum_{k=0}^{\infty} \mathcal{E}\|z_k\|^2.$$

If the above pair (u_∞^*, v_∞^*) exist, we say that the infinite horizon H_2/H_∞ control problem of (5.33) is solvable.

Obviously, when there is no jump in system (5.34), the zero state equilibrium of system (5.34) is SESMS iff it is ESMS.

5.3.1 Definitions and preliminaries

We first provide some useful definitions and preliminaries. Consider the following stochastic control system

$$\begin{cases} x_{k+1} = A_1^{k,\theta_k} x_k + B_1^{k,\theta_k} u_k + \sum_{j=1}^{r} (\bar{A}_j^{k,\theta_k} x_k + \bar{B}_j^{k,\theta_k} u_k) w_j^k, \\ y_k = C_{k,\theta_k} x_k, \\ x_0 \in \mathcal{R}^n, \theta_0 \in \overline{N}, k \in \mathcal{N}. \end{cases} \tag{5.35}$$

Now, we state the notions of stabilizability and detectability of [50, 52] as follows:

DEFINITION 5.4 *We say*

$$x_{k+1} = A_1^{k,\theta_k} x_k + B_1^{k,\theta_k} u_k + \sum_{j=1}^{r} (\bar{A}_j^{k,\theta_k} x_k + \bar{B}_j^{k,\theta_k} u_k) w_j^k$$

or $(A_1, B_1; \bar{A}, \bar{B})$ is stochastically stabilizable if there exists a feedback control $u_k = K^{k,\theta_k} x_k$, such that the closed-loop system

$$x_{k+1} = (A_1^{k,\theta_k} + B_1^{k,\theta_k} K^{k,\theta_k}) x_k + \sum_{j=1}^{r} (\bar{A}_j^{k,\theta_k} + \bar{B}_j^{k,\theta_k} K^{k,\theta_k}) x_k w_j^k \tag{5.36}$$

is SESMS, where $K^{k,\theta_k} \in \mathcal{R}^{n \times n_u}$ is called a stabilizing feedback gain matrix.

DEFINITION 5.5   *The following system*

$$\begin{cases} x_{k+1} = A_1^{k,\theta_k} x_k + \sum_{j=1}^{r} \bar{A}_j^{k,\theta_k} x_k w_j^k, \\ y_k = C_{k,\theta_k} x_k, \ x_0 \in \mathcal{R}^n, \ \theta_0 = i \in \overline{N} \end{cases} \tag{5.37}$$

or $(A_1, \bar{A}|C)$, is called stochastically detectable if there exists a uniformly bounded sequence of $\{H^{k,\theta_k}\}_{k\in\mathcal{N}}$ such that

$$x_{k+1} = (A_1^{k,\theta_k} + H^{k,\theta_k} C_{k,\theta_k}) x_k + \sum_{j=1}^{r} \bar{A}_j^{k,\theta_k} x_k w_j^k$$

is SESMS.

REMARK 5.2   There are other definitions for detectability of stochastic system (5.37). For instance, [170] introduced detectability in conditional mean for a countably infinite state Markov chain, while weak detectability was defined in [171]. It was shown in [171] that stochastic detectability implies weak detectability, but the converse is not true. For simplicity, we now apply stochastic detectability to our concerned H_2/H_∞ problem. Obviously, for the main results of this section, there exists room for improvement. We note that generalizing uniform detectability and exact detectability [218] of linear discrete time-varying systems without Markov jump to system (5.37) is valuable and deserves further study.   ∎

The following lemma generalizes Lemma 2.11 to discrete-time Markov jump systems. It can be easily shown by analogous arguments, so the details are omitted.

LEMMA 5.2
Suppose that C_1^{k,θ_k}, C_{k,θ_k}, B_1^{k,θ_k}, K_1^{k,θ_k}, K_2^{k,θ_k}, K_3^{k,θ_k} and $\hat{H}_{k,\theta_k}^1 > 0$ for $\theta_k \in \overline{N}$ are matrices of suitable dimensions. Set

$$\tilde{A}_2^{k,\theta_k} = \begin{bmatrix} C_{k,\theta_k} \\ (\hat{H}_{k,\theta_k}^1)^{-\frac{1}{2}} K_3^{k,\theta_k} \\ K_2^{k,\theta_k} \end{bmatrix}, \quad \tilde{A}_3^{k,\theta_k} = \begin{bmatrix} C_{k,\theta_k} \\ K_2^{k,\theta_k} \end{bmatrix}.$$

Then we have

(1) If $(A_1, \bar{A}|C)$ is stochastically detectable, so is $(A_1 + B_1 K_2, \bar{A}|\tilde{A}_2)$.

(2) If $(A_1 + C_1 K_1, \bar{A}|C)$ is stochastically detectable, so is $(A_1 + C_1 K_1 + B_1 K_2, \bar{A}|\tilde{A}_3)$.

5.3.2   An SBRL

Consider the following stochastic perturbed system:

$$\begin{cases} x_{k+1} = A_{11}^{k,\theta_k} x_k + B_{11}^{k,\theta_k} v_k + \sum_{j=1}^{r} (\bar{A}_{j1}^{k,\theta_k} x_k + \bar{B}_{j1}^{k,\theta_k} v_k) w_j^k, \\ z_1^k = C_{11}^{k,\theta_k} x_k, \ x_0 \in \mathcal{R}^n, \ \theta_0 \in \overline{N}, \ k \in \mathcal{N}. \end{cases} \tag{5.38}$$

In system (5.38), for any disturbance $v \in l_w^2(\mathcal{N}, \mathcal{R}^{n_v})$ and the corresponding output $z_1^k \in l_w^2(\mathcal{N}, \mathcal{R}^{n_{z_1}})$, a linear perturbed operator $\tilde{\mathcal{L}}_\infty : l_w^2(\mathcal{N}, \mathcal{R}^{n_v}) \mapsto l_w^2(\mathcal{N}, \mathcal{R}^{n_{z_1}})$ can be defined as

$$\tilde{\mathcal{L}}_\infty(v_k) = z_1^k \big|_{x_0 = 0, \theta_0 \in \overline{N}}, \quad \forall v \in l_w^2(\mathcal{N}, \mathcal{R}^{n_v})$$

with its H_∞ norm given by

$$\|\tilde{\mathcal{L}}_\infty\| := \sup_{v \in l_w^2(\mathcal{N}, \mathcal{R}^{n_v}), v \neq 0, x_0 = 0, \theta_0 \in \overline{N}} \frac{\| z_1 \|_{l_w^2(\mathcal{N}, \mathcal{R}^{n_{z_1}})}}{\| v \|_{l_w^2(\mathcal{N}, \mathcal{R}^{n_v})}}.$$

Similarly, for any $\mathcal{T} \in \mathbb{Z}_{1+} = \{1, 2, \cdots\}$, we can define another linear operator $\tilde{\mathcal{L}}_\mathcal{T} : l_w^2(N_\mathcal{T}, \mathcal{R}^{n_v}) \mapsto l_w^2(N_\mathcal{T}, \mathcal{R}^{n_{z_1}})$. Obviously, $\|\tilde{\mathcal{L}}_\mathcal{T}\| \leq \|\tilde{\mathcal{L}}_\infty\|$ for all $\mathcal{T} \in \mathbb{Z}_{1+}$.

To simplify the expressions, introduce the following notations:

$$\Pi_i^1(P^{k+1}) := (A_{11}^{k,i})^T \Psi_i(P^{k+1}) A_{11}^{k,i} + \sum_{j=1}^r (\bar{A}_{j1}^{k,i})^T \Psi_i(P^{k+1}) \bar{A}_{j1}^{k,i},$$

$$\Pi_i^2(P^{k+1}) := (A_{11}^{k,i})^T \Psi_i(P^{k+1}) B_{11}^{k,i} + \sum_{j=1}^r (\bar{A}_{j1}^{k,i})^T \Psi_i(P^{k+1}) \bar{B}_{j1}^{k,i},$$

$$\Pi_i^3(P^{k+1}) := (B_{11}^{k,i})^T \Psi_i(P^{k+1}) B_{11}^{k,i} + \sum_{j=1}^r (\bar{B}_{j1}^{k,i})^T \Psi_i(P^{k+1}) \bar{B}_{j1}^{k,i} - \gamma^2 I,$$

where $\Psi_i(P^{k+1})$ is defined in (5.3).

PROPOSITION 5.1

If system (5.38) is internally stable and $\|\tilde{\mathcal{L}}_\infty\| < \gamma$ for $\gamma > 0$, then the following discrete-time Riccati equation admits a solution $\{P_\mathcal{T}^k\} \in \mathcal{S}_n^{N+}$ for any $\mathcal{T} \in \mathbb{Z}_{1+}$:

$$\begin{cases} P_\mathcal{T}^{k,i} = \Pi_i^1(P^{k+1}) + (C_{11}^{k,i})^T C_{11}^{k,i} - \Pi_i^2(P^{k+1}) \Pi_i^3(P^{k+1})^{-1} \Pi_i^2(P^{k+1})^T, \\ P_\mathcal{T}^{\mathcal{T}+1,i} = 0, \\ \Pi_i^3(P_\mathcal{T}^{k+1}) < -\varepsilon_0 I_{n_v}, \quad \varepsilon_0 \in (0, \gamma^2 - \|\tilde{\mathcal{L}}_\mathcal{T}\|^2). \end{cases} \quad (5.39)$$

Moreover, $0 \leq P_\mathcal{T}^{k,i} \leq P_{\mathcal{T}+1}^{k,i} \leq \zeta I_n$, where $(k, i) \in N_\mathcal{T} \times \overline{N}$, and $\zeta > 0$ is a constant independent of i, k, and $\mathcal{T}$.

Proof. By the same procedure as in Lemma 8.14 [52], there is no difficulty to prove that for all $k_0 \in \mathcal{N}$, $x_{k_0} \in \mathcal{R}^n$ and $v \in l_w^2(\mathcal{N}, \mathcal{R}^{n_v})$,

$$-J_1^\infty(x, v; x_{k_0}, \theta_{k_0}, k_0) := \sum_{k=k_0}^\infty \mathcal{E}(\|z_1^k\|^2 - \gamma^2 \|v_k\|^2) \leq \rho \|x_{k_0}\|^2,$$

where $\rho > 0$ is a constant. Moreover, $\|\tilde{\mathcal{L}}_\infty\| < \gamma$ implies that $\|\tilde{\mathcal{L}}_\mathcal{T}\| < \gamma$ for all $\mathcal{T} \in \mathbb{Z}_{1+}$. By Theorem 5.1, the solution $P_\mathcal{T}^k \in \mathcal{S}_n^{N+}$ to (5.39) is well defined. Similar to the proof of Proposition 8.5 [52], it is easy to show that $P_\mathcal{T}^{k,i} \leq P_{\mathcal{T}+1}^{k,i}$ for $(k, i) \in N_\mathcal{T} \times \overline{N}$. Furthermore, for any $(k_0, i) \in [0, \mathcal{T}+1] \times \overline{N}$, $x_{k_0} \in \mathcal{R}^n$ and $\mathcal{T} \in \mathbb{Z}_{1+}$, it follows from Lemma 5.1 that

$$\pi_{k_0}(i) x_{k_0}^T P_\mathcal{T}^{k_0, i} x_{k_0} \leq \mathcal{E}(x_{k_0}^T P_\mathcal{T}^{k_0, \theta_{k_0}} x_{k_0}) = \sum_{k=k_0}^\mathcal{T} \mathcal{E}(\|z_1^k\|^2 - \gamma^2 \|v_k\|^2)$$

$$\leq -J_1^\infty(x, \bar{v}; x_{k_0}, \theta_{k_0}, k_0) \leq \rho \|x_{k_0}\|^2,$$

where $\bar{v}_k = v_k$ for $0 \le k \le \mathcal{T}$ and $\bar{v}_k = 0$ if $k \ge \mathcal{T} + 1$. Under the Assumption-(A2), we have that $x_{k_0}^T P_{\mathcal{T}}^{k_0,i} x_{k_0} \le \zeta \|x_{k_0}\|^2$ with $\zeta = \rho / \inf_{k \in \mathcal{N}} \pi_k(i) > 0$. This completes the proof. $\square$

Based on Proposition 5.1, we now present an SBRL for system (5.38), which can be regarded as a time-varying version of Theorem 8.13-(i)$\Leftrightarrow$(iii) [52].

THEOREM 5.4 (SBRL)

For a prescribed $\gamma > 0$, if system (5.38) is internally stable and $\|\tilde{\mathcal{L}}_\infty\| < \gamma$, then the following discrete-time Riccati equation has a bounded stabilizing solution $P^k = (P^{k,1}, \cdots, P^{k,N}) \in \mathcal{S}_n^{N+}$ on $\mathcal{N} \times \overline{N}$:

$$\begin{cases} P^{k,i} = \Pi_i^1(P^{k+1}) + (C_{11}^{k,i})^T C_{11}^{k,i} - \Pi_i^2(P^{k+1})\Pi_i^3(P^{k+1})^{-1}\Pi_i^2(P^{k+1})^T \\ \Pi_i^3(P^{k+1}) \le -\varepsilon_0 I_{n_v}, \ \varepsilon_0 \in (0, \gamma^2 - \|\tilde{\mathcal{L}}_\infty\|^2). \end{cases} \quad (5.40)$$

Conversely, if (5.40) admits a bounded stabilizing solution $P^k \in \mathcal{S}_n^{N+}$, then system (5.38) is internally stable and $\|\tilde{\mathcal{L}}_\infty\| < \gamma$.

Here, P^k is said to be a stabilizing solution if $v_k = F^{k,\theta_k} x_k$ guarantees that the closed-loop system is SESMS, where

$$F^{k,\theta_k} = -\Pi_{\theta_k}^3(P^{k+1})^{-1}\Pi_{\theta_k}^2(P^{k+1})^T.$$

Proof. Obviously, Proposition 5.1 implies that the sequence

$$P_{\mathcal{T}}^k = (P_{\mathcal{T}}^{k,1}, \cdots, P_{\mathcal{T}}^{k,N}) \in \mathcal{S}_n^{N+}$$

converges as $\mathcal{T} \to +\infty$. Denoting $P^{k,i} = \lim_{\mathcal{T} \to +\infty} P_{\mathcal{T}}^{k,i}$ and taking $\mathcal{T} \to +\infty$ on both sides of (5.39), we obtain that $P^k \in \mathcal{S}_n^{N+}$ is a bounded solution of (5.40). The remainder is to show that P^k is a stabilizing solution. To this end, we define another linear operator $L_\infty^\alpha : l_w^2(\mathcal{N}, \mathcal{R}^{n_v}) \mapsto l_w^2(\mathcal{N}, \mathcal{R}^{n+n_{z_1}})$ as

$$L_\infty^\alpha(v_k) = [(C_{11}^{k,\theta_k})^T \quad \alpha I_n]^T x_{(k,v;0,\theta_0)}$$

where $x_{(k,v;0,\theta_0)}$ is the solution of (5.38) with $x_0 = 0$ and θ_0. By Corollary 3.9 [52], $\|L_\infty^\alpha\| < \gamma$ still holds for sufficiently small $\alpha > 0$ and so does $\|L_{\mathcal{T}}^\alpha\| < \gamma$, where $L_{\mathcal{T}}^\alpha(v_k) = L_\infty^\alpha(v_k)$ for $k \in N_{\mathcal{T}}$. Then, from Proposition 5.1, it can be deduced that the following difference equation

$$\begin{cases} P_{(\alpha,\mathcal{T})}^{k,i} = \Pi_i^1(P^{k+1}) + (C_{11}^{k,i})^T C_{11}^{k,i} + \alpha^2 I_n - \Pi_i^2(P_{(\alpha,\mathcal{T})}^{k+1})\Pi_i^3(P_{(\alpha,\mathcal{T})}^{k+1})^{-1}\Pi_i^2(P_{(\alpha,\mathcal{T})}^{k+1})^T, \\ P_{(\alpha,\mathcal{T})}^{\mathcal{T}+1,i} = 0, \\ \Pi_i^3(P_{(\alpha,\mathcal{T})}^{k+1}) < -\varepsilon_0 I_{n_v}, \ \varepsilon_0 \in (0, \gamma^2 - \|L_{\mathcal{T}}^\alpha\|^2) \end{cases}$$

(5.41)

admits a bounded solution $P_{(\alpha,\mathcal{T})}^k \in \mathcal{S}_n^{N+}$ for $(k,i) \in N_{\mathcal{T}} \times \overline{N}$. Repeating the above procedure, we will derive that there exists a bounded sequence $P_\alpha^{k,i} = \lim_{\mathcal{T} \to +\infty} P_{(\alpha,\mathcal{T})}^{k,i}$, which solves

$$\begin{cases} P_\alpha^{k,i} = \Pi_i^1(P^{k+1}) + (C_{11}^{k,i})^T C_{11}^{k,i} + \alpha^2 I_n - \Pi_i^2(P_\alpha^{k+1})\Pi_i^3(P_\alpha^{k+1})^{-1}\Pi_i^2(P_\alpha^{k+1})^T, \\ \Pi_i^3(P_\alpha^{k+1}) < -\varepsilon_0 I_{n_v}, \ \varepsilon_0 \in (0, \gamma^2 - \|L_\infty^\alpha\|^2). \end{cases}$$

(5.42)

Comparing (5.39) with (5.41) and utilizing a comparison theorem (Theorem 5.1 of [52]), we have that $P_T^{k,i} \le P_{(\alpha,T)}^{k,i}$ for all $(k,i) \in N_T \times \overline{N}$. Then, letting $T \to +\infty$, it yields that

$$P^{k,i} \le P_\alpha^{k,i}, \ (k,i) \in \mathcal{N} \times \overline{N}. \tag{5.43}$$

Subtracting (5.40) from (5.42) yields

$$P_\alpha^{k,i} - P^{k,i} = (A_{11}^{k,i} + B_{11}^{k,i}F^{k,i})^T \Psi_i(P_\alpha^{k+1} - P^{k+1})(A_{11}^{k,i} + B_{11}^{k,i}F^{k,i})$$
$$+ \sum_{j=1}^{r}(\bar{A}_{j1}^{k,i} + \bar{B}_{j1}^{k,i}F^{k,i})^T \Psi_i(P_\alpha^{k+1} - P^{k+1})$$
$$\cdot (\bar{A}_{j1}^{k,i} + \bar{B}_{j1}^{k,i}F^{k,i}) + M^{k,i}, \tag{5.44}$$

where

$$M^{k,i} = \alpha^2 I_{n_v} - (F^{k,i} - F_\alpha^{k,i})^T \Pi_i^3(P_\alpha^{k+1})(F^{k,i} - F_\alpha^{k,i})$$

and $F_\alpha^{k,i} := F^{k,i}|_{Pk=P_\alpha^k}$. In view of the second inequality of (5.42) and Assumption-(A3), the above indicates that $M^{k,i} > \alpha^2 I_n$ is uniformly bounded for all $i \in \overline{N}$. Therefore, (5.44) admits a bounded solution $Y^k := P_\alpha^k - P^k$. In view of (5.43), $Y^k \in \mathcal{S}_n^{N+}$. Applying (vi)$\Rightarrow$(i) of Theorem 2.4 [52], (5.38) is SESMS for $v_k = F^{k,\theta_k}x_k$, which verifies the first part of this theorem.

To prove the converse, we denote by P^k the stabilizing solution of (5.40). A simple calculation shows that $\hat{P}^k = -P^k$ is the stabilizing solution of

$$\begin{cases} \hat{P}^{k,i} = \Pi_i^1(\hat{P}^{k+1}) - (C_{11}^{k,i})^T C_{11}^{k,i} - \Pi_i^2(\hat{P}^{k+1})\bar{\Pi}_i^3(\hat{P}^{k+1})^{-1}\Pi_i^2(\hat{P}^{k+1})^T \\ \bar{\Pi}_i^3(\hat{P}^{k+1}) = -\Pi_i^3(P^{k+1}) \ge \varepsilon_0 I_{n_v}, \\ \varepsilon_0 \in (0, \gamma^2 - \|\tilde{\mathcal{L}}_\infty\|^2). \end{cases} \tag{5.45}$$

Hence, according to Theorem 5.12 [52], there exists $\bar{Y}^k \in \mathcal{S}_n^N$ satisfying

$$\begin{bmatrix} \Gamma_i(\bar{Y}^{k+1}) - (C_{11}^{k,i})^T C_{11}^{k,i} & \Pi_i^2(\bar{Y}^{k+1}) \\ \Pi_i^2(\bar{Y}^{k+1})^T & \bar{\Pi}_i^3(\bar{Y}^{k+1}) \end{bmatrix} \ge \mu I_{n+n_v}, \tag{5.46}$$

where $\Gamma_i(\bar{Y}^{k+1}) = \Pi_i^1(\bar{Y}^{k+1}) - \bar{Y}^{k,i}$ and $\mu > 0$ is a constant independent of k and i. On the other hand, by Proposition 5.1 [52], $\hat{P}^{k,i}$ is also the maximal solution of (5.45). Therefore, $\bar{Y}^{k,i} \le \hat{P}^{k,i} = -P^{k,i} \le 0$. Note that the $(1,1)$-block of (5.46) implies

$$\Gamma_i(\bar{Y}^{k+1}) - (C_{11}^{k,i})^T C_{11}^{k,i} \ge \mu I_n.$$

So $\bar{Y}^{k,i} \le -\mu I_n$. Now, we get $\Pi_i^1(\bar{Y}^{k+1}) - \bar{Y}^{k,i} \ge \mu I_n$ with $\bar{Y}^{k,i} \le -\mu I_n$. By Theorem 3.7 [52], system (5.38) is internally stable. Applying Lemma 5.1 and completing squares technique, it is easy to show that $\|\tilde{\mathcal{L}}_\infty\| < \gamma$ following the line of Lemma 3.7. $\square$

As pointed out by Corollary 5.2 [52], the discrete-time Riccati equation (5.40) admits at most one bounded stabilizing solution.

REMARK 5.3 The proof above indicates that if the GARE (5.40) admits a bounded stabilizing solution, then (5.46) is solvable. Indeed, we can further show that the solvability of (5.46) also guarantees that GARE (5.40) has a bounded stabilizing solution. So, there is no essential difficulty to provide a SBRL in terms of difference LMIs [78]. In the special case that the coefficients of system (5.38) are time-invariant and involve no jump parameters, the aforementioned conclusion is precisely Theorem 2.5 (i)$\Leftrightarrow$(iv) [65]. ∎

5.3.3 Main result

Next, we give our main result on the infinite horizon stochastic H_2/H_∞ control based on the above preliminaries.

THEOREM 5.5
For system (5.33), assume that $(A_1, \bar{A}|C)$ and $(A_1 + C_1 K_1, \bar{A}|C)$ are stochastically detectable. If the following CDMREs admit a bounded solution $(P_{1,\infty}^{k,i}, K_1^{k,i}; P_{2,\infty}^{k,i}, K_2^{k,i}) \in \mathcal{S}_n^{N+} \times \mathcal{R}_{n \times n_v}^N \times \mathcal{S}_n^{N+} \times \mathcal{R}_{n \times n_u}^N$ on $\mathcal{N} \times \overline{N}$:

$$
\begin{cases}
P_{1,\infty}^{k,i} = (A_1^{k,i} + B_1^{k,i} K_2^{k,i})^T \Psi_i(P_{1,\infty}^{k+1})(A_1^{k,i} + B_1^{k,i} K_2^{k,i}) \\
\quad + \sum\limits_{j=1}^{r} (\bar{A}_j^{k,i})^T \Psi_i(P_{1,\infty}^{k+1}) \bar{A}_j^{k,i} + C_{k,i}^T C_{k,i} \\
\quad + (K_2^{k,i})^T K_2^{k,i} + K_3^{k,i} H_i^1(P_{1,\infty}^{k+1})^{-1}(K_3^{k,i})^T, \\
H_i^1(P_{1,\infty}^{k+1}) \geq \varepsilon_0 I_{n_v}, \ \varepsilon_0 \in (0, \gamma^2 - \|\mathcal{L}_\infty\|^2),
\end{cases} \tag{5.47}
$$

$$
K_1^{k,i} = H_i^1(P_{1,\infty}^{k+1})^{-1}(K_3^{k,i})^T, \tag{5.48}
$$

$$
P_{2,\infty}^{k,i} = (A_1^{k,i} + C_1^{k,i} K_1^{k,i})^T \Psi_i(P_{2,\infty}^{k+1})(A_1^{k,i} + C_1^{k,i} K_1^{k,i}) + C_{k,i}^T C_{k,i}
$$

$$
\quad + \sum_{j=1}^{r} (\bar{A}_j^{k,i})^T \Psi_i(P_{2,\infty}^{k+1}) \bar{A}_j^{k,i} - K_4^{k,i} H_i^2(P_{2,\infty}^{k+1})^{-1}(K_4^{k,i})^T, \tag{5.49}
$$

$$
K_2^{k,i} = -H_i^2(P_{2,\infty}^{k+1})^{-1}(K_4^{k,i})^T, \tag{5.50}
$$

where

$$
H_i^1(P_{1,\infty}^{k+1}) = \gamma^2 I_{n_v} - (C_1^{k,i})^T \Psi_i(P_{1,\infty}^{k+1}) C_1^{k,i},
$$

$$
H_i^2(P_{2,\infty}^{k+1}) = I_{n_u} + (B_1^{k,i})^T \Psi_i(P_{2,\infty}^{k+1}) B_1^{k,i},
$$

$$
K_3^{k,i} = (A_1^{k,i})^T \Psi_i(P_{1,\infty}^{k+1}) C_1^{k,i} + (K_2^{k,i})^T (B_1^{k,i})^T \Psi_i(P_{1,\infty}^{k+1}) C_1^{k,i},
$$

$$
K_4^{k,i} = (A_1^{k,i})^T \Psi_i(P_{2,\infty}^{k+1}) B_1^{k,i} + (K_1^{k,i})^T (C_1^{k,i})^T \Psi_i(P_{2,\infty}^{k+1}) B_1^{k,i},
$$

then the H_2/H_∞ control problem admits a pair of solutions

$$
(u_\infty^*, v_\infty^*) = \{u_k^* = u_{\infty,k}^* = K_2^{k,\theta_k} x_k, \quad v_k^* = v_{\infty,k}^* = K_1^{k,\theta_k} x_k\}_{k \in \mathcal{N}}
$$

with $(K_1^{k,i}, K_2^{k,i})$ given by (5.48) and (5.50).
Conversely, if $(A_1 + C_1 K_1, \bar{A}|C)$ is stochastically detectable and the H_2/H_∞ control problem for system (5.33) is solved by

$$
(u_{\infty,k}^* = K_2^{k,\theta_k} x_k, \ v_{\infty,k}^* = K_1^{k,\theta_k} x_k),
$$

then the CDMREs (5.47)–(5.50) admit a unique solution $(P_{1,\infty}^{k,i}, K_1^{k,i}; P_{2,\infty}^{k,i}, K_2^{k,i}) \in \mathcal{S}_n^{N+} \times \mathcal{R}_{n \times n_v}^N \times \mathcal{S}_n^{N+} \times \mathcal{R}_{n \times n_u}^N$ on $\mathcal{N} \times \overline{N}$. Moreover, $(K_1^{k,i}, K_2^{k,i})$ are determined by (5.48) and (5.50).

Proof. Sufficiency. Observe that (5.47) and (5.49) can be rewritten as

$$
P_{1,\infty}^{k,i} = (A_1^{k,i} + B_1^{k,i} K_2^{k,i})^T \Psi_i(P_{1,\infty}^{k+1})(A_1^{k,i} + B_1^{k,i} K_2^{k,i})
$$

$$
\quad + \sum_{j=1}^{r} (\bar{A}_j^{k,i})^T \Psi_i(P_{1,\infty}^{k+1}) \bar{A}_j^{k,i} + (\tilde{A}_2^{k,i})^T \tilde{A}_2^{k,i} \tag{5.51}
$$

and

$$
\begin{aligned}
P_{2,\infty}^{k,i} &= (A_1^{k,i} + C_1^{k,i}K_1^{k,i} + B_1^{k,i}K_2^{k,i})^T \Psi_i(P_{2,\infty}^{k+1}) \\
&\quad \cdot (A_1^{k,i} + C_1^{k,i}K_1^{k,i} + B_1^{k,i}K_2^{k,i}) \\
&\quad + \sum_{j=1}^{r}(\bar{A}_j^{k,i})^T \Psi_i(P_{2,\infty}^{k+1})\bar{A}_j^{k,i} + (\tilde{A}_3^{k,i})^T \tilde{A}_3^{k,i},
\end{aligned}
$$

$$(5.52)$$

respectively, where $\tilde{A}_2^{k,i}$ and $\tilde{A}_3^{k,i}$ are defined in Lemma 5.2. Since $(A_1 + C_1K_1, \bar{A}\,|C)$ is stochastically detectable, Lemma 5.2 shows that $(A_1 + C_1K_1 + B_1K_2, \bar{A}|\tilde{A}_3)$ is also stochastically detectable. Then, (5.52) together with Theorem 4.1 [52] implies that $(A + C_1K_1 + B_1K_2, \bar{A})$ is SESMS. Thus, $(u_\infty^*, v_\infty^*) \in l_w^2(\mathcal{N}, \mathcal{R}^{n_u}) \times l_w^2(\mathcal{N}, \mathcal{R}^{n_v})$. By a similar argument, we can prove that $(A_1 + B_1K_2, \bar{A})$ is SESMS, i.e., $u_{\infty,k}^* = K_2^{k,\theta_k}x_k$ stabilizes the system (5.33) internally.

Next, let us verify that $\|\mathcal{L}_\infty\| < \gamma$. Substituting $u_{\infty,k}^* = K_2^{k,\theta_k}x_k$ into system (5.33), it follows that

$$
\begin{cases}
x_{k+1} = (A_1^{k,\theta_k} + B_1^{k,\theta_k}K_2^{k,\theta_k})x_k + C_1^{k,\theta_k}v_k + \sum_{j=1}^{r}\bar{A}_j^{k,\theta_k}x_kw_j^k, \\
z_k = \begin{bmatrix} C_{k,\theta_k} \\ D_{k,\theta_k}K_2^{k,\theta_k} \end{bmatrix} x_k, \\
D_{k,\theta_k}^T D_{k,\theta_k} = I_{n_u}, \\
x_0 \in \mathcal{R}^n, \ \theta_0 \in \bar{N}, \ k \in \mathcal{N}.
\end{cases}
$$

$$(5.53)$$

From the preceding derivation, $(A_1 + C_1K_1 + B_1K_2, \bar{A})$ is SESMS, which reveals that (5.47) admits a bounded stabilizing solution $P_{1,\infty}^k \in \mathcal{S}_n^{N+}$. By Theorem 5.4, $\|\mathcal{L}_\infty\| < \gamma$.

Finally, it remains to show that u_∞^* minimizes the output energy $J_{2,\infty}(u, v_\infty^*)$. To this end, we employ the technique of completing squares to $J_{1,\infty}(u_\infty^*, v)$ associated with (5.47) and (5.53). It can be computed that

$$
J_{1,\infty}(u_\infty^*, v) = \sum_{i=1}^{N} \pi_0(i)x_0^T P_{1,\infty}^{0,i}x_0
$$

$$(5.54)$$

$$
- \sum_{k=0}^{\infty} \mathcal{E}\{[v_k - (H_{\theta_k}^1)^{-1}(K_3^{k,\theta_k})^T x_k]^T H_{\theta_k}^1 [v_k - (H_{\theta_k}^1)^{-1}(K_3^{k,\theta_k})^T x_k]\}
$$

$$
\leq \sum_{i=1}^{N} \pi_0(i)x_0^T P_{1,\infty}^{0,i}x_0 = J_{1,\infty}(u_\infty^*, v_\infty^*),
$$

$$(5.55)$$

which indicates that $v_{\infty,k}^* = K_1^{k,\theta_k}x_k$ is just the worst-case disturbance corresponding to u_∞^*. By substituting v_∞^* into system (5.33), we obtain

$$
\begin{cases}
x_{k+1} = (A_1^{k,\theta_k} + C_1^{k,\theta_k}K_1^{k,\theta_k})x_k + B_1^{k,\theta_k}u_k + \sum_{j=1}^{r}\bar{A}_j^{k,\theta_k}x_kw_j^k, \\
z_k = \begin{bmatrix} C_{k,\theta_k}x_k \\ D_{k,\theta_k}u_k \end{bmatrix}, \ D_{k,\theta_k}^T D_{k,\theta_k} = I_{n_u}, \\
x_0 \in \mathcal{R}^n, \ \theta_0 \in \bar{N}, \ k \in \mathcal{N}.
\end{cases}
$$

$$(5.56)$$

The rest is to show that u_∞^* is an optimal solution to the following LQ problem:

$$
\min_{u \in l_w^2(\mathcal{N}, \mathcal{R}^{n_u})} J_{2,\infty}(u, v_\infty^*),
$$
subject to (5.56),

$$(5.57)$$

which is a standard LQ problem discussed in [52]. Note that $(A_1 + C_1 K_1 + B_1 K_2, \bar{A})$ is SESMS. So, (5.49) has a bounded stabilizing solution $P_{2,\infty}(t) \in \mathcal{S}_n^{N+}$. In view of (5.49) and the stochastic detectability of $(A_1 + C_1 K_1, \bar{A}|C)$, Theorem 5.14 and Proposition 6.3 [52] lead directly to

$$
\min_{u \in l_w^2(\mathcal{N}, \mathcal{R}^{n_u})} J_{2,\infty}(u, v^*) = \sum_{i=1}^{N} \pi_0(i) x_0^T P_{2,\infty}^{0,i} x_0 = J_{2,\infty}(u_\infty^*, v_\infty^*),
$$

where $u_{\infty,k}^* = K_2^{k,\theta_k} x_k$. This completes the sufficiency proof.

Necessity: Firstly, substitute $u_{\infty,k}^*$ into (5.33) to obtain system (5.53). From the notion of H_2/H_∞ control, $u_{\infty,k}^*$ stabilizes (5.53) internally, and $\|\mathcal{L}_\infty\| < \gamma$. Thus, by Theorem 5.4 , (5.47) admits a stabilizing solution $P_{1,\infty}^k \in \mathcal{S}_n^{N+}$, which means that $(A_1 + C_1 K_1 + B_1 K_2, \bar{A})$ is SESMS. Moreover, according to Corollary 3.9 of [52], the internal stability of (5.53) ensures $x_k \in l_w^2(\mathcal{N}, \mathcal{R}^n)$ for any $v_k \in l_w^2(\mathcal{N}, \mathcal{R}^{n_v})$. Then, similar to the derivation as in the sufficiency proof, (5.53) together with (5.47) gives rise to (5.55), which justifies that

$$
K_1^{k,\theta_k} = H_{\theta_k}^1 (P_{1,\infty}^{k+1})^{-1} (K_3^{k,\theta_k})^T.
$$

To proceed, substituting $v_{\infty,k}^*$ into (5.33) gives system (5.56). Recalling Definition 5.3, $u_{\infty,k}^*$ is optimal for the LQ optimization (5.57). Taking into account that $(A_1 + C_1 K_1 + B_1 K_2, \bar{A})$ is SESMS, we have that $(A_1 + C_1 K_1, B_1; \bar{A})$ is stochastically stabilizable. Using Theorem 5.14 [52], (5.49) admits a stabilizing solution $P_{2,\infty} \in \mathcal{S}_n^{N+}$. Furthermore, making use of the completing squares technique, we have

$$
\begin{aligned}
J_{2,\infty}(u, v_\infty^*) =& \sum_{i=1}^{N} \pi_0(i) x_0^T P_{2,\infty}^{0,i} x_0 \\
&+ \sum_{k=0}^{\infty} \mathcal{E}\{[u_k + (H_{\theta_k}^2 (P_{2,\infty}^{k+1}))^{-1} (K_4^{k,\theta_k})^T x_k]^T H_{\theta_k}^2 (P_{2,\infty}^{k+1}) \\
&\quad \cdot [u_k + (H_{\theta_k}^2 (P_{2,\infty}^{k+1}))^{-1} (K_4^{k,\theta_k})^T x_k]\} \\
\geq& \sum_{i=1}^{N} \pi_0(i) x_0^T P_{2,\infty}^{0,i} x_0 = J_{2,\infty}(u_\infty^*, v_\infty^*),
\end{aligned}
$$

which verifies that $K_2^{k,\theta_k} = -H_{\theta_k}^2 (P_{2,\infty}^{k+1})^{-1} (K_4^{k,\theta_k})^T$. The proof is completed. $\square$

REMARK 5.4　Lemma 5.2 only holds for systems with state-dependent noise under stochastic detectability. As said in Remark 4.4 [206], to extend Theorem 5.5 to general stochastic systems with (x, u, v)-dependent noise, we have to replace stochastic detectability with other notions of detectability such as uniform detectability or exact detectability. We believe, by generalizing the recently developed uniform detectability and exact detectability [218] to the system (5.37), one can succeed in dealing with the infinite horizon stochastic H_2/H_∞ control of discrete time-varying Markov jump systems with (x, u, v)-dependent noise. ∎

5.3.4   An economic example

Consider the following multiplier-accelerator macroeconomic system [166]:

$$\begin{cases} \mathcal{C}_k = (1 - \phi)\mathcal{Y}_{k-1}, \\ \mathcal{I}_k = \psi(\mathcal{Y}_{k-1} - \mathcal{Y}_{t-2}), \\ \mathcal{Y}_k = \mathcal{C}_k + \mathcal{I}_k + \mathcal{G}_k + \mathcal{E}_k, \ k \in \mathcal{N}, \end{cases} \tag{5.58}$$

where $\mathcal{C}_k$ is the consumption expenditure, $\mathcal{Y}_k$ is the national income, $\mathcal{I}_k$ is the induced private investment, $\mathcal{G}_k$ is the government expenditure, $\mathcal{E}_k$ is the net export, ϕ is the marginal propensity to save and ψ is the accelerator coefficient. It has long been recognized that the economy may be in either a fast or slow growth phase with the switching between the two phases governed by the outcome of a Markov process [95]. Thus, a homogeneous Markov chain with the state space $\overline{N} = \{1, 2\}$ and the transition probability matrix $\mathcal{P}_{2\times2} = (p_{ij})_{2\times2}$ is introduced to describe the economic situations: "$i = 1$" stands for "fast growth phase" and "$i = 2$" means "slow growth phase." Let the parameters $\phi = \phi_{0,k} + \sigma_1 w_1^k$ and $\psi = \psi_{0,k} + \sigma_2 w_2^k$ where w_1^k and w_2^k are independent standard Gaussian white noises specifying statistical bias caused by inflation or deflation, the nominal values $\phi_{0,k}$ and $\psi_{0,k}$ are time-varying due to the periodic fluctuation of economy in each growth phase. By denoting $x_{1,k} = \mathcal{Y}_{k-1}, x_{2,k} = \mathcal{Y}_k$, $x_k = [x_{1,k} \ x_{2,k}]^T$, $u_k = \mathcal{G}_k$, $v_k = \mathcal{E}_k$, $z_k = (\mathcal{Y}_k \ \mathcal{G}_k)^T$ and $\xi_k = 1 - \phi_{0,k} + \psi_{0,k}$, we obtain the following state-space representation for system (5.58):

$$\begin{cases} x_{k+1} = \begin{bmatrix} 0 & 1 \\ -\psi_{0,k} & \xi_k \end{bmatrix} x_k + \begin{bmatrix} 0 \\ 1 \end{bmatrix} u_k + \begin{bmatrix} 0 \\ 1 \end{bmatrix} v_k \\ \qquad\quad + \left(\begin{bmatrix} 0 & 0 \\ 0 & -\sigma_1 \end{bmatrix} w_1^k + \begin{bmatrix} 0 & 0 \\ -\sigma_2 & \sigma_2 \end{bmatrix} w_2^k \right) x_k, \\ z_k = \begin{bmatrix} [0 \ 1]x_k \\ u_k \end{bmatrix}, \ k \in \mathcal{N}. \end{cases} \tag{5.59}$$

In reality, the government needs to regulate the reliance of the national economy on external trade by means of government purchases. On the other hand, the risk of high inflation induced by high money growth rates and possible budget deficits arising from governmental overspending must be carefully taken into consideration, namely, the total quantity of the national income and the government expenditure, which is assessed by $J_{2,\infty}(u, v) = \sum_{k=0}^{\infty} \mathcal{E}[x_{2,k}^2 + u_k^2]$, has to be minimized.

Given a disturbance attenuation level $\gamma = 1.97$, we set $(x_{1,0}, x_{2,0}) = (10, 12)$, $(\sigma_1, \sigma_2) = (0.02, 0.01)$, the initial mode $\theta_0 = 1$ and the accuracy $\epsilon = 1 \times 10^{-4}$. Based on the data of [43], $p_{11} = 0.5$, $p_{12} = 0.5$, $p_{21} = 0.2$ and $p_{22} = 0.8$. According to the acceleration principle [12], we may take $\phi_{0,k} = 0.84 + 0.03 \cos \pi k$, $\psi_{0,k} = 0.73 + 0.01 \cos \pi k$ if $\theta_k = 1$ (fast growth phase) and $\phi_{0,k} = 0.9 + 0.03 \cos \pi k$, $\psi_{0,k} = 0.54 + 0.01 \cos \pi k$ if $\theta_k = 2$ (slow growth phase). By means of Theorem 5.5, we obtain the worst-case disturbance (the adverse consequence of the international trade cycle) and H_2/H_∞ control design (the fiscal policy of government):

$$v_{\infty,k}^* = -(0.13 + 0.005 \cos \pi k)x_{1,k} + (0.129 - 0.001 \cos \pi k)x_{2,k},$$

$$u_{\infty,k}^* = (0.507 + 0.001 \cos \pi k)x_{1,k} - 0.505x_{2,k}$$

when $\theta_k = 1$. When $\theta_k = 2$,

$$v_{\infty,k}^* = -(0.096 + 0.005 \cos \pi k)x_{1,k} + 0.078x_{2,k},$$

$$u_{\infty,k}^* = (0.375 + 0.001 \cos \pi k)x_{1,k} - 0.303x_{2,k}.$$

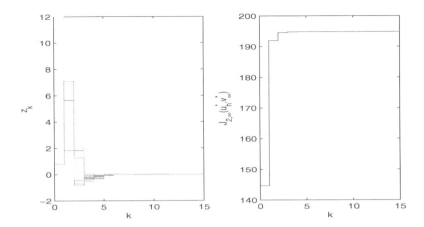

FIGURE 5.1

z_k and $J_{2,\infty}$ corresponding to $(\tilde{u}_\infty^*, v_\infty^*)$.

As discussed previously, it is easy to retrieve the H_2 optimal control

$$u_{s,k}^* = -(1.006 + 0.007 \cos \pi k)x_{1,k} + (1.024 - 0.008 \cos \pi k)x_{2,k},$$

and H_∞ optimal control

$$\tilde{u}_{\infty,k}^* = (0.711 + 0.002 \cos \pi k)x_{1,k} - (0.657 - 0.004 \cos \pi k)x_{2,k}$$

for $\theta_k = 1$. When $\theta_k = 2$,

$$u_{s,k}^* = -(0.744 + 0.008 \cos \pi k)x_{1,k} + (0.624 - 0.005 \cos \pi k)x_{2,k},$$

$$\tilde{u}_{\infty,k}^* = 0.526x_{1,k} - (0.374 - 0.007 \cos \pi k)x_{2,k}.$$

It is clearly shown in Figures 5.1–5.2 that $J_{2,\infty}(u_\infty^*, v_\infty^*) < J_{2,\infty}(\tilde{u}_\infty^*, v_\infty^*)$, which demonstrates the advantage of the H_2/H_∞ control design. Besides, when H_2 control strategy u_s^* is also applied to (5.59), it fails to stabilize the closed-loop system in the presence of v_∞^* and so $J_{2,\infty}(u_s^*, v_\infty^*)$ tends to infinity, which is displayed in Figure 5.3.

5.4 Infinite Horizon Discrete Time-Invariant H_2/H_∞ Control

Up to now, we have only dealt with the H_2/H_∞ control for discrete time-varying stochastic Markov jump systems with state-dependent noise. In this section, under exact detectability, we consider the H_2/H_∞ control for discrete time-invariant Markov jump systems with (x, u, v)-dependent noise, which can be viewed as an extension of Section 3.3. The system under

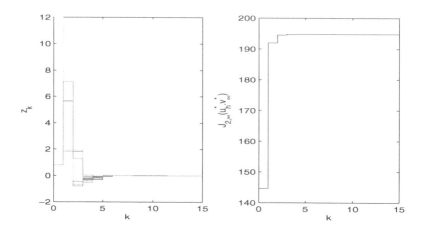

FIGURE 5.2

z_k and $J_{2,\infty}$ corresponding to $\left(u_\infty^*, v_\infty^*\right)$.

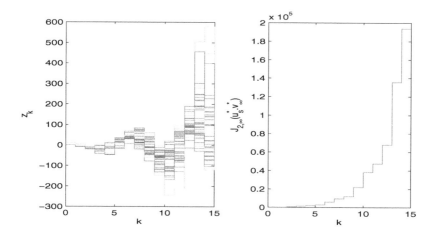

FIGURE 5.3

z_k and $J_{2,\infty}$ corresponding to $\left(u_s^*, v_\infty^*\right)$.

consideration is described as

$$
\begin{cases}
x_{k+1} = A_1^{\theta_k} x_k + B_1^{\theta_k} u_k + C_1^{\theta_k} v_k + (A_2^{\theta_k} x_k + B_2^{\theta_k} u_k + C_2^{\theta_k} v_k) w_k, \\[2mm]
z_k = \begin{bmatrix} C_{\theta_k} x_k \\ D_{\theta_k} u_k \end{bmatrix}, \ D_{\theta_k}^T D_{\theta_k} = I_{n_u}, \\[4mm]
x_0 \in \mathcal{R}^n, \ \theta_0 \in \overline{N}, \ k \in \mathcal{N},
\end{cases}
\tag{5.60}
$$

where, different from previous sections, in (5.60), θ_k is a homogeneous Markov process with the transition probability matrix $\mathcal{P}_{N \times N} = (p_{ij})_{N \times N}$, $p_{ij} = \mathcal{P}(\theta_{k+1} = j | \theta_k = i)$. To study the H_2/H_∞ control of (5.60), we should introduce some preliminaries.

5.4.1 Stability, stabilization, and SBRL

For the Hilbert space S_n^N, its inner product is defined as

$$
\langle U, V \rangle = \sum_{i=1}^{N} \text{Trace}(U_i V_i)
$$

for any $U, V \in S_n^N$. Associated with the system

$$
\begin{cases}
x_{k+1} = A_1^{\theta_k} x_k + A_2^{\theta_k} x_k w_k, \\
x_0 \in \mathcal{R}^n, \ \theta_0 \in \overline{N}, \ k \in \mathcal{N},
\end{cases}
\tag{5.61}
$$

define a Lyapunov operator $\mathcal{D}^{A_1, A_2} : U = (U_1, U_2, \cdots, U_N) \in S_n^N \mapsto S_n^N$ as

$$
\mathcal{D}^{A_1, A_2}(U) = (\mathcal{D}_1^{A_1, A_2}(U), \cdots, \mathcal{D}_N^{A_1, A_2}(U))
$$

with

$$
\mathcal{D}_i^{A_1, A_2}(U) = \sum_{j=1}^{N} p_{ji} A_1^j U_j (A_1^j)^T + \sum_{j=1}^{N} p_{ji} A_2^j U_j (A_2^j)^T.
$$

Obviously, $\mathcal{D}^{A_1, A_2}$ is a linear positive operator, and its adjoint operator is given by

$$
(\mathcal{D}^{A_1, A_2})^*(U) = ((\mathcal{D}_1^{A_1, A_2})^*(U), \cdots, (\mathcal{D}_N^{A_1, A_2})^*(U)),
$$

where

$$
(\mathcal{D}_i^{A_1, A_2})^*(U) = \sum_{j=1}^{N} p_{ij} (A_1^i)^T U_j A_1^i + \sum_{j=1}^{N} p_{ij} (A_2^i)^T U_j A_2^i.
$$

DEFINITION 5.6 [52]   *The discrete-time linear stochastic system*

$$
x_{k+1} = A_1^{\theta_k} x_k + A_2^{\theta_k} x_k w_k
\tag{5.62}
$$

is called ASMS if for all $x_0 \in \mathcal{R}^n$ and $\theta_0 \in \overline{N}$,

$$
\lim_{k \to \infty} \mathcal{E} \|x_k\|^2 = \lim_{k \to \infty} \mathcal{E} \|\Phi(k, 0) x_0\|^2 = 0,
$$

where $\Phi(k, 0)$ is the fundamental matrix solution of (5.62). In this case, $(A_1, A_2; \theta)$ is also called Schur stable.

Below, we introduce the notion of stabilizability for the following linear stochastic time-invariant control system

$$\begin{cases} x_{k+1} = A_1^{\theta_k} x_k + B_1^{\theta_k} u_k + (A_2^{\theta_k} x_k + B_2^{\theta_k} u_k) w_k, \\ x_0 \in \mathcal{R}^n, \ \theta_0 \in \overline{N}, \ k \in \mathcal{N}. \end{cases} \qquad (5.63)$$

DEFINITION 5.7   *System (5.63) is stochastically stabilizable or stabilizable in short if there exists a sequence $\{F_{\theta_k}\}_{k \in \mathcal{N}} \in \mathcal{R}^{n \times n_u}$ such that the closed-loop system*

$$x_{k+1} = (A_1^{\theta_k} + B_1^{\theta_k} F_{\theta_k}) x_k + (A_2^{\theta_k} + B_2^{\theta_k} F_{\theta_k}) x_k w_k$$

is ASMS for any $(x_0, \theta_0) \in \mathcal{R}^n \times \overline{N}$, where $u_k = F_{\theta_k} x_k$ is called a stabilizing feedback.

By the finite dimensional Krein–Rutman Theorem [164], we have the following lemma.

LEMMA 5.3
Let $\rho(\mathcal{D}^{A_1, A_2})$ be the spectral radius of $\mathcal{D}^{A_1, A_2}$. Then, there exists a non-zero $X \in S_n^{N+}$ such that $\mathcal{D}^{A_1, A_2}(X) = \rho(\mathcal{D}^{A_1, A_2}) X$.

Parallel to Theorem 3.6, we have the following result.

THEOREM 5.6 [52]
System (5.62) is ASMS iff $\rho(\mathcal{D}^{A_1, A_2}) < 1$, where $\rho(\cdot)$ denotes the spectral radius of $\mathcal{D}^{A_1, A_2}$.

Now, consider the following perturbed system

$$\begin{cases} x_{k+1} = A_{11}^{\theta_k} x_k + B_{11}^{\theta_k} v_k + (A_{12}^{\theta_k} x_k + B_{12}^{\theta_k} v_k) w_k, \\ z_1^k = C_{11}^{\theta_k} x_k, \ x_0 \in \mathcal{R}^n, \ \theta_0 \in \overline{N}, \ k \in \mathcal{N}. \end{cases} \qquad (5.64)$$

Corresponding to Theorem 5.4, for system (5.64), we have the following SBRL.

LEMMA 5.4 [52]
For a prescribed $\gamma > 0$, if system (5.64) is internally stable (i.e., system (5.64) is ASMS when $v \equiv 0$) and $\|\tilde{\mathcal{L}}_\infty\| < \gamma$, then the following GARE has a stabilizing solution $P = (P^1, P^2, \cdots, P^N) \in S_n^{N+}$:

$$\begin{cases} P^i = \Pi_i^1(P) + (C_{11}^i)^T C_{11}^i - \Pi_i^2(P) \Pi_i^3(P)^{-1} \Pi_i^2(P)^T, \\ \Pi_i^3(P) < 0. \end{cases} \qquad (5.65)$$

Conversely, if (5.65) admits a stabilizing solution $P \in S_n^{N+}$, then system (5.64) is internally stable and $\|\tilde{\mathcal{L}}_\infty\| < \gamma$. Here, in (5.65),

$$\Psi_i(P) = \sum_{j=1}^N p_{ij} P^j,$$

$$\Pi_i^1(P) = (A_{11}^i)^T \Psi_i(P) A_{11}^i + (A_{12}^i)^T \Psi_i(P) A_{12}^i,$$
$$\Pi_i^2(P) = (A_{11}^i)^T \Psi_i(P) B_{11}^i + (A_{12}^i)^T \Psi_i(P) B_{12}^i,$$
$$\Pi_i^3(P) = (B_{11}^i)^T \Psi_i(P) B_{11}^i + (B_{12}^i)^T \Psi_i(P) B_{12}^i - \gamma^2 I.$$

5.4.2 Exact detectability and extended Lyapunov theorem

DEFINITION 5.8 *Consider the following discrete-time Markov jump system:*

$$\begin{cases} x_{k+1} = A_1^{\theta_k} x_k + A_2^{\theta_k} x_k w_k, \\ y_k = C_{\theta_k} x_k, \ x_0 \in \mathcal{R}^n, \ \theta_0 \in \overline{N}, \ k \in \mathcal{N}, \end{cases} \tag{5.66}$$

where y_k is the measurement output.

For arbitrary $x_0 \in \mathcal{R}^n$ and $\theta_0 \in \overline{N}$, if $y_k \equiv 0$ a.s. for $k \in \mathcal{N}$, implies $\lim_{k\to\infty} \mathcal{E}\|x_k\|^2 = 0$, then system (5.66) or $(A_1, A_2|C)$ is said to be exactly detectable. It is said to be exactly observable if $y_k \equiv 0$ a.s., $k \in \mathcal{N} \Rightarrow x_0 = 0$.

REMARK 5.5 We note that [42] introduced the so-called "W-detectability" for the system

$$x_{k+1} = A_1^{\theta_k} x_k, \ y_k = C_{\theta_k} x_k,$$

which has been generalized to the system (5.66) by [161, 217] and shown to be equivalent to Definition 5.8. ∎

In what follows, we present a PBH criterion for the exact detectability of $(A_1, A_2|C)$.

THEOREM 5.7
$(A_1, A_2|C)$ is exactly detectable iff there does not exist non-zero $X \in S_n^N$ such that

$$\mathcal{D}^{A_1,A_2}(X) = \lambda X, \ CX := (C_1 X_1, \cdots, C_N X_N) = 0, \ |\lambda| \geq 1. \tag{5.67}$$

Proof. Denote $X_i^k := \mathcal{E}[x_k x_k^T I_{\{\theta_k = i\}}]$, $X^k := (X_1^k, \ X_2^k, \ \cdots, X_N^k)$; then it is easy to show that $X_i^{k+1} = \mathcal{D}_i^{A_1,A_2}(X^k)$. Obviously,

$$\lim_{k\to\infty} \|x_k\|^2 = 0 \Leftrightarrow \lim_{k\to\infty} X_i^k = 0, \forall i \in \overline{N} \Leftrightarrow \lim_{k\to\infty} X^k = 0,$$

while

$$y_k = 0, \ a.s. \Leftrightarrow Y_i^k := \mathcal{E}[y_k y_k^T I_{\{\theta_k = i\}}] = 0 \Leftrightarrow C_i X_i^k = 0, \forall i \in \overline{N} \Leftrightarrow Y^k = 0$$

where $Y^k = (Y_1^k, Y_2^k, \cdots, Y_N^k)$. Hence, $(A_1, A_2|C)$ is exactly detectable iff the deterministic matrix-valued system

$$X^{k+1} = \mathcal{D}^{A_1,A_2}(X^k), \ Y^k = CX^k$$

is exactly detectable. By means of the $\mathcal{H}$-representation technique and the PBH criterion for the complete detectability of deterministic vector-valued systems, this theorem is easily proved. □

Stochastic detectability of $(A_1, A_2|C)$ was given in [50, 52], which said that $(A_1, A_2|C)$ is stochastically detectable if there are matrices K^{θ_k}, $\theta_k \in \overline{N}$, $k \in \mathcal{N}$, such that

$$x_{k+1} = (A_1^{\theta_k} + K^{\theta_k} C_{\theta_k}) x_k + A_2^{\theta_k} x_k w_k \tag{5.68}$$

is ASMS.

PROPOSITION 5.2

If $(A_1, A_2|C)$ is stochastically detectable, then it is also exactly detectable.

Proof. Because $(A_1, A_2|C)$ is stochastically detectable, we can find a feedback gain $K(\theta_k)$ such that (5.68) is ASMS. Provided that $C_{\theta_k} x_k \equiv 0$, then system (5.68) coincides with (5.66). That is, $C_{\theta_k} x_k \equiv 0$ implies that the state trajectory of system (5.66) is ASMS. Hence, $(A_1, A_2|C)$ is exactly detectable. $\square$

The converse of Proposition 5.2 does not hold, i.e., exact detectability is weaker than stochastic detectability; see the following counterexample.

Example 5.2
Let

$$A_1^1 = A_1^2 = A_1 = \begin{bmatrix} 2 & 0 \\ 0 & 1 \end{bmatrix}, \quad A_2^1 = A_2^2 = A_2 = \begin{bmatrix} 1 & 0 \\ -1 & 0 \end{bmatrix},$$

$$C_1 = C_2 = C = [0\ 1], \quad \overline{N} = \{1, 2\}.$$

$\mathcal{P}_{2\times2} = (p_{ij})_{2\times2}$ is an arbitrary stochastic matrix. It is straightforward to test that there does not exist a non-zero $X \in \mathcal{S}_n$ satisfying (5.67). Thus, $(A_1, A_2|C)$ is exactly detectable. By Corollary 4.3 [52], a necessary and sufficient condition for $(A_1, A_2|C)$ to be stochastically detectable is that there are matrices $Y > 0$ and $Z \in \mathcal{R}^{n \times n_z}$ satisfying the following LMI:

$$\begin{bmatrix} -Y & A_1^T Y + C^T Z^T & A_2^T Y \\ Y A_1 + ZC & -Y & 0 \\ Y A_2 & 0 & -Y \end{bmatrix} < 0. \tag{5.69}$$

However, by using the LMI control toolbox in MATLAB, (5.69) is not strictly feasible, which means that $(A_1, A_2|C)$ is not stochastically detectable. $\quad\square$

THEOREM 5.8
Assume that $(A_1, A_2|C)$ is exactly detectable. Then, $(A_1, A_2; \theta)$ is Schur stable iff the following GLE has a unique solution $X = (X_1, \cdots, X_N) \in \mathcal{S}_n^{N+}$:

$$X = (\mathcal{D}^{A_1, A_2})^*(X) + \tilde{C}, \quad \tilde{C} = (\tilde{C}_1, \cdots, \tilde{C}_N), \quad \tilde{C}_i = C_i^T C_i, \tag{5.70}$$

or equivalently,

$$X_i = (A_1^i)^T \Psi_i(X) A_1^i + (A_2^i)^T \Psi_i(X) A_2^i + C_i^T C_i, \quad i \in \overline{N}.$$

Proof. Necessity: Since $(A_1, A_2; \theta)$ is Schur stable, from Theorem 2.5 [52] and the non-negativity of $\tilde{C}$, the equation (5.70) admits a unique solution $X \in \mathcal{S}_n^{N+}$.

Sufficiency: Assume that the equation (5.70) admits a solution $X \in S_n^{N+}$, but $(A_1, A_2; \theta)$ is not Schur stable, then $\rho(\mathcal{D}^{A_1,A_2}) \geq 1$. Moreover, by Lemma 5.3, there exists non-zero $\bar{X} \in S_n^{N+}$ satisfying $\mathcal{D}^{A_1,A_2}(\bar{X}) = \rho(\mathcal{D}^{A_1,A_2})\bar{X}$. Taking into account that $(A_1, A_2|C)$ is exactly detectable, by Theorem 5.7, we deduce $C_i\bar{X}_i \neq 0$, $i \in \overline{N}$. On the other hand, by standard inner product manipulations, we have

$$0 \leq \sum_{i=1}^N \mathrm{Trace}(C_i\bar{X}_iC_i^T) = \sum_{i=1}^N \mathrm{Trace}(C_i^TC_i\bar{X}_i) = \langle \tilde{C}, \bar{X} \rangle$$

$$= \langle X - (\mathcal{D}^{A_1,A_2})^*(X), \bar{X} \rangle = \langle X, \bar{X} \rangle - \langle (\mathcal{D}^{A_1,A_2})^*(X), \bar{X} \rangle$$

$$= \langle X, \bar{X} \rangle - \langle X, (\mathcal{D}^{A_1,A_2})(\bar{X}) \rangle = \langle X, (1 - \rho((\mathcal{D}^{A_1,A_2})(X)))\bar{X} \rangle \leq 0,$$

which reveals that $C_i\bar{X}_iC_i^T = 0$ or further $C_i\bar{X}_i = 0$ for all $i \in \overline{N}$ due to $\bar{X} \geq 0$. This is a contradiction. So $(A_1, A_2; \theta)$ must be Schur stable. $\square$

When $A_2^{\theta_k} \equiv 0$ for $k \in \mathcal{N}$, Theorem 5.8 yields Theorem 13 of [42]. Theorem 5.8 also generalizes Lemma 3.5-(iii) to Markov jump systems.

5.4.3 Main result and numerical algorithm

Similar to the proof of Theorem 3.8, we present the following theorem without proof. Define

$$P_{1,\infty} = (P_{1,\infty}^1, P_{1,\infty}^2, \cdots, P_{1,\infty}^N) \in S_n^{N+}, \ P_{2,\infty} = (P_{2,\infty}^1, P_{2,\infty}^2, \cdots, P_{2,\infty}^N) \in S_n^{N+},$$

$$K_1 = (K_1^1, K_1^2, \cdots, K_1^N) \in \mathcal{R}_{n_v \times n}^N, \ K_2 = (K_2^1, K_2^2, \cdots, K_2^N) \in \mathcal{R}_{n_u \times n}^N.$$

THEOREM 5.9

For system (5.60), assume that the following CDMREs admit a solution $(P_{1,\infty}, K_1; P_{2,\infty}, K_2) \in S_n^{N+} \times \mathcal{R}_{n_v \times n}^N \times S_n^{N+} \times \mathcal{R}_{n_u \times n}^N$ *for* $i \in \overline{N}$:

$$\begin{cases} P_{1,\infty}^i = \sum_{j=1}^2 (A_j^i + B_j^iK_2^i)^T\Psi_i(P_{1,\infty})(A_j^i + B_j^iK_2^i) + C_i^TC_i \\ \qquad + (K_2^i)^TK_2^i + K_3^i\hat{H}_i^1(P_{1,\infty})^{-1}(K_3^i)^T, \\ H_i^1(P_{1,\infty}) > 0, \end{cases} \tag{5.71}$$

$$K_1^i = H_i^1(P_{1,\infty})^{-1}(K_3^i)^T, \tag{5.72}$$

$$P_{2,\infty}^i = \sum_{j=1}^2 (A_j^i + B_j^iK_1^i)^T\Psi_i(P_{2,\infty})(A_j^i + B_j^iK_1^i) + C_i^TC_i$$

$$\qquad - K_4^iH_i^2(P_{2,\infty})^{-1}(K_4^i)^T, \tag{5.73}$$

$$K_2^i = -H_i^2(P_{2,\infty})^{-1}(K_4^i)^T, \tag{5.74}$$

where

$$H_i^1(P_{1,\infty}) = \gamma^2 I_{n_v} - (C_1^i)^T\Psi_i(P_{1,\infty})C_1^i - (C_2^i)^T\Psi_i(P_{1,\infty})C_2^i,$$

$$H_i^2(P_{2,\infty}) = I_{n_u} + (B_1^i)^T\Psi_i(P_{2,\infty})B_1^i + (B_2^i)^T\Psi_i(P_{2,\infty})B_2^i,$$

$$K_3^i = \sum_{j=1}^2 (A_j^i + B_j^iK_2^i)^T\Psi_i(P_{1,\infty})C_j^i,$$

$$K_4^i = \sum_{j=1}^2 (A_j^i + C_j^iK_1^i)^T\Psi_i(P_{2,\infty})B_j^i.$$

If $(A_1, A_2|C)$ and $(A_1 + C_1K_1, A_2 + C_2K_1|C)$ are exactly detectable, then the H_2/H_∞ control problem admits a pair of solutions $(u^*_{\infty,k} = K_2^{\theta_k} x_k, v^*_{\infty,k} = K_1^{\theta_k} x_k)$ with K_1 and K_2 given by (5.72) and (5.74), respectively.
Conversely, if $(A_1 + C_1K_1, A_2 + C_2K_1|C)$ is exactly detectable and the H_2/H_∞ control problem of (5.60) is solved by

$$u^*_{\infty,k} = K_2^{\theta_k} x_k, \quad v^*_{\infty,k} = K_1^{\theta_k} x_k,$$

then the CDMREs (5.71)–(5.74) have a unique quaternion solution $(P_{1,\infty}, K_1; P_{2,\infty}, K_2) \in S_n^{N+} \times \mathcal{R}_{n \times n_v}^N \times S_n^{N+} \times \mathcal{R}_{n \times n_u}^N$.

REMARK 5.6   From the development of this section, we can see that most results are simple extensions of discrete-time stochastic systems without Markov jumps. However, how to establish the H_2/H_∞ theory for infinite state nonhomogeneous Markov jump systems deserves further investigation.   ∎

From Theorem 5.9, we see that the key to the design of a state feedback H_2/H_∞ controller lies in computing the solutions of CDMREs (5.71)–(5.74). In the following, we will focus on seeking an iterative algorithm to solve (5.71)–(5.74). By Theorem 5.2, associated with the finite horizon H_2/H_∞ control problem, if the following CDMREs

$$\begin{cases} P_{1,\mathcal{T}}^{k,i} = \sum_{j=1}^2 (A_j^i + B_j^i K_{2,\mathcal{T}}^{k,i})^T \Psi_i(P_{1,\mathcal{T}}^{k+1})(A_j^i + B_j^i K_{2,\mathcal{T}}^{k,i}) \\ \qquad + C_i^T C_i + (K_{2,\mathcal{T}}^{k,i})^T K_{2,\mathcal{T}}^{k,i} + K_{3,\mathcal{T}}^{k,i} H_i^1(P_{1,\mathcal{T}}^{k+1})^{-1}(K_{3,\mathcal{T}}^{k,i})^T, \\ H_i^1(P_{1,\mathcal{T}}^{k+1}) = \gamma^2 I_{n_v} - \sum_{j=1}^2 (C_j^i)^T \Psi_i(P_{1,\mathcal{T}}^{k+1})C_j^i > 0, \\ P_{1,\mathcal{T}}^{\mathcal{T}+1} = 0, \end{cases} \qquad (5.75)$$

$$K_{1,\mathcal{T}}^{k,i} = H_i^1(P_{1,\mathcal{T}}^{k+1})^{-1}(K_{3,\mathcal{T}}^{k,i})^T, \qquad (5.76)$$

$$\begin{cases} P_{2,\infty}^{\mathcal{T},i} = \sum_{j=1}^2 (A_j^i + C_j^i K_{1,\mathcal{T}}^{k,i})^T \Psi_i(P_{2,\mathcal{T}}^{k+1})(A_j^i + C_j^i K_{1,\mathcal{T}}^{k,i}) \\ \qquad + C_i^T C_i - K_{4,\mathcal{T}}^{k,i} H_i^2(P_{2,\mathcal{T}}^{k+1})^{-1}(K_{4,\mathcal{T}}^{k,i})^T, \\ H_i^2(P_{2,\mathcal{T}}^{k+1}) = \sum_{j=1}^2 (B_j^i)^T \Psi_i(P_{2,\mathcal{T}}^{k+1})B_j^i + I_{n_u}, \\ P_{2,\mathcal{T}}^{\mathcal{T}+1} = 0, \end{cases} \qquad (5.77)$$

$$K_{2,\mathcal{T}}^{k,i} = -H_i^2(P_{2,\mathcal{T}}^{k+1})^{-1}(K_{4,\mathcal{T}}^{k,i})^T \qquad (5.78)$$

admit a quaternion solution $(P_{1,\mathcal{T}}^k, K_{1,\mathcal{T}}^k; P_{2,\mathcal{T}}^k, K_{2,\mathcal{T}}^k) \in S_n^{N+} \times \mathcal{R}_{n_v \times n}^N \times S_n^{N+} \times \mathcal{R}_{n_u \times n}^N$ on $N_{\mathcal{T}+1} \times \overline{N}$, then

$$J_{1,\mathcal{T}}(u^*_{\mathcal{T}}, v^*_{\mathcal{T}}; \theta_0 = i) := \min_{v \in l_w^2(N_{\mathcal{T}}, \mathcal{R}^{n_v}), u = u^*_{\mathcal{T}}} \sum_{k=0}^{\mathcal{T}} \mathcal{E}(\gamma^2 \|v_k\|^2 - \|z_k\|^2 | \theta_0 = i)$$

$$= -x_0^T P_{1,\mathcal{T}}^{0,i} x_0, \qquad (5.79)$$

$$J_{2,\mathcal{T}}(u^*_{\mathcal{T}}, v^*_{\mathcal{T}}; \theta_0 = i) := \min_{u \in l_w^2(N_{\mathcal{T}}, \mathcal{R}^{n_u}), v = v^*_{\mathcal{T}}} \sum_{k=0}^{\mathcal{T}} \mathcal{E}(\|z_k\|^2 | r_0 = i)$$

$$= x_0^T P_{2,\mathcal{T}}^{0,i} x_0, \qquad (5.80)$$

where $u^*_{\mathcal{T},k} = K^{k,\theta_k}_{2,\mathcal{T}} x_k$, $v^*_{\mathcal{T},k} = K^{k,\theta_k}_{1,\mathcal{T}} x_k$, and

$$K^{k,i}_{3,\mathcal{T}} = \sum_{j=1}^{2}(A^i_j + B^i_j K^{k+1,i}_{2,\mathcal{T}})^T \Psi_i(P^{k+1}_{1,\mathcal{T}})C^i_j, \tag{5.81}$$

$$K^{k,i}_{4,\mathcal{T}} = \sum_{j=1}^{2}(A^i_j + C^i_j K^{k+1,i}_{1,\mathcal{T}})^T \Psi_i(P^{k+1}_{2,\mathcal{T}})B^i_j. \tag{5.82}$$

By the approximation analysis of the LQ optimal control problem (see Chapter 6 of [52]), for any $x_0 \in \mathcal{R}^n$, we have

$$\lim_{\mathcal{T}\to\infty} x_0^T P^{0,i}_{1,\mathcal{T}} x_0 = \lim_{\mathcal{T}\to\infty} J_{1,\mathcal{T}}(u^*_\mathcal{T}, v^*_\mathcal{T}; \theta_0 = i) = J_{1,\infty}(u^*_\infty, v^*_\infty; \theta_0 = i)$$
$$= x_0^T P^i_{1,\infty} x_0, \tag{5.83}$$

and

$$\lim_{\mathcal{T}\to\infty} x_0^T P^{0,i}_{2,\mathcal{T}} x_0 = \lim_{\mathcal{T}\to\infty} J_{2,\mathcal{T}}(u^*_\mathcal{T}, v^*_\mathcal{T}; \theta_0 = i) = J_{2,\infty}(u^*_\infty, v^*_\infty; \theta_0 = i)$$
$$= x_0^T P^i_{2,\infty} x_0 \tag{5.84}$$

which implies

$$\lim_{\mathcal{T}\to\infty} P^{0,i}_{1,\mathcal{T}} = P^i_{1,\infty}, \quad \lim_{\mathcal{T}\to\infty} P^{0,i}_{2,\mathcal{T}} = P^i_{2,\infty},$$

and further leads to

$$\lim_{\mathcal{T}\to\infty} K^{0,i}_{1,\mathcal{T}} = K^i_1, \quad \lim_{\mathcal{T}\to\infty} K^{0,i}_{2,\mathcal{T}} = K^i_2,$$

where $(P_{1,\infty}, K_1; P_{2,\infty}, K_2)$ are the solutions of (5.71)–(5.74). Therefore, a backward iterative algorithm for solving CDMREs (5.71)–(5.74) can be summarized as follows:

(i) Take a large $\mathcal{T} \in \mathcal{N}$ with the terminal values $P^{\mathcal{T}+1,i}_{1,\mathcal{T}} = 0$, $P^{\mathcal{T}+1,i}_{2,\mathcal{T}} = 0$; we have

$$K^{\mathcal{T},i}_{3,\mathcal{T}} = 0, \quad K^{\mathcal{T},i}_{4,\mathcal{T}} = 0,$$

$$H^1_i(P^{\mathcal{T}+1}_{1,\mathcal{T}}) = \gamma^2 I, \quad H^2_i(P^{\mathcal{T}+1}_{2,\mathcal{T}}) = I.$$

(ii) Substitute $K^{\mathcal{T},i}_{3,\mathcal{T}}$, $K^{\mathcal{T},i}_{4,\mathcal{T}}$, $H^1_i(P^{\mathcal{T}+1}_{1,\mathcal{T}})$ and $H^2_i(P^{\mathcal{T}+1}_{2,\mathcal{T}})$ into (5.76) and (5.78), respectively, $K^{\mathcal{T},i}_{1,\mathcal{T}}$ and $K^{\mathcal{T},i}_{2,\mathcal{T}}$ are in turn obtained. Furthermore, $P^{\mathcal{T},i}_{1,\mathcal{T}}$ and $P^{\mathcal{T},i}_{2,\mathcal{T}}$ can be computed from (5.75) and (5.77), respectively.

(iii) Plug the obtained

$$P^{\mathcal{T}}_{1,\mathcal{T}}, \ P^{\mathcal{T}}_{2,\mathcal{T}}, \ K^{\mathcal{T}}_{1,\mathcal{T}}, \ K^{\mathcal{T}}_{2,\mathcal{T}}, H^1(P^{\mathcal{T}}_{1,\mathcal{T}}), \ H^2(P^{\mathcal{T}}_{2,\mathcal{T}})$$

into (5.81) and (5.82); $K^{\mathcal{T}-1}_{3,\mathcal{T}}$ and $K^{\mathcal{T}-1}_{4,\mathcal{T}}$ can be respectively computed.

(iv) Proceed with the steps (ii)–(iii) for $k = \mathcal{T} - 1, \mathcal{T} - 2, \cdots, 0$, then $(P^0_{1,\mathcal{T}}, P^0_{2,\mathcal{T}})$ and $(K^0_{1,\mathcal{T}}, K^0_{2,\mathcal{T}})$ are obtained.

TABLE 5.5
Parameters of system (5.85).

i	A_1^i	A_2^i	C_1^i	C_2^i	B_1^i	B_2^i
$i = 1$	$\begin{bmatrix} 0.55 & 0 \\ 0 & 0.6 \end{bmatrix}$	$\begin{bmatrix} 0.4 & 0 \\ 0 & 0.45 \end{bmatrix}$	$\begin{bmatrix} 0.65 \\ 0.45 \end{bmatrix}$	$\begin{bmatrix} 0.7 \\ 0.5 \end{bmatrix}$	$\begin{bmatrix} 0.65 \\ 0.5 \end{bmatrix}$	$\begin{bmatrix} 0.5 \\ 0.3 \end{bmatrix}$
$i = 2$	$\begin{bmatrix} 0.95 & 0 \\ 0 & 0.64 \end{bmatrix}$	$\begin{bmatrix} 0.75 & 0 \\ 0 & 0.5 \end{bmatrix}$	$\begin{bmatrix} 0.8 \\ 0.3 \end{bmatrix}$	$\begin{bmatrix} 0.75 \\ 0.45 \end{bmatrix}$	$\begin{bmatrix} 0.85 \\ 0.25 \end{bmatrix}$	$\begin{bmatrix} 0.8 \\ 0.4 \end{bmatrix}$

TABLE 5.6
Approximate solutions for $T = 39$.

i	$P_{1,\infty}^i$	$P_{2,\infty}^i$	K_1^i	$-K_2^i$
$i = 1$	$\begin{bmatrix} 0.4094 & 0.4849 \\ 0.4849 & 0.7424 \end{bmatrix}$	$\begin{bmatrix} 0.4261 & 0.4960 \\ 0.4960 & 0.7623 \end{bmatrix}$	$[0.0576 \ 0.0723]$	$[0.3129 \ 0.3715]$
$i = 2$	$\begin{bmatrix} 0.6574 & 0.5700 \\ 0.5700 & 0.7703 \end{bmatrix}$	$\begin{bmatrix} 0.7166 & 0.6007 \\ 0.6007 & 0.8020 \end{bmatrix}$	$[0.0875 \ 0.0678]$	$[0.5060 \ 0.3335]$

(v) Repeat the above procedures to calculate the solutions of (5.75)–(5.78) at $k = 0$ to obtain $P_{1,T}^{0,i}$, $P_{2,T}^{0,i}$, $K_{1,T}^{0,i}$, $K_{2,T}^{0,i}$.

(vi) For a prescribed accuracy $\epsilon > 0$, if the maximal error

$$\max_{i \in \overline{N}} \left\{ \| P_{1,T}^{0,i} - P_{1,T+1}^{0,i} \|, \| P_{2,T}^{0,i} - P_{2,T+1}^{0,i} \| \right\} \le \epsilon,$$

then stop.

Example 5.3
Consider the following second-order discrete-time Markov jump system with (x, u, v)-dependent noise:

$$\begin{cases} x_{k+1} = A_1^{\theta_k} x_k + B_1^{\theta_k} u_k + C_1^{\theta_k} v_k + (A_2^{\theta_k} x_k + B_2^{\theta_k} u_k + C_2^{\theta_k} v_k) w_k, \\ z_k = \begin{bmatrix} [0.5 \ 0.7] x_k \\ u_k \end{bmatrix}, \\ x_0 \in \mathcal{R}^n, \ k \in \mathcal{N}, \ \theta_0 \in \overline{N}. \end{cases} \quad (5.85)$$

In (5.85), the state space of θ_k is $\overline{N} = \{1, 2\}$ and its transition possibility matrix $\mathcal{P}_{2\times 2} = (p_{ij})_{2\times 2}$ is defined via $p_{11} = 0.2$, $p_{12} = 0.8$, $p_{21} = 0.3$, $p_{22} = 0.7$. Moreover, the coefficients corresponding to the two modes are given in Table 5.3.

Set $\gamma = 2.3$ and select an accuracy $\epsilon = 1 \times 10^{-4}$. By applying the proposed iterative algorithm, the approximate solutions of (5.71)–(5.74) can be obtained after 39 iterations; see Table 5.3. The maximal estimation error generated in the numerical experiment is given by 1.3718×10^{-5}. For clarity, we plot the evolution of the backward iterations in Figure 5.4.

□

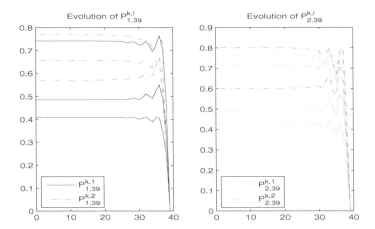

FIGURE 5.4

Backward iterations of $P_{1,39}^{k,i}$ and $P_{2,39}^{k,i}$.

5.5 Finite Horizon H_2/H_∞ Control of Continuous-Time Systems

In this section, we generalize some results of Chapter 2 to stochastic Markovian systems. Consider the following continuous time-varying stochastic Markov jump system with state- and disturbance-dependent noise:

$$
\begin{cases}
dx(t) = [A_1(t,\theta(t))x(t) + B_1(t,\theta(t))u(t) + C_1(t,\theta(t))v(t)]\,dt \\
\quad + [A_2(t,\theta(t))x(t) + C_2(t,\theta(t))v(t)]\,dB(t), \\
z(t) = \begin{bmatrix} C(t,\theta(t))x(t) \\ D(t,\theta(t))u(t) \end{bmatrix}, \\
D^T(t,\theta(t))D(t,\theta(t)) = I,\ x(0) = x_0 \in \mathcal{R}^n,\ t \in [0,T],\ \theta(0) \in \overline{N},
\end{cases}
\tag{5.86}
$$

where, as in Chapter 2, $B(t)$ is the one-dimensional standard Brownian motion, and the jumping process $\{\theta(t), t \geq 0\}$ is a continuous-time discrete-state Markov process taking values in $\overline{N}$ with transition probability described by

$$
\mathcal{P}(\theta(t+h) = j|\theta(t) = i) = \begin{cases} \pi_{ij}h + o(h), & \text{if } i \neq j, \\ 1 + \pi_{ii}h + o(h), & \text{if } i = j, \end{cases}
\tag{5.87}
$$

where $h > 0$, $\lim_{h \to 0} o(h)/h = 0$ and $\pi_{ij} \geq 0$ for $i, j \in \overline{N}, i \neq j$, determine the switching rate from mode i at time t to mode j at time $t + h$, and $\pi_{ii} = -\sum_{j=1, j\neq i}^{N} \pi_{ij}$ for all $i \in \overline{N}$. The processes $\theta(t)$ and $B(t)$ are assumed to be independent. All coefficients of (5.86) are assumed to be continuous matrix-valued functions of suitable dimensions. $B(t)$ and $\theta(t)$ are defined on the complete probability space $(\Omega, \mathcal{F}, \mathcal{P})$ with the natural filter $\mathcal{F}_t$ generated by $B(\cdot)$ and $\theta(\cdot)$ up to time t.

5.5.1   Definitions and lemmas

To give our main results, we need the following definitions and lemmas. Given a disturbance attenuation level $\gamma > 0$, associated with (5.86), define two performance indices as follows:

$$J_{1,T}(u,v) := \gamma^2 \|v(t)\|_{[0,T]}^2 - \|z(t)\|_{[0,T]}^2 = \mathcal{E}\left\{ \int_0^T (\gamma^2\|v(t)\|^2 - \|z(t)\|^2)\, dt \right\},$$

$$J_{2,T}(u,v) := \|z(t)\|_{[0,T]}^2 = \mathcal{E}\left\{ \int_0^T \|z(t)\|^2\, dt \right\}.$$

DEFINITION 5.9   *For system (5.86) and a given $\gamma > 0$, $0 < T < \infty$, find, if it exists, a state feedback control $u_T^*(t) \in \mathcal{L}_{\mathcal{F}}^2([0,T], \mathcal{R}^{n_u})$ such that*

(i) $\|\mathcal{L}_T\| < \gamma$ with

$$\|\mathcal{L}_T\| = \sup_{\substack{v \in \mathcal{L}_{\mathcal{F}}^2([0,T],\mathcal{R}^{n_v}) \\ x_0 = 0,\, v \neq 0,\, \theta(0) \in \overline{N}}} \frac{\mathcal{E}\left\{ \int_0^T (\|C(t,\theta(t))x(t)\|^2 + \|u_T^*(t)\|^2)\, dt \right\}^{1/2}}{\mathcal{E}\left\{ \int_0^T \|v(t)\|^2\, dt \right\}^{1/2}}$$

where $\mathcal{L}_T$ is an operator associated with system (5.86) which is defined by

$$\mathcal{L}_T : \mathcal{L}_{\mathcal{F}}^2([0,T], \mathcal{R}^{n_v}) \mapsto \mathcal{L}_{\mathcal{F}}^2([0,T], \mathcal{R}^{n_z}),$$
$$\mathcal{L}_T(v(t)) = z(t)|_{x_0=0,\, \theta(0)=\theta_0 \in \overline{N}},\ t \in [0,T].$$

(ii) When the worst-case disturbance $v_T^*(t) \in \mathcal{L}_{\mathcal{F}}^2([0,T], \mathcal{R}^{n_v})$, if it exists, is applied to (5.86), $u_T^*(t)$ minimizes the output energy

$$J_{2,T}(u, v_T^*) = \mathcal{E}\left\{ \int_0^T (\|C(t,\theta(t))x(t)\|^2 + \|u(t)\|^2)\, dt \right\}$$

where $v_T^*(t)$ is defined as

$$v_T^*(t) = \arg\min\left\{ J_{1,T}(u_T^*, v) = \mathcal{E}\left\{ \int_0^T (\gamma^2\|v(t)\|^2 - \|z(t)\|^2)\, dt \right\} \right\}.$$

If the above (u_T^*, v_T^*) exist, then we say that the finite horizon H_2/H_∞ control of system (5.86) is solvable and has a pair of solutions (u_T^*, v_T^*).

DEFINITION 5.10   *$(u_T^*, v_T^*) \in \mathcal{L}_{\mathcal{F}}^2([0,T], \mathcal{R}^{n_u}) \times \mathcal{L}_{\mathcal{F}}^2([0,T], \mathcal{R}^{n_v})$ are called the Nash equilibrium strategies of a two-person non-zero sum LQ game corresponding to cost functionals $J_{1,T}(u,v)$ and $J_{2,T}(u,v)$ if*

$$J_{1,T}(u_T^*, v_T^*) \le J_{1,T}(u_T^*, v),\ J_{2,T}(u_T^*, v_T^*) \le J_{2,T}(u, v_T^*),$$
$$\forall (u(t), v(t)) \in \mathcal{L}_{\mathcal{F}}^2([0,T], \mathcal{R}^{n_u}) \times \mathcal{L}_{\mathcal{F}}^2([0,T], \mathcal{R}^{n_v}),\ \theta(0) \in \overline{N}.$$

LEMMA 5.5 (Generalized Itô formula)[131]
Let $\alpha(t, x, i)$ and $\beta(t, x, i)$ be given $\mathcal{R}^n$-valued, $\mathcal{F}_t$-adapted processes, $i \in \overline{N}$, and $dx(t) = \alpha(t, x(t), \theta(t)) \, dt + \beta(t, x(t), \theta(t)) \, dB(t)$. Then for given $\phi(t, x, i) \in \mathcal{C}^{1,2}([0, T] \times \mathcal{R}^n; \mathcal{R})$, $i \in \overline{N}$, we have

$$\mathcal{E}\left\{ \phi(T, x(T), \theta(T)) - \phi(s, x(s), \theta(s)) | \theta(s) = i \right\}$$
$$= \mathcal{E}\left\{ \int_s^T \Gamma\phi(t, x(t), \theta(t)) \, dt | \theta(s) = i \right\}$$

with $\Gamma\phi(t, x, i) = \phi_t(t, x, i) + \alpha^T(t, x, i)\phi_x(t, x, i) + \frac{1}{2}\beta^T(t, x, i)\phi_{x,x}(t, x, i)\beta(t, x, i) + \sum_{j=1}^N \pi_{ij}\phi(t, x, j)$.

To study the finite horizon H_2/H_∞ control problem, we need to establish an SBRL, which is the key in developing the H_∞ control theory. Consider the following stochastic perturbed system with Markov jump parameters

$$\begin{cases} dx(t) = [A_{11}(t, \theta(t))x(t) + B_{11}(t, \theta(t))v(t)] \, dt \\ \qquad + [A_{12}(t, \theta(t))x(t) + B_{12}(t, \theta(t))v(t)] \, dB(t), \\ z_1(t) = C_{11}(t, \theta(t))x(t), \ x(0) = x_0 \in \mathcal{R}^n, \ t \in [0, T], \ \theta(0) \in \overline{N}. \end{cases} \quad (5.88)$$

Associated with system (5.88), the perturbed operator

$$\tilde{\mathcal{L}}_T : \mathcal{L}_{\mathcal{F}}^2([0, T], \mathcal{R}^{n_v}) \mapsto \mathcal{L}_{\mathcal{F}}^2([0, T], \mathcal{R}^{n_{z_1}})$$

is defined as $\tilde{\mathcal{L}}_T(v(t)) = z_1(t)|_{x_0=0} = C_{11}(t, \theta(t))x(t)|_{x_0=0}, \ t \in [0, T]$, and

$$\|\tilde{\mathcal{L}}_T\| = \sup_{v \in \mathcal{L}_{\mathcal{F}}^2([0,T],\mathcal{R}^{n_v}), v \neq 0, x_0=0, \theta(0) \in \overline{N}} \frac{\mathcal{E}\left\{ \int_0^T \|C_{11}(t, \theta(t))x(t)\|^2 \, dt \right\}^{1/2}}{\mathcal{E}\left\{ \int_0^T \|v(t)\|^2 \, dt \right\}^{1/2}}.$$

LEMMA 5.6
For system (5.88) and a given disturbance attenuation $\gamma > 0$, $\|\tilde{\mathcal{L}}_T\| < \gamma$ iff the following coupled GDREs (the time variable t is suppressed)

$$\begin{cases} \dot{P}_i + P_i A_{11}(\cdot, i) + A_{11}^T(\cdot, i)P_i - (P_i B_{11}(\cdot, i) + A_{12}^T(\cdot, i)P_i B_{12}(\cdot, i)) \\ \quad \cdot(\gamma^2 I + B_{12}^T(\cdot, i)P_i B_{12}(\cdot, i))^{-1}(P_i B_{11}(\cdot, i) + A_{12}^T(\cdot, i)P_i B_{12}(\cdot, i))^T \\ \quad + A_{12}^T(\cdot, i)P_i A_{12}(\cdot, i) - C_{11}^T(\cdot, i)C_{11}(\cdot, i) + \sum_{j=1}^N \pi_{ij}P_j = 0, \\ P_i(T) = 0, \\ \gamma^2 I + B_{12}^T(\cdot, i)P_i B_{12}(\cdot, i) > 0, \quad t \in [0, T], \quad i \in \overline{N} \end{cases} \quad (5.89)$$

have a bounded solution $(P_1(t), \ldots, P_N(t)) \leq 0 \in \mathcal{C}([0, T]; \mathcal{S}_n^N)$.

Proof. The proof is very similar to that of Lemma 2.1 except that we only need to replace usual Itô's formula with Lemma 5.5 and

$$J_1^T(x, v; x(t_0), t_0) := \mathcal{E} \int_{t_0}^T (\gamma^2 \|v\|^2 - \|z_1\|^2) \, dt$$

therein with the following

$$J_1^T(x, v; x(t_0), t_0, \theta(0) = i) := \mathcal{E}\left\{\int_{t_0}^T (\gamma^2\|v\|^2 - \|z_1\|^2)\, dt | \theta(0) = i\right\}.$$

Hence, the proof is omitted. $\square$

We now consider the finite horizon stochastic LQ control for Markov jump systems. More specifically, under the constraint of

$$\begin{cases} dx(t) = [A_{11}(t, \theta(t))x(t) + B_{11}(t, \theta(t))u(t)]\, dt \\ \qquad + [A_{12}(t, \theta(t))x(t) + B_{12}(t, \theta(t))u(t)]\, dB(t), \\ x(0) = x_0 \in \mathcal{R}^n,\ t \in [0, T],\ \theta(0) \in \overline{N}, \end{cases} \tag{5.90}$$

we consider the optimization problem:

$$\min_{u \in \mathcal{L}_{\mathcal{F}}^2([0,T], \mathcal{R}^{n_u})} \left\{ J_T(x_0, \theta(0) = i; u) := \mathcal{E}\left\{\int_0^T [x^T(t)Q_{\theta(t)}(t)x(t) \right.\right.$$

$$\left.\left. + u^T(t)R_{\theta(t)}(t)u(t)]\, dt | \theta(0) = i \right\}\right\} \tag{5.91}$$

where $Q_i \in \mathcal{C}([0, T]; \mathcal{S}_n)$, $R_i \in \mathcal{C}([0, T]; \mathcal{S}_{n_u})$, $i \in \overline{N}$. Note that Q_i and R_i in (5.91) are indefinite, and hence the above is an indefinite stochastic LQ control problem.

The indefinite LQ control is associated with the following GDRE

$$\begin{cases} \dot{P}_i + P_i A_{11}(\cdot, i) + A_{11}^T(\cdot, i)P_i - (P_i B_{11}(\cdot, i) + A_{12}^T(\cdot, i)P_i B_{12}(\cdot, i)) \\ \quad \cdot(R_i + B_{12}^T(\cdot, i)P_i B_{12}(\cdot, i))^+ (P_i B_{11}(\cdot, i) + A_{12}^T(\cdot, i)P_i B_{12}(\cdot, i))^T \\ \quad + A_{12}^T(\cdot, i)P_i A_{12}(\cdot, i) + Q_i + \sum_{j=1}^N \pi_{ij}P_j = 0, \\ P_i(T) = 0, \\ (R_i + B_{12}^T(\cdot, i)P_i B_{12}(\cdot, i))(R_i + B_{12}^T(\cdot, i)P_i B_{12}(\cdot, i))^+ \\ \quad \cdot(P_i B_{11}(\cdot, i) + A_{12}^T(\cdot, i)P_i B_{12}(\cdot, i))^T - (P_i B_{11}(\cdot, i) + A_{12}^T(\cdot, i)P_i B_{12}(\cdot, i))^T = 0, \\ R_i + B_{12}^T(\cdot, i)P_i B_{12}(\cdot, i) \geq 0, \quad t \in [0, T], \quad i \in \overline{N}. \end{cases} \tag{5.92}$$

Note that when $R_i(t) + B_{12}^T(t, i)P_i(t)B_{12}(t, i) > 0$ for $i \in \overline{N}$ and $t \in [0, T]$, GDRE (5.92) reduces to

$$\begin{cases} \dot{P}_i + P_i A_{11}(\cdot, i) + A_{11}^T(\cdot, i)P_i - (P_i B_{11}(\cdot, i) + A_{12}^T(\cdot, i)P_i B_{12}(\cdot, i)) \\ \quad \cdot(R_i + B_{12}^T(\cdot, i)P_i B_{12}(\cdot, i))^{-1}(P_i B_{11}(\cdot, i) + A_{12}^T(\cdot, i)P_i B_{12}(\cdot, i))^T \\ \quad + A_{12}^T(\cdot, i)P_i A_{12}(\cdot, i) + Q_i + \sum_{j=1}^N \pi_{ij}P_j = 0, \\ P_i(T) = 0, \\ R_i + B_{12}^T(\cdot, i)P_i B_{12}(\cdot, i) > 0, \quad \forall t \in [0, T], \quad i \in \overline{N}. \end{cases} \tag{5.93}$$

We state respectively the indefinite and standard stochastic LQ control results as follows.

LEMMA 5.7 [122]

1) If GDRE (5.92) admits a solution $(P_1(t), \ldots, P_N(t)) \in \mathcal{C}([0, T]; \mathcal{S}_n^N)$, then the stochastic LQ control problem (5.90)–(5.91) is well posed. In particular, the optimal cost performance

$$V(x_0, \theta(0) = i) := \min_{u \in \mathcal{L}_{\mathcal{F}}^2([0,T], \mathcal{R}^{n_u})} J_T(x_0, \theta(0) = i; u) = x_0^T P_i(0)x_0$$

and all the optimal control laws can be parameterized by

$$u^*(\mathcal{Y}_i, \mathcal{Z}_i; t)$$

$$= -\sum_{i=1}^{N} \left\{ [(R_i + B_{12}^T(\cdot, i)P_i B_{12}(\cdot, i))^+(P_i B_{11}(\cdot, i) + A_{12}^T(\cdot, i)P_i B_{12}(\cdot, i))^T + \mathcal{Y}_i \right.$$

$$-(R_i + B_{12}^T(\cdot, i)P_i B_{12}(\cdot, i))^+(R_i + B_{12}^T(\cdot, i)P_i B_{12}(\cdot, i))\mathcal{Y}_i]x(t) + \mathcal{Z}_i$$

$$\left. -(R_i + B_{12}^T(\cdot, i)P_i B_{12}(\cdot, i))^+(R_i + B_{12}^T(\cdot, i)P_i B_{12}(\cdot, i))\mathcal{Z}_i \right\} I_{\{\theta(t)=i\}}(t),$$

for any $\mathcal{Y}_i \in \mathcal{L}_{\mathcal{F}}^2([0,T], \mathcal{R}^{n_u \times n})$ and $\mathcal{Z}_i \in \mathcal{L}_{\mathcal{F}}^2([0,T], \mathcal{R}^{n_u})$.

2) If there is an optimal state feedback control $u^*(t)$ to the LQ control problem (5.90)–(5.91) with respect to (x_0, r_0), then GDRE (5.92) must have a solution $(P_1(t), \ldots, P_N(t)) \in \mathcal{C}([0,T]; \mathcal{S}_N^n)$, and $u^*(t) = \sum_{i=1}^{N} K(t, i) I_{\{\theta(t)=i\}}(t)x(t)$ with

$$K(\cdot, i) = -[(R_i + B_{12}^T(\cdot, i)P_i B_{12}(\cdot, i))^+(P_i B_{11}(\cdot, i) + A_{12}^T(\cdot, i)P_i B_{12}(\cdot, i))^T$$

$$+\mathcal{Y}_i - (R_i + B_{12}^T(\cdot, i)P_i B_{12}(\cdot, i))^+(R_i + B_{12}^T(\cdot, i)P_i B_{12}(\cdot, i))\mathcal{Y}_i].$$

LEMMA 5.8 [53]
For the standard stochastic LQ problem (5.90)-(5.91), where $Q_i(t) \geq 0$ and $R_i(t) > 0$ on $[0, T] \times \overline{N}$, GDRE (5.93) admits a unique global solution $(P_1(t), \cdots, P_N(t)) \geq 0 \in \mathcal{C}([0,T]; \mathcal{S}_n^N)$. The optimal cost value and the unique optimal control law are respectively given by

$$V(x_0, \theta(0) = i) = x_0^T P_i(0)x_0,$$

and

$$u^*(t) =$$

$$-\sum_{i=1}^{N} \left\{ (R_i + B_{12}^T(\cdot, i)P_i B_{12}(\cdot, i))^{-1}(P_i B_{11}(\cdot, i) + A_{12}^T(\cdot, i)P_i B_{12}(\cdot, i))^T x(t) \right\} I_{\{\theta(t)=i\}}(t).$$

5.5.2 Nash equilibrium strategy and H_2/H_∞ control

We first consider the following system with only state-dependent noise:

$$\begin{cases} dx(t) = [A_1(t, \theta(t))x(t) + B_1(t, \theta(t))u(t) + C_1(t, \theta(t))v(t)] \, dt \\ \qquad + A_2(t, \theta(t))x(t) \, dB(t), \\ z(t) = \begin{bmatrix} C(t, \theta(t))x(t) \\ D(t, \theta(t))u(t) \end{bmatrix}, \\ D^T(t, \theta(t))D(t, \theta(t)) = I, \ x(0) = x_0 \in \mathcal{R}^n, \theta(0) \in \overline{N}, t \in [0, T]. \end{cases} \quad (5.94)$$

For system (5.94), we present a necessary and sufficient condition for the existence of two-person non-zero sum Nash equilibrium strategies. In our subsequent analysis, we define $M_i(t) := M(t, i)$, $M_{1i}(t) := M_1(t, i)$, $M_{2i}(t) := M_2(t, i)$, $i \in \overline{N}$ for convenience.

THEOREM 5.10

For system (5.94), there exist linear memoryless state feedback Nash equilibrium strategies

$$u_T^*(t) = \sum_{i=1}^N K_2(t,i) I_{\{\theta(t)=i\}}(t)x(t) = K_2(t,\theta(t))x(t) \in \mathcal{L}_{\mathcal{F}}^2([0,T],\mathcal{R}^{n_u})$$

and

$$v_T^*(t) = \sum_{i=1}^N K_1(t,i) I_{\{\theta(t)=i\}}(t)x(t) = K_1(t,\theta(t))x(t) \in \mathcal{L}_{\mathcal{F}}^2([0,T],\mathcal{R}^{n_v}),$$

i.e.,

$$J_{1,T}(u_T^*, v_T^*) \le J_{1,T}(u_T^*, v), \ \forall v(t) \in \mathcal{L}_{\mathcal{F}}^2([0,T],\mathcal{R}^{n_v}), \tag{5.95}$$

and

$$J_{2,T}(u_T^*, v_T^*) \le J_{2,T}(u, v_T^*), \ \forall u(t) \in \mathcal{L}_{\mathcal{F}}^2([0,T],\mathcal{R}^{n_u}) \tag{5.96}$$

iff the following GDREs

$$\begin{cases} -\dot{P}_i^{1,T} = P_i^{1,T} A_{1i} + A_{1i}^T P_i^{1,T} + A_{2i}^T P_i^{1,T} A_{2i} - C_i^T C_i \\ \quad - [P_i^{1,T} \ P_i^{2,T}] \begin{bmatrix} \gamma^{-2} C_{1i} C_{1i}^T & B_{1i} B_{1i}^T \\ B_{1i} B_{1i}^T & B_{1i} B_{1i}^T \end{bmatrix} \begin{bmatrix} P_i^{1,T} \\ P_i^{2,T} \end{bmatrix} \\ \quad + \sum_{j=1}^N \pi_{ij} P_j^{1,T}, \\ P_i^{1,T}(T) = 0, \ t \in [0,T], \ i \in \overline{N} \end{cases} \tag{5.97}$$

and

$$\begin{cases} -\dot{P}_i^{2,T} = P_i^{2,T} A_{1i} + A_{1i}^T P_i^{2,T} + A_{2i}^T P_i^{2,T} A_{2i} + C_i^T C_i \\ \quad - [P_i^{1,T} \ P_i^{2,T}] \begin{bmatrix} 0 & \gamma^{-2} C_{1i} C_{1i}^T \\ \gamma^{-2} C_{1i} C_{1i}^T & B_{1i} B_{1i}^T \end{bmatrix} \begin{bmatrix} P_i^{1,T} \\ P_i^{2,T} \end{bmatrix} \\ \quad + \sum_{j=1}^N \pi_{ij} P_j^{2,T}, \\ P_i^{2,T}(T) = 0, \ t \in [0,T], \ i \in \overline{N} \end{cases} \tag{5.98}$$

have solutions $P^{1,T}(t) = (P_1^{1,T}(t),\cdots,P_N^{1,T}(t))$, $P^{2,T}(t) = (P_1^{2,T}(t),\cdots,P_N^{2,T}(t)) \in \mathcal{C}([0,T];\mathcal{S}_n^N)$. *If the solutions of (5.97) and (5.98) exist, then*

(i)

$$u_T^*(t) = -B_1^T(t,\theta(t))P_{\theta(t)}^{2,T}(t)x(t), \ v_T^*(t) = -\gamma^{-2} C_1^T(t,\theta(t))P_{\theta(t)}^{1,T}(t)x(t).$$

(ii)

$$J_{1,T}(u_T^*, v_T^*) = \mathcal{E}[x_0^T P_{\theta(0)}^{1,T}(0)x_0] = x_0^T \left[\sum_{i=1}^N P_i^{1,T}(0)\mathcal{P}(\theta(0)=i) \right] x_0,$$

$$J_{2,T}(u_T^*, v_T^*) = x_0^T P_{\theta(0)}^{2,T}(0)x_0 = x_0^T \left[\sum_{i=1}^N P_i^{2,T}(0)\mathcal{P}(\theta(0)=i) \right] x_0.$$

(iii)

$$P^{1,T}(t) \le 0, \ P^{2,T}(t) \ge 0, \ \forall t \in [0,T].$$

Proof. Sufficiency: Applying Lemma 5.5 and considering the constraint of (5.94), we have

$$J_{1,T}(u, v; \theta(0) = i) := \mathcal{E}\left\{ x_0^T P_{\theta(0)}^{1,T}(0)x_0 - x^T(T)P_{\theta(T)}^{1,T}x(T)|\theta(0) = i \right\}$$

$$+ \mathcal{E}\left\{ \int_0^T (\gamma^2 \|v(t)\|^2 - \|z(t)\|^2)\, dt \right.$$

$$\left. + \Gamma(x^T(t)P_{\theta(t)}^{1,T}(t)x(t))|\theta(0) = i \right\}$$

$$= x_0^T P_i^{1,T}(0)x_0 + \mathcal{E}\left\{ \int_0^T (\gamma^2 \|v(t)\|^2 - \|z(t)\|^2)\, dt \right.$$

$$+ x^T(t)\dot{P}_{\theta(t)}^{1,T}(t)x(t) + dx^T(t)P_{\theta(t)}^{1,T}(t)x(t)$$

$$+ x^T(t)P_{\theta(t)}^{1,T}(t)dx(t) + dx^T(t)P_{\theta(t)}^{1,T}(t)dx(t)$$

$$\left. + \sum_{j=1}^N \pi_{\theta(t)j}x^T(t)P_j^{1,T}(t)x(t)|\theta(0) = i \right\}.$$

Similar to Lemma 2.4, by the standard completion of squares argument and considering (5.97), it is easy to see that

$$J_{1,T}(u_T^*, v; \theta(0) = i) \geq J_{1,T}(u_T^*, v_T^*; \theta(0) = i) = x_0^T P_i^{1,T}(0)x_0.$$

So

$$J_{1,T}(u_T^*, v) \geq J_{1,T}(u_T^*, v_T^*) = x_0^T \left[\sum_{i=1}^N P_i^{1,T}(0)\mathcal{P}(\theta(0) = i) \right] x_0.$$

The first Nash inequality (5.95) is derived. The rest is almost the same with the derivation of Lemma 2.4, and hence is omitted. $\square$

Because it is very similar to the results in Chapter 2, below, we only present results but without proofs. First, parallel to Theorem 2.4, for system (5.86), we have the following results.

THEOREM 5.11

For system (5.86), there exist linear memoryless state feedback Nash equilibrium strategies $u_T^*(t)$ and $v_T^*(t)$ for (5.95)–(5.96) iff $\forall t \in [0, T]$, $i \in \overline{N}$, the following GDREs

$$\begin{cases}
-\dot{P}_i^{1,T} = P_i^{1,T}(A_{1i} + B_{1i}K_{2i}) + (A_{1i} + B_{1i}K_{2i})^T P_i^{1,T} \\
\quad + A_{2i}^T P_i^{1,T} A_{2i} - C_i^T C_i + \sum_{j=1}^N \pi_{ij}P_j^{1,T} - K_{2i}^T K_{2i} \\
\quad - (P_i^{1,T}C_{1i} + A_{2i}^T P_i^{1,T} C_{2i})(\gamma^2 I + C_{2i}^T P_i^{1,T} C_{2i})^+ \\
\quad \cdot (C_{1i}^T P_i^{1,T} + C_{2i}^T P_i^{1,T} A_{2i}), \\
(\gamma^2 I + C_{2i}^T P_i^{1,T} C_{2i})(\gamma^2 I + C_{2i}^T P_i^{1,T} C_{2i})^+ \\
\quad \cdot (C_{1i}^T P_i^{1,T} + C_{2i}^T P_i^{1,T} A_{2i}) - (C_{1i}^T P_i^{1,T} + C_{2i}^T P_i^{1,T} A_{2i}) = 0, \\
K_{2i} = -B_{1i}^T P_i^{2,T}, \\
P_i^{1,T}(T) = 0, \\
\gamma^2 I + C_{2i}^T P_i^{1,T} C_{2i} \geq 0
\end{cases} \quad (5.99)$$

and

$$\begin{cases} -\dot{P}_i^{2,T} = P_i^{2,T}(A_{1i} + C_{1i}K_{1i}) + (A_{1i} + C_{1i}K_{1i})^T P_i^{2,T} \\ \qquad + C_i^T C_i + \sum_{j=1}^N \pi_{ij} P_j^{2,T} - P_i^{2,T} B_{1i} B_{1i}^T P_i^{2,T} \\ \qquad + (A_{2i} + C_{2i}K_{1i})^T P_i^{2,T}(A_{2i} + C_{2i}K_{1i}), \\ K_{1i} = -(\gamma^2 I + C_{2i}^T P_i^{1,T} C_{2i})^{-1}(C_{1i}^T P_i^{1,T} + C_{2i}^T P_i^{1,T} A_{2i}), \\ P_i^{2,T}(T) = 0 \end{cases} \quad (5.100)$$

have solutions $P^{1,T}(t) \in \mathcal{C}([0,T]; \mathcal{S}_n^N) \leq 0$, $P^{2,T}(t) \geq 0 \in \mathcal{C}([0,T]; \mathcal{S}_n^N)$.

Secondly, parallel to Theorem 2.1, the relationship between the existence of Nash equilibrium strategies and the solvability of H_2/H_∞ control for system (5.94) is clarified as follows.

THEOREM 5.12
The following three statements are equivalent:

1) *The finite horizon H_2/H_∞ control of (5.94) has the solution (u_T^*, v_T^*) with $u_T^*(t) = \sum_{i=1}^N K_2(t,i) I_{\{\theta(t)=i\}}(t)x(t)$, and $v_T^*(t) = \sum_{i=1}^N K_1(t,i) I_{\{\theta(t)=i\}}(t)x(t)$.*

2) *There exist linear memoryless state feedback Nash equilibrium strategies $u_T^*(t)$ and $v_T^*(t)$ for (5.95) and (5.96).*

3) *The GDREs (5.97) and (5.98) have a solution $(P^{1,T}(t) \leq 0, P^{2,T}(t) \geq 0) \in \mathcal{C}([0,T]; \mathcal{S}_n^N) \times \mathcal{C}([0,T]; \mathcal{S}_n^N)$.*

Finally, parallel to Theorem 2.3 on the stochastic H_2/H_∞ control, we have the following result.

THEOREM 5.13
For system (5.86), its finite horizon H_2/H_∞ control has a solution (u_T^*, v_T^*) with $u_T^*(t) = \sum_{i=1}^N K_2(t,i) I_{\{\theta(t)=i\}}(t)x(t)$, $v_T^*(t) = \sum_{i=1}^N K_1(t,i) I_{\{\theta(t)=i\}}(t)x(t)$, iff $\forall t \in [0,T]$, $i \in \bar{N}$, the following coupled GDREs

$$\begin{cases} -\dot{P}_i^{1,T} = P_i^{1,T}(A_{1i} + B_{1i}K_{2i}) + (A_{1i} + B_{1i}K_{2i})^T P_i^{1,T} \\ \qquad + A_{2i}^T P_i^{1,T} A_{2i} - C_i^T C_i + \sum_{j=1}^N \pi_{ij} P_j^{1,T} - K_{2i}^T K_{2i} \\ \qquad - (P_i^{1,T} C_{1i} + A_{2i}^T P_i^{1,T} C_{2i})(\gamma^2 I + C_{2i}^T P_i^{1,T} C_{2i})^{-1} \\ \qquad \cdot (C_{1i}^T P_i^{1,T} + C_{2i}^T P_i^{1,T} A_{2i}), \\ K_{2i} = -B_{1i}^T P_i^{2,T}, \\ P_i^{1,T}(T) = 0, \\ \gamma^2 I + C_{2i}^T P_i^{1,T} C_{2i} > 0 \end{cases} \quad (5.101)$$

and

$$\begin{cases} -\dot{P}_i^{2,T} = P_i^{2,T}(A_{1i} + C_{1i}K_{1i}) + (A_{1i} + C_{1i}K_{1i})^T P_i^{2,T} \\ \qquad + C_i^T C_i + \sum_{j=1}^N \pi_{ij} P_j^{2,T} - P_i^{2,T} B_{1i} B_{1i}^T P_i^{2,T} \\ \qquad + (A_{2i} + C_{2i}K_{1i})^T P_i^{2,T}(A_{2i} + C_{2i}K_{1i}), \\ K_{1i} = -(\gamma^2 I + C_{2i}^T P_i^{1,T} C_{2i})^{-1}(C_{1i}^T P_i^{1,T} + C_{2i}^T P_i^{1,T} A_{2i}), \\ P_i^{2,T}(T) = 0 \end{cases} \quad (5.102)$$

have a solution $(P^{1,T}, P^{2,T}) \in \mathcal{C}([0,T]; \mathcal{S}_n^N) \times \mathcal{C}([0,T]; \mathcal{S}_n^N)$. In this case, we have

1)

$$u_T^*(t) = \sum_{i=1}^{N} K_{2i}(t) I_{\{\theta(t)=i\}}(t) x(t) = K_2(t, \theta(t)) x(t),$$

$$v_T^*(t) = \sum_{i=1}^{N} K_{1i}(t) I_{\{\theta(t)=i\}}(t) x(t) = K_2(t, \theta(t)) x(t).$$

2)

$$J_{1,T}(u_T^*, v_T^*) = \mathcal{E}[x_0^T P_{\theta(0)}^{1,T}(0) x_0] = x_0^T \left[\sum_{i=1}^{N} P_i^{1,T}(0) \mathcal{P}(\theta(0) = i) \right] x_0,$$

$$J_{2,T}(u_T^*, v_T^*) = x_0^T P_{\theta(0)}^{2,T}(0) x_0 = x_0^T \left[\sum_{i=1}^{N} P_i^{2,T}(0) \mathcal{P}(\theta(0) = i) \right] x_0.$$

3)

$$P^{1,T}(t) \leq 0, \quad P^{2,T}(t) \geq 0.$$

5.6 Infinite Horizon Continuous-Time H_2/H_∞ Control

In [98], under stochastic detectability [52], the infinite horizon H_2/H_∞ control was studied for Itô-type time-invariant Markovian jump systems with state-dependent noise, which has been improved by [143] under the weaker assumption of exact detectability. Because this section is similar to Section 5.4, we omit most proofs in the subsequent discussions.

5.6.1 A moment equation

Consider the time-invariant homogeneous Markov jump system

$$\begin{cases} dx(t) = A_1(\theta(t))x(t)\,dt + A_2(\theta(t))x(t)\,dB(t), \\ x(0) = x_0, \ \theta(0) \in \overline{N}, \end{cases} \tag{5.103}$$

where the Markov chain $\{\theta(t)\}_{t \geq 0}$ is defined as in (5.87) with the stationary transition rate matrix $\Lambda = [(\pi_{ij})]$. Let

$$X_i(t) = \mathcal{E}[x(t)x^T(t) I_{\{\theta(t)=i\}}], \ X(t) = (X_1(t) \ X_2(t) \ \cdots \ X_N(t)).$$

The following lemma comes from Lemma 9.2 of [115].

LEMMA 5.9
The indicator process $\Phi(t) = [I_{\{\theta(t)=1\}}, \ \cdots, I_{\{\theta(t)=N\}}]^T$ satisfies the following stochastic integral equation

$$\Phi(t) = \Phi(0) + \int_0^t \Lambda^T \Phi(s)\,ds + m(t) \tag{5.104}$$

where $m(t)$ is an $\mathcal{R}^N$-valued square integrable martingale with respect to $\mathcal{F}_t$, which is independent of the Brownian motion $B(\cdot)$.

THEOREM 5.14

For system (5.103), $X_i(t), i \in \overline{N}$ satisfies the GLE

$$\begin{cases} \dot{X}_i(t) = A_{1i}X_i(t) + X_i(t)A_{1i}^T + A_{2i}X_i(t)A_{2i}^T + \sum_{j=1}^N \pi_{ji}X_j(t), \\ X_i(0) = [x(0)x^T(0)I_{\{\theta(0)=i\}}], \end{cases} \qquad (5.105)$$

where A_{1i} and A_{2i} are defined respectively as $A_{1i} := A_1(\theta(t))$ and $A_{2i} := A_2(\theta(t))$ for $\theta(t) = i$.

Proof. Let $Y(t) = x(t)x^T(t)$. By using Itô's formula, we get

$$dY(t) = [A_1(\theta(t))Y(t) + Y(t)A_1^T(\theta(t)) + A_2(\theta(t))Y(t)A_2^T(\theta(t))]\,dt$$
$$+[A_2(\theta(t))Y(t) + Y(t)A_2^T(\theta(t))]\,dB(t).$$

According to (5.104),

$$\Phi_i(t) = \Phi_i(0) + \int_0^t [\pi_{1i}\Phi_1(s) + \cdots + \pi_{Ni}\Phi_N(s)]ds + m_i(t).$$

So by the formula of integration-by-parts of the semimartingale, and the fact that $Y(t)$ is continuous with respect to t, we have

$$Y(t)\Phi_i(t) = Y_0\Phi_i(0) + \int_0^t Y(s)d\Phi_i(s) + \int_0^t \Phi_i(s_-)\,dY(s) + \langle Y, \Phi_i^c \rangle_t$$

$$= Y_0\Phi_i(0) + \langle Y, \Phi_i^c \rangle_t + \int_0^t \sum_{j=1}^N \pi_{ji}Y(s)\Phi_j(s)ds$$

$$+\int_0^t [A_1(\theta)Y + YA_1^T(\theta) + A_2(\theta)YA_2^T(\theta)](s)\Phi_i(s_-)ds$$

$$+\int_0^t Y(s)dm_i(s) + \int_0^t [A_2(\theta)Y + YA_2^T(\theta)](s)\Phi_i(s_-)\,dB(s), \qquad (5.106)$$

where $\Phi_i(s_-)$ is the left-limitation process of $\Phi_i(s)$, and Φ_i^c is the continuous martingale part of semimartingale Φ_i, which is equivalent to m_i^c. Because $m(\cdot)$ is independent of $B(\cdot)$, by the continuity of $Y(t)$,

$$\langle Y, \Phi_i^c \rangle_t = \left\langle \int_0^t [A_2(\theta(s))Y(s) + Y(s)A_2^T(\theta(s))]\,dB(s),\ m_i \right\rangle_t$$

$$= \int_0^t [A_2(\theta(s))Y(s) + Y(s)A_2^T(\theta(s))]d\langle B, m_i^c \rangle_s \equiv 0, \qquad (5.107)$$

where m_i^c is the continuous martingale part of m_i, which is also a square integral martingale independent of $B(\cdot)$. Because of $\Phi_i(s) = I_{\{\theta(s)=i\}}$, by the regularity of the Markov chain, it follows that

$$\mu\{s : \Phi_i(s_-) \neq \Phi_i(s)\} = 0,$$

where $\mu(\cdot)$ is the Lebesgue measure. Hence,

$$\int_0^t [A_1(\theta)Y + YA_1^T(\theta) + A_2(\theta)YA_2^T(\theta)](s)\Phi_i(s_-)\,ds$$
$$= \int_0^t [A_1(\theta)Y + YA_1^T(\theta) + A_2(\theta)YA_2^T(\theta)](s)\Phi_i(s)\,ds. \qquad (5.108)$$

Taking the expectation on both sides of (5.106), by (5.107)–(5.108), the properties of stochastic integral, and the Fubini theorem, we get

$$X_i(t) = X_i(0) + \int_0^t \left[A_{1i}X_i + X_iA_{1i}^T + A_{2i}X_iA_{2i}^T + \sum_{j=1}^N \pi_{ji}X_j \right](s)\,ds, (5.109)$$

which is the integral form of (5.105). The proof is completed. $\square$

REMARK 5.7   As can be seen in the proof of Theorem 5.14, the independence of $B(\cdot)$ and $\theta(\cdot)$ is a very important assumption, otherwise, (5.107) is not identically zero. $\blacksquare$

REMARK 5.8   References [41, 67] considered (5.103)-like equations for Markov jump linear systems without multiplicative noise. A similar result to Theorem 5.14 for infinite dimensional stochastic differential systems with multiplicative noise but no jumps can be found in [172]. $\blacksquare$

Now, we define the following operators associated with $A_1 = [A_{11}, A_{12}, \cdots, A_{1N}]$ and $A_2 = [A_{21}, A_{22}, \cdots, A_{2N}]$:

$$\begin{cases} \mathcal{L}_i^{A_1,A_2}(X) = A_{1i}X_i + X_iA_{1i}^T + A_{2i}X_iA_{2i}^T + \sum_{j=1}^N \pi_{ji}X_j, \ X \in \mathcal{S}_n^N \\ \mathcal{L}^{A_1,A_2}(X) = (\mathcal{L}_1^{A_1,A_2}(X), \ \mathcal{L}_2^{A_1,A_2}(X), \ \cdots, \mathcal{L}_N^{A_1,A_2}(X)). \end{cases} \qquad (5.110)$$

Obviously, $\mathcal{L}^{A_1,A_2}$ is a bounded linear operator on the Hilbert space $\mathcal{S}_n^N$ with the inner product defined as $\langle A, B \rangle = \sum_i^N Tr(A_iB_i)$ for $A = [A_1, A_2, \cdots, A_N], B = [B_1, B_2, \cdots, B_N] \in \mathcal{S}_n^N$, and its adjoint operator is

$$\begin{cases} (\mathcal{L}_i^{A_1,A_2})^*(X) = X_iA_{1i} + A_{1i}^TX_i + A_{2i}^TX_iA_{2i} + \sum_{j=1}^N \pi_{ij}X_j, \\ (\mathcal{L}^{A_1,A_2})^*(X) = ((\mathcal{L}_1^{A_1,A_2})^*(X), \ (\mathcal{L}_2^{A_1,A_2})^*(X), \ \cdots, (\mathcal{L}_N^{A_1,A_2})^*(X)). \end{cases}$$

Similar to Theorem 1.7, it is easy to show the following:

THEOREM 5.15
System (5.103) is ASMS iff $\sigma(\mathcal{L}^{A_1,A_2}) \subset \mathcal{C}^-$.

5.6.2   Exact observability and detectability

Consider the time-invariant unforced Markovian jump system

$$\begin{cases} dx(t) = A_1(\theta(t))x(t)\,dt + A_2(\theta(t))x(t)\,dB(t), \\ y(t) = C(\theta(t))x(t) \\ x(0) = x_0 \in \mathcal{R}^n, \ \theta(0) \in \overline{N} \end{cases} \qquad (5.111)$$

whose exact observability and detectability can be defined as in Definition 1.8 and Definition 1.13.

DEFINITION 5.11   *Considering system (5.111), we call $x(0) \in \mathcal{R}^n$ an unobservable state, if for some $T > 0$, the corresponding output response always equals zero, i.e., $y(t) \equiv 0, \forall t \in [0, T]$. $(A_1, A_2|C)$ is called exactly observable, if there is no unobservable state except for the zero initial state.*

DEFINITION 5.12 $(A_1, A_2|C)$ *is said to be exactly detectable if $y(t) \equiv 0$ a.s., $t \in [0, T], \forall T \geq 0$, implies $\lim_{t \to \infty} \mathcal{E}\|x(t)\|^2 = 0$.*

Because it is very similar to the results of Itô-type systems without Markov jumps, we only state our results but omit their proofs.

THEOREM 5.16 (PBH criterion)
$(A_1, A_2|C)$ *is exactly observable iff*

$$(C_1 X_1, C_2 X_2, \cdots, C_N X_N) \neq 0$$

for every eigenvector $X = (X_1, X_2, \cdots, X_N) \in \mathcal{S}_N^n$ of $\mathcal{L}^{A_1, A_2}$ corresponding to any eigenvalue $\lambda \in \mathcal{C}$, where $C_i = C(\theta(t))$ for $\theta(t) = i$.

THEOREM 5.17 (PBH criterion)
$(A_1, A_2|C)$ *is exactly detectable iff*

$$(C_1 X_1, C_2 X_2, \cdots, C_N X_N) \neq 0$$

for every eigenvector $X = (X_1, X_2, \cdots, X_N) \in \mathcal{S}_N^n$ of $\mathcal{L}^{A_1, A_2}$ corresponding to any eigenvalue λ with $Re(\lambda) \geq 0$.

Another concept called "stochastic detectability" was introduced in [53, 98].

DEFINITION 5.13   *System $(A_1, A_2|C)$ is called stochastically detectable, if there is a gain matrix function $H(\theta(t))$ which is constant for $\theta(t) = i \in \overline{N}$, such that $(A_1 + HC, A_2)$ is ASMS for any $x_0 \in \mathcal{R}^n$ and $\theta(0) \in \overline{N}$, i.e., the following system*

$$dx(t) = [A_1(\theta(t)) + H(\theta(t))C(\theta(t))]x(t)\, dt + A_2(\theta(t))x(t)\, dB(t)$$

satisfies

$$\lim_{t \to \infty} \mathcal{E}[x^T(t)x(t)] = 0.$$

Parallel to Theorem 1.14, for Markov jump systems, the following still holds.

PROPOSITION 5.3
If $(A_1, A_2|C)$ is stochastically detectable, then it is also exactly detectable.

Proof. Let $H(\theta(t))$ be the gain as in Definition 5.13. If $(A_1, A_2|C)$ is not exactly detectable, then by Theorem 5.17, there exists an eigenvector $X \in \mathcal{S}_n^N$ of $\mathcal{L}^{A_1,A_2}$ corresponding to some eigenvalue $\lambda \in \sigma(\mathcal{L}^{A_1,A_2})$, $Re(\lambda) \geq 0$, such that $(C_1X_1, C_2X_2, \cdots, C_N X_N) = 0$. So

$$\mathcal{L}_i^H(X) := (A_{1i} + H_i C_i)X_i + X_i(A_{1i} + H_i C_i)^T + A_{2i}X_i A_{2i}^T + \sum_{j=1}^N \pi_{ji} X_j$$

$$= [A_{1i}X_i + X_i A_{1i}^T + A_{2i}X_i A_{2i}^T + \sum_{j=1}^N \pi_{ji} X_j] + [H_i C_i X_i + (H_i C_i X_i)^T]$$

$$= \mathcal{L}_i^{A_1,A_2}(X) = \lambda X_i.$$

Equivalently,

$$\mathcal{L}^H(X) = \mathcal{L}^{A_1,A_2}(X) = \lambda X, \ Re(\lambda) \geq 0,$$

which contradicts the stability of $(A_1 + HC, A_2)$ according to Theorem 5.15. So $(A_1, A_2|C)$ is exactly detectable. $\square$

Parallel to Theorem 2.7, for Markov jump systems, we have

THEOREM 5.18
Assume $Y \in \mathcal{S}_n^N$, $Y \geq 0$, and $(A_1, A_2|Y)$ is exactly detectable. Then (A_1, A_2) is ASMS iff the GLE

$$(\mathcal{L}^{A_1,A_2})^*(X) = -Y \tag{5.112}$$

has a solution $X \geq 0$.

Proof. The necessity proof can be found in Theorem 15 of [53]. The sufficiency can be shown by repeating the same line of Theorem 2.7. $\square$

Theorem 2.6 can also be extended to stochastic Markov jump systems.

THEOREM 5.19
Assume $Y \in \mathcal{S}_n^N$, $Y \geq 0$, $(A_1, A_2|Y)$ is exactly observable. Then (A_1, A_2) is ASMS iff the GLE

$$(\mathcal{L}^{A_1,A_2})^*(X) = -Y \tag{5.113}$$

has a solution $X > 0$.

Proof. By Theorem 5.18, we only need to show that, under the exact observability of $(A_1, A_2|Y)$, the GLE (5.113) has a strictly positive definite solution $X > 0$. Otherwise, there exists $X_i \geq 0$ which is not strictly positive definite. Suppose ς is an eigenvector of X_i corresponding to zero eigenvalue. By Lemma 5.5, for any $T > 0$, under the constraint of

$$\begin{cases} dx(t) = A_1(\theta(t))x(t)\, dt + A_2(\theta(t))x(t)\, dB(t), \\ y(t) = Y(\theta(t))x(t), \\ x(0) = \varsigma, \theta(0) \in \overline{N}, \end{cases}$$

we have for any $\theta(0) = i \in \overline{N}$ that

$$0 \leq \mathcal{E}\left\{\int_0^T x^T(s)Y(\theta(s))x(s)\,ds|\theta(0) = i\right\}$$
$$= \varsigma^T X_i \varsigma - \mathcal{E}[x^T(T)Y(\theta(T))x(T)|\theta(0) = i]$$
$$= -\mathcal{E}[x^T(T)Y(\theta(T))x(T)|\theta(0) = i] \leq 0,$$

which implies $y(s) \equiv 0$ a.s. for $\forall s \in [0, T]$ corresponding to a non-zero initial state $x(0) = \varsigma$, this contradicts the exact observability of $(A_1, A_2|Y)$. $\square$

5.6.3 Comments on the H_2/H_∞ control

Fundamental to the H_2/H_∞ control theory, an infinite horizon SBRL and an indefinite LQ control result of linear Itô-type Markov jump systems were obtained in Theorem 10 of [53] and Theorems 4.1-5.1 of [6], respectively. Additionally, based on Sections 5.6.1-5.6.2, it is easy to generalize the results of Section 2.3.4 and Section 2.4 to stochastic Markov jump systems. For instance, consider the following time-invariant system with state-dependent noise

$$\begin{cases} dx(t) = [A_1(\theta(t))x(t) + B_1(\theta(t))u(t) + C_1(\theta(t))v(t)]\,dt \\ \qquad\quad + A_2(\theta(t))x(t)\,dB(t), \\ z(t) = \begin{bmatrix} C(\theta(t))x(t) \\ D(\theta(t))u(t) \end{bmatrix}, \\ D^T(\theta(t))D(\theta(t)) = I, \ x(0) = x_0 \in \mathcal{R}^n, \theta(0) \in \overline{N}, t \in [0, \infty). \end{cases} \quad (5.114)$$

Theorem 2.15 in Section 2.3.4 still holds for system (5.114) if we replace the coupled GAREs (2.128)–(2.129) with

$$P_i^1 A_{1i} + A_{1i}^T P_i^1 + A_{2i}^T P_i^1 A_{2i} - C_i^T C_i + \sum_{j=1}^N \pi_{ij} P_j^1$$

$$- [P_i^1 \ P_i^2]\begin{bmatrix} \gamma^{-2}C_{1i}C_{1i}^T & B_{1i}B_{1i}^T \\ B_{1i}B_{1i}^T & B_{1i}B_{1i}^T \end{bmatrix}\begin{bmatrix} P_i^1 \\ P_i^2 \end{bmatrix} = 0, \ i \in \overline{N}$$

and

$$P_i^2 A_{1i} + A_{1i}^T P_i^2 + A_{2i}^T P_i^2 A_{2i} + C_i^T C_i + \sum_{j=1}^N \pi_{ij} P_j^2$$

$$- [P_i^1 \ P_i^2]\begin{bmatrix} 0 & \gamma^{-2}C_{1i}C_{1i}^T \\ \gamma^{-2}C_{1i}C_{1i}^T & B_{1i}B_{1i}^T \end{bmatrix}\begin{bmatrix} P_i^1 \\ P_i^2 \end{bmatrix} = 0, \ i \in \overline{N}.$$

The detailed proof can be found in Theorem 4.1 of [143]. The reference [98] studied the infinite horizon H_2/H_∞ control of (5.114) under a stronger assumption of stochastic detectability. The main result of [98] was improved by [143] under a weaker assumption of exact detectability. We can also discuss a relationship between the infinite horizon H_2/H_∞ control and the existence of two-person non-zero sum Nash equilibrium strategy as in Theorem 2.21 and Corollary 2.2.

At the end of this section, we make the following comments:

- In our viewpoint, for the finite state homogenous Markov jump process $\{\theta(t)\}$, there is no difficulty in generalizing all the results of Chapter 2 to Itô-type Markov jump systems, which can be viewed as a trivial extension.

- Up to now, there have been few papers to deal with the H_2/H_∞ control of Itô-type infinite state non-homogenous Markov jump systems, which is a valuable research topic. Some essential differences from previous references on finite state homogenous Markov jump systems are expected. Reference [72] presented a unified approach for stochastic and mean square stability of continuous-time infinite state Markov jump systems with additive disturbance, where the method may be utilized to deal with multiplicative disturbance.

5.7   Notes and References

There are many excellent books and papers concerned with Markov jump systems; we refer the reader to the new monographs [52, 53, 131]. Both continuous- and discrete-time Markovian jumping systems have been extensively investigated by many researchers, for example, various definitions on observability and detectability can be found in [41, 42, 91, 143, 161, 172, 217], while [43] and [122] studied the LQ optimal control problem. Stochastic stability of Markov jump systems is also an attractive direction; see [72, 170, 171]. The references [44, 89, 90, 97, 137, 139, 165] which studied the stochastic H_2/H_∞ control of linear Markov jump systems are closely related to this chapter.

In our viewpoint, most results of this chapter can be viewed as trivial extensions of Chapters 2–3 except for those in Section 5.3. A challenging problem is to consider the H_2/H_∞ control when $\theta(t)$ is a countable nonhomogeneous infinite state Markov chain. In this case, to solve the H_2/H_∞ control problem, we have to search for new methods and introduce new concepts that are different from the existing ones. Theorem 5.14 was first proved in [143], which plays an important role in the study of stochastic mean square stability.

The materials of this chapter mainly come from [88, 89, 90, 139, 143, 165].

6

Nonlinear Continuous-Time Stochastic H_∞ and H_2/H_∞ Controls

The aim of this chapter is to generalize the linear stochastic H_2/H_∞ control in Chapter 2 to nonlinear stochastic Itô's systems. Nonlinear H_∞ control of deterministic continuous-time systems was a popular research topic in the 1990s. We refer the reader to the well-known works [31, 92, 101, 173]. In [101, 173], the differential geometric approach was employed to study the strict relation between the Hamilton–Jacobi equation (HJE) and invariant manifolds of Hamiltonian vector fields, and the existence of local solution to the primal nonlinear H_∞ control was discussed. However, the differential geometric approach has seldom been applied to stochastic control systems. We mainly use the method of completion of squares together with stochastic dissipative theory to discuss global solutions to nonlinear stochastic H_∞ and mixed H_2/H_∞ control problems.

6.1 Dissipative Stochastic Systems

Consider the following nonlinear stochastic control system governed by the Itô-type differential equation

$$\begin{cases} dx(t) = [f(x(t)) + g(x(t))u(t)]\, dt + [h(x(t)) + l(x(t))u(t)]\, dB(t), \\ f(0) = 0,\ h(0) = 0, \\ z = m(x(t)),\ m(0) = 0. \end{cases} \quad (6.1)$$

In the above, the matrix functions f, g, h, l and m are uniformly continuous and Lipschitz, satisfying a linear growth condition, which guarantee that (6.1) has a unique strong solution [196] on $[0, T]$ for any $T > 0$. $x(t) \in \mathcal{R}^n$ is the system state, $u(t) \in \mathcal{R}^{n_u}$ is the control input, and $z(t) \in \mathcal{R}^{n_z}$ is the regulated output. $B(t)$ is one-dimensional standard Brownian process defined on the complete probability space $(\Omega, \mathcal{F}, \mathcal{P})$, with the natural filter $\mathcal{F}_t$ generated by $B(\cdot)$ up to time t. The control input $u(t)$ is an adapted process with respect to $\{\mathcal{F}_t\}_{t \geq 0}$ such that system (6.1) has a unique strong solution under the above conditions. The dissipative dynamic system theory was established in [178], which has become an important tool in studying stability and stabilization of nonlinear systems; see [93], [94] and [30]. In recent years, the dissipative theory developed by [178] has been extended to stochastic systems in various ways by [70], [28], [167] and [179]. In the following, we shall develop a dissipative theory for stochastic systems which is slightly different from those in the previous references and makes the book self-contained in developing the H_∞ control theory.

Following the terminology of [178], a function $w(\cdot, \cdot) : \mathcal{R}^{n_u} \times \mathcal{R}^{n_z} \mapsto \mathcal{R}$ associated with system (6.1) is called the supply rate on $[s, \infty)$, if it has the following property: for any $u \in \mathcal{L}_{\mathcal{F}}^2([s, T], \mathcal{R}^{n_u})$, the deterministic initial state $x(s) \in \mathcal{R}^n$, the controlled output $z(t) = m(x(t))$ of (6.1) is such that

$$\mathcal{E} \int_s^T |w(u(t), z(t))| \, dt < \infty \ \ for \ all \ T \geq s \geq 0.$$

DEFINITION 6.1   *System (6.1) with supply rate w is said to be dissipative on $[s, \infty)$, $s \geq 0$, if there exists a nonnegative continuous function $V : \mathcal{R}^n \mapsto \mathcal{R}^+$, called the storage function, such that for all $t \geq s \geq 0$, the deterministic initial state $x(s) \in \mathcal{R}^n$,*

$$\mathcal{E}V(x(t)) - V(x(s)) \leq \mathcal{E} \int_s^t w(u(\tau), z(\tau)) \, d\tau. \qquad (6.2)$$

As in deterministic systems [178], (6.2) can be called the dissipative inequality.

In [179], dissipativeness was defined in a slightly more general form with (6.2) replaced by

$$\mathcal{E}[V(x(t))|\mathcal{F}_s] - V(x(s)) \leq \mathcal{E} \left(\int_s^t w(u(\tau), z(\tau)) \, d\tau \middle| \mathcal{F}_s \right), \ a.s..$$

PROPOSITION 6.1
If there exists a positive definite Lyapunov function $V \in \mathcal{C}^2(\mathcal{R}^n; \mathcal{R}^+)$ satisfying

$$\mathcal{L}_u V(x) \leq w(u, z), \quad \forall (u, z) \in \mathcal{R}^{n_u} \times \mathcal{R}^{n_z},$$

then system (6.1) is dissipative with supply rate w on $[s, \infty)$ for any $s \geq 0$, where $\mathcal{L}_u$ is the infinitesimal generator of the equation

$$dx(t) = [f(x(t)) + g(x(t))u(t)] \, dt + [h(x(t)) + l(x(t))u(t)] \, dB(t). \qquad (6.3)$$

Proof. By Itô's formula, for any $t \geq s \geq 0$, $x(s) \in \mathcal{R}^n$,

$$V(x(t)) - V(x(s)) = \int_s^t \mathcal{L}_u V(x) \, d\tau + \int_s^t \frac{\partial V^T(x)}{\partial x}[h(x) + l(x)u] \, dB(\tau).$$

By taking the expectation in both sides of the above equation, we get

$$\mathcal{E}V(x(t)) - V(x(s)) = \mathcal{E} \int_s^t \mathcal{L}_u V(x) \, d\tau \leq \mathcal{E} \int_s^t w(u(\tau), z(\tau)) \, d\tau.$$

This ends the proof. □

DEFINITION 6.2 *An available storage function with supply rate w on $[s, \infty)$, $s \geq 0$, is defined by*

$$V_{a,s}(x) = - \inf_{u \in \mathcal{L}_{\mathcal{F}}^2([s,t],\mathcal{R}^{n_u}),t \geq s, x(s)=x \in \mathcal{R}^n} \mathcal{E} \int_s^t w(u(\tau), z(\tau)) \, d\tau$$

$$= \sup_{u \in \mathcal{L}_{\mathcal{F}}^2([s,t],\mathcal{R}^{n_u}),t \geq s, x(s)=x \in \mathcal{R}^n} -\mathcal{E} \int_s^t w(u(\tau), z(\tau)) \, d\tau. \quad (6.4)$$

A stochastic version of Proposition 2.3 [30] is as follows.

PROPOSITION 6.2
If system (6.1) with supply rate w is dissipative on $[s, \infty)$, $s \geq 0$, then the available storage function $V_{a,s}(x)$ is finite for each $x \in \mathcal{R}^n$. Moreover, for any other possible storage function V_s,

$$0 \leq V_{a,s}(x) \leq V_s(x), \quad \forall x \in \mathcal{R}^n. \quad (6.5)$$

Conversely, if $V_{a,s}$ is finite for each $x \in \mathcal{R}^n$, then system (6.1) is dissipative on $[s, \infty)$.

Proof. $V_{a,s} \geq 0$ is obvious. Next, by Definition 6.1, if system (6.1) with supply rate w is dissipative on $[s, \infty)$, then (6.2) holds for some storage function V_s. So for any $x(s) = x \in \mathcal{R}^n, t \geq s \geq 0$,

$$V_s(x) \geq -\mathcal{E} \int_s^t w(u(\tau), z(\tau)) \, d\tau + \mathcal{E}V_s(x(t)) \geq -\mathcal{E} \int_s^t w(u(\tau), z(\tau)) \, d\tau,$$

which yields

$$V_s(x) \geq \sup_{t \geq s, u \in \mathcal{L}_{\mathcal{F}}^2([s,t],\mathcal{R}^{n_u}),x(s)=x} -\mathcal{E} \int_s^t w(u(\tau), z(\tau)) \, d\tau$$

$$= - \inf_{t \geq s, u \in \mathcal{L}_{\mathcal{F}}^2([s,t],\mathcal{R}^{n_u}),x(s)=x} \mathcal{E} \int_s^t w(u(\tau), z(\tau)) \, d\tau = V_{a,s}(x).$$

Therefore, $V_{a,s}$ is finite and (6.5) holds. The rest of the proof can be carried out along the line of [178] by using the following relation

$$V_{a,s}(x) + \mathcal{E} \int_s^t w(u(\tau), z(\tau)) \, d\tau \geq \mathcal{E}V_{a,s}(x(t)). \quad (6.6)$$

□

We make the following assumption, which is necessary.

Assumption 6.1. The storage function (6.4), if it exists, belongs to $\mathcal{C}^2(\mathcal{R}^n; \mathcal{R}^+)$.

The following theorem with $w(u, z) = z^T Q z + 2 z^T S u + u^T R u$ will be used in proving the infinite horizon SBRL, where $Q \in \mathcal{S}_{n_z}$, $S \in \mathcal{R}^{n_z \times n_u}$, and $R \in \mathcal{S}_{n_u}$ are constant matrices.

THEOREM 6.1

A necessary and sufficient condition for system (6.1) to be dissipative on $[s, \infty)$ with respect to a supply rate $w(\cdot, \cdot)$ is that there exists a storage function $V_s(x)$ with $V_s(0) = 0$, $\tilde{l}(x) : \mathcal{R}^n \mapsto \mathcal{R}^q$, and $\tilde{w}(x) : \mathcal{R}^n \mapsto \mathcal{R}^{q \times n_u}$ for some integer $q > 0$, such that

$$m^T Q m - \frac{\partial V_s^T}{\partial x} f - \frac{1}{2} h^T \frac{\partial^2 V_s}{\partial x^2} h = \tilde{l}^T \tilde{l}, \tag{6.7}$$

$$R - \frac{1}{2} l^T \frac{\partial^2 V_s}{\partial x^2} l = \tilde{w}^T \tilde{w}, \tag{6.8}$$

$$2 S^T m - g^T \frac{\partial V_s}{\partial x} - l^T \frac{\partial^2 V_s}{\partial x^2} h = 2 \tilde{w}^T \tilde{l}. \tag{6.9}$$

Proof. If system (6.1) is dissipative on $[s, \infty)$ with respect to a supply rate $w(\cdot, \cdot)$, by Proposition 6.1, $V_{a,s}$ is a possible storage function, which satisfies (6.6). By (6.6) with any $x(s) = x \in \mathcal{R}^n$ and Assumption 6.1, we have

$$-\frac{\mathcal{E} V_{a,s}(x(t)) - V_{a,s}(x)}{t - s} + \frac{\mathcal{E} \int_s^t w(u(\tau), z(\tau))\, d\tau}{t - s} \geq 0, \ t > s.$$

Let $t \downarrow s$ in the above and note that (applying Itô's formula)

$$\mathcal{E} V_{a,s}(x(t)) = V_{a,s}(x) + \mathcal{E} \int_s^t \left[\frac{\partial V_{a,s}^T}{\partial x}(f + gu) + \frac{1}{2}(h + lu)^T \frac{\partial^2 V_{a,s}}{\partial x^2}(h + lu) \right] d\tau,$$

it follows that

$$J(x, u) := m^T Q m + 2 m^T S u + u^T R u - \frac{\partial V_{a,s}^T}{\partial x}(f + gu)$$

$$- \frac{1}{2}(h + lu)^T \frac{\partial^2 V_{a,s}}{\partial x^2}(h + lu) \geq 0 \tag{6.10}$$

for all x and u. Obviously, by the fact that the right-hand side of (6.10) is quadratic in u, there exist $\tilde{l} : \mathcal{R}^n \mapsto \mathcal{R}^q$ and $\tilde{w} : \mathcal{R}^n \mapsto \mathcal{R}^{q \times n_u}$ (not necessarily unique), such that

$$J(x, u) = [\tilde{l}(x) + \tilde{w}(x)u]^T [\tilde{l}(x) + \tilde{w}(x)u].$$

By comparing the coefficients corresponding to the same powers of u, (6.7), (6.8) and (6.9) are derived. The converse can be very easily shown by noting that for any

$x(s) = x \in \mathcal{R}^n$, we have

$$\mathcal{E} \int_s^t w(u(\tau), z(\tau)) \, d\tau = \mathcal{E} \int_s^t [\tilde{l}(x) + \tilde{w}(x)u]^T [\tilde{l}(x) + \tilde{w}(x)u] \, d\tau$$
$$+ \mathcal{E} V_s(x(t)) - V_s(x) \geq \mathcal{E} V_s(x(t)) - V_s(x).$$

The proof of this theorem is complete. $\square$

Theorem 6.1 is an important result, which extends Theorem 1 of [94]. Similar to the application of Theorem 1 [94] in [93], [94] and [30], Theorem 6.1 can be applied to nonlinear stochastic stability analysis and stabilization [126]. In the following, we mainly study dissipativity with $w(u, z) = \gamma^2 u^T u - z^T z$, i.e., the finite $\mathcal{L}_2$ gain problem [173]. When $w(u, z) = u^T z$, it is called a passive system [30, 70], which is very useful in the study of stability of stochastic nonlinear systems. If (6.2) is replaced by

$$\mathcal{E} V(x(t)) - V(x(s)) = \mathcal{E} \int_s^t w(u(\tau), z(\tau)) \, d\tau, \quad \forall (u, z) \in \mathcal{R}^{n_u} \times \mathcal{R}^{n_z}, \quad (6.11)$$

system (6.1) is said to be lossless; see [29]. In [151], (6.7)–(6.9) is called a nonlinear Lure equation.

REMARK 6.1   For stochastic system $dx(t) = m(x, u) \, dt + \sigma(x) \, dB(t)$, a more general definition for stochastic dissipativeness can be found in Definitions 4.1–4.2 of [28]. However, the above Definition 6.1 is sufficient for our purpose. In particular, when $w(u, z) = u^T z$, by using the well-known Dynkin's formula, it can be seen that Definition 6.1 extends Definition 4.1 of [70] about stochastic passive systems.   ∎

REMARK 6.2   If we let $s = 0$ and t be any bounded stopping time in (6.2) and (6.4), then Definitions 6.1–6.2 have been introduced in [167] for the following general nonlinear stochastic system

$$dx(t) = f(x, u) \, dt + g(x, u) \, dB(t), \quad x(0) = x \in \mathcal{R}^n.$$

Here, we take the terminal time t to be any fixed scalar only for technicality. ∎

6.2   Observability and Detectability

The following definition can be considered as an extension of the exact observability and detectability of linear stochastic systems introduced in Chapter 1.

DEFINITION 6.3   *We say that the following system*

$$dx = f(x)\, dt + l(x)\, dB, \ \ f(0) = 0, \ \ l(0) = 0, \ \ z = h(x) \tag{6.12}$$

or $(f, l|h)$ for short, is locally zero-state detectable if there is a neighborhood $\tilde{N}_0$ of the origin such that $\forall x(0) = x_0 \in \tilde{N}_0$,

$$z(t) = h(x(t)) = 0, a.s. \ \forall t \geq 0 \Rightarrow \mathcal{P}(\lim_{t \to \infty} x(t) = 0) = 1.$$

If $\tilde{N}_0 = \mathcal{R}^n$, (6.12) is called zero-state detectable. (6.12) is locally (respectively, globally) zero-state observable, if there is a neighborhood $\tilde{N}_0$ of the origin such that $\forall x_0 \in \tilde{N}_0$ (respectively, $\mathcal{R}^n$), $z(t) \equiv 0$ implies $x_0 \equiv 0$.

Obviously, for the linear time-invariant stochastic system

$$dx = Fx\, dt + Lx\, dB, \ \ z = Hx, \tag{6.13}$$

there is no difference among local zero-state observability, global zero-state observability and exact observability. Although there is no difference between local zero-state detectability and global zero-state detectability, they are not equivalent to exact detectability of $(F, L|H)$, because $\mathcal{P}(\lim_{t \to \infty} x(t) = 0) = 1$ is not equivalent to $\lim_{t \to \infty} \mathcal{E}\|x(t)\|^2 = 0$.

The following lemma, which is called the stochastic version of LaSalle's invariance principle [104] will be used.

LEMMA 6.1

Assume there exists a Lyapunov function V such that

$$\mathcal{L}_{u \equiv 0} V(x) \leq 0$$

for any $x \in \mathcal{R}^n$; then the solution $x(t)$ of the system

$$dx = f(x)\, dt + h(x)\, dB, \ x(0) = x_0 \in \mathcal{R}^n, \ \ f(0) = h(0) = 0 \tag{6.14}$$

tends in probability one to the largest invariant set whose support is contained in the locus $\Upsilon := \{x : \mathcal{L}_{u \equiv 0} V(x) = 0\}$ for any $t \geq 0$.

There are other LaSalle-type invariance principles; see, e.g., [132]. A discrete-time LaSalle-type invariance principle for systems with multiplicative noise can be found in [221].

6.3 Infinite Horizon H_∞ Control

Consider the nonlinear time-invariant stochastic system (the time variable t is suppressed)

$$\begin{cases} dx = [f(x) + g(x)u + k(x)v]\, dt + [h(x) + l(x)v]\, dB, \\ f(0) = 0,\ h(0) = 0, \\ z = \begin{bmatrix} m(x) \\ u \end{bmatrix},\ m(0) = 0, \end{cases} \tag{6.15}$$

where $v(t)$ still stands for the exogenous disturbance, which is an adapted process with respect to $\mathcal{F}_t$. Under very mild conditions, (6.15) has a unique strong solution $x(t)$ or for clarity $x(t, u, v, x(t_0), t_0)$ [196] on any finite interval $[t_0, T]$ under initial state $x(t_0) \in \mathcal{R}^n$.

DEFINITION 6.4 (Infinite horizon nonlinear state feedback H_∞ control). *Given $\gamma > 0$, we want to find, if it exits, an admissible control $\tilde{u}_\infty^* \in \mathcal{L}_\mathcal{F}^2(\mathcal{R}^+, \mathcal{R}^{n_u})$, such that for any non-zero $v \in \mathcal{L}_\mathcal{F}^2(\mathcal{R}^+, \mathcal{R}^{n_v})$, when $x(0) = 0$, the following inequality holds.*

$$\|z\|_{[0,\infty)} := \left\{ \mathcal{E} \int_0^\infty \|z\|^2\, dt \right\}^{1/2} \le \gamma \|v\|_{[0,\infty)} := \gamma \left\{ \mathcal{E} \int_0^\infty \|v\|^2\, dt \right\}^{1/2}. \tag{6.16}$$

(6.16) is equivalent to $\|\mathcal{L}_\infty\| \le \gamma$, where the perturbation operator $\mathcal{L}_\infty$ is defined by $\mathcal{L}_\infty : v \in \mathcal{L}_\mathcal{F}^2(\mathcal{R}^+, \mathcal{R}^{n_v}) \mapsto z \in \mathcal{L}_\mathcal{F}^2(\mathcal{R}^+, \mathcal{R}^{n_z})$ subject to (6.15) with

$$\|\mathcal{L}_\infty\| = \sup_{v \in \mathcal{L}_\mathcal{F}^2(\mathcal{R}^+, \mathcal{R}^{n_v}), u = \tilde{u}_\infty^*, v \neq 0, x(0) = 0} \frac{\|z\|_{[0,\infty)}}{\|v\|_{[0,\infty)}}.$$

In Definition 2.7 of linear time-invariant stochastic H_∞ control, the internal mean square stability is required. We, of course, expect the closed-loop system to be internally stable in a certain sense, which will be guaranteed by (6.16) together with zero-state observability or zero-state detectability. As pointed out by [173], it is easier to first consider an infinite horizon nonlinear state feedback H_∞ control as in Definition 6.4. More specifically, if we let $u \equiv 0$ in (6.15), $\tilde{\mathcal{L}}_\infty(v) := m(x(t, 0, v, 0, 0))$, $v \in \mathcal{L}_\mathcal{F}^2(\mathcal{R}^+, \mathcal{R}^{n_v})$, then when $\|\tilde{\mathcal{L}}_\infty\| \le \gamma$ for some $\gamma > 0$, the nonlinear system

$$\begin{cases} dx = [f(x) + k(x)v]\, dt + [h(x) + l(x)v]\, dB, \\ z_1 = m(x) \end{cases} \tag{6.17}$$

is said to be externally stable or $\mathcal{L}_2$ input-output stable. Another point that should be emphasized is that Definition 6.4 adopts $\|\mathcal{L}_\infty\| \le \gamma$ rather than $\|\mathcal{L}_\infty\| < \gamma$ as in linear time-invariant systems. This is because it is not easy to guarantee $\|\mathcal{L}_\infty\| <$

γ in nonlinear stochastic systems. The following theorem extends Theorem 16 of [173].

THEOREM 6.2

Suppose there exists a nonnegative solution $V \in \mathcal{C}^2(\mathcal{R}^n; \mathcal{R}^+)$ with $V(0) = 0$ to the HJE

$$
\begin{cases}
\mathcal{H}^1_\infty(V(x)) := \frac{\partial V^T}{\partial x} f + \frac{1}{2}\left(\frac{\partial V^T}{\partial x}k + h^T \frac{\partial^2 V}{\partial x^2}l\right)\left(\gamma^2 I - l^T \frac{\partial^2 V}{\partial x^2}l\right)^{-1}\left(k^T \frac{\partial V}{\partial x} + l^T \frac{\partial^2 V}{\partial x^2}h\right) \\
\quad - \frac{1}{2}\frac{\partial V^T}{\partial x}gg^T\frac{\partial V}{\partial x} + \frac{1}{2}m^T m + \frac{1}{2}h^T \frac{\partial^2 V}{\partial x^2}h = 0, \\
\gamma^2 I - l^T \frac{\partial^2 V}{\partial x^2}l > 0, V(0) = 0,
\end{cases}
\tag{6.18}
$$

then

$$
\tilde{u}^*_\infty = -g^T \frac{\partial V}{\partial x}
\tag{6.19}
$$

is an H_∞ control for system (6.15).

Proof. By Itô's formula,

$$
dV(x) = \frac{\partial V^T}{\partial x}(f + gu + kv) + \frac{1}{2}(h + lv)^T \frac{\partial^2 V}{\partial x^2}(h + lv)\,dt
$$
$$
+ \frac{\partial V^T}{\partial x}(h + lv)\,dB(t).
\tag{6.20}
$$

By completing the squares and taking into account (6.18), we have for any $T > 0$,

$$
\mathcal{E}V(x(T)) - V(0) = \mathcal{E}V(x(T)) = \mathcal{E}\int_0^T \left[\frac{\partial V^T}{\partial x}(f + gu + kv)\right.
$$
$$
\left. + \frac{1}{2}(h + lv)^T \frac{\partial^2 V}{\partial x^2}(h + lv)\right]dt
$$
$$
= \frac{1}{2}\mathcal{E}\int_0^T \left(\left\|u + g^T\frac{\partial V}{\partial x}\right\|^2 + 2\mathcal{H}^1_\infty(V(x))\right.
$$
$$
- \left\|v - \left(\gamma^2 I - l^T\frac{\partial^2 V}{\partial x^2}l\right)^{-1}\left(k^T\frac{\partial V}{\partial x} + l^T\frac{\partial^2 V}{\partial x^2}h\right)\right\|^2_{\gamma,l,V}
$$
$$
\left. - \|z\|^2 + \gamma^2\|v\|^2\right)dt
$$
$$
= \frac{1}{2}\mathcal{E}\int_0^T \left(\left\|u + g^T\frac{\partial V}{\partial x}\right\|^2 - \|z\|^2 + \gamma^2\|v\|^2\right.
$$
$$
\left. - \left\|v - \left(\gamma^2 I - l^T\frac{\partial^2 V}{\partial x^2}l\right)^{-1}\left(k^T\frac{\partial V}{\partial x} + l^T\frac{\partial^2 V}{\partial x^2}h\right)\right\|^2_{\gamma,l,V}\right)dt, \tag{6.21}
$$

where $\|Z(x)\|^2_{\gamma,l,V} := Z^T(x)(\gamma^2 I - l^T \frac{\partial^2 V}{\partial x^2} l)Z(x)$. Obviously, when $u = \tilde{u}^*_\infty$, (6.21) leads to

$$\mathcal{E}\int_0^T \|z\|^2\, dt = -\mathcal{E}\int_0^T \left\| v - \left(\gamma^2 I - l^T \frac{\partial^2 V}{\partial x^2} l\right)^{-1} \left(k^T \frac{\partial V}{\partial x} + l^T \frac{\partial^2 V}{\partial x^2} h\right) \right\|^2_{\gamma,l,V}\, dt$$

$$-2\mathcal{E}V(x(T)) + 2V(0) + \gamma^2 \mathcal{E}\int_0^T \|v\|^2\, dt$$

$$\leq \gamma^2 \mathcal{E}\int_0^T \|v\|^2\, dt. \tag{6.22}$$

Let $T \to \infty$ in (6.22), then it follows that $\|\mathcal{L}_\infty\| \leq \gamma$ because of $V \geq 0$ and $V(0) = 0$. This ends the proof of Theorem 6.2. $\square$

REMARK 6.3 From the proof of Theorem 6.2, it can be seen that we have in fact obtained the following identity

$$\mathcal{L}_{u,v}V(x) := \frac{\partial V^T}{\partial x}(f + gu + kv) + \frac{1}{2}(h + lv)^T \frac{\partial^2 V}{\partial x^2}(h + lv)$$

$$= \frac{1}{2}(\|u - \tilde{u}^*_\infty\|^2 - \|v - \tilde{v}^*_\infty\|^2_{\gamma,l,V} + 2\mathcal{H}^1_\infty(V(x))$$

$$- \|z\|^2 + \gamma^2 \|v\|^2), \tag{6.23}$$

where $\mathcal{L}_{u,v}$ is the infinitesimal generator of

$$dx = [f(x) + g(x)u + k(x)v]\, dt + [h(x) + l(x)v]\, dB$$

and

$$\tilde{v}^*_\infty = \left(\gamma^2 I - l^T \frac{\partial^2 V}{\partial x^2} l\right)^{-1} \left(k^T \frac{\partial V}{\partial x} + l^T \frac{\partial^2 V}{\partial x^2} h\right).$$

We can also see that Theorem 6.2 still holds if HJE (6.18) is replaced by HJI

$$\mathcal{H}^1_\infty(V(x)) \leq 0,\ \gamma^2 I - l^T \frac{\partial^2 V}{\partial x^2} l > 0, V(0) = 0,$$

which is convenient to be used in practice. $\blacksquare$

REMARK 6.4 From the inequality (6.22), it immediately follows that for any $v \in \mathcal{L}^2_{\mathcal{F}}(\mathcal{R}^+, \mathcal{R}^{n_v})$, we have $z \in \mathcal{L}^2_{\mathcal{F}}(\mathcal{R}^+, \mathcal{R}^{n_z})$, $\tilde{u}^*_\infty \in \mathcal{L}^2_{\mathcal{F}}(\mathcal{R}^+, \mathcal{R}^{n_u})$. However, we cannot assert $\tilde{v}^*_\infty \in \mathcal{L}^2_{\mathcal{F}}(\mathcal{R}^+, \mathcal{R}^{n_v})$. $\blacksquare$

The following result generalizes Corollary 17 of [173] to the stochastic case.

COROLLARY 6.1

Under the condition of Theorem 6.2, if $(f, h|m)$ is zero-state observable, then any solution to HJE (6.18) satisfies $V(x) > 0$ for $x \neq 0$, and the closed-loop system (with $v \equiv 0$)

$$dx = [f(x) + g(x)\tilde{u}_\infty^*] \, dt + h(x) \, dB \qquad (6.24)$$

is locally asymptotically stable in probability one. If V is also proper (i.e., for each $a > 0$, $V^{-1}[0, a]$ is compact), then it is globally asymptotically stable in probability one. Moreover, $\lim_{t \to \infty} \mathcal{E}V(x(t)) = 0$.

Proof. By (6.23), we have

$$\mathcal{L}_{u=\tilde{u}_\infty^*, v=0} V(x) = -\frac{1}{2} \left(\|\tilde{v}_\infty^*\|_{\gamma, l, V}^2 + \|m(x)\|^2 + \|\tilde{u}_\infty^*\|^2 \right)$$

$$\leq -\frac{1}{2} \begin{bmatrix} m^T(x) & (\tilde{u}_\infty^*)^T \end{bmatrix} \begin{bmatrix} m(x) \\ \tilde{u}_\infty^* \end{bmatrix}. \qquad (6.25)$$

If $V(x)$ is not strictly positive definite in the sense of Lyapunov, then there exists $x_0 \neq 0$, such that $V(x_0) = 0$. Integrating from 0 to T, and then taking expectation on both sides of (6.25), it follows that

$$0 \leq \mathcal{E}V(x(T)) = -\frac{1}{2}\mathcal{E}\int_0^T \left(\|m\|^2 + \|\tilde{u}_\infty^*\|^2 \right) dt \leq 0. \qquad (6.26)$$

Equation (6.26) concludes $z(t)|_{u=\tilde{u}_\infty^*} \equiv 0, t \in [0, T]$ for any $T > 0$. From the zero-state observability of $(f, h|m)$, it is easy to prove the zero-state observability of $(f + g\tilde{u}_\infty^*, h|[m^T \ (\tilde{u}_\infty^*)^T]^T)$. According to the definition of zero-state observability, we must have $x(t) \equiv 0$ from $z(t)|_{u=\tilde{u}_\infty^*, v=0} \equiv 0$, a.s., which contradicts $x_0 \neq 0$. $V > 0$ is proved.

In addition, by the above analysis, we have

$$\Upsilon = \{x : \mathcal{L}_{u=\tilde{u}_\infty^*, v=0} V(x) = 0\} \subset \{x : m(x) = 0\} = \{0\}.$$

Hence, the asymptotic stability is proved by use of Lemma 6.1.

Finally, to show $\lim_{t \to \infty} \mathcal{E}V(x(t)) = 0$, we apply Itô's formula to system (6.24) and obtain that for any $t > s > 0$,

$$V(x(t)) = V(x(s)) + \int_s^t \mathcal{L}_{u=\tilde{u}_\infty^*, v=0} V(x(\tau)) \, d\tau + \int_s^t h^T \frac{\partial V}{\partial x} \, dB(\tau)$$

$$= V(x(s)) - \frac{1}{2} \int_s^t \left(\|\tilde{v}_\infty^*\|_{\gamma, l, V}^2 + \|m(x)\|^2 + \|\tilde{u}_\infty^*\|^2 \right) dt$$

$$+ \int_s^t h^T \frac{\partial V}{\partial x} \, dB(\tau).$$

So

$$\mathcal{E}[V(x(t))|\mathcal{F}_s] = \mathcal{E}[V(x(s))|\mathcal{F}_s] + \mathcal{E}\left[\int_s^t h^T \frac{\partial V}{\partial x}\, dB(\tau)\Big|\mathcal{F}_s\right]$$

$$-\frac{1}{2}\mathcal{E}\left[\int_s^t (\|\tilde{v}_\infty^*\|_{\gamma,l,V}^2 + \|m(x)\|^2 + \|\tilde{u}_\infty^*\|^2)\, d\tau\Big|\mathcal{F}_s\right]$$

$$= V(x(s)) - \frac{1}{2}\mathcal{E}\left[\int_s^t (\|\tilde{v}_\infty^*\|_{\gamma,l,V}^2 + \|m(x)\|^2 + \|\tilde{u}_\infty^*\|^2)\, d\tau\Big|\mathcal{F}_s\right]$$

$$\leq V(x(s)),$$

which shows that $\{V(x(t)), \mathcal{F}_t\}$ is a nonnegative supermartingale. By Doob's convergence theorem and asymptotic stability, $V(x(\infty)) = \lim_{t\to\infty} V(x(t)) = 0$ a.s.. Moreover, $\lim_{t\to\infty} \mathcal{E}V(x(t)) = \mathcal{E}V(x(\infty)) = 0$. The proof of this corollary is complete. $\square$

REMARK 6.5 By analogous discussions as in Corollary 6.1, if we replace zero-state observability with zero-state detectability, then Corollary 6.1 still holds with $V > 0$ replaced by $V \geq 0$. ∎

We attempt to show the necessity of Theorem 6.2, however, there are some technical problems that cannot be overcome at present. Nonetheless, we believe that the following lemma, which can be called the "infinite horizon nonlinear SBRL," would contribute to establishing the necessity of Theorem 6.2.

LEMMA 6.2
Given $\gamma > 0$, consider system (6.17). If there exists a nonnegative solution $V \in \mathcal{C}^2(\mathcal{R}^n; \mathcal{R}^+)$ to the HJE

$$\begin{cases} \frac{\partial V^T}{\partial x} f + \frac{1}{2}\left(\frac{\partial V^T}{\partial x}k + h^T \frac{\partial^2 V}{\partial x^2}l\right)\left(\gamma^2 I - l^T \frac{\partial^2 V}{\partial x^2}l\right)^{-1} \\ \qquad \cdot \left(k^T \frac{\partial V}{\partial x} + l^T \frac{\partial^2 V}{\partial x^2}h\right) + \frac{1}{2}m^T m + \frac{1}{2}h^T \frac{\partial^2 V}{\partial x^2}h = 0, \\ \gamma^2 I - l^T \frac{\partial^2 V}{\partial x^2}l > 0, \forall x \in \mathcal{R}^n, V(0) = 0, \end{cases} \quad (6.27)$$

then $\|\tilde{\mathcal{L}}_\infty\| \leq \gamma$ for $x(0) = 0$.
Conversely, assume that the following conditions hold:
(i) There exists a positive definite function $q(x) : \mathcal{R}^n \mapsto \mathcal{R}^+$, $q(0) = 0$, such that for $\forall x(0) = x \in \mathcal{R}^n, v \in \mathcal{L}_{\mathcal{F}}^2(\mathcal{R}^+, \mathcal{R}^{n_v})$,

$$\|z\|_{[0,\infty)}^2 \leq \gamma^2 \|v\|_{[0,\infty)}^2 + q(x), v \neq 0. \quad (6.28)$$

(ii) The storage function $V_{a,0} \in \mathcal{C}^2(\mathcal{R}^n, \mathcal{R}^+)$ exists and satisfies that $\gamma^2 I - l^T \frac{\partial^2 V_{a,0}}{\partial x^2}l > 0$ for all $x \in \mathcal{R}^n$, where

$$V_{a,0}(x) = -\inf_{v \in \mathcal{L}_{\mathcal{F}}^2([0,T],\mathcal{R}^{n_v}),T\geq 0,x(0)=x\in\mathcal{R}^n} \mathcal{E}\int_0^T w(v,z)\, dt$$

with $w(v,z) = \frac{1}{2}\gamma^2\|v\|^2 - \frac{1}{2}\|z\|^2$.

Then $V_{a,0}$ solves HJE (6.27). Moreover, for any solution V of (6.27),

$$V \geq V_{a,0} \geq 0, V_{a,0}(0) = 0. \tag{6.29}$$

Proof. The first part is an immediate corollary of Theorem 6.2 ($g \equiv 0, u \equiv 0$). As for the converse part, we first note that (6.28) concludes, for any $T \geq 0$, that

$$\|z\|^2_{[0,T]} \leq \gamma^2\|v\|^2_{[0,T]} + q(x), \ \forall v \in \mathcal{L}^2_\mathcal{F}([0,T], \mathcal{R}^{n_v}).$$

Actually, for any $v \in \mathcal{L}^2_\mathcal{F}([0,T], \mathcal{R}^{n_v})$, if we let

$$\hat{v}(t) = \begin{cases} v(t), \ t \in [0,T], \\ 0, \ t \in (T,\infty), \end{cases}$$

then $\hat{v} \in \mathcal{L}^2_\mathcal{F}([0,T], \mathcal{R}^{n_v})$. By (6.28),

$$\begin{aligned} \|z\|^2_{[0,T]} &\leq \|z\|^2_{[0,\infty)} = \|z\|^2_{[0,T]} + \|z\|^2_{(T,\infty)} \\ &\leq \gamma^2\|\hat{v}\|^2_{[0,\infty)} + q(x) \\ &= \gamma^2\|v\|^2_{[0,T]} + q(x). \end{aligned} \tag{6.30}$$

So

$$0 \leq V_{a,0}(x) \leq \frac{1}{2}q(x), \ V_{a,0}(0) = 0.$$

Take $R = \frac{1}{2}\gamma^2 I, S = 0, Q = -\frac{1}{2}I, V_s = V_{a,0}$, and

$$\tilde{w} = \frac{\sqrt{2}}{2}\left(\gamma^2 I - l^T \frac{\partial^2 V_{a,0}}{\partial x^2} l\right)^{1/2}.$$

HJE (6.27) is derived from Theorem 6.1. It is easy to show that any solution V of (6.27) is a possible storage function with supply rate w. Therefore, (6.29) is derived from Proposition 6.2. $\square$

For the linear time-invariant system

$$\begin{cases} dx = (A_1 x + B_1 u + C_1 v)\, dt + (A_2 x + C_2 v)\, dB, \\ z = \begin{bmatrix} Cx \\ u \end{bmatrix}. \end{cases} \tag{6.31}$$

Let $V(x) = \frac{1}{2}x^T P x$. Then Theorem 6.2 and Corollary 6.1 lead to the following corollary.

COROLLARY 6.2

Suppose there exists a solution $P \geq 0$ to the GARE

$$\begin{cases} PA_1 + A_1^T P + A_2^T PA_2 + (PC_1 + A_2^T PC_2)(\gamma^2 I - C_2^T PC_2)^{-1} \\ \qquad \cdot (C_1^T P + C_2^T PA_2) - PB_1 B_1^T P + C^T C = 0, \\ \gamma^2 I - C_2^T PC_2 > 0 \end{cases} \tag{6.32}$$

for some $\gamma > 0$, then $\tilde{u}_\infty^*(x) = -B_1^T P x$ is an H_∞ control, which makes the closed-loop system satisfy $\|\mathcal{L}_\infty\| \leq \gamma$. Additionally, if $(A_1, A_2|C)$ is observable (respectively, detectable), then

(i) *$P > 0$ (respectively, $P \geq 0$);*

(ii) *system*

$$dx = (A_1 - B_1 B_1^T P)x\, dt + A_2 x\, dB$$

is not only asymptotically stable in probability one, but also ASMS.

Proof. The first part is an immediate corollary of Theorem 6.2. As for the second part, (i) and (ii) are concluded from Corollary 6.1 and Remark 6.5. $\square$

The following example is given to illustrate the design procedure of the proposed nonlinear stochastic H_∞ controller.

Example 6.1

Consider the following nonlinear stochastic system with state-dependent noise:

$$dx = \left(\begin{bmatrix} x_1^3 - 2x_1 - 4x_2 \\ x_2^3 - 2x_2 \end{bmatrix} + \begin{bmatrix} 2x_1 \\ 2x_2 \end{bmatrix} u + \begin{bmatrix} 1 \\ 1 \end{bmatrix} v \right) dt + \begin{bmatrix} x_2^2 \\ x_1 x_2 \end{bmatrix} dB(t),$$

$$z = \begin{bmatrix} 2(x_1 + x_2) \\ u(t) \end{bmatrix}.$$

Let the desired disturbance attenuation level $\gamma = 1$. Then, by Theorem 6.2 and Remark 6.3, we need to solve the following HJI

$$\mathcal{H}_\infty^1(V(x)) = \frac{\partial V^T}{\partial x} \begin{bmatrix} x_1^3 - 2x_1 - 4x_2 \\ x_2^3 - 2x_2 \end{bmatrix} + \frac{1}{2} \frac{\partial V^T}{\partial x} \begin{bmatrix} 1 \\ 1 \end{bmatrix} [1\ 1] \frac{\partial V}{\partial x}$$

$$- \frac{1}{2} \frac{\partial V^T}{\partial x} \begin{bmatrix} 2x_1 \\ 2x_2 \end{bmatrix} [2x_1\ 2x_2] \frac{\partial V}{\partial x} + 2(x_1 + x_2)^2 + \frac{1}{2} [x_2^2\ x_1 x_2] \frac{\partial^2 V}{\partial x^2} \begin{bmatrix} x_2^2 \\ x_1 x_2 \end{bmatrix} \leq 0.$$

If we take $V(x) = x_1^2 + x_2^2$, then

$$\mathcal{H}_\infty^1(V(x)) = -6x_1^4 - 5x_2^4 - 15x_1^2 x_2^2 \leq 0,$$

i.e., if we choose $\tilde{u}_\infty^* = -g^T \frac{\partial V}{\partial x} = -[2x_1\ 2x_2] \begin{bmatrix} 2x_1 \\ 2x_2 \end{bmatrix} = -4x_1^2 - 4x_2^2$, then the desired H_∞ performance is achieved. $\square$

REMARK 6.6 As stated in the linear case, all the above results can be extended to systems with multiple noises:

$$\begin{cases} dx = [f(x) + g(x)u + k(x)v]\, dt + \sum_{i=1}^N [h_i(x) + l_i(x)v]\, dB_i, \\ z = \begin{bmatrix} m(x) \\ u \end{bmatrix}, \end{cases}$$

where $B = [B_1, B_2, \cdots, B_N]^T$ is a multi-dimensional Brownian motion. In this case, HJE (6.18) becomes

$$\begin{cases} \frac{\partial V^T}{\partial x} f + \frac{1}{2}\left(\frac{\partial V^T}{\partial x} k + \sum_{i=1}^N h_i^T \frac{\partial^2 V}{\partial x^2} l_i\right)\left(\gamma^2 I - \sum_{i=1}^N l_i^T \frac{\partial^2 V}{\partial x^2} l_i\right)^{-1} \\ \quad \cdot \left(k^T \frac{\partial V}{\partial x} + \sum_{i=1}^N l_i^T \frac{\partial^2 V}{\partial x^2} h_i\right) - \frac{1}{2}\frac{\partial V^T}{\partial x} g g^T \frac{\partial V}{\partial x} + \frac{1}{2} m^T m + \frac{1}{2}\sum_{i=1}^N h_i^T \frac{\partial^2 V}{\partial x^2} h_i = 0, \\ \gamma^2 I - \sum_{i=1}^N l_i^T \frac{\partial^2 V}{\partial x^2} l_i > 0, V(0) = 0. \end{cases}$$

∎

REMARK 6.7   In (6.15), if $l(x)v$ is replaced by $l(x)u$, then by the same discussion as in Theorem 6.2, we can show that Theorem 6.2 still holds if we replace (6.18) and (6.19) by

$$\begin{cases} \frac{\partial V^T}{\partial x} f + \frac{1}{2\gamma^2}\frac{\partial V^T}{\partial x} k k^T \frac{\partial V}{\partial x} - \frac{1}{2}\left(\frac{\partial V^T}{\partial x} g + h^T \frac{\partial^2 V}{\partial x^2} l\right)(I + l^T \frac{\partial^2 V}{\partial x^2} l)^{-1} \\ \quad \cdot \left(g^T \frac{\partial V}{\partial x} + l^T \frac{\partial^2 V}{\partial x^2} h\right) + \frac{1}{2} m^T m + \frac{1}{2} h^T \frac{\partial^2 V}{\partial x^2} h = 0, \\ I + l^T \frac{\partial^2 V}{\partial x^2} l > 0, V(0) = 0 \end{cases}$$

and

$$\tilde{u}_\infty^*(x) = -\left(I + l^T \frac{\partial^2 V}{\partial x^2} l\right)^{-1}\left(g^T \frac{\partial V}{\partial x} + l^T \frac{\partial^2 V}{\partial x^2} h\right),$$

respectively.   ∎

6.4   Finite Horizon Nonlinear H_∞ Control

In this section, we study the finite horizon H_∞ control problem. Suppose the system is governed by the following stochastic time-varying equation

$$\begin{cases} dx = [f(t,x) + g(t,x)u + k(t,x)v] \, dt + [h(t,x) + l(t,x)v] \, dB, \\ z(t) = \begin{bmatrix} m(t,x) \\ u \end{bmatrix}, \ t \in [0,T]. \end{cases} \tag{6.33}$$

The finite horizon H_∞ control is not associated with stochastic stability, so different from (6.15), we do not need to assume $f(t,0) = 0$, $h(t,0) = 0$ and $m(t,0) = 0$ in (6.33). The so-called finite horizon H_∞ control is to find, if it exists, a $\tilde{u}_T^* \in \mathcal{L}_{\mathcal{F}}^2([0,T], \mathcal{R}^{n_u})$, such that for any given $\gamma > 0$, and all non-zero $v \in \mathcal{L}_{\mathcal{F}}^2([0,T], \mathcal{R}^{n_v})$, $x(0) = 0$, the closed-loop system satisfies

$$\|z\|_{[0,T]} \le \gamma \|v\|_{[0,T]}. \tag{6.34}$$

Similar to the definition of $\mathcal{L}_\infty$, a perturbation operator $\mathcal{L}_T : \mathcal{L}_{\mathcal{F}}^2([0,T], \mathcal{R}^{n_v}) \mapsto \mathcal{L}_{\mathcal{F}}^2([0,T], \mathcal{R}^{n_z})$ can be defined associated with (6.33) with $u = \tilde{u}_T^*$ as follows:

$$\mathcal{L}_T(v(t)) = z(t)|_{x(0)=0}.$$

As such, (6.34) is equivalent to

$$\|\mathcal{L}_T\| = \sup_{v \in \mathcal{L}_{\mathcal{F}}^2([0,T],\mathcal{R}^{n_v}), v \neq 0, x(0)=0} \frac{\|z\|_{[0,T]}}{\|v\|_{[0,T]}} \leq \gamma.$$

In particular, if we set $\tilde{\mathcal{L}}_T = \mathcal{L}_T|_{u \equiv 0}$, then when $\|\tilde{\mathcal{L}}_T\| \leq \gamma$ for any given $\gamma > 0$, the system is said to have $\mathcal{L}_2$-gain less than or equal to γ.

In analogy with the proof of Theorem 6.2, the following result is easily obtained.

THEOREM 6.3
Assume $V_T(t,x) \in \mathcal{C}^{1,2}([0,T] \times \mathcal{R}^n; \mathcal{R}^+)$ satisfies the following HJE

$$\begin{cases} \mathcal{H}_T^1(t,x) := \frac{\partial V_T}{\partial t} + \frac{\partial V_T^T}{\partial x} f + \frac{1}{2} \left(\frac{\partial V_T^T}{\partial x} k + h^T \frac{\partial^2 V_T}{\partial x^2} l \right) \left(\gamma^2 I - l^T \frac{\partial^2 V_T}{\partial x^2} l \right)^{-1} \\ \quad \cdot \left(k^T \frac{\partial V_T}{\partial x} + l^T \frac{\partial^2 V_T}{\partial x^2} h \right) - \frac{1}{2} \frac{\partial V_T^T}{\partial x} g g^T \frac{\partial V_T}{\partial x} + \frac{1}{2} m^T m + \frac{1}{2} h^T \frac{\partial^2 V_T}{\partial x^2} h = 0, \\ \gamma^2 I - l^T \frac{\partial^2 V_T}{\partial x^2} l > 0, V_T(T,x) = 0, V_T(t,0) = 0, \forall (t,x) \in [0,T] \times \mathcal{R}^n, \end{cases}$$
$$(6.35)$$

then $(\tilde{u}_T^*, \tilde{v}_T^*)$ is a saddle point for the following stochastic two-person zero-sum game problem:

$$\min_{u \in \mathcal{L}_{\mathcal{F}}^2([0,T],\mathcal{R}^{n_u})} \max_{v \in \mathcal{L}_{\mathcal{F}}^2([0,T],\mathcal{R}^{n_v})} \mathcal{E} \int_0^T (\|z\|^2 - \gamma^2 \|v\|^2) \, dt,$$

where $\tilde{u}_T^*$ and $\tilde{v}_T^*$ are defined respectively as

$$\tilde{u}_T^* = -g^T \frac{\partial V_T}{\partial x}$$

and

$$\tilde{v}_T^* = \left(\gamma^2 I - l^T \frac{\partial^2 V_T}{\partial x^2} l \right)^{-1} \left(k^T \frac{\partial V_T}{\partial x} + l^T \frac{\partial^2 V_T}{\partial x^2} h \right).$$

Moreover, $\tilde{u}_T^*$ is an H_∞ control for system (6.33), and $\tilde{v}_T^*$ is the corresponding worst-case disturbance.

Now, we highlight the relationship between the solutions of finite and infinite horizon HJEs. Let $V(x)$ and $V_T(t,x,Q(x))$ stand for the solutions of (6.18) and (6.35) with terminal condition $V_T(T,x) = Q(x) \geq 0$ for all $x \in \mathcal{R}^n$, respectively. A generalized version of Lemma 2.6 of [114] is as follows.

PROPOSITION 6.3
If $V(x) \geq Q(x)$ for all $x \in \mathcal{R}^n$, then $V(x) \geq V_T(t,x,Q(x)) \geq V_T(t,x,0) = V_T(t,x) \geq 0$ for all $(t,x) \in [0,T] \times \mathcal{R}^n$.

Proof. For any initial time $t \geq 0$ and state $x(t) := x \in \mathcal{R}^n$, one only needs to note the following identities:

$$\frac{1}{2}\mathcal{E}\int_t^T (\|z\|^2 - \gamma^2\|v\|^2)\,d\tau = V(x) - \mathcal{E}V(x(T))$$

$$+\frac{1}{2}\mathcal{E}\int_t^T (\|u - \tilde{u}_T^*\|^2 - \|v - \tilde{v}_T^*\|_{\gamma,l,V}^2)$$

$$+2\mathcal{H}_T^1(V(x)))\,d\tau \qquad (6.36)$$

and

$$\frac{1}{2}\mathcal{E}\int_t^T (\|z\|^2 - \gamma^2\|v\|^2)\,d\tau = V_T(t,x,Q(x)) - \mathcal{E}V_T(T,x(T),Q(x))$$

$$+\frac{1}{2}\mathcal{E}\int_t^T (\|u - \tilde{u}_T^*\|^2 - \|v - \tilde{v}_T^*\|_{\gamma,l,V_T(t,x,Q(x))}^2)$$

$$+2\mathcal{H}_T^1(t,x))\,d\tau. \qquad (6.37)$$

The rest is similar to the proof of Lemma 2.6 of [114], and is omitted. $\square$

PROPOSITION 6.4
There is at most one solution to (6.35).

Proof. By contradiction, let $V_T^{(1)}(\cdot,\cdot)$ and $V_T^{(2)}(\cdot,\cdot)$ be two solutions of (6.35). Set

$$J_T(x,u,v,x(t_0),t_0) = \frac{1}{2}\mathcal{E}\int_{t_0}^T (\|z\|^2 - \gamma^2\|v\|^2)\,dt,$$

$$u_{i,T}^* = -g^T \frac{\partial V_T^{(i)}}{\partial x}$$

and

$$v_{i,T}^* = \left(\gamma^2 I - l^T \frac{\partial^2 V_T^{(i)}}{\partial x^2} l\right)^{-1}\left(k^T \frac{\partial V_T^{(i)}}{\partial x} + l^T \frac{\partial^2 V_T^{(i)}}{\partial x^2} h\right), \quad i = 1,2.$$

For any $x(s) = y$, $(s,y) \in [0,T) \times \mathcal{R}^n$, from (6.37), we have

$$J_T(x,u_{1,T}^*,v_{1,T}^*,y,s) = V_T^{(1)}(s,y) \leq J_T(x,u_{2,T}^*,v_{1,T}^*,y,s)$$

$$\leq J_T(x,u_{2,T}^*,v_{2,T}^*,y,s) = V_T^{(2)}(s,y).$$

Similarly, we can have $V_T^{(2)}(s,y) \leq V_T^{(1)}(s,y)$. Hence, $V_T^{(2)}(s,y) = V_T^{(1)}(s,y)$. $\square$

PROPOSITION 6.5
$V_T(\cdot,\cdot)$ is monotonically increasing with respect to $T > 0$.

Proof. For any $0 \le s \le T_0 \le T_1 < \infty$, $x(s) = y \in \mathcal{R}^n$, by (6.37), we have

$$
\begin{aligned}
J_{T_0}(x, u_{T_0}^*, v_{T_0}^*, y, s) = V_{T_0}(s, y) &\le J_{T_0}(x, u_{T_1}^*, v_{T_0}^*, y, s) \\
&\le J_{T_1}(x, u_{T_1}^*, v_{T_0}^*, y, s) \le J_{T_1}(x, u_{T_1}^*, v_{T_1}^*, y, s), \\
&= V_{T_1}(s, y).
\end{aligned}
$$

The proof of this proposition is complete. $\square$

If system (6.33) is time-invariant, and

$$
\bar{V}(t, x) := \lim_{T \to \infty} V_T(t, x, Q(x))
$$

exists, then $\bar{V}$ only depends on x, and is a solution of (6.18) (we refer the reader to the proof of Corollary 2.7 of [114]). In particular, if there exists $(\tilde{u}_\infty^*, \tilde{v}_\infty^*) \in \mathcal{L}_{\mathcal{F}}^2(\mathcal{R}^+, \mathcal{R}^{n_u}) \times \mathcal{L}_{\mathcal{F}}^2(\mathcal{R}^+, \mathcal{R}^{n_v})$, such that $J_\infty(x, \tilde{u}_\infty^*, \tilde{v}_\infty^*, y, s) < \infty$, then by making use of Propositions 6.3–6.5, $\bar{V}(x)$ exists due to the monotonicity and uniform boundedness of $V_T(t, x)$.

In general, the converse of Theorem 6.3 is not true, i.e., $\|\mathcal{L}_T\| \le \gamma$ does not necessarily imply that HJE (6.35) has a solution. A converse result will be presented in the following under some other conditions. To this end, assume $u = \tilde{k}(t, x)$ is an H_∞ control law of (6.33) and define $\tilde{V}_{T,\tilde{k}}(s, x) : [0, T] \times \mathcal{R}^n \mapsto \mathcal{R}^+$ as

$$
\begin{aligned}
\tilde{V}_{T,\tilde{k}}(s, x) &= -\frac{1}{2} \inf_{v \in \mathcal{L}_{\mathcal{F}}^2([s,T], \mathcal{R}^{n_v}), u = \tilde{k}, x(s) = x} \mathcal{E} \int_s^T (\gamma^2 \|v\|^2 - \|z\|^2) \, dt \\
&= \sup_{v \in \mathcal{L}_{\mathcal{F}}^2([s,T], \mathcal{R}^{n_v}), u = \tilde{k}, x(s) = x} -\frac{1}{2} \mathcal{E} \int_s^T (\gamma^2 \|v\|^2 - \|z\|^2) \, dt.
\end{aligned}
$$

It is easy to test the following properties of $\tilde{V}_{T,\tilde{k}}$: (1) $\tilde{V}_{T,\tilde{k}} \ge 0$; (2) $\tilde{V}_{T,\tilde{k}}(T, x) = 0$ for all $x \in \mathcal{R}^n$. The following proposition can also be shown in the same way as in [30] and [116].

PROPOSITION 6.6

(i) $\|\mathcal{L}_T\| \le \gamma$ *implies* $\tilde{V}_{T,\tilde{k}}(s, 0) = 0, \forall s \in [0, T]$.

(ii) $\tilde{V}_{T,\tilde{k}}$ *is finite on* $[0, T] \times \mathcal{R}^n$ *if and only if there exists a nonnegative function* $V(s, x) : [0, T] \times \mathcal{R}^n \mapsto \mathcal{R}^+$, *satisfying the following integral dissipation inequality (IDI):*

$$
\mathcal{E}V(T, x(T)) - V(s, x) \le \frac{1}{2} \mathcal{E} \int_s^T (\gamma^2 \|v\|^2 - \|z\|^2) \, dt. \tag{6.38}
$$

Moreover, when $\tilde{V}_{T,\tilde{k}}(s, x)$ *is finite,* $\tilde{V}_{T,\tilde{k}}$ *is itself a solution of (6.38).*

In the literature, such as [173] and [116], to guarantee the finiteness of $\tilde{V}_{T,\tilde{k}}(s, x)$, an essential concept called "reachability" in system theory was introduced.

LEMMA 6.3

If $\tilde{V}_{T,\tilde{k}}(s,x) \in \mathcal{C}^{1,2}([0,T] \times \mathcal{R}^n; \mathcal{R}^+)$ is finite with $\gamma^2 I - l^T \frac{\partial^2 \tilde{V}_{T,\tilde{k}}}{\partial x^2} l > 0$ for some $\gamma > 0$, and $\|\mathcal{L}_T\| \leq \gamma$, then $\tilde{V}_{T,\tilde{k}}$ solves the HJE

$$
\begin{cases}
\mathcal{H}(V_{T,\tilde{k}}) := \frac{\partial V_{T,\tilde{k}}}{\partial t} + \frac{\partial V_{T,\tilde{k}}^T}{\partial x}(f + g\tilde{k}) + \frac{1}{2}\left(\frac{\partial V_{T,\tilde{k}}^T}{\partial x}k + h^T \frac{\partial^2 V_{T,\tilde{k}}}{\partial x^2}l\right)\left(\gamma^2 I - l^T \frac{\partial^2 V_{T,\tilde{k}}}{\partial x^2}l\right)^{-1} \\
\qquad \cdot \left(k^T \frac{\partial V_{T,\tilde{k}}}{\partial x} + l^T \frac{\partial^2 V_{T,\tilde{k}}}{\partial x^2}h\right) + \frac{1}{2}(m^T m + \tilde{k}^T \tilde{k}) + \frac{1}{2}h^T \frac{\partial^2 V_{T,\tilde{k}}}{\partial x^2}h = 0, \\
\gamma^2 I - l^T \frac{\partial^2 V_{T,\tilde{k}}}{\partial x^2}l > 0, V_{T,\tilde{k}}(T,x) = 0, V_{T,\tilde{k}}(t,0) = 0, \forall (t,x) \in [0,T] \times \mathcal{R}^n.
\end{cases}
$$

$$(6.39)$$

Proof. We have shown that $\tilde{V}_{T,\tilde{k}}$ satisfies the boundary conditions of (6.39) above. Now, let $\hat{V} = -\tilde{V}_{T,\tilde{k}}$, then by the dynamic programming principle, $\hat{V}$ solves the following HJE [196]

$$
-\frac{\partial \hat{V}}{\partial t} + \max_{v \in U} H\left(t, x, v, -\frac{\partial \hat{V}}{\partial x}, -\frac{\partial^2 \hat{V}}{\partial x^2}\right) = 0, \qquad (6.40)
$$

where (U, ρ) is a polish space, $U \subset \mathcal{R}^{n_v}$, and the generalized Hamiltonian function

$$
H\left(t, x, v, -\frac{\partial \hat{V}}{\partial x}, -\frac{\partial^2 \hat{V}}{\partial x^2}\right) := -\frac{1}{2}\gamma^2\|v\|^2 + \frac{1}{2}\|z\|^2
$$

$$
-\frac{\partial \hat{V}^T}{\partial x}(f + g\tilde{k} + kv) - \frac{1}{2}(h + lv)^T \frac{\partial^2 \hat{V}}{\partial x^2}(h + lv)
$$

$$
= \mathcal{H}(\tilde{V}_{T,\tilde{k}}) - \frac{\partial \tilde{V}_{T,\tilde{k}}}{\partial t} - \frac{1}{2}\|v - \hat{v}_T\|_{\gamma,l,\tilde{V}_{T,\tilde{k}}}^2
$$

with $\hat{v}_T = (\gamma^2 I - l^T \frac{\partial^2 \tilde{V}_{T,\tilde{k}}}{\partial x^2}l)^{-1}(k^T \frac{\partial \tilde{V}_{T,\tilde{k}}}{\partial x} + l^T \frac{\partial^2 \tilde{V}_{T,\tilde{k}}}{\partial x^2}h)$. Obviously,

$$
\max_{v \in U} H\left(t, x, v, -\frac{\partial \hat{V}}{\partial x}, -\frac{\partial^2 \hat{V}}{\partial x^2}\right) = H\left(t, x, \hat{v}_T, \frac{\partial \tilde{V}_{T,\tilde{k}}}{\partial x}, \frac{\partial^2 \tilde{V}_{T,\tilde{k}}}{\partial x^2}\right)
$$

$$
= \mathcal{H}(\tilde{V}_{T,\tilde{k}}) - \frac{\partial \tilde{V}_{T,\tilde{k}}}{\partial t}.
$$

Therefore, (6.40) is equivalent to $\mathcal{H}(\tilde{V}_{T,\tilde{k}}) = 0$ and the proof of this lemma is complete. $\square$

THEOREM 6.4

If there exists an H_∞ control $u = \tilde{k}(t,x)$ for system (6.33), such that the conditions of Lemma 6.3 hold, then HJE (6.35) admits a unique solution.

Proof. By applying Lemma 6.3 and noting the identity (6.37), we have

$$\mathcal{E}\int_t^T (\|z\|^2 - \gamma^2\|v\|^2)\, d\tau = \mathcal{E}\int_t^T \left(\|\tilde{k} + g^T\frac{\partial \tilde{V}_{T,\tilde{k}}}{\partial x}\|^2 - \|v - \hat{v}_T\|_{\gamma,l,\tilde{V}_{T,\tilde{k}}}^2 \right.$$
$$\left. + \mathcal{H}(\tilde{V}_{T,\tilde{k}}) \right) d\tau$$
$$= \mathcal{E}\int_t^T \left(\|\tilde{k} + g^T\frac{\partial \tilde{V}_{T,\tilde{k}}}{\partial x}\|^2 - \|v - \hat{v}_T\|_{\gamma,l,\tilde{V}_{T,\tilde{k}}}^2 \right) d\tau.$$

Obviously, to have $\|\mathcal{L}_T\| \leq \gamma$, we must take $\tilde{k} = -g^T\frac{\partial \tilde{V}_{T,\tilde{k}}}{\partial x}$ as an H_∞ control. Substituting $\tilde{k}$ into (6.39), (6.35) is derived. The uniqueness of the solution follows from Proposition 6.4. □

REMARK 6.8   Note that the above derivation does not shed much light on the extension to the case of (x, u, v)-dependent noise. One reason for this is that u and v are no longer separable in the HJE. In the next section, we shall deal with general nonlinear stochastic H_∞ control with (x, u, v)-dependent noise. ∎

6.5 H_∞ Control of More General Stochastic Nonlinear Systems

Consider the following general nonlinear stochastic time-varying system with (x, u, v)-dependent noise:

$$\begin{cases} dx(t) = [f(t, x) + g(t, x)u(t) + k(t, x)v(t)]\, dt \\ \qquad\quad + [h(t, x) + s(t, x)u(t) + l(t, x)v(t)]\, dB(t), \\ x(0) = x_0 \in \mathcal{R}^n,\ t \in [0, T] \end{cases} \tag{6.41}$$

together with the regulated output

$$z(t) = Col(m(t, x), u(t)) := \begin{bmatrix} m(t, x) \\ u(t) \end{bmatrix}. \tag{6.42}$$

We first discuss the finite-time H_∞ control problem and obtain the following result:

THEOREM 6.5

For a given disturbance attenuation level $\gamma > 0$, if there exists a function $V(t, x) \in \mathcal{C}^{1,2}([0, T] \times \mathcal{R}^n; \mathcal{R})$ with $\frac{\partial^2 V}{\partial x^2}(t, x) \leq 0$ for $\forall(t, x) \in [0, T] \times \mathcal{R}^n$, *which solves the following constrained HJI*

$$
\begin{cases}
\mathcal{H}(t,x) := \frac{\partial V}{\partial t} + \frac{\partial V^T}{\partial x} f + \frac{1}{2} h^T \frac{\partial^2 V}{\partial x^2} h - m^T m - \frac{1}{4}(h^T \frac{\partial^2 V}{\partial x^2} l + \frac{\partial V^T}{\partial x} k) \\
\qquad \cdot (\gamma^2 I + l^T \frac{\partial^2 V}{\partial x^2} l)^{-1} (l^T \frac{\partial^2 V}{\partial x^2} h + k^T \frac{\partial V}{\partial x}) - \frac{1}{4}(h^T \frac{\partial^2 V}{\partial x^2} s + \frac{\partial V^T}{\partial x} g) \\
\qquad \cdot (-I + s^T \frac{\partial^2 V}{\partial x^2} s)^{-1} (s^T \frac{\partial^2 V}{\partial x^2} h + g^T \frac{\partial V}{\partial x}) \geq 0, \\
\gamma^2 I + l^T \frac{\partial^2 V}{\partial x^2} l > 0, \\
V(0,0) = 0, V(T,x) = 0, \forall x \in \mathcal{R}^n,
\end{cases}
\tag{6.43}
$$

then

$$
\tilde{u}_T^* = -\frac{1}{2} \left(-I + s^T \frac{\partial^2 V}{\partial x^2} s \right)^{-1} \left(s^T \frac{\partial^2 V}{\partial x^2} h + g^T \frac{\partial V}{\partial x} \right)
$$

is an H_∞ control of (6.41)–(6.42).

To prove Theorem 6.5, we give the following useful lemma whose proof is straightforward and omitted.

LEMMA 6.4
Given $x, b \in \mathcal{R}^n$ and $A \in \mathcal{S}_n$. If A^{-1} exists, then

$$
x^T A x + x^T b + b^T x = (x + A^{-1}b)^T A(x + A^{-1}b) - b^T A^{-1} b. \tag{6.44}
$$

REMARK 6.9   Lemma 6.4 is very useful in the application of the completing squares technique, which was first introduced in [226]; see Lemma 2.4 of [226].   ∎

Proof of Theorem 6.5. Applying Itô's formula to $V(t,x)$, we have

$$
dV(t,x) = \left\{ \frac{\partial V}{\partial t} + \frac{\partial V^T}{\partial x}(f + gu + kv) + \frac{1}{2}(h + su + lv)^T \frac{\partial^2 V}{\partial x^2} \right.
$$
$$
\left. \cdot (h + su + lv) \right\} dt + \frac{\partial V^T}{\partial x}(h + su + lv)\, dB(t). \tag{6.45}
$$

By integrating from 0 to T and taking mathematical expectation on both sides of (6.45), we have

$$
\mathcal{E}V(T, x(T)) - V(0, x(0))
$$
$$
= \mathcal{E} \int_0^T \left\{ \frac{\partial V}{\partial t} + \frac{\partial V^T}{\partial x}(f + gu + kv) + \frac{1}{2}(h + su + lv)^T \frac{\partial^2 V}{\partial x^2}(h + su + lv) \right\} dt.
$$

Hence,

$$
\gamma^2 \|v(t)\|_{[0,T]}^2 - \|z(t)\|_{[0,T]}^2 = \mathcal{E} \int_0^T (\gamma^2 \|v\|^2 - \|z\|^2)\, dt
$$
$$
= \mathcal{E} \int_0^T \{(\gamma^2 \|v\|^2 - \|z\|^2)\, dt + dV(t, x(t))\} + V(0, x_0) - \mathcal{E}V(T, x(T))
$$

$$= \mathcal{E} \int_0^T \left(\Delta_1(v,x) + \Delta_2(x) + \Delta_3(u,x) \right) dt$$

$$+ \mathcal{E} \int_0^T \frac{1}{2} \left(u^T s^T \frac{\partial^2 V}{\partial x^2} lv + v^T l^T \frac{\partial^2 V}{\partial x^2} su \right) dt$$

$$+ V(0,x_0) - \mathcal{E} V(T, x(T)), \tag{6.46}$$

where

$$\Delta_1(v,x) = v^T \left(\gamma^2 I + \frac{1}{2} l^T \frac{\partial^2 V}{\partial x^2} l \right) v + \frac{1}{2} \left(h^T \frac{\partial^2 V}{\partial x^2} l + \frac{\partial V}{\partial x}^T k \right) v$$

$$+ \frac{1}{2} v^T \left(l^T \frac{\partial^2 V}{\partial x^2} h + k^T \frac{\partial V}{\partial x} \right),$$

$$\Delta_2(x) = \frac{\partial V}{\partial t} + \frac{\partial V}{\partial x}^T f + \frac{1}{2} h^T \frac{\partial^2 V}{\partial x^2} h - m^T m,$$

$$\Delta_3(u,x) = u^T \left(-I + \frac{1}{2} s^T \frac{\partial^2 V}{\partial x^2} s \right) u + \frac{1}{2} \left(h^T \frac{\partial^2 V}{\partial x^2} s + \frac{\partial V}{\partial x}^T g \right) u$$

$$+ \frac{1}{2} u^T \left(s^T \frac{\partial^2 V}{\partial x^2} h + g^T \frac{\partial V}{\partial x} \right).$$

In addition, the assumption of $\frac{\partial^2 V}{\partial x^2} \leq 0$ leads to

$$\frac{1}{2}(-u^T s^T + v^T l^T) \frac{\partial^2 V}{\partial x^2} (-su + lv) \leq 0,$$

which yields

$$\frac{1}{2} \left(u^T s^T \frac{\partial^2 V}{\partial x^2} lv + v^T l^T \frac{\partial^2 V}{\partial x^2} su \right) \geq \frac{1}{2} u^T s^T \frac{\partial^2 V}{\partial x^2} su + \frac{1}{2} v^T l^T \frac{\partial^2 V}{\partial x^2} lv.$$

So

$$\gamma^2 \|v(t)\|_{[0,T]}^2 - \|z(t)\|_{[0,T]}^2 \geq \mathcal{E} \int_0^T \left(\Delta_1(v,x) + \Delta_2(x) + \Delta_3(u,x) \right) dt$$

$$+ \frac{1}{2} \mathcal{E} \int_0^T \left(u^T s^T \frac{\partial^2 V}{\partial x^2} su + v^T l^T \frac{\partial^2 V}{\partial x^2} lv \right) dt$$

$$+ V(0,x_0) - \mathcal{E} V(T, x(T))$$

$$= \mathcal{E} \int_0^T \left(\tilde{\Delta}_1(v,x) + \Delta_2(x) + \tilde{\Delta}_3(u,x) \right) dt$$

$$+ V(0,x_0) - \mathcal{E} V(T, x(T)), \tag{6.47}$$

where

$$\tilde{\Delta}_3(u,x) = u^T \left(-I + s^T \frac{\partial^2 V}{\partial x^2} s \right) u + \frac{1}{2} \left(h^T \frac{\partial^2 V}{\partial x^2} s + \frac{\partial V}{\partial x}^T g \right) u$$
$$+ \frac{1}{2} u^T \left(s^T \frac{\partial^2 V}{\partial x^2} h + g^T \frac{\partial V}{\partial x} \right),$$

$$\tilde{\Delta}_1(v,x) = v^T \left(\gamma^2 I + l^T \frac{\partial^2 V}{\partial x^2} l \right) v + \frac{1}{2} \left(h^T \frac{\partial^2 V}{\partial x^2} l + \frac{\partial V}{\partial x}^T k \right) v$$
$$+ \frac{1}{2} v^T \left(l^T \frac{\partial^2 V}{\partial x^2} h + k^T \frac{\partial V}{\partial x} \right).$$

Applying Lemma 6.4 to $\tilde{\Delta}_3(u,x)$ and $\tilde{\Delta}_1(v,x)$, we have

$$\tilde{\Delta}_3(u,x) = \left[u + \frac{1}{2} \left(-I + s^T \frac{\partial^2 V}{\partial x^2} s \right)^{-1} \left(s^T \frac{\partial^2 V}{\partial x^2} h + g^T \frac{\partial V}{\partial x} \right) \right]^T \left(-I + s^T \frac{\partial^2 V}{\partial x^2} s \right)$$
$$\cdot \left[u + \frac{1}{2} \left(-I + s^T \frac{\partial^2 V}{\partial x^2} s \right)^{-1} \left(s^T \frac{\partial^2 V}{\partial x^2} h + g^T \frac{\partial V}{\partial x} \right) \right]$$
$$- \frac{1}{4} \left(h^T \frac{\partial^2 V}{\partial x^2} s + \frac{\partial V}{\partial x}^T g \right) \left(-I + s^T \frac{\partial^2 V}{\partial x^2} s \right)^{-1} \left(s^T \frac{\partial^2 V}{\partial x^2} h + g^T \frac{\partial V}{\partial x} \right) \quad (6.48)$$

and

$$\tilde{\Delta}_1(v,x) = \left[v + \frac{1}{2} \left(\gamma^2 I + l^T \frac{\partial^2 V}{\partial x^2} l \right)^{-1} \left(l^T \frac{\partial^2 V}{\partial x^2} h + k^T \frac{\partial V}{\partial x} \right) \right]^T \left(\gamma^2 I + l^T \frac{\partial^2 V}{\partial x^2} l \right)$$
$$\cdot \left[v + \frac{1}{2} \left(\gamma^2 I + l^T \frac{\partial^2 V}{\partial x^2} l \right)^{-1} \left(l^T \frac{\partial^2 V}{\partial x^2} h + k^T \frac{\partial V}{\partial x} \right) \right]$$
$$- \frac{1}{4} \left(h^T \frac{\partial^2 V}{\partial x^2} l + \frac{\partial V}{\partial x}^T k \right) \left(\gamma^2 I + l^T \frac{\partial^2 V}{\partial x^2} l \right)^{-1}$$
$$\cdot \left(l^T \frac{\partial^2 V}{\partial x^2} h + k^T \frac{\partial V}{\partial x} \right). \quad (6.49)$$

We note that $-I + s^T \frac{\partial^2 V}{\partial x^2} s < 0$ due to $\frac{\partial^2 V}{\partial x^2} \leq 0$. Substituting (6.48) and (6.49) into (6.47), and considering (6.43) with its constraint conditions, we have that

$$\gamma^2 \|v(t)\|^2_{[0,T]} - \|z(t)\|^2_{[0,T]} \geq \mathcal{E} \int_0^T (u + \tilde{K}_2)^T \left(-I + s^T \frac{\partial^2 V}{\partial x^2} s \right) (u + \tilde{K}_2)\, dt$$
$$+ \mathcal{E} \int_0^T (v + \tilde{K}_1)^T \left(\gamma^2 I + l^T \frac{\partial^2 V}{\partial x^2} l \right) (v + \tilde{K}_1)\, dt$$
$$+ V(0, x_0), \quad (6.50)$$

where

$$\tilde{K}_2 = \frac{1}{2}\left(-I + s^T \frac{\partial^2 V}{\partial x^2} s\right)^{-1} \left(s^T \frac{\partial^2 V}{\partial x^2} h + g^T \frac{\partial V}{\partial x}\right),$$

$$\tilde{K}_1 = \frac{1}{2}\left(\gamma^2 I + l^T \frac{\partial^2 V}{\partial x^2} l\right)^{-1} \left(l^T \frac{\partial^2 V}{\partial x^2} h + k^T \frac{\partial V}{\partial x}\right).$$

In view of the constraints of (6.43), for $x_0 = 0$, if we take

$$u(t) = \tilde{u}_T^*(t) = -\tilde{K}_2(t, x) = -\frac{1}{2}\left(-I + s^T \frac{\partial^2 V}{\partial x^2} s\right)^{-1} \left(s^T \frac{\partial^2 V}{\partial x^2} h + g^T \frac{\partial V}{\partial x}\right),$$

then it follows from (6.50) that

$$\|z(t)\|_{[0,T]} \leq \gamma \|v(t)\|_{[0,T]}. \tag{6.51}$$

The proof is complete. $\square$

Obviously, in (6.43), if $V(T, x) = 0$ is relaxed to $V(T, x) \leq 0$, Theorem 6.5 still holds. The proof of Theorem 6.5 is based on the elementary identity (6.44).

Now, we give a similar result for the infinite horizon H_∞ control of the following nonlinear stochastic time-invariant system

$$\begin{cases} dx(t) = [f(x) + g(x)u(t) + k(x)v(t)] \, dt + [h(x) + s(x)u(t) + l(x)v(t)] \, dB(t), \\ x(0) = x_0 \in \mathcal{R}^n \end{cases}$$
$$\tag{6.52}$$

with the controlled output as

$$z(t) = Col(m(x), u(x)). \tag{6.53}$$

THEOREM 6.6
Assume $V(x) \in \mathcal{C}^2(\mathcal{R}^n; \mathcal{R}^-)$ is proper with $V(0) = 0$, $V(x) < 0$ and $\frac{\partial^2 V}{\partial x^2}(x) \leq 0$ for all non-zero $x \in \mathcal{R}^n$. If $V(x)$ solves the following HJI

$$\begin{cases} \mathcal{H}_1(x) := \frac{\partial V^T}{\partial x} f + \frac{1}{2} h^T \frac{\partial^2 V}{\partial x^2} h - m^T m - \frac{1}{4}(h^T \frac{\partial^2 V}{\partial x^2} l + \frac{\partial V^T}{\partial x} k)(\gamma^2 I + l^T \frac{\partial^2 V}{\partial x^2} l)^{-1} \\ \quad \cdot (l^T \frac{\partial^2 V}{\partial x^2} h + k^T \frac{\partial V}{\partial x}) - \frac{1}{4}(h^T \frac{\partial^2 V}{\partial x^2} s + \frac{\partial V^T}{\partial x} g)(-I + s^T \frac{\partial^2 V}{\partial x^2} s)^{-1} \\ \quad \cdot (s^T \frac{\partial^2 V}{\partial x^2} h + k^T \frac{\partial V}{\partial x}) > 0, \\ \gamma^2 I + l^T \frac{\partial^2 V}{\partial x^2} l > 0, \forall x \in \mathcal{R}^n, \end{cases}$$
$$\tag{6.54}$$

then

$$\tilde{u}_\infty^* = -\frac{1}{2}\left(-I + s^T \frac{\partial^2 V}{\partial x^2} s\right)^{-1} \left(s^T \frac{\partial^2 V}{\partial x^2} h + g^T \frac{\partial V}{\partial x}\right)$$

is an infinite horizon H_∞ control law of (6.52)–(6.53).

Proof. Repeating the same procedure as in Theorem 6.5, it is easy to show (6.16). Now we only need to show that (6.52) is internally stable, i.e., the following system

$$dx(t) = [f(x) + g(x)\tilde{u}_\infty^*] \, dt + [h(x) + s(x)\tilde{u}_\infty^*] \, dB(t) \tag{6.55}$$

is globally asymptotically stable in probability. Let $\mathcal{L}_{\tilde{u}_\infty^*}$ be the infinitesimal generator of the system (6.55) and $V_1(x) := -V(x) > 0$ for $x \neq 0$, then

$$
\begin{aligned}
\mathcal{L}_{\tilde{u}_\infty^*} V_1 &= \frac{\partial V_1^T}{\partial x}(f + g\tilde{u}_\infty^*) + \frac{1}{2}(h + s\tilde{u}_\infty^*)^T \frac{\partial^2 V_1}{\partial x^2}(h + s\tilde{u}_\infty^*) \\
&= -\frac{\partial V^T}{\partial x}f - \frac{1}{2}h^T \frac{\partial^2 V}{\partial x^2}h - \frac{\partial V^T}{\partial x}g(x)\tilde{u}_\infty^* \\
&\quad -\frac{1}{2}h^T \frac{\partial^2 V}{\partial x^2}s\tilde{u}_\infty^* - \frac{1}{2}(s\tilde{u}_\infty^*)^T \frac{\partial^2 V}{\partial x^2}h - \frac{1}{2}(s\tilde{u}_\infty^*)^T \frac{\partial^2 V}{\partial x^2}(s\tilde{u}_\infty^*) \\
&= -\frac{\partial V^T}{\partial x}f - \frac{1}{2}h^T \frac{\partial^2 V}{\partial x^2}h + \Pi_1(x) + \Pi_2(x),
\end{aligned} \tag{6.56}
$$

where

$$
\begin{aligned}
\Pi_1(x) &= -\frac{\partial V^T}{\partial x}g(x)\tilde{u}_\infty^* - \frac{1}{2}h^T \frac{\partial^2 V}{\partial x^2}s\tilde{u}_\infty^* - \frac{1}{2}(s\tilde{u}_\infty^*)^T \frac{\partial^2 V}{\partial x^2}h \\
&= \frac{1}{2}\frac{\partial V^T}{\partial x}g(x)\left(-I + s^T \frac{\partial^2 V}{\partial x^2}s\right)^{-1}\left(s^T \frac{\partial^2 V}{\partial x^2}h + g^T \frac{\partial V}{\partial x}\right) \\
&\quad + \frac{1}{4}h^T \frac{\partial^2 V}{\partial x^2}s\left(-I + s^T \frac{\partial^2 V}{\partial x^2}s\right)^{-1}\left(s^T \frac{\partial^2 V}{\partial x^2}h + g^T \frac{\partial V}{\partial x}\right) \\
&\quad + \frac{1}{4}\left(h^T \frac{\partial^2 V}{\partial x^2}s + \frac{\partial V^T}{\partial x}g\right)\left(-I + s^T \frac{\partial^2 V}{\partial x^2}s\right)^{-1}s^T \frac{\partial^2 V}{\partial x^2}h
\end{aligned}
$$

and

$$
\begin{aligned}
\Pi_2(x) &= -\frac{1}{2}(s\tilde{u}_\infty^*)^T \frac{\partial^2 V}{\partial x^2}(s\tilde{u}_\infty^*) \\
&= -\frac{1}{8}\left(h^T \frac{\partial^2 V}{\partial x^2}s + \frac{\partial V^T}{\partial x}g\right)\left(-I + s^T \frac{\partial^2 V}{\partial x^2}s\right)^{-1} \\
&\quad \cdot s^T \frac{\partial^2 V}{\partial x^2}s\left(-I + s^T \frac{\partial^2 V}{\partial x^2}s\right)^{-1}\left(s^T \frac{\partial^2 V}{\partial x^2}h + g^T \frac{\partial V}{\partial x}\right).
\end{aligned}
$$

It is easy to verify that

$$
\begin{aligned}
&\Pi_1(x) \\
&= \frac{1}{2}\left(h^T \frac{\partial^2 V}{\partial x^2}s + \frac{\partial V^T}{\partial x}g\right)\left(-I + s^T \frac{\partial^2 V}{\partial x^2}s\right)^{-1}\left(s^T \frac{\partial^2 V}{\partial x^2}h + g^T \frac{\partial V}{\partial x}\right)
\end{aligned} \tag{6.57}
$$

and

$$
\Pi_2(x) \leq -\frac{1}{8}\left(h^T \frac{\partial^2 V}{\partial x^2}s + \frac{\partial V^T}{\partial x}g\right)(-I + s^T \frac{\partial^2 V}{\partial x^2}s)^{-1}\left(s^T \frac{\partial^2 V}{\partial x^2}h + g^T \frac{\partial V}{\partial x}\right). \tag{6.58}
$$

Substituting (6.57) and (6.58) into (6.56) and considering (6.54), it follows that

$$
\mathcal{L}_{\tilde{u}_\infty^*} V_1 \leq -\frac{\partial V^T}{\partial x} f - \frac{1}{2} h^T \frac{\partial^2 V}{\partial x^2} h + \frac{3}{8} \left(h^T \frac{\partial^2 V}{\partial x^2} s + \frac{\partial V^T}{\partial x} g \right) \left(-I + s^T \frac{\partial^2 V}{\partial x^2} s \right)^{-1}
$$
$$
\cdot \left(s^T \frac{\partial^2 V}{\partial x^2} h + g^T \frac{\partial V}{\partial x} \right)
$$

$$
\leq -\frac{\partial V^T}{\partial x} f - \frac{1}{2} h^T \frac{\partial^2 V}{\partial x^2} h + \frac{1}{4} \left(h^T \frac{\partial^2 V}{\partial x^2} s + \frac{\partial V^T}{\partial x} g \right) \left(-I + s^T \frac{\partial^2 V}{\partial x^2} s \right)^{-1}
$$
$$
\cdot \left(s^T \frac{\partial^2 V}{\partial x^2} h + g^T \frac{\partial V}{\partial x} \right)
$$
$$
< -m^T m - \frac{1}{4} \left(h^T \frac{\partial^2 V}{\partial x^2} l + \frac{\partial V^T}{\partial x} k \right) \left(\gamma^2 I + l^T \frac{\partial^2 V}{\partial x^2} l \right)^{-1} \left(l^T \frac{\partial^2 V}{\partial x^2} h + k^T \frac{\partial V}{\partial x} \right)
$$
$$
\leq 0, \tag{6.59}
$$

which implies that (6.55) is globally asymptotically stable in probability by the result of [87]. This theorem is proved. □

REMARK 6.10 Theorems 6.5–6.6 transform the nonlinear stochastic H_∞ problem into the solvability of a single HJI. An approximate method of designing a nonlinear stochastic H_∞ controller is the fuzzy linearized technique based on the Takagi–Sugeno (T-S) model, which avoids solving HJIs; see [40]. However, in the stochastic case, it is not easy to show that the linearized fuzzy system converges to the original nonlinear system. ∎

REMARK 6.11 Reference [15] presented a single-degree-of-freedom inverted pendulum example modeled by a nonlinear stochastic control system with state-dependent noise. A benchmark mechanical system example can be found in [190]. In [38, 39, 40, 184], it was shown that in the field of systems biology, nonlinear stochastic Itô systems with state-dependent noise are ideal models. In practice, the control input and external disturbance may also be subject to noise, giving rise to the general nonlinear stochastic system (6.41). For example, in quantized feedback, the input quantization error can be considered as a multiplicative noise [75]. ∎

Example 6.2
Consider the RLC electric circuit with nonlinear resistance. Assume that the volt-ampere characteristics of the nonlinear resistance is $u(t) = \frac{1}{3} i^3(t) - i(t)$, where $u(t)$ and $i(t)$ are voltage and current, respectively. The charge $Q(t)$ at time t in the nonlinear RLC circuit satisfies the second-order differential

equation

$$HÜ{Q}(t) + \left(\frac{1}{3}\dot{Q}^3(t) - \dot{Q}(t)\right) + \frac{1}{C}Q(t) = F(t), \qquad (6.60)$$

where H is the inductance, C is the capacitance, and $F(t)$ is the potential source [130]. Suppose that the potential source is subject to the environmental noise and is described by $F(t) = G(t) + \alpha(t)\dot{B}(t)$, where $\dot{B}(t)$ is a one-dimensional white noise and $\alpha(t)$ is the intensity of the noise. Then (6.60) becomes

$$H\ddot{Q}(t) + \left(\frac{1}{3}\dot{Q}^3(t) - \dot{Q}(t)\right) + \frac{1}{C}Q(t) = G(t) + \alpha(t)\dot{B}(t). \qquad (6.61)$$

In this situation, if we introduce a two-dimensional process

$$x(t) = (x_1(t), x_2(t))^T = (Q(t), \dot{Q}(t))^T,$$

then (6.61) can be expressed as an Itô-type equation

$$\begin{cases} dx_1(t) = x_2(t)\, dt, \\ dx_2(t) = \frac{1}{H}\left(-\frac{1}{3}x_2^3(t) + x_2(t) - \frac{1}{C}x_1(t) + G(t)\right) dt + \frac{\alpha(t)}{H}\, dB(t). \end{cases} \qquad (6.62)$$

In (6.62), a control device is introduced and the controlled system may be described by

$$\begin{cases} dx_1(t) = x_2(t)\, dt, \\ dx_2(t) = \frac{1}{H}\left(-\frac{1}{3}x_2^3(t) + x_2(t) - \frac{1}{C}x_1(t) + bu(t) + G(t)\right) dt \\ \qquad + \frac{\alpha(t)}{H}\, dB(t). \end{cases} \qquad (6.63)$$

To make the calculation simple, let us specify the parameters as $H = 1$, $C = 1$, $b = 2$, $\alpha(t) = \frac{1}{2}(x_2(t) + u(t))$. If the controlled output is assumed to be $(x_2(t), u(t))^T$, then (6.63) takes the form of (6.52)–(6.53) with

$$f(x) = \begin{bmatrix} x_2 \\ -\frac{1}{3}x_2^3 + x_2 - x_1 \end{bmatrix}, \quad g(x) = \begin{bmatrix} 0 \\ 2 \end{bmatrix}, \quad k(x) = \begin{bmatrix} 0 \\ 1 \end{bmatrix},$$

$$h(x) = \begin{bmatrix} 0 \\ \frac{1}{2}x_2 \end{bmatrix}, \quad s(x) = \begin{bmatrix} 0 \\ \frac{1}{2} \end{bmatrix}, \quad l(x) = 0, \quad m(x) = x_2.$$

Take $V(x) = x^T P x$ with $P = \mathrm{diag}\{p_1, p_2\} < 0$ to be determined. For a given disturbance attenuation level $\gamma = \sqrt{3}$, $P = \mathrm{diag}\{-2, -2\}$ is a solution to (6.54). According to Theorem 6.6, $\tilde{u}_\infty^* = -\frac{9}{4}x_2$ is an H_∞ control of system (6.63).

The initial condition is chosen as $x_0 = [0.2, \ 0.5]^T$ and $G(t)$ is assumed to be 1. The state responses of unforced system (6.62) and controlled system (6.63) are shown in Fig. 6.1 and Fig. 6.2, respectively. From Fig. 6.2, one can find that the nonlinear stochastic system can achieve stability. Note that since $G(t) \equiv 1$, the equilibrium point of the system is $[1 \ 0]^T$. ☐

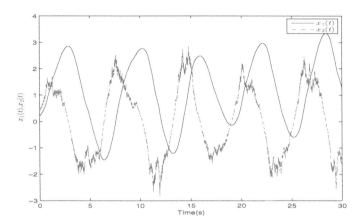

FIGURE 6.1

State response of the unforced system (6.62).

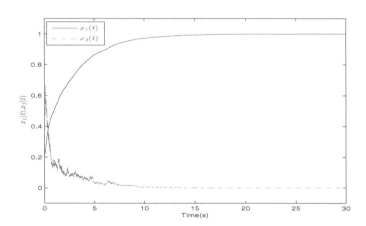

FIGURE 6.2

State response of the controlled system (6.63).

6.6   Finite Horizon H_2/H_∞ Control

Consider the system (6.41)–(6.42). The finite horizon H_2/H_∞ control is defined as follows:

DEFINITION 6.5 (Finite horizon nonlinear stochastic H_2/H_∞ control)   *For any given $\gamma > 0$, $0 < T < \infty$, $v \in \mathcal{L}^2_{\mathcal{F}}([0,T],\mathcal{R}^{n_v})$, find, if possible, a state feedback control law $u = u_T^* \in \mathcal{L}^2_{\mathcal{F}}([0,T],\mathcal{R}^{n_u})$, such that*

  (i) *The trajectory of the closed-loop system (6.41) starting from $x_0 = 0$ satisfies*

$$\|z(t)\|_{[0,T]} \le \gamma \|v\|_{[0,T]} \quad or \quad \|\mathcal{L}_T\| \le \gamma \tag{6.64}$$

  for $\forall v \ne 0 \in \mathcal{L}^2_{\mathcal{F}}([0,T],\mathcal{R}^{n_v})$.

 (ii) *When the worst-case disturbance $v_T^* \in \mathcal{L}^2_{\mathcal{F}}([0,T],\mathcal{R}^{n_v})$, if it exists, is implemented in (6.41), u_T^* minimizes the quadratic performance*

$$\mathcal{E}\int_0^T \|z(t)\|^2\,dt$$

  simultaneously.

Definition 6.5 extends the definition of deterministic nonlinear H_2/H_∞ control [127]. If we define

$$J_{1,T}(u,v) := \mathcal{E}\int_0^T (\gamma^2\|v\|^2 - \|z\|^2)\,dt, \quad J_{2,T}(u,v) := \mathcal{E}\int_0^T \|z\|^2\,dt,$$

then, different from linear stochastic H_2/H_∞ control, the nonlinear mixed H_2/H_∞ control problem is equivalent to finding the Nash equilibria (u_T^*, v_T^*) of

$$J_{1,T}(u_T^*, v_T^*) \le J_{1,T}(u_T^*, v), \quad \forall v \in \mathcal{L}^2_{\mathcal{F}}([0,T],\mathcal{R}^{n_v}), \tag{6.65}$$

$$J_{2,T}(u_T^*, v_T^*) \le J_{2,T}(u, v_T^*), \quad \forall u \in \mathcal{L}^2_{\mathcal{F}}([0,T],\mathcal{R}^{n_u}). \tag{6.66}$$

This is because, in nonlinear H_2/H_∞ control, one typically adopts $\|\mathcal{L}_T\| \le \gamma$ instead of $\|\mathcal{L}_T\| < \gamma$. v_T^* is a worst-case disturbance corresponding to (6.65), while u_T^* minimizes the H_2 performance corresponding to (6.66). Conversely, if (6.65) and (6.66) hold, then (u_T^*, v_T^*) solves the mixed H_2/H_∞ control if $J_{1,T}(u_T^*, v_T^*) \ge 0$.

THEOREM 6.7
If there exists a pair of solutions $(V_1 \le 0, V_2 \ge 0)$ to the following four

cross-coupled HJEs (the variables t and x are suppressed)

$$
\begin{cases}
H_1(V_1) := \frac{\partial V_1}{\partial t} + \frac{\partial V_1^T}{\partial x}\tilde{f} - \frac{1}{4}\left(\frac{\partial V_1^T}{\partial x}k + \tilde{h}^T\frac{\partial^2 V_1}{\partial x^2}l\right)\left(\gamma^2 I + \frac{1}{2}l^T\frac{\partial^2 V_1}{\partial x^2}l\right)^{-1} \\
\qquad \cdot \left(k^T\frac{\partial V_1}{\partial x} + l^T\frac{\partial^2 V_1}{\partial x^2}\tilde{h}\right) - \tilde{m}^T\tilde{m} + \frac{1}{2}\tilde{h}^T\frac{\partial^2 V_1}{\partial x^2}\tilde{h} = 0, \\
\gamma^2 I + \frac{1}{2}l^T\frac{\partial^2 V_1}{\partial x^2}l > 0, V_1(T,x) = 0, \\
\forall (t,x) \in [0,T] \times \mathcal{R}^n,
\end{cases}
$$

$$(6.67)$$

$$
K_{1,T} = -\frac{1}{2}\left(\gamma^2 I + \frac{1}{2}l^T\frac{\partial^2 V_1}{\partial x^2}l\right)^{-1}\left(k^T\frac{\partial V_1}{\partial x} + l^T\frac{\partial^2 V_1}{\partial x^2}\tilde{h}\right), \qquad (6.68)
$$

$$
\begin{cases}
H_2(V_2) := \frac{\partial V_2^T}{\partial t} + \frac{\partial V_2^T}{\partial x}\tilde{f}_1 + \frac{1}{2}\tilde{h}_1^T\frac{\partial^2 V_2}{\partial x^2}\tilde{h}_1 + m^T m - \frac{1}{4}\left(\frac{\partial V_2^T}{\partial x}g + \tilde{h}_1^T\frac{\partial^2 V_2}{\partial x^2}s\right) \\
\qquad \cdot \left(I + \frac{1}{2}s^T\frac{\partial^2 V_2}{\partial x^2}s\right)^{-1}\left(g^T\frac{\partial V_2}{\partial x} + s^T\frac{\partial^2 V_2}{\partial x^2}\tilde{h}_1\right) = 0, \\
I + \frac{1}{2}s^T\frac{\partial^2 V_2}{\partial x^2}s > 0, V_2(T,x) = 0, \\
\forall (t,x) \in [0,T] \times \mathcal{R}^n,
\end{cases}
$$

$$(6.69)$$

$$
K_{2,T} = -\frac{1}{2}\left(I + \frac{1}{2}s^T\frac{\partial^2 V_2}{\partial x^2}s\right)^{-1}\left(g^T\frac{\partial V_2}{\partial x} + s^T\frac{\partial^2 V_2}{\partial x^2}\tilde{h}_1\right), \qquad (6.70)
$$

then the finite horizon H_2/H_∞ control has a pair of solutions (u_T^*, v_T^*) with

$$
u_T^* = K_{2,T}, \ v_T^* = K_{1,T}, \ J_{2,T}(u_T^*, v_T^*) = V_2(0,x_0),
$$

where in (6.67)–(6.70),

$$
\tilde{f} = f + gK_{2,T}, \ \tilde{h} = h + sK_{2,T}, \ \tilde{m} = Col(m, K_{2,T}),
$$

$$
\tilde{f}_1 = f + kK_{1,T}, \ \tilde{h}_1 = h + lK_{1,T}.
$$

To prove Theorem 6.7, we should present some preliminaries. Consider the following stochastic time-varying perturbed system

$$
\begin{cases}
dx(t) = [f(t,x) + k(t,x)v(t)]\,dt + [h(t,x) + l(t,x)v(t)]\,dB(t), \\
z(t) = m(t,x).
\end{cases} \qquad (6.71)
$$

For any $0 < T < \infty$, define the perturbation operator $\tilde{\mathcal{L}}_T : \mathcal{L}_{\mathcal{F}}^2([0,T], \mathcal{R}^{n_v}) \mapsto \mathcal{L}_{\mathcal{F}}^2([0,T], \mathcal{R}^{n_z})$ as

$$
\tilde{\mathcal{L}}_T(v) = m(t,x)|_{x_0=0}, \ t \geq 0, \ v \in \mathcal{L}_{\mathcal{F}}^2([0,T], \mathcal{R}^{n_v})
$$

with its norm

$$
\|\tilde{\mathcal{L}}_T\| := \sup_{\mathcal{L}_{\mathcal{F}}^2([0,T], \mathcal{R}^{n_z}), v \neq 0, x_0 = 0} \frac{\left\{\mathcal{E}\int_0^T \|m(t,x)\|^2\,dt\right\}^{1/2}}{\left\{\mathcal{E}\int_0^T \|v\|^2\,dt\right\}^{1/2}}.
$$

Obviously, $\tilde{\mathcal{L}}_T$ is a nonlinear operator. When $\|\tilde{\mathcal{L}}_T\| \leq \gamma$, system (6.71) is said to have $\mathcal{L}_2$-gain no larger than γ.

LEMMA 6.5 (Finite horizon nonlinear SBRL)
Given system (6.71) and disturbance attenuation level $\gamma > 0$, if there exists a function $V \in \mathcal{C}^{1,2}([0,T] \times \mathcal{R}^n; \mathcal{R}^-)$ satisfying the following HJE

$$
\begin{cases}
H_3(V) := \frac{\partial V}{\partial t} + \frac{\partial V^T}{\partial x} f - \frac{1}{4} \left(\frac{\partial V^T}{\partial x} k + h^T \frac{\partial^2 V}{\partial x^2} l \right) \left(\gamma^2 I + \frac{1}{2} l^T \frac{\partial^2 V}{\partial x^2} l \right)^{-1} \\
\qquad \cdot \left(k^T \frac{\partial V}{\partial x} + l^T \frac{\partial^2 V}{\partial x^2} h \right) - m^T m + \frac{1}{2} h^T \frac{\partial^2 V}{\partial x^2} h = 0, \\
\gamma^2 I + \frac{1}{2} l^T \frac{\partial^2 V}{\partial x^2} l > 0, V(T,x) = 0, \\
\forall (t,x) \in [0,T] \times \mathcal{R}^n,
\end{cases}
\tag{6.72}
$$

then $\|\tilde{\mathcal{L}}_T\|_{[0,T]} \leq \gamma$.

Proof. One only needs to note the following identity:

$$
\mathcal{E} \int_0^T (\gamma^2 \|v\|^2 - \|z\|^2)\, dt = V(0,x_0) - \mathcal{E}V(T,x(T))
$$

$$
+ \mathcal{E} \int_0^T \left[\|(v - \hat{v}_T^*(x))\|_{(\gamma,s,V)}^2 + H_3(V(t,x)) \right] dt, \tag{6.73}
$$

where

$$
\|Z\|_{(\gamma,l,V)}^2 = Z^T \left(\gamma^2 I + \frac{1}{2} l^T \frac{\partial^2 V}{\partial x^2} l \right) Z
$$

and

$$
\hat{v}_T^* = -\frac{1}{2} \left(\gamma^2 I + \frac{1}{2} l^T \frac{\partial^2 V}{\partial x^2} l \right)^{-1} \left(k^T \frac{\partial V}{\partial x} + l^T \frac{\partial^2 V}{\partial x^2} h \right).
$$

From (6.73), we can also see that $\hat{v}_T^*$ is the corresponding worst-case disturbance. $\square$

LEMMA 6.6 (Nonlinear quadratic optimal control)
Consider the nonlinear stochastic system

$$
\begin{cases}
dx(t) = [f(t,x) + g(t,x)u(t)]\, dt + [h(t,x) + s(t,x)u(t)]\, dB(t), \\
z(t) = Col(m(t,x), u(t)).
\end{cases}
\tag{6.74}
$$

If the following HJE

$$
\begin{cases}
H_4(V_2) := \frac{\partial V_2}{\partial t} + \frac{\partial V_2^T}{\partial x} f + \frac{1}{2} h^T \frac{\partial^2 V_2}{\partial x^2} h + m^T m - \frac{1}{4} \left(\frac{\partial V_2^T}{\partial x} g + h^T \frac{\partial^2 V_2}{\partial x^2} s \right) \\
\qquad \cdot \left(I + \frac{1}{2} s^T \frac{\partial^2 V_2}{\partial x^2} s \right)^{-1} \left(g^T \frac{\partial V_2}{\partial x} + s^T \frac{\partial^2 V_2}{\partial x^2} h \right) = 0, \\
I + \frac{1}{2} s^T \frac{\partial^2 V_2}{\partial x^2} s > 0, V_2(T,x) = 0, \forall (t,x) \in [0,T] \times \mathcal{R}^n
\end{cases}
$$

$$
\tag{6.75}
$$

admits a nonnegative solution $V_2 \in C^{1,2}([0,T] \times \mathcal{R}^n; \mathcal{R}^+)$, then we have

$$V_T(x_0) := \inf_{u \in \mathcal{L}_{\mathcal{F}}^2([0,T], \mathcal{R}^{n_u})} \left\{ J_T(0, x_0; u) := \mathcal{E} \int_0^T \|z(t)\|^2 \, dt \right\}$$

$$= J_T(0, x_0; \bar{u}_T^*) = V_2(0, x_0) \tag{6.76}$$

with the optimal control

$$\bar{u}_T^* = -\frac{1}{2} \left(I + \frac{1}{2} s^T \frac{\partial^2 V_2}{\partial x^2} s \right)^{-1} \left(g^T \frac{\partial V_2}{\partial x} + s^T \frac{\partial^2 V_2}{\partial x^2} h \right). \tag{6.77}$$

Proof. Using the Itô's formula and completion of squares technique and taking into consideration the HJE (6.75), we have

$$J_T(0, x_0; u) = \mathcal{E} \int_0^T \|z(t)\|^2 \, dt$$

$$= \mathcal{E} \int_0^T [(m^T m + u^T u) \, dt + dV_2] + V_2(0, x_0) - \mathcal{E} V_2(T, x(T))$$

$$= \mathcal{E} \int_0^T \left\{ (m^T m + u^T u + \frac{\partial V_2}{\partial t} + \frac{\partial V_2^T}{\partial x}(f + gu) + \frac{1}{2}(h + su)^T \frac{\partial^2 V_2}{\partial x^2}(h + su) \right\} dt$$

$$+ V_2(0, x_0)$$

$$= \mathcal{E} \int_0^T H_4(V_2(t)) \, dt + \mathcal{E} \int_0^T (u - \bar{u}_T^*)^T \tilde{R}(u - \bar{u}_T^*) \, dt + V_2(0, x_0)$$

$$= \mathcal{E} \int_0^T (u - \bar{u}_T^*)^T \tilde{R}(u - \bar{u}_T^*) \, dt + V_2(0, x_0), \tag{6.78}$$

where $\tilde{R} = (I + \frac{1}{2} s^T \frac{\partial^2 V_2}{\partial x^2} s)$, and the third equality holds because $V_2(T, \cdot) = 0$. From (6.78), it follows that

$$V_T(x_0) = \inf_{u \in \mathcal{L}_{\mathcal{F}}^2([0,T], \mathcal{R}^{n_u})} J_T(0, x_0; u) = V_2(0, x_0)$$

with the optimal control given by (6.77). $\square$

Proof of Theorem 6.7. Substituting $u = K_{2,T}(t, x)$ with $K_{2,T}$ defined by (6.70) into (6.41) and (6.42), it follows that

$$\begin{cases} dx = (\tilde{f} + kv) \, dt + (\tilde{h} + lv) \, dB, \\ z = \text{Col}(m, K_{2,T}). \end{cases} \tag{6.79}$$

By applying Lemma 6.5 to the system (6.79), (6.67) follows. Furthermore, it can be seen from (6.73) that $v_T^* = K_{1,T}$ with $K_{1,T}$ given by (6.68) is the worst-case disturbance. In addition, if we substitute $v = v_T^* = K_{1,T}(t, x)$ into (6.41), it yields

$$\begin{cases} dx = (\tilde{f}_1 + gu) \, dt + (\tilde{h}_1 + su) \, dB, \\ x(0) = x_0 \in R^n. \end{cases} \tag{6.80}$$

Minimizing $J_{2,T}(u, v_T^*)$ under the constraint of (6.80) is a standard nonlinear regulator problem, and a direct application of Lemma 6.6 concludes that $J_{2,T}(u, v_T^*)$ achieves its minimum at $u_T^* = K_{2,T}(t,x)$ with $J_{2,T}(u_T^*, v_T^*) = V_2(0, x_0)$. The proof of this theorem is complete. $\square$

In some special cases, it only needs to solve two coupled HJEs. For instance, if in (6.41), $s \equiv 0, l \equiv 0$, i.e., only the state-dependent noise is considered, we have the following corollary.

COROLLARY 6.3

For system

$$\begin{cases} dx = (f + gu + kv)\, dt + h\, dB, \\ z = Col(m, u), x(0) = x_0 \in \mathcal{R}^n, \end{cases} \tag{6.81}$$

if there exist a non-positive solution $V_1 \in \mathcal{C}^{1,2}([0,T] \times \mathcal{R}^n; \mathcal{R}^-)$, and a non-negative solution $V_2 \in \mathcal{C}^{1,2}([0,T] \times \mathcal{R}^n; \mathcal{R}^+)$ solving the following two coupled HJEs

$$\begin{cases} \frac{\partial V_1}{\partial t} + \frac{\partial V_1^T}{\partial x} f - \frac{1}{2} \frac{\partial V_1^T}{\partial x} gg^T \frac{\partial V_2}{\partial x} - \frac{1}{4} \gamma^{-2} \frac{\partial V_1^T}{\partial x} kk^T \frac{\partial V_1}{\partial x} - m^T m + \frac{1}{2} h^T \frac{\partial^2 V_1}{\partial x^2} h \\ \quad - \frac{1}{4} \frac{\partial V_2^T}{\partial x} gg^T \frac{\partial V_2}{\partial x} = 0, \\ V_1(T, x) = 0, \forall (t, x) \in [0, T] \times \mathcal{R}^n \end{cases} \tag{6.82}$$

and

$$\begin{cases} \frac{\partial V_2}{\partial t} + \frac{\partial V_2^T}{\partial x} f - \frac{1}{2} \gamma^{-2} \frac{\partial V_2^T}{\partial x} kk^T \frac{\partial V_1}{\partial x} + \frac{1}{2} h^T \frac{\partial^2 V_2}{\partial x^2} h + m^T m - \frac{1}{4} \frac{\partial V_2^T}{\partial x} gg^T \frac{\partial V_2}{\partial x} = 0, \\ V_2(T, x) = 0, \forall (t, x) \in [0, T] \times \mathcal{R}^n, \end{cases} \tag{6.83}$$

then the finite horizon stochastic H_2/H_∞ control has a solution (u_T^*, v_T^*) with $u_T^* = K_{2,T} = -\frac{1}{2} g^T \frac{\partial V_2}{\partial x}$ and $v_T^* = K_{1,T} = -\frac{1}{2} \gamma^{-2} k^T \frac{\partial V_1}{\partial x}$.

In addition, it is easy to test that for the case of (x, u) or (x, v)-dependent noise, HJEs (6.67)–(6.70) still reduce to two coupled HJEs with respect to V_1 and V_2.

REMARK 6.12   Observe that it is challenging to solve the HJEs (6.67)–(6.70) or their special form such as (6.82)–(6.83) analytically. So it is necessary to search for numerical algorithms in the stochastic H_2/H_∞ controller design.

∎

In general, the converse of Theorem 6.7 is not true, i.e., the existence of finite horizon H_2/H_∞ control does not necessarily imply the solvability of HJEs (6.67)–(6.70). In fact, similar to the nonlinear state feedback H_∞ case [92], HJEs (6.67)–(6.70) hold in the viscosity solution sense.

To present a converse result, we define respectively two extreme value functions $\tilde{V}_T^1(s, x) : [0, T] \times \mathcal{R}^n \mapsto \mathcal{R}^-$ and $\tilde{V}_T^2(s, x) : [0, T] \times \mathcal{R}^n \mapsto \mathcal{R}^+$ associated with

(6.41) as follows:

$$\tilde{V}_T^1(s,x) = \inf_{v \in \mathcal{L}_{\mathcal{F}}^2([s,T],\mathcal{R}^{n_v}), x(s)=x} J_{1,T}(u_T^*, v) \tag{6.84}$$

and

$$\tilde{V}_T^2(s,x) = \inf_{u \in \mathcal{L}_{\mathcal{F}}^2([s,T],\mathcal{R}^{n_u}), x(s)=x} J_{2,T}(u, v_T^*). \tag{6.85}$$

The following result is obvious.

PROPOSITION 6.7
(i) $\tilde{V}_T^1 \le 0$; $\tilde{V}_T^1(T,x) = 0$ for all $x \in \mathcal{R}^n$. (ii) $\tilde{V}_T^2 \ge 0$; $\tilde{V}_T^2(T,x) = 0$ for all $x \in \mathcal{R}^n$.

THEOREM 6.8
Assume that there exists a pair of solutions $(u_T^*, v_T^*) \in \mathcal{L}_{\mathcal{F}}^2([0,T],\mathcal{R}^{n_u}) \times \mathcal{L}_{\mathcal{F}}^2([0,T],\mathcal{R}^{n_v})$ to the H_2/H_∞ control problem and, moreover, assume that $\tilde{V}_T^1, \tilde{V}_T^2 \in \mathcal{C}^{1,2}([0,T] \times \mathcal{R}^n; \mathcal{R})$ satisfy $\gamma^2 I + \frac{1}{2} l^T \frac{\partial^2 \tilde{V}_T^1}{\partial x^2} l > 0$ and $I + \frac{1}{2} s^T \frac{\partial^2 \tilde{V}_T^2}{\partial x^2} s > 0$. Then the four cross-coupled HJEs (6.67)-(6.70) admit a pair of solutions $(\tilde{V}_T^1 \le 0, \tilde{V}_T^2 \ge 0)$.

Proof. For notational simplicity, we still write $v_T^* = K_{1,T}(t,x), u_T^* = K_{2,T}(t,x)$ with $K_{1,T}$ and $K_{2,T}$ to be determined. Because (u_T^*, v_T^*) solves the mixed H_2/H_∞ control and (6.64) holds, by Lemma 6.3, $\tilde{V}_T^1$ solves the HJE

$$\begin{cases} \frac{\partial \tilde{V}_T^1}{\partial t} + \left(\frac{\partial \tilde{V}_T^1}{\partial x}\right)^T (f + g u_T^*) + \frac{1}{2}(h + s u_T^*)^T \frac{\partial^2 \tilde{V}_T^1}{\partial x^2}(h + s u_T^*) - \|m\|^2 - \|K_{2,T}\|^2 \\ \quad - \frac{1}{4}\left(\frac{\partial \tilde{V}_T^1}{\partial x}k + \tilde{h}^T \frac{\partial^2 \tilde{V}_T^1}{\partial x^2} l\right)\left(\gamma^2 I + \frac{1}{2} l^T \frac{\partial^2 \tilde{V}_T^1}{\partial x^2} l\right)^{-1} \left(k^T \frac{\partial \tilde{V}_T^1}{\partial x} + l^T \frac{\partial^2 \tilde{V}_T^1}{\partial x^2} \tilde{h}\right) = 0, \\ \gamma^2 I + \frac{1}{2} l^T \frac{\partial^2 \tilde{V}_T^1}{\partial x^2} l > 0, \ \tilde{V}_T^1(T,x) = 0, \forall (t,x) \in [0,T] \times \mathcal{R}^n. \end{cases} \tag{6.86}$$

Now, applying the identity (6.73), it follows from (6.86) and Proposition 6.7 that

$$J_{1,T}(u_T^*, v) = \tilde{V}_T^1(0,x_0) + \mathcal{E} \int_0^T \|(v - \tilde{v}_T^*)\|_{(\gamma,l,\tilde{V}_T^1)}^2 \, dt, \tag{6.87}$$

where

$$\tilde{v}_T^* = -\frac{1}{2}\left(\gamma^2 I + \frac{1}{2} l^T \frac{\partial^2 \tilde{V}_T^1}{\partial x^2} l\right)^{-1} \left(k^T \frac{\partial \tilde{V}_T^1}{\partial x} + l^T \frac{\partial^2 \tilde{V}_T^1}{\partial x^2} \tilde{h}\right).$$

From (6.87), we see that $\tilde{v}_T^*$ is the worst-case disturbance, so

$$v_T^* = K_{1,T} = \tilde{v}_T^* = -\frac{1}{2}\left(\gamma^2 I + \frac{1}{2} l^T \frac{\partial^2 \tilde{V}_T^1}{\partial x^2} l\right)^{-1} \left(k^T \frac{\partial \tilde{V}_T^1}{\partial x} + l^T \frac{\partial^2 \tilde{V}_T^1}{\partial x^2} \tilde{h}\right). \tag{6.88}$$

Now, by substituting $v = v_T^* = K_{1,T}$ into (6.41), we obtain

$$dx = (\tilde{f}_1 + gu)\, dt + (\tilde{h}_1 + su)\, dB. \tag{6.89}$$

By our definition, $u_T^* = K_{2,T}(t, x)$ is the optimal solution to (6.85). Additionally, by the stochastic dynamic programming principle [196], $(\tilde{V}_T^2, u_T^*)$ solves the following HJE

$$-\frac{\partial \tilde{V}_T^2}{\partial t} + \max_{u \in \mathcal{L}_{\mathcal{F}}^2([0,T],\mathcal{R}^{n_u})} H\left(t, x, u, -\frac{\partial \tilde{V}_T^2}{\partial x}, -\frac{\partial^2 \tilde{V}_T^2}{\partial x^2}\right) = 0, \tag{6.90}$$

i.e.,

$$-\frac{\partial \tilde{V}_T^2}{\partial t} + H\left(t, x, u_T^*, -\frac{\partial \tilde{V}_T^2}{\partial x}, -\frac{\partial^2 \tilde{V}_T^2}{\partial x^2}\right) = 0, \tag{6.91}$$

where the generalized Hamiltonian function is defined as

$$H\left(t, x, u, -\frac{\partial \tilde{V}_T^2}{\partial x}, -\frac{\partial^2 \tilde{V}_T^2}{\partial x^2}\right)$$
$$:= -\|u\|^2 - \|m\|^2 - \frac{\partial \tilde{V}_T^2}{\partial x}^T (\tilde{f}_1 + gu) - \frac{1}{2}(\tilde{h}_1 + su)^T \frac{\partial^2 \tilde{V}_T^2}{\partial x^2}(\tilde{h}_1 + su). \tag{6.92}$$

By a simple computation, (6.92) is equivalent to

$$H\left(t, x, u, -\frac{\partial \tilde{V}_T^2}{\partial x}, -\frac{\partial^2 \tilde{V}_T^2}{\partial x^2}\right) = -H_2(\tilde{V}_T^2) - \frac{\partial \tilde{V}_T^2}{\partial t} - \|u - \tilde{u}_T^*\|_{(h,s,\tilde{V}_T^2)}^2 \tag{6.93}$$

with $\tilde{u}_T^* = -\frac{1}{2}(I + \frac{1}{2}s^T \frac{\partial^2 \tilde{V}_T^2}{\partial x^2}s)^{-1}(g^T \frac{\partial \tilde{V}_T^2}{\partial x} + s^T \frac{\partial^2 \tilde{V}_T^2}{\partial x^2}\tilde{h}_1)$. From (6.93), we conclude that

$$u_T^* = K_{2,T} = \tilde{u}_T^* = -\frac{1}{2}\left(I + \frac{1}{2}s^T \frac{\partial^2 \tilde{V}_T^2}{\partial x^2}s\right)^{-1}\left(g^T \frac{\partial \tilde{V}_T^2}{\partial x} + s^T \frac{\partial^2 \tilde{V}_T^2}{\partial x^2}\tilde{h}_1\right). \tag{6.94}$$

Substituting the above u_T^* into (6.91) and considering Proposition 6.7, we obtain the following HJE

$$\begin{cases} \frac{\partial \tilde{V}_T^2}{\partial t} + (\frac{\partial \tilde{V}_T^2}{\partial x})^T \tilde{f}_1 + \frac{1}{2}\tilde{h}_1^T \frac{\partial^2 \tilde{V}_T^2}{\partial x^2}\tilde{h}_1 + m^T m - \frac{1}{4}[(\frac{\partial \tilde{V}_T^2}{\partial x})^T g + \tilde{h}_1^T \frac{\partial^2 \tilde{V}_T^2}{\partial x^2}s] \\ \quad \cdot(I + \frac{1}{2}s^T \frac{\partial^2 \tilde{V}_T^2}{\partial x^2}s)^{-1}(g^T \frac{\partial \tilde{V}_T^2}{\partial x} + s^T \frac{\partial^2 \tilde{V}_T^2}{\partial x^2}\tilde{h}_1) = 0, \\ I + \frac{1}{2}s^T \frac{\partial^2 \tilde{V}_T^2}{\partial x^2}s > 0, \tilde{V}_T^2(T, x) = 0, \forall (t, x) \in [0, T] \times \mathcal{R}^n. \end{cases} \tag{6.95}$$

It is easy to test that HJEs (6.86), (6.88), (6.95) and (6.94) are the same as HJEs (6.67)–(6.70), respectively. Hence the proof of this theorem is complete. $\square$

Theorem 6.8 explains why the solvability of the coupled HJEs (6.67)–(6.70) is not equivalent to the existence of H_2/H_∞ control, because $\tilde{V}_1^1$ and $\tilde{V}_1^2$ do not necessarily belong to $\mathcal{C}^{1,2}([0, T] \times \mathcal{R}^n, \mathcal{R})$. For the linear stochastic system (2.54), if we take $V_1(t, x) = x^T P_1(t)x$ and $V_2(t, x) = x^T P_2(t)x$ with $P_1(t)$ and $P_2(t)$ being continuously differentiable with respect to $t \in [0, T]$, then the coupled HJEs (6.67)–(6.70) reduce to the four coupled GDREs (2.55)–(2.58). When $h \equiv 0$, $s \equiv 0$ and $l \equiv 0$, the coupled HJEs (6.67)–(6.70) become those of [127].

REMARK 6.13   We note that even for deterministic systems, to design an H_2/H_∞ controller, one has no effective way to solve the associated coupled HJEs [127], which are special forms of (6.67)–(6.70). In practice, one may resort to a linearization approach or fuzzy T-S model as in [37]. A main difficulty in applying the T-S fuzzy approach is to prove that the fuzzy system approaches the original nonlinear stochastic system in some probability sense. ∎

Because it is tedious, we do not plan to introduce the results of infinite horizon nonlinear stochastic H_2/H_∞ control. The reader can refer to [214].

6.7   Notes and References

The affine nonlinear stochastic H_∞ control and mixed H_2/H_∞ control problems have been studied by using the completion of squares method. Global solutions to nonlinear stochastic H_∞ and finite horizon mixed H_2/H_∞ control are obtained, which is different from [173], where only local solution was presented via differential geometric approach. At this stage, it is not clear how to use the differential geometric approach to study the relationship between HJEs and invariant manifolds of Hamiltonian vector fields in nonlinear stochastic H_∞ control.

There are at least a few open research problems in nonlinear stochastic H_2/H_∞ control design: First, under what conditions, does a local solution exist for nonlinear H_∞ (H_2/H_∞) control and what is the relationship between the H_2/H_∞ control of the primal nonlinear system and that of its linearized system? Secondly, up to now, there has been no effective way to solve HJEs (6.18), (6.35), and (6.67)–(6.70), which is a key difficulty in nonlinear H_∞ control design. Finally, when the system state is not completely available, one needs to consider H_2/H_∞ controller design under measurement feedback [101], which remains open.

The materials of this chapter mainly come from [213], [214], [209] and [219]. We refer the reader to [163] for H_∞ analysis of nonlinear stochastic time-delay systems.

7

Nonlinear Stochastic H_∞ and H_2/H_∞ Filtering

The H_∞ filtering problem is to design an estimator for the state or a combination of state variables using measurement output such that the $\mathcal{L}_2$ gain from the external disturbance to the estimation error is less than a prescribed level $\gamma > 0$; see [17] for the deterministic nonlinear H_∞ filtering. In general, an H_∞ filter is not unique. Unlike the Kalman filter, where the statistical properties of the disturbance signal are known a priori, an H_∞ filter only assumes that the disturbance has bounded energy. On the other hand, the H_2/H_∞ filtering selects one among all the H_∞ filters to minimize the estimation error variance; see [13, 76, 112] for deterministic nonlinear H_2/H_∞ filtering. Note that SBRL plays an important role not only in the H_∞ control but also in the H_∞ filtering. A linear stochastic Itô-type H_∞ filtering was studied in [80, 191] by means of an SBRL in [84]. Moreover, a suboptimal linear H_2/H_∞ filtering problem was discussed in [80]. As an application of nonlinear SBRL developed in Chapter 5, this chapter aims to present approaches for the design of an H_∞ filter and a mixed H_2/H_∞ filter for general nonlinear stochastic systems based on HJIs and LMIs. For a class of quasi-linear systems, an LMI-based approach is presented.

7.1 Nonlinear H_∞ Filtering: Delay-Free Case

Consider the following affine nonlinear stochastic system

$$\begin{cases} dx(t) = [f(x(t)) + k(x(t))v(t)] \, dt + [h(x(t)) + l(x(t))v(t)] \, dB(t), \\ y(t) = n(x(t)) + r(x(t))v(t), \\ z(t) = m(x(t)), \ t \in [0, \infty). \end{cases} \tag{7.1}$$

In the above, $x(t) \in \mathcal{R}^n$ and $v \in \mathcal{R}^{n_v}$ stand for, as in previous chapters, the system state and external disturbance, $z(t) \in \mathcal{R}^{n_z}$ is the state combination to be estimated, $y(t) \in \mathcal{R}^{n_y}$ is the measurement output. f, k, h, l, n, r and m satisfy definite conditions such as Lipschitz conditions and linear growth conditions, which guarantee that the system (7.1) has a unique strong solution on any interval $[0, T]$ for $v \in \mathcal{L}_{\mathcal{F}}^2(\mathcal{R}^+, \mathcal{R}^{n_v})$. $f(0) = h(0) = 0$, $n(0) = 0$, $m(0) = 0$. When the system state is not completely available, to estimate $z(t)$ from the observed information

$\sigma(y(s) : 0 \le s \le t)$ with an $\mathcal{L}_2$-gain (from v to the estimation error) less than a prescribed level $\gamma > 0$, we have to construct a nonlinear stochastic H_∞ filter in general.

7.1.1 Lemmas and definitions

As commented in Remark 6.3, the first part of Lemma 6.2 still holds if the HJE (6.27) is replaced by the corresponding HJI. Below, we present an SBRL in terms of HJIs, which is more convenient to be used in practice, because it is easier to solve an HJI than an HJE.

LEMMA 7.1 (SBRL)
Consider system (7.1) and a given $\gamma > 0$. If there exists a positive function $V(x) \in \mathcal{C}^2(\mathcal{R}^n; \mathcal{R}^+)$ solving the following HJI

$$\begin{cases} \frac{\partial V^T}{\partial x} f + \frac{1}{2} \left(\frac{\partial V^T}{\partial x} k + h^T \frac{\partial^2 V}{\partial x^2} l \right) \left(\gamma^2 I - l^T \frac{\partial^2 V}{\partial x^2} l \right)^{-1} \\ \qquad \cdot \left(k^T \frac{\partial V}{\partial x} + l^T \frac{\partial^2 V}{\partial x^2} h \right) + \frac{1}{2} m^T m + \frac{1}{2} h^T \frac{\partial^2 V}{\partial x^2} h \le 0, \qquad (7.2) \\ \gamma^2 I - l^T \frac{\partial^2 V}{\partial x^2} l > 0, \forall x \in \mathcal{R}^n, V(0) = 0, \end{cases}$$

then $\|z(t)\|_{[0,\infty)} \le \gamma \|v(t)\|_{[0,\infty)}$ for $\forall v \in \mathcal{L}_\mathcal{F}^2(\mathcal{R}^+, \mathcal{R}^{n_v})$, $x(0) = 0$.

REMARK 7.1   From the proof of Theorem 6.2, we can find that for the system (7.1),

$$v^* = \left(\gamma^2 I - l^T \frac{\partial^2 V}{\partial x^2} l \right)^{-1} \left(k^T \frac{\partial V}{\partial x} + l^T \frac{\partial^2 V}{\partial x^2} h \right)$$

is the worst-case disturbance in the sense that v^* achieves the maximal possible energy gain from the disturbance input v to the controlled output z.

∎

In what follows, we construct an estimator for $z(t)$ of the form:

$$\begin{cases} d\hat{x} = \hat{f}(\hat{x}) \, dt + \hat{G}(\hat{x})y \, dt, \\ \hat{z} = \hat{m}(\hat{x}), \\ \hat{f}(0) = 0, \ \hat{m}(0) = 0, \ \hat{x}(0) = 0, \end{cases} \qquad (7.3)$$

where $\hat{f}(\hat{x})$, $\hat{G}(\hat{x})$ and $\hat{m}(\hat{x})$ are matrices of appropriate dimensions to be determined. By setting $\eta = [x^T \ \hat{x}^T]^T$, we get the following augmented system

$$d\eta = [f_e(\eta) + k_e(\eta)v] \, dt + [h_e(\eta) + l_e(\eta)v] \, dB, \qquad (7.4)$$

where

$$f_e(\eta) = \begin{bmatrix} f(x) \\ \hat{f}(\hat{x}) + \hat{G}(\hat{x})n(x)) \end{bmatrix}, \ k_e(\eta) = \begin{bmatrix} k(x) \\ \hat{G}(\hat{x})r(x) \end{bmatrix}$$

$$h_e(\eta) = \begin{bmatrix} h(x) \\ 0 \end{bmatrix}, \; l_e(\eta) = \begin{bmatrix} l(x) \\ 0 \end{bmatrix}.$$

In addition, let

$$\tilde{z} = z - \hat{z} = m(x) - \hat{m}(\hat{x})$$

denote the estimator error, then the nonlinear stochastic H_∞ filtering problem can be stated as follows:

DEFINITION 7.1 (Nonlinear stochastic H_∞ filtering) *Find the filter matrices $\hat{f}(\hat{x})$, $\hat{G}(\hat{x})$ and $\hat{m}(\hat{x})$ in (7.3), such that*

(i) The equilibrium point $\eta \equiv 0$ of the augmented system (7.4) is globally asymptotically stable in probability in the case of $v = 0$.

(ii) For a given disturbance attenuation level $\gamma > 0$, the following relation holds.

$$\|\tilde{z}\|_{[0,\infty)} \leq \gamma \|v\|_{[0,\infty)}, \; \forall v \in \mathcal{L}_\mathcal{F}^2(\mathcal{R}^+, \mathcal{R}^{n_v}), \; v \neq 0. \tag{7.5}$$

REMARK 7.2 According to practical needs, Definition 7.1-(i) can be replaced by other stability requirements such as exponential stability and mean square stability. (7.3) is more general than a Luenberger-type observer which only has the observer gain matrix as a design parameter and is more conservative. ∎

7.1.2 Main results

The main result of this subsection is the following theorem.

THEOREM 7.1
For a given disturbance attenuation level $\gamma > 0$, if there exists a positive Lyapunov function $V(\eta) \in \mathcal{C}^2(\mathcal{R}^{2n}; \mathcal{R}^+)$ solving the following HJI

$$\begin{cases} \Gamma_0(\eta) := V_x^T f(x) + V_{\hat{x}}^T [\hat{f}(\hat{x}) + \hat{G}(\hat{x})n(x)] + \frac{1}{2}\Theta^T(\eta)\left[\gamma^2 I - l^T(x)V_{xx}l(x)\right]^{-1} \\ \quad \cdot \Theta(\eta) + \frac{1}{2}\|\tilde{z}\|^2 + \frac{1}{2}h^T(x)V_{xx}h(x) < 0 \\ \gamma^2 I - l^T(x)V_{xx}l(x) > 0, \; \forall x \in \mathcal{R}^n, \; V(0) = 0 \end{cases}$$

$$\tag{7.6}$$

for some matrices $\hat{f}, \hat{G}$ and $\hat{m}$ of suitable dimensions, then the stochastic H_∞ filtering problem is solved by (7.3), where

$$\Theta^T(\eta) = V_x^T k(x) + V_{\hat{x}}^T \hat{G}(\hat{x})r(x) + h^T(x)V_{xx}l(x).$$

To prove Theorem 7.1, the following lemma is needed for the global asymptotic stability of the equilibrium point $x \equiv 0$ of

$$\begin{cases} dx(t) = f(x(t))\,dt + g(x(t))\,dB(t), x(0) = x_0 \in \mathcal{R}^n, \\ f(0) = g(0) = 0. \end{cases} \tag{7.7}$$

LEMMA 7.2 [87]
Assume there exists a positive Lyapunov function $V(x) \in C^2(\mathcal{R}^n; \mathcal{R}^+)$ satisfying $\mathcal{L}V(x) < 0$ for all non-zero $x \in \mathcal{R}^n$, then the equilibrium point $x \equiv 0$ of (7.7) is globally asymptotically stable in probability. Here, $\mathcal{L}$ is the infinitesimal generator of (7.7) defined by

$$\mathcal{L}V(x) = \frac{\partial V^T}{\partial x} f + \frac{1}{2} g^T \frac{\partial^2 V}{\partial x^2} g.$$

Proof of Theorem 7.1. We first show that (7.5) holds. Applying Lemma 7.1 to the system (7.4), we immediately have that (7.5) holds if

$$\begin{cases} \frac{\partial V^T}{\partial \eta} f_e + \frac{1}{2}(\frac{\partial V^T}{\partial \eta} k_e + h_e^T \frac{\partial^2 V}{\partial \eta^2} l_e)(\gamma^2 I - l_e^T \frac{\partial^2 V}{\partial \eta^2} l_e)^{-1} \\ \qquad \cdot (k_e^T \frac{\partial V}{\partial \eta} + l_e^T \frac{\partial^2 V}{\partial \eta^2} h_e) + \frac{1}{2}\|\tilde{z}\|^2 + \frac{1}{2}h_e^T \frac{\partial^2 V}{\partial \eta^2} h_e < 0, \\ \gamma^2 I - l_e^T \frac{\partial^2 V}{\partial \eta^2} l_e > 0, \ \forall \eta \in \mathcal{R}^{2n}, \ V(0) = 0 \end{cases} \qquad (7.8)$$

admits solutions $V(\eta) \in C^2(\mathcal{R}^{2n}; \mathcal{R}^+)$, $V(\eta) > 0$ for $\eta \neq 0$, $\hat{f}(\hat{x})$, $\hat{G}(\hat{x})$ and $\hat{m}(\hat{x})$. By a series of computations, (7.8) is equivalent to (7.6).

Secondly, we show that $\eta \equiv 0$ of the augmented system (7.4) is globally asymptotically stable in probability in the case when $v = 0$. By Lemma 7.2, we only need to prove $\tilde{\mathcal{L}}_{v=0}V(\eta) < 0$ for some $V(\eta) \in C^2(\mathcal{R}^{2n}; \mathcal{R}^+)$, i.e.,

$$\frac{\partial V^T}{\partial \eta} f_e(\eta) + \frac{1}{2} h_e^T(\eta) \frac{\partial^2 V}{\partial \eta^2} h_e(\eta) < 0, \qquad (7.9)$$

where $\tilde{\mathcal{L}}_v$ is defined as the infinitesimal generator of the system (7.4).

$$d\eta = f_e(\eta) \, dt + h_e(\eta) \, dB.$$

Noting that (7.9) is obvious because of (7.6), the proof of Theorem 7.1 is complete. □

Theorem 7.1 can lead to the following corollaries.

COROLLARY 7.1
For $l \equiv 0$ in (7.1), if the following HJI

$$\frac{\partial V^T}{\partial x} f(x) + \frac{\partial V^T}{\partial \hat{x}}(\hat{f}(\hat{x}) + \hat{G}(\hat{x})n(x)) + \frac{1}{2}\gamma^{-2}\Theta^T(\eta)\Theta(\eta) + \frac{1}{2}\|\tilde{z}\|^2$$
$$+ \frac{1}{2}h^T(x)\frac{\partial^2 V}{\partial x^2}h(x) < 0$$

admits solutions $V \in C^2(\mathcal{R}^{2n}; \mathcal{R}^+)$, $V(\eta) > 0$ for $\eta \neq 0$, $V(0) = 0$, $\hat{f}, \hat{G}$ and $\hat{m}$ for some $\gamma > 0$, then the stochastic H_∞ filtering problem is solved by (7.3), where

$$\Theta^T(\eta) = \frac{\partial V^T}{\partial x} k(x) + \frac{\partial V^T}{\partial \hat{x}} \hat{G}(\hat{x})r(x).$$

COROLLARY 7.2

Given a disturbance attenuation $\gamma > 0$, the stochastic H_∞ filtering problem is solved by (7.3) if there exist a positive constant $\mu > 0$ and a positive Lyapunov function $V(\eta) \in C^2(\mathcal{R}^{2n}; \mathcal{R}^+)$ such that the following conditions hold:

(H1)

$$\gamma^2 I - l^T(x)\frac{\partial^2 V}{\partial x^2} l(x) > \mu I; \tag{7.10}$$

(H2)

$$\Gamma_1(\eta) := \frac{\partial V^T}{\partial x} f(x) + \frac{1}{2} h^T(x)\frac{\partial^2 V}{\partial x^2} h(x) + \frac{3}{2}\mu^{-1} \left\| \frac{\partial V^T}{\partial x} k(x) \right\|^2$$

$$+ \frac{3}{2}\mu^{-1} \left\| h^T(x)\frac{\partial^2 V}{\partial x^2} l(x) \right\|^2 + \|m(x)\|^2 + \frac{1}{2}n^T(x)n(x)$$

$$< 0; \tag{7.11}$$

(H3)

$$\Gamma_2(\eta) := \frac{\partial V^T}{\partial \hat{x}} \hat{f}(\hat{x}) + \frac{1}{2} \left\| \frac{\partial V^T}{\partial \hat{x}} \hat{G}(\hat{x}) \right\|^2$$

$$+ \frac{3}{2}\mu^{-1} \left\| \frac{\partial V^T}{\partial \hat{x}} \hat{G}(\hat{x}) \right\|^2 + \|\hat{m}(\hat{x})\|^2 < 0. \tag{7.12}$$

Proof of Corollary 7.2. By applying the fact that

$$X^T Y + Y^T X \le \varepsilon X^T X + \varepsilon^{-1} Y^T Y, \ \forall \varepsilon > 0, \tag{7.13}$$

it follows that (take $\varepsilon = 1$)

$$\frac{\partial V^T}{\partial \hat{x}} \hat{G}(\hat{x})n(x) = \frac{1}{2}\frac{\partial V^T}{\partial \hat{x}} \hat{G}(\hat{x})n(x) + \frac{1}{2}n^T(x)\hat{G}^T(\hat{x})\frac{\partial V}{\partial \hat{x}}$$

$$\le \frac{1}{2}\frac{\partial V^T}{\partial \hat{x}} \hat{G}(\hat{x})\hat{G}^T(\hat{x})\frac{\partial V}{\partial \hat{x}} + \frac{1}{2}n^T(x)n(x),$$

$$\frac{1}{2}\|\tilde{z}\|^2 = \frac{1}{2}\|m(x) - \hat{m}(\hat{x})\|^2 \le \|m(x)\|^2 + \|\hat{m}(\hat{x})\|^2.$$

By condition $(H1)$,

$$\frac{1}{2}\Theta^T(\eta)\left[\gamma^2 I - l^T(x)\frac{\partial^2 V}{\partial \hat{x}^2} l(x)\right]^{-1}\Theta(\eta) \le \frac{1}{2}\mu^{-1}\|\Theta(\eta)\|^2$$

$$\le \frac{3}{2}\mu^{-1}\left(\left\| \frac{\partial V^T}{\partial x} k(x) \right\|^2 + \left\| \frac{\partial V^T}{\partial \hat{x}} \hat{G}(\hat{x})r(x) \right\|^2 + \left\| h^T(x)\frac{\partial^2 V}{\partial x^2} l(x) \right\|^2\right). \tag{7.14}$$

Since it is easy to test that $\Gamma_0(\eta) \leq \Gamma_1(\eta) + \Gamma_2(\eta) < 0$ by $(H2)$ and $(H3)$, this corollary is proved. $\square$

In general, (7.11) and (7.12) are a pair of coupled HJIs. However, if we take $V(\eta)$ in the form of $V(\eta) = V_1(x) + V_2(\hat{x})$, then (7.11) and (7.12) become decoupled and can be solved independently. Specifically, for $l \equiv 0$, Corollary 7.2 yields the following:

COROLLARY 7.3
The results of Corollary 7.2 still hold if:

(H4) There exists a positive Lyapunov function $V(\eta) \in C^2(\mathcal{R}^{2n}; \mathcal{R}^+)$ solving the HJI

$$
\Gamma_3(\eta) := \frac{\partial V^T}{\partial x} f(x) + \frac{1}{2} h^T(x) \frac{\partial^2 V}{\partial x^2} h(x) + \frac{1}{2} n^T(x) n(x) + \gamma^{-2} \left\| \frac{\partial V^T}{\partial x} k(x) \right\|^2
$$
$$
+ \| m(x) \|^2 < 0. \tag{7.15}
$$

(H5) $V(\eta)$ also solves HJI

$$
\Gamma_4(\eta) := \frac{\partial V^T}{\partial \hat{x}} \hat{f}(\hat{x}) + \frac{1}{2} \left\| \frac{\partial V^T}{\partial \hat{x}} \hat{G}(\hat{x}) \right\|^2 + \gamma^{-2} \left\| \frac{\partial V^T}{\partial \hat{x}} \hat{G}(\hat{x}) \right\|^2 + \| \hat{m}(\hat{x}) \|^2 < 0 \tag{7.16}
$$

for some matrices $\hat{f}$, $\hat{G}$ and $\hat{m}$ of suitable dimensions.

Proof. Note that in this case, we can take $\mu = \gamma^2$. Additionally,

$$
\frac{1}{2} \gamma^{-2} \Theta^T(\eta) \Theta(\eta) = \frac{1}{2} \gamma^{-2} \left\| \frac{\partial V^T}{\partial x} k(x) + \frac{\partial V^T}{\partial \hat{x}} \hat{G}(\hat{x}) r(x) \right\|^2
$$
$$
\leq \gamma^{-2} \left\| \frac{\partial V^T}{\partial x} k(x) \right\|^2 + \gamma^{-2} \left\| \frac{\partial V^T}{\partial \hat{x}} \hat{G}(\hat{x}) \right\|^2.
$$

The rest of the proof is omitted. $\square$

If $dim(y) = dim(v)$, under a standard assumption $rr^T = r^T r \equiv I$ [103], by repeating the same procedure as in Corollary 7.2, we have the following result.

COROLLARY 7.4
If (H1), (H2) and (H3) of Corollary 7.2 are replaced by the following ($\bar{H}1$), ($\bar{H}2$) and ($\bar{H}3$), respectively, then Corollary 7.2 still holds.

($\bar{H}1$)

$$
\frac{\partial V^T}{\partial x} k(x) r^T(x) + h^T(x) \frac{\partial^2 V}{\partial x^2} l(x) r^T(x) = -\frac{\partial V^T}{\partial \hat{x}} \hat{G}(\hat{x}). \tag{7.17}
$$

($\bar{H}2$)

$$\Gamma_5(\eta) := \frac{\partial V^T}{\partial x} f(x) + \frac{1}{2} h^T(x) \frac{\partial^2 V}{\partial x^2} h(x) + \frac{1}{2} n^T(x) n(x) + \|m(x)\|^2 < 0. \tag{7.18}$$

($\bar{H}3$)

$$\Gamma_6(\eta) := \frac{\partial V^T}{\partial \hat{x}} \hat{f}(\hat{x}) + \frac{1}{2} \| \frac{\partial V^T}{\partial \hat{x}} \hat{G}(\hat{x})\|^2 + \|\hat{m}(\hat{x})\|^2 < 0. \tag{7.19}$$

Proof. We only need to note that under the condition of $(\bar{H}1)$, $\Theta^T(\eta) = 0$ because of the assumption $r^T r \equiv I$. $\Box$

REMARK 7.3 In most literature (e.g., [101]) on deterministic nonlinear H_∞ control or filtering, one often assumes $k(x) r^T(x) \equiv 0$ for simplicity. Under the above assumption, (7.17) reduces to

$$h^T(x) \frac{\partial^2 V}{\partial x^2} l(x) r^T(x) = -\frac{\partial V^T}{\partial \hat{x}} \hat{G}(\hat{x}).$$

∎

Example 7.1

For the one-dimensional nonlinear stochastic system

$$\begin{cases} dx = [(-6x^5 - 2x^3) + x^2 v] \, dt + (\sqrt{2}x^3 + \frac{1}{\sqrt{2}} v) dB, \\ y = -4x^4 + xv, \\ z = x^2, \end{cases}$$

we can construct an H_∞ filter to estimate $z(t)$ from the measurement information $y(s)$, $s \in [0, t]$. Assume that the disturbance attenuation level is given by $\gamma = \sqrt{2}$, and the state estimator is of the form of (7.3), then the augmented system is given by (7.4) with

$$f_e(\eta) = \begin{bmatrix} -6x^5 - 2x^3 \\ \hat{f}(\hat{x}) - 4x^4 \hat{G}(\hat{x}) \end{bmatrix}, \quad k_e(\eta) = \begin{bmatrix} x^2 \\ x\hat{G}(\hat{x}) \end{bmatrix},$$

$$h_e(\eta) = \begin{bmatrix} \sqrt{2}x^3 \\ 0 \end{bmatrix}, \quad l_e(\eta) = \begin{bmatrix} \frac{1}{\sqrt{2}} \\ 0 \end{bmatrix}.$$

By setting $V(x, \hat{x}) = x^2 + \hat{x}^2$, it is easy to verify that $\hat{f} = -\hat{x}^3$, $\hat{G} = \frac{1}{4}$ and $\hat{m} = \hat{x}^2$ solve HJI (7.6). Hence, the H_∞ filter can be taken as

$$d\hat{x} = -\hat{x}^3 \, dt + \frac{1}{4} y \, dt, \quad \hat{z} = \hat{x}^2.$$

Clearly, there may be more than one solution to HJI (7.6). Therefore, the H_∞ state estimator is not unique. ▯

7.2   Suboptimal Mixed H_2/H_∞ Filtering

Because the H_∞ filter is not unique, we now seek one from the set of the H_∞ filters to minimize the total error energy $\mathcal{E} \int_0^\infty \|\tilde{z}\|^2 \, dt$ when the worst-case disturbance $\hat{v}$ (from v to $\tilde{z}$) is considered under the initial state $(x(0), \hat{x}(0)) = (x(0), 0) \in \mathcal{R}^n \times \{0\}$, which is referred to as the mixed H_2/H_∞ filtering problem. By Remark 7.1, the worst-case disturbance from v to $\tilde{z}$ is of the following form

$$
\begin{aligned}
\hat{v} &= \left(\gamma^2 I - l_e^T \frac{\partial^2 V}{\partial \eta^2} l_e\right)^{-1} \left(k_e^T \frac{\partial V}{\partial \eta} + l_e^T \frac{\partial^2 V}{\partial \eta^2} h_e\right) \\
&= \left[\gamma^2 I - l^T(x)\frac{\partial^2 V}{\partial x^2} l(x)\right]^{-1} \left[k^T(x)\frac{\partial V}{\partial x} + r^T(x)\hat{G}^T(\hat{x})\frac{\partial V}{\partial \hat{x}} \right.\\
&\quad \left. + l^T(x)\frac{\partial^2 V}{\partial x^2} h(x)\right],
\end{aligned}
\tag{7.20}
$$

where $V(\eta)$ is an admissible solution of (7.8) or (7.6). Substituting the above $\hat{v}$ into (7.4) yields

$$
d\eta = [f_e(\eta) + k_e(\eta)\hat{v}(\eta)] \, dt + [h_e(\eta) + l_e(\eta)\hat{v}(\eta)] \, dB.
\tag{7.21}
$$

THEOREM 7.2

For a prescribed disturbance attenuation level $\gamma > 0$, if there exists a positive Lyapunov function $V(\eta) \in \mathcal{C}^2(\mathcal{R}^{2n}; \mathcal{R}^+)$ that solves the following HJI

$$
\begin{cases}
\Theta_1(\eta) := \frac{1}{2}\gamma^2 \left(\frac{\partial V^T}{\partial \eta}k_e + h_e^T \frac{\partial^2 V}{\partial \eta^2} l_e\right)\left(\gamma^2 I - l_e^T \frac{\partial^2 V}{\partial \eta^2} l_e\right)^{-1}\left(k_e^T \frac{\partial V}{\partial \eta} + l_e^T \frac{\partial^2 V}{\partial \eta^2} h_e\right) \\
\quad + \Gamma_0(\eta) + \frac{1}{2}\|\tilde{z}\|^2 < 0, \\
\gamma^2 I - l_e^T \frac{\partial^2 V}{\partial \eta^2} l_e > 0, \ \forall \eta \in \mathcal{R}^{2n}
\end{cases}
\tag{7.22}
$$

for some filter gain matrices $\hat{f}(\hat{x})$, $\hat{G}(\hat{x})$ and $\hat{m}(\hat{x})$ of suitable dimensions, then a suboptimal mixed H_2/H_∞ filter can be synthesized by solving the following constrained optimization problem:

$$
\min_{\hat{f}(\hat{x}),\hat{G}(\hat{x}),\hat{m}(\hat{x}),\ (7.22)} V(\eta(0)).
\tag{7.23}
$$

Proof. First, we have $\Gamma_0(\eta) < 0$ from (7.22). Hence, by Theorem 7.1, the H_∞ filtering problem is solved by (7.3).

Secondly, we assert that $\lim_{T\to\infty} \mathcal{E}V(\eta(T)) = 0$ when $\hat{v}$ of (7.20) is implemented in (7.21), i.e., $\hat{v} \in \mathcal{L}^2(\mathcal{R}^+; \mathcal{R}^{n_v}) \cap \tilde{\Omega}$ with $\tilde{\Omega} := \{v : \lim_{t\to\infty} \mathcal{E}V(\eta(t)) = 0\}$. To prove this assertion, we first note that for any $T > 0$, by Itô's formula, we have

$$
0 \le \mathcal{E}\int_0^T \|\tilde{z}\|^2 \, dt = \mathcal{E}\int_0^T [\|\tilde{z}\|^2 \, dt + dV(\eta)] + V(\eta(0)) - \mathcal{E}V(\eta(T))
$$

$$= V(\eta(0)) - \mathcal{E}V(\eta(T)) + \mathcal{E}\int_0^T \Theta_1(\eta(t))\, dt + \mathcal{E}\int_0^T (h_e + l_e(\eta)\hat{v})^T \frac{\partial V}{\partial \eta}\, dB$$

$$= V(\eta(0)) - \mathcal{E}V(\eta(T)) + \mathcal{E}\int_0^T \Theta_1(\eta(t))\, dt$$

$$\leq V(\eta(0)) - \mathcal{E}V(\eta(T)). \tag{7.24}$$

In addition, by using Itô's formula, for $T \geq s \geq 0$, it follows that

$$V(\eta(T)) = V(\eta(s)) + \int_s^T \tilde{\mathcal{L}}_{v=\hat{v}}V(\eta(\tau))\, d\tau + \int_s^T [h_e(\eta(\tau)) + l_e(\eta(\tau))\hat{v}]^T \frac{\partial V}{\partial \eta}\, dB(\tau)$$

$$= V(\eta(s)) + \int_s^T \left\{ \frac{\partial V^T}{\partial \eta}[f_e(\eta(\tau)) + k_e(\eta(\tau))\hat{v}] \right.$$

$$\left. + \frac{1}{2}[h_e(\eta(\tau)) + l_e(\eta(\tau))\hat{v}]^T \frac{\partial^2 V}{\partial \eta^2}[h_e(\eta(\tau)) + l_e(\eta(\tau))\hat{v}] \right\} d\tau$$

$$+ \int_s^T [h_e(\eta(\tau)) + l_e(\eta(\tau))\hat{v}]^T \frac{\partial V}{\partial \eta}\, dB(\tau). \tag{7.25}$$

By letting $\tilde{\mathcal{F}}_t = \sigma(B(s), y(s), 0 \leq s \leq t)$, it follows from (7.24) that $\mathcal{E}|V(\eta(t))| < \infty$ for $\forall t \geq 0$. Moreover,

$$\mathcal{E}[V(\eta(T))|\tilde{\mathcal{F}}_s] \leq \mathcal{E}[V(\eta(s))|\tilde{\mathcal{F}}_s] - \mathcal{E}\left[\int_s^T \|\tilde{z}(s)\|^2 ds|\tilde{\mathcal{F}}_s\right]$$

$$+ \mathcal{E}\left[\int_s^T (h_e(\eta(s)) + l_e(\eta(s))\hat{v})^T \frac{\partial V}{\partial \eta}dB(s)|\tilde{\mathcal{F}}_s\right]$$

$$\leq V(\eta(s)) \ a.s.. \tag{7.26}$$

Thus $\{V(\eta(t)), \tilde{\mathcal{F}}_t\}$ is a nonnegative supermartingale with respect to $\{\tilde{\mathcal{F}}_t\}_{t\geq 0}$. Additionally, from (7.22), we have

$$\tilde{\mathcal{L}}_{v=\hat{v}}V(\eta) = \frac{\partial V^T}{\partial \eta}[f_e(\eta) + k_e(\eta)\hat{v}] + \frac{1}{2}[h_e(\eta) + l_e(\eta)\hat{v}]^T \frac{\partial^2 V}{\partial \eta^2}[h_e(\eta) + l_e(\eta)\hat{v}]$$

$$< -\|\tilde{z}\|^2 \leq 0. \tag{7.27}$$

Hence, (7.21) is globally asymptotically stable in probability. By Doob's convergence theorem [145, 130], $V(\eta(\infty)) = \lim_{t\to\infty} V(\eta(t)) = 0 \ a.s.$ Moreover,

$$\lim_{t\to\infty} \mathcal{E}V(\eta(t)) = \mathcal{E}V(\eta(\infty)) = 0.$$

Finally, taking $T \to \infty$ in (7.24) and applying the above assertion yield that

$$\min_{\hat{f}(\hat{x}), \hat{G}(\hat{x}), \hat{m}(\hat{x}), (7.22)} \mathcal{E}\int_0^\infty \|\tilde{z}\|^2 dt \leq \min_{\hat{f}(\hat{x}), \hat{G}(\hat{x}), \hat{m}(\hat{x}), (7.22)} V(\eta(0)).$$

Theorem 7.2 is accordingly concluded, i.e., by solving (7.23), a suboptimal mixed H_2/H_∞ filter is obtained. $\square$

Obviously, HJI (7.6) for nonlinear H_∞ filtering is implied by (7.22). Hence, for a suboptimal mixed H_2/H_∞ filtering problem, we only need to minimize $V(\eta(0))$ subject to (7.22). From Theorems 7.1–7.2, to synthesize a nonlinear H_∞ or mixed H_2/H_∞ filter, one should solve HJI (7.6) or constrained optimization problem (7.23), neither of which, however, is easy except for some special cases, such as for linear time-invariant systems. Moreover, from the proof of Theorem 7.2, $V_1(\eta(0))$ is a tight upper bound of $\mathcal{E}\int_0^\infty \|\tilde{z}\|^2\, dt$ due to $\lim_{T\to\infty} \mathcal{E}V(\eta(T)) = 0$. In particular, when $\Theta_1(\eta) = 0$, it follows from (7.24) that

$$\min_{\hat{f}(\hat{x}),\ \hat{G}(\hat{x}),\ \hat{m}(\hat{x}),\ (7.22)} \mathcal{E}\int_0^\infty \|\tilde{z}\|^2\, dt = \min_{s.t.\ \hat{f}(\hat{x}),\ \hat{G}(\hat{x}),\ \hat{m}(\hat{x}),\ (7.22)} V(\eta(0)),$$

and an optimal H_2/H_∞ filtering is obtained.

7.3   LMI-Based Approach for Quasi-Linear H_∞ Filter Design

From the previous section, it is noted that for general nonlinear stochastic system (7.1), to design an H_∞ filter, one needs to solve HJI (7.6), which is not an easy task. This section shows that for a class of quasi-linear stochastic systems, the above-mentioned problem can be converted into that of solving LMIs.

We consider the following quasi-linear stochastic system

$$\begin{cases} dx = [Ax + F_0(x) + B_0 v]\, dt + [Cx + F_1(x)]\, d\tilde{B}_0, \\ \|F_i(x)\| \le \lambda\|x\|,\ i = 0, 1,\ \lambda > 0,\ \forall x \in \mathcal{R}^n \end{cases} \tag{7.28}$$

with the linear measurement output

$$dy = (A_1 x + B_1 v)\, dt + C_1 x\, d\tilde{B}_1,\quad z = Dx, \tag{7.29}$$

where all the coefficients are constant matrices of suitable dimensions and $F_i(0) = 0$, $i = 0, 1$, $\tilde{B}_0(t)$ and $\tilde{B}_1(t)$ are independent one-dimensional standard Brownian motions. (7.28) is a special case of the state equation of (7.1) with, for the purpose of simplicity, only a state-dependent noise. As a matter of fact, in (7.1), if one takes $k(x) = B_0$, $l(x) \equiv 0$, and regards $Ax + F_0(x)$ and $Cx + F_1(x)$ as the Taylor series expansions of $f(x)$ and $h(x)$, respectively, then the state equation (7.1) reduces to (7.28), which is called a quasi-linear stochastic system. In (7.29), we assume that the measurement output, as in [80], is governed by a stochastic differential equation.

For the special nonlinear system (7.28) with the measurement equation (7.29), we adopt the following linear filter for the estimation of $z(t)$ (see [200] for the treatment of discrete-time nonlinear systems):

$$d\hat{x} = A_f \hat{x}\, dt + B_f\, dy,\quad \hat{x}(0) = 0,\quad \hat{z} = D\hat{x}, \tag{7.30}$$

where $\hat{x} \in \mathcal{R}^n$. Again, let $\eta = \begin{bmatrix} x \\ \hat{x} \end{bmatrix}$, $\tilde{z} = z - \hat{z}$, then

$$\begin{cases} d\eta = \tilde{A}\eta \, dt + \tilde{D}_1\eta d\tilde{B}_0 + \tilde{D}_2\eta d\tilde{B}_1 + \tilde{F}_1 \, dt + \tilde{F}_2 \, d\tilde{B}_0 + \tilde{F}_3 v \, dt, \\ \|F_i(x)\| \le \lambda\|x\|, \ i = 0, 1, \ \lambda > 0, \ \forall x \in \mathcal{R}^n, \end{cases} \tag{7.31}$$

where

$$\tilde{A} = \begin{bmatrix} A & 0 \\ B_f A_1 & A_f \end{bmatrix}, \ \tilde{D}_1 = \begin{bmatrix} C & 0 \\ 0 & 0 \end{bmatrix}, \ \tilde{D}_2 = \begin{bmatrix} 0 & 0 \\ B_f C_1 & 0 \end{bmatrix},$$

$$\tilde{F}_1 = \begin{bmatrix} F_0(x) \\ 0 \end{bmatrix}, \ \tilde{F}_2 = \begin{bmatrix} F_1(x) \\ 0 \end{bmatrix}, \ \tilde{F}_3 = \begin{bmatrix} B_0 \\ B_f B_1 \end{bmatrix}. \tag{7.32}$$

For a prescribed disturbance attenuation level $\gamma > 0$, we want to find constant matrices A_f and B_f, such that

$$\|\tilde{z}(t)\|_{[0,\infty)}^2 < \gamma^2 \|v(t)\|_{[0,\infty)}^2 \tag{7.33}$$

holds for any $v \in \mathcal{L}_\mathcal{F}^2(\mathcal{R}^+, \mathcal{R}^{n_v})$. Define the H_∞ performance index as

$$J_s = \|\tilde{z}(t)\|_{[0,\infty)}^2 - \gamma^2 \|v(t)\|_{[0,\infty)}^2. \tag{7.34}$$

Obviously, the H_∞ filtering performance (7.33) is achieved iff $J_s < 0$.

As in [80], the H_∞-based state estimation problem is formulated as follows:

(i) Given a prescribed value $\gamma > 0$, find an estimator (7.30) such that (7.31) is exponentially mean square stable in the case of $v \equiv 0$. That is, for the system

$$d\eta = \tilde{A}\eta \, dt + \tilde{D}_1\eta d\tilde{B}_0 + \tilde{D}_2\eta d\tilde{B}_1 + \tilde{F}_1 \, dt + \tilde{F}_2 \, d\tilde{B}_0,$$

there are some positive constants ρ and ϱ such that

$$\mathcal{E}\|\eta(t)\|^2 \le \rho\|\eta(0)\|^2 \exp(-\varrho t), \ t \ge 0.$$

(ii) $J_s < 0$ for all non-zero $v \in \mathcal{L}_\mathcal{F}^2(\mathcal{R}^+, \mathcal{R}^{n_v})$ with $\eta(0) = 0$.

LEMMA 7.3

If the following matrix inequalities

$$P\tilde{A} + \tilde{A}^T P + \tilde{D}_1^T P \tilde{D}_1 + \tilde{D}_2^T P \tilde{D}_2 + P + 3\lambda^2 \alpha I + Q + \frac{1}{\gamma^2} P \tilde{F}_3 \tilde{F}_3^T P < 0 \tag{7.35}$$

and

$$P \le \alpha I \tag{7.36}$$

with

$$Q = \begin{bmatrix} D^T D & -D^T D \\ -D^T D & D^T D \end{bmatrix} = \begin{bmatrix} D^T \\ -D^T \end{bmatrix} \begin{bmatrix} D & -D \end{bmatrix} \ge 0$$

have solutions $P > 0$ and $\alpha > 0$, then (7.31) is exponentially mean square stable when $v \equiv 0$, and the H_∞ performance $J_s < 0$ when $v(t) \neq 0$.

REMARK 7.4   In (7.28), $\|F_i(x)\| \leq \lambda \|x\|$ for $\forall x \in \mathcal{R}^n$, $i = 0, 1$, imply that $F_i(x)$ satisfies the global Lipschitz condition at the origin. If in the definition of stochastic H_∞ filtering, we require that (7.31) be locally exponentially mean square stable ($v \equiv 0$), then $\|F_i(x)\| \leq \lambda \|x\|$ is only required to be satisfied in the neighborhood of the origin.   ∎

Proof of Lemma 7.3. We first prove that (7.31) is exponentially mean square stable when $v \equiv 0$. By taking the Lyapunov function candidate as $V(\eta) = \eta^T P \eta$ with $P > 0$ a solution to (7.35) and (7.36), and letting $\hat{\mathcal{L}}_v$ be the infinitesimal generator of (7.31), then

$$\hat{\mathcal{L}}_{v\equiv 0}V(\eta) = \eta^T(P\tilde{A} + \tilde{A}^T P + \tilde{D}_1^T P \tilde{D}_1 + \tilde{D}_2^T P \tilde{D}_2)\eta$$
$$+ 2\tilde{F}_1^T P\eta + 2\eta^T \tilde{D}_1^T P \tilde{F}_2 + \tilde{F}_2^T P \tilde{F}_2. \qquad (7.37)$$

Applying (7.13) (set $\varepsilon = 1$) and (7.36), we have the following

$$2\tilde{F}_1^T P\eta \leq \tilde{F}_1^T P \tilde{F}_1 + \eta^T P\eta \leq \alpha\lambda^2\|x\|^2 + \eta^T P\eta \leq \lambda^2\alpha\|\eta\|^2 + \eta^T P\eta, \quad (7.38)$$

$$2\eta^T \tilde{D}_1^T P \tilde{F}_2 \leq \eta^T \tilde{D}_1^T P \tilde{D}_1 \eta + \tilde{F}_2^T P \tilde{F}_2 \leq \eta^T \tilde{D}_1^T P \tilde{D}_1 \eta + \lambda^2\alpha\|\eta\|^2, \quad (7.39)$$

$$\tilde{F}_2^T P \tilde{F}_2 \leq \lambda^2\alpha\|\eta\|^2. \qquad (7.40)$$

Substituting (7.38)–(7.40) into (7.37), we have

$$\hat{\mathcal{L}}_{v\equiv 0}V(\eta) \leq \eta^T(P\tilde{A} + \tilde{A}^T P + \tilde{D}_1^T P \tilde{D}_1 + \tilde{D}_2^T P \tilde{D}_2 + P + 3\lambda^2\alpha I)\eta. \; (7.41)$$

Obviously, if (7.35) holds, then there exists $k_3 > 0$ such that

$$P\tilde{A} + \tilde{A}^T P + \tilde{D}_1^T P \tilde{D}_1 + \tilde{D}_2^T P \tilde{D}_2 + P + 3\lambda^2\alpha I < -k_3 I.$$

Hence, $\hat{\mathcal{L}}_{v\equiv 0}V(\eta) \leq -k_3\|\eta\|^2$, which yields that (7.31) is exponentially mean square stable for $v \equiv 0$ by Theorem 1.6-(4).

Secondly, we prove $J_s < 0$ for all non-zero $v \in \mathcal{L}_{\mathcal{F}}^2(\mathcal{R}^+, \mathcal{R}^{n_v})$ with $\eta(0) = 0$. Note that for any $T > 0$,

$$J_s(T) \; := \; \mathcal{E}\int_0^T (\|\tilde{z}\|^2 - \gamma^2\|v\|^2)\, dt = \mathcal{E}\int_0^T [(\eta^T Q\eta - \gamma^2\|v\|^2)\, dt + d(\eta^T P\eta)$$
$$-d(\eta^T P\eta)]$$
$$= -\mathcal{E}[\eta^T(T)P\eta(T)] + \mathcal{E}\int_0^T [(\eta^T Q\eta - \gamma^2\|v\|^2) + \hat{\mathcal{L}}_v V(\eta)]\, dt$$
$$\leq \mathcal{E}\int_6^T [(\eta^T Q\eta - \gamma^2\|v\|^2) + \hat{\mathcal{L}}_{v\equiv 0}V(\eta) + v^T \tilde{F}_3^T P\eta + \eta^T P\tilde{F}_3 v]\, dt$$

$$\leq \mathcal{E} \int_0^T [\eta^T (P\tilde{A} + \tilde{A}^T P + \tilde{D}_1^T P \tilde{D}_1 + \tilde{D}_2^T P \tilde{D}_2 + P + 3\lambda^2 \alpha I + Q)\eta$$

$$-\gamma^2 \|v\|^2] \, dt + \mathcal{E} \int_0^T (v^T \tilde{F}_3^T P \eta + \eta^T P \tilde{F}_3 v) \, dt$$

$$= \mathcal{E} \int_0^T \begin{bmatrix} \eta \\ v \end{bmatrix}^T \begin{bmatrix} \{P\tilde{A} + \tilde{A}^T P + \tilde{D}_1^T P \tilde{D}_1 + P & P\tilde{F}_3 \\ + \tilde{D}_2^T P \tilde{D}_2 + 3\lambda^2 \alpha I + Q\} & \\ \tilde{F}_3^T P & -\gamma^2 I \end{bmatrix} \begin{bmatrix} \eta \\ v \end{bmatrix} \, dt.$$

Therefore, if

$$\begin{bmatrix} \{P\tilde{A} + \tilde{A}^T P + \tilde{D}_1^T P \tilde{D}_1 + P & P\tilde{F}_3 \\ + \tilde{D}_2^T P \tilde{D}_2 + 3\lambda^2 \alpha I + Q\} & \\ \tilde{F}_3^T P & -\gamma^2 I \end{bmatrix} < 0, \qquad (7.42)$$

then there exists $\epsilon > 0$ such that $J_s(T) \leq -\epsilon^2 \mathcal{E} \int_0^T \|v\|^2 \, dt < 0$ for any non-zero $v \in \mathcal{L}_{\mathcal{F}}^2(\mathcal{R}^+, \mathcal{R}^{n_v})$, which yields $J_s \leq -\epsilon^2 \mathcal{E} \int_0^\infty \|v\|^2 \, dt < 0$ by taking $T \to \infty$. As (7.42) is equivalent to (7.35) according to Schur's complement, the proof of this lemma is completed. $\square$

Lemma 7.3 is inconvenient to use in designing an H_∞ filter, because (7.35) is not an LMI. The following result is more useful in practice.

THEOREM 7.3
If the following LMIs

$$\begin{bmatrix} P_{11} - \alpha I & 0 \\ 0 & P_{22} - \alpha I \end{bmatrix} < 0, \qquad (7.43)$$

and

$$\begin{bmatrix} \Upsilon_{11} & A_1^T Z_1^T - D^T D & C^T P_{11} & C_1^T Z_1^T & P_{11} B_0 \\ Z_1 A_1 - D^T D & \Upsilon_{22} & 0 & 0 & Z_1 B_1 \\ P_{11} C & 0 & -P_{11} & 0 & 0 \\ Z_1 C_1 & 0 & 0 & -P_{22} & 0 \\ B_0^T P_{11} & B_1^T Z_1^T & 0 & 0 & -\gamma^2 I \end{bmatrix} < 0 \qquad (7.44)$$

with

$$\Upsilon_{11} = P_{11} A + A^T P_{11} + 3\lambda^2 \alpha I + P_{11} + D^T D$$

and

$$\Upsilon_{22} = Z + Z^T + 3\lambda^2 \alpha I + P_{22} + D^T D$$

have solutions $P_{11} > 0$, $P_{22} > 0$, $\alpha > 0$, $Z_1 \in \mathcal{R}^{n \times n_y}$, and $Z \in \mathcal{R}^{n \times n}$, then (7.31) is exponentially mean square stable for $v \equiv 0$ and the H_∞ filtering performance $J_s < 0$ is achieved with the following filter

$$d\hat{x} = P_{22}^{-1} Z \hat{x} \, dt + P_{22}^{-1} Z_1 dy, \quad \hat{z} = D\hat{x}. \qquad (7.45)$$

Proof. By Schur's complement, (7.35) is equivalent to

$$
\begin{bmatrix}
P\tilde{A} + \tilde{A}^T P + P + 3\lambda^2\alpha I + Q & \tilde{D}_1^T P & \tilde{D}_2^T P & P\tilde{F}_3 \\
P\tilde{D}_1 & -P & 0 & 0 \\
P\tilde{D}_2 & 0 & -P & 0 \\
\tilde{F}_3^T P & 0 & 0 & -\gamma^2 I
\end{bmatrix} < 0. \tag{7.46}
$$

If we take $P = diag(P_{11}, P_{22})$, then (7.36) is equivalent to (7.43). By substituting (7.32) into (7.46), we have

$$
\begin{bmatrix}
\Psi_{11} & \Psi_{12}^T & \Psi_{13}^T & \Psi_{14}^T \\
\Psi_{12} & \Psi_{22} & 0 & 0 \\
\Psi_{13} & 0 & \Psi_{33} & 0 \\
\Psi_{14} & 0 & 0 & \Psi_{44}
\end{bmatrix} < 0, \tag{7.47}
$$

where

$$
\Psi_{11} = \begin{bmatrix} \Upsilon_{11} & A_1^T B_f^T P_{22} - D^T D \\ P_{22} B_f A_1 - D^T D & \Psi_{11}^{22} \end{bmatrix},
$$

$$
\Psi_{11}^{22} = P_{22} A_f + A_f^T P_{22} + 3\lambda^2\alpha I + P_{22} + D^T D,
$$

$$
\Psi_{22} = \Psi_{33} = \begin{bmatrix} -P_{11} & 0 \\ 0 & -P_{22} \end{bmatrix}, \quad \Psi_{44} = -\gamma^2 I,
$$

$$
\Psi_{12}^T = \begin{bmatrix} C^T P_{11} & 0 \\ 0 & 0 \end{bmatrix}, \quad \Psi_{13}^T = \begin{bmatrix} 0 & C_1^T B_f^T P_{22} \\ 0 & 0 \end{bmatrix}, \quad \Psi_{14}^T = \begin{bmatrix} P_{11} B_0 \\ P_{22} B_f B_1 \end{bmatrix}.
$$

Then (7.47) is equivalent to

$$
\begin{bmatrix}
\Upsilon_{11} & A_1^T B_f^T P_{22} - D^T D & C^T P_{11} & C_1^T B_f^T P_{22} & P_{11} B_0 \\
P_{22} B_f A_1 - D^T D & \Psi_{11}^{22} & 0 & 0 & P_{22} B_f B_1 \\
P_{11} C & 0 & -P_{11} & 0 & 0 \\
P_{22} B_f C_1 & 0 & 0 & -P_{22} & 0 \\
B_0^T P_{11} & B_1^T B_f^T P_{22} & 0 & 0 & -\gamma^2 I
\end{bmatrix} < 0. \tag{7.48}
$$

By letting $P_{22} A_f = Z$ and $P_{22} B_f = Z_1$, (7.48) becomes (7.44). By our assumption, $A_f = P_{22}^{-1} Z$ and $B_f = P_{22}^{-1} Z_1$. Hence, an H_∞ filter is constructed in the form of (7.45), and the proof is completed. □

Based on the above discussion, we summarize the following design algorithm.

Design Algorithm:

Step i. Obtain solutions $P_{11} > 0$, $P_{22} > 0$, $\alpha > 0$, Z_1 and Z by solving LMIs (7.43)–(7.44).

Step ii. Set $A_f = P_{22}^{-1} Z$, $B_f = P_{22}^{-1} Z_1$, and substitute the obtained A_f, B_f into (7.30). Then, (7.30) is the desired H_∞ filter.

We should point out that we can also set the augmented state vector as $\eta = \begin{bmatrix} x \\ x - \hat{x} \end{bmatrix}$, and then (7.31) is changed into

$$
\begin{cases}
d\eta = \tilde{A}\eta \, dt + \tilde{D}_1\eta d\tilde{B}_0 + \tilde{D}_2\eta d\tilde{B}_1 + \tilde{F}_1 \, dt + \tilde{F}_2 \, d\tilde{B}_0 + \tilde{F}_3 v \, dt, \\
\|F_i(x)\| \le \lambda\|x\|, \ i = 0, 1, \ \lambda > 0, \ \forall x \in \mathcal{R}^n,
\end{cases} \tag{7.49}
$$

where

$$\tilde{A} = \begin{bmatrix} A & 0 \\ A - B_f A_1 - A_f & -A_f \end{bmatrix}, \quad \tilde{D}_1 = \begin{bmatrix} C & 0 \\ C & 0 \end{bmatrix}, \quad \tilde{D}_2 = \begin{bmatrix} 0 & 0 \\ -B_f C_1 & 0 \end{bmatrix}, \quad (7.50)$$

$$\tilde{F}_1 = \begin{bmatrix} F_0(x) \\ F_0(x) \end{bmatrix}, \quad \tilde{F}_2 = \begin{bmatrix} F_1(x) \\ F_1(x) \end{bmatrix}, \quad \tilde{F}_3 = \begin{bmatrix} B_0 \\ B_0 - B_f B_1 \end{bmatrix}. \quad (7.51)$$

Repeating the same procedure as in Theorem 7.3, it is easy to obtain the following theorem.

THEOREM 7.4
If the following LMIs

$$\begin{bmatrix} P_{11} - \alpha I & 0 \\ 0 & P_{22} - \alpha I \end{bmatrix} < 0 \qquad (7.52)$$

and

$$\begin{bmatrix} a_{11} & \Delta_{12} & C^T P_{11} & C^T P_{22} & -C_1^T Z_1^T & P_{11} B_0 \\ * & a_{22} & 0 & 0 & 0 & \Delta_{26} \\ * & 0 & -P_{11} & 0 & 0 & 0 \\ * & 0 & 0 & -P_{22} & 0 & 0 \\ * & 0 & 0 & 0 & -P_{22} & 0 \\ * & * & 0 & 0 & 0 & -\gamma^2 I \end{bmatrix} < 0 \qquad (7.53)$$

have solutions $P_{11} > 0, P_{22} > 0, \alpha > 0, Z_1 \in \mathcal{R}^{n \times n_y}$, and $Z \in \mathcal{R}^{n \times n}$, then (7.49) is internally stable and $J_s < 0$. Moreover,

$$d\hat{x} = P_{22}^{-1} Z \hat{x}\, dt + P_{22}^{-1} Z_1 dy \qquad (7.54)$$

is the corresponding H_∞ filter. In (7.53), $a_{11} = P_{11} A + A^T P_{11} + 3\lambda^2 \alpha I + P_{11}$, $a_{22} = -Z - Z^T + 3\lambda^2 \alpha I + D^T D + P_{22}$, $\Delta_{12} = A^T P_{22} - A_1^T Z_1^T - Z^T$, $\Delta_{26} = P_{22} B_0 - Z_1 B_1$, and $*$ is derived by symmetry.

REMARK 7.5 (7.33) or (7.34) is a standard requirement for H_∞ performance, which is weaker than that which requires the $\mathcal{L}_2$-gain from the external disturbance to the estimation error being strictly less than $\gamma > 0$, i.e.,

$$\sup_{v \in \mathcal{L}_{\mathcal{F}}^2(\mathcal{R}^+, \mathcal{R}^{n_v}), v \neq 0, x_0 = 0} \frac{\|\tilde{z}\|_{[0,\infty)}}{\|v\|_{[0,\infty)}} < \gamma. \qquad (7.55)$$

However, from the proof of Theorem 7.3, Theorems 7.3–7.4 in fact guarantee that (7.55) holds. ∎

7.4   Suboptimal Mixed H_2/H_∞ Filtering of Quasi-Linear Systems

In this section, we follow the line of [80] to discuss the mixed H_2/H_∞ filtering for the quasi-linear system (7.28)-(7.29). The mixed H_2/H_∞ filtering requires choosing one from the set of all H_∞ filters to minimize the estimation error variance:

$$
\begin{aligned}
J_2 : &= \lim_{t\to\infty} \mathcal{E}[\tilde{z}^T(t)\tilde{z}(t)] = \lim_{t\to\infty} \mathcal{E}[\eta^T(t)(0\ I)^T D^T D(0\ I)\eta(t)] \\
&= \lim_{t\to\infty} \text{Trace}\{D(0\ I)\mathcal{E}[\eta(t)\eta^T(t)](0\ I)^T D^T\},
\end{aligned}
\tag{7.56}
$$

where $\eta^T = \begin{bmatrix} x^T & (x-\hat{x})^T \end{bmatrix}$. Different from Section 7.3, the external disturbance signal v in (7.28) and (7.29) is considered as a white noise when minimizing J_2. The two indices J_s in (7.34) and J_2 in (7.56) are associated with H_∞ robustness and H_2 performance, respectively. If $v(t) = \dot{\tilde{B}}_2(t)$ is a white noise independent of $\tilde{B}_0$ and $\tilde{B}_1$, then (7.49) accordingly becomes

$$
\begin{cases}
d\eta = \tilde{A}\eta\, dt + \tilde{D}_1\eta d\tilde{B}_0 + \tilde{D}_2\eta d\tilde{B}_1 + \tilde{F}_1\, dt + \tilde{F}_2\, d\tilde{B}_0 + \tilde{F}_3\, d\tilde{B}_2, \\
\|F_i(x)\| \le \lambda\|x\|,\ i=0,1,\ \lambda>0,\ \forall x \in \mathcal{R}^n,
\end{cases}
\tag{7.57}
$$

where the coefficients of (7.57) are defined in (7.50) and (7.51). Letting $X(t) = \mathcal{E}[\eta(t)\eta^T(t)]$ in (7.57), it then follows from Itô's formula that

$$
\begin{aligned}
\dot{X}(t) &= \tilde{A}X(t) + X(t)\tilde{A}^T + \mathcal{E}(\tilde{F}_1\eta^T + \eta\tilde{F}_1^T) + \tilde{D}_1 X \tilde{D}_1^T \\
&\quad + \mathcal{E}(\tilde{D}_1\eta\tilde{F}_2^T + \tilde{F}_2\eta^T\tilde{D}_1^T) + \mathcal{E}(\tilde{F}_2\tilde{F}_2^T) + \tilde{D}_2 X(t)\tilde{D}_2^T + \tilde{F}_3\tilde{F}_3^T.
\end{aligned}
\tag{7.58}
$$

By means of

$$
\mathcal{E}(\tilde{F}_1\eta^T + \eta\tilde{F}_1^T) \le \mathcal{E}(\tilde{F}_1\tilde{F}_1^T) + X(t)
$$

and

$$
\mathcal{E}(\tilde{D}_1\eta\tilde{F}_2^T + \tilde{F}_2\eta^T\tilde{D}_1^T) \le \tilde{D}_1 X \tilde{D}_1^T + \mathcal{E}(\tilde{F}_2\tilde{F}_2^T),
$$

we have

$$
\begin{aligned}
\dot{X}(t) &\le \tilde{A}X(t) + X(t)\tilde{A}^T + 2\tilde{D}_1 X(t)\tilde{D}_1^T + \tilde{D}_2 X(t)\tilde{D}_2^T \\
&\quad + X(t) + 2\mathcal{E}(\tilde{F}_2\tilde{F}_2^T) + \mathcal{E}(\tilde{F}_1\tilde{F}_1^T) + \tilde{F}_3\tilde{F}_3^T.
\end{aligned}
\tag{7.59}
$$

By assuming that $F_i(x)(i=0,1)$ satisfy

$$
F_i(x)F_i^T(x) \le G_i xx^T G_i^T,\ i=0,1, \forall x \in \mathcal{R}^n,
\tag{7.60}
$$

where G_1 and G_2 are constant matrices of suitable dimensions, we have

$$
\begin{aligned}
\tilde{F}_i\tilde{F}_i^T &= \begin{bmatrix} I & 0 \\ I & I \end{bmatrix}\begin{bmatrix} F_i F_i^T & 0 \\ 0 & 0 \end{bmatrix}\begin{bmatrix} I & I \\ 0 & I \end{bmatrix} \\
&\le \begin{bmatrix} I & 0 \\ I & I \end{bmatrix}\begin{bmatrix} G_i xx^T G_i^T & 0 \\ 0 & 0 \end{bmatrix}\begin{bmatrix} I & I \\ 0 & I \end{bmatrix}
\end{aligned}
$$

$$= \begin{bmatrix} I & 0 \\ I & I \end{bmatrix} \begin{bmatrix} G_i & 0 \\ 0 & 0 \end{bmatrix} \eta \eta^T \begin{bmatrix} G_i^T & 0 \\ 0 & 0 \end{bmatrix} \begin{bmatrix} I & I \\ 0 & I \end{bmatrix}$$

$$= \begin{bmatrix} G_i & 0 \\ G_i & 0 \end{bmatrix} \eta \eta^T \begin{bmatrix} G_i^T & G_i^T \\ 0 & 0 \end{bmatrix}$$

$$:= \tilde{G}_i \eta \eta^T \tilde{G}_i^T, \quad i = 0, 1, \tag{7.61}$$

where

$$\tilde{G}_i = \begin{bmatrix} G_i & 0 \\ G_i & 0 \end{bmatrix}.$$

So (7.59) becomes

$$\dot{X}(t) \leq \tilde{A}X(t) + X(t)\tilde{A}^T + 2\tilde{D}_1 X(t)\tilde{D}_1^T + \tilde{D}_2 X(t)\tilde{D}_2^T + X(t)$$
$$+ 2\tilde{G}_2 X(t)\tilde{G}_2^T + \tilde{G}_1 X(t)\tilde{G}_1^T + \tilde{F}_3 \tilde{F}_3^T. \tag{7.62}$$

In addition, if $X_1(t)$ solves

$$\begin{cases} \dot{X}_1(t) = \tilde{A}X_1(t) + X_1(t)\tilde{A}^T + 2\tilde{D}_1 X_1(t)\tilde{D}_1^T + \tilde{D}_2 X_1(t)\tilde{D}_2^T + X_1(t) \\ \qquad + 2\tilde{G}_2 X_1(t)\tilde{G}_2^T + \tilde{G}_1 X_1(t)\tilde{G}_1^T + \tilde{F}_3 \tilde{F}_3^T, \\ X_1(0) = X(0), \end{cases}$$

$$\tag{7.63}$$

it is easy to prove that $X(t) \leq X_1(t)$. Denote $\bar{X}_1 := \lim_{t \to \infty} X_1(t)$, where $\bar{X}_1$ satisfies

$$\tilde{A}\bar{X}_1 + \bar{X}_1 \tilde{A}^T + 2\tilde{D}_1 \bar{X}_1 \tilde{D}_1^T + \tilde{D}_2 \bar{X}_1 \tilde{D}_2^T + 2\tilde{G}_2 \bar{X}_1 \tilde{G}_2^T + \tilde{G}_1 \bar{X}_1 \tilde{G}_1^T + \bar{X}_1 + \tilde{F}_3 \tilde{F}_3^T$$
$$= 0.$$

Obviously, $\lim_{t \to \infty} X(t) \leq \bar{X}_1$, and so

$$J_2 \leq \mathrm{Trace}[D(0 \; I)\bar{X}_1(0 \; I)^T D^T] = \mathrm{Trace}[\bar{X}_1 Q],$$

where $Q = (0 \; D)^T(0 \; D)$. As in [80], the following is easily obtained.

LEMMA 7.4
If $\hat{P}$ is a solution of

$$\tilde{A}^T \hat{P} + \hat{P}\tilde{A} + 2\tilde{D}_1^T \hat{P}\tilde{D}_1 + \tilde{D}_2^T \hat{P}\tilde{D}_2 + 2\tilde{G}_2^T \hat{P}\tilde{G}_2 + \tilde{G}_1^T \hat{P}\tilde{G}_1 + Q + \hat{P} = 0, \tag{7.64}$$

then $\mathrm{Trace}(\bar{X}_1 Q) = \mathrm{Trace}(\hat{P}\tilde{F}_3 \tilde{F}_3^T).$

Secondly, suppose $P > 0$ satisfies

$$\tilde{A}^T P + P\tilde{A} + 2\tilde{D}_1^T P\tilde{D}_1 + \tilde{D}_2^T P\tilde{D}_2 + Q + P + 2\tilde{G}_2^T P\tilde{G}_2 + \tilde{G}_1^T P\tilde{G}_1 < 0. \tag{7.65}$$

By applying Theorem 1.5, it is easy to show that $P > \hat{P}$. So we have the following lemma.

LEMMA 7.5
If there exist positive definite solutions P and $\hat{P}$ to (7.65) and (7.64), respectively, then $P > \hat{P}$.

By Lemmas 7.4–7.5, we have that

$$
\begin{aligned}
J_2 &= \lim_{t\to\infty} \text{Trace}[D(0\ I)X(t)(0\ I)^T D^T] \\
&\le \lim_{t\to\infty} \text{Trace}[D(0\ I)X_1(t)(0\ I)^T D^T] \\
&= \text{Trace}[D(0\ I)\bar{X}_1(0\ I)^T D^T] \\
&= \text{Trace}[\bar{X}_1 Q] = \text{Trace}(\hat{P}\tilde{F}_3\tilde{F}_3^T) \\
&= \text{Trace}(\tilde{F}_3^T \hat{P}\tilde{F}_3) \\
&\le \hat{J}_2 := \text{Trace}(\tilde{F}_3^T P\tilde{F}_3).
\end{aligned}
\tag{7.66}
$$

That (7.65) has a positive definite solution $P > 0$ is equivalent to

$$
\begin{bmatrix}
P\tilde{A} + \tilde{A}^T P + P + Q & \sqrt{2}\tilde{D}_1^T P & \tilde{D}_2^T P & \tilde{G}_1^T P & \sqrt{2}\tilde{G}_2^T P \\
\sqrt{2}P\tilde{D}_1 & -P & 0 & 0 & 0 \\
P\tilde{D}_2 & 0 & -P & 0 & 0 \\
P\tilde{G}_1 & 0 & 0 & -P & 0 \\
\sqrt{2}P\tilde{G}_2 & 0 & 0 & 0 & -P
\end{bmatrix} < 0.
\tag{7.67}
$$

If we still take $P = diag(P_{11}, P_{22}) > 0$, a suboptimal H_2/H_∞ filter can be obtained by minimizing $\text{Trace}(H)$ subject to (7.52), (7.53), (7.67) and

$$
H - \tilde{F}_3^T P\tilde{F}_3 > 0.
\tag{7.68}
$$

(7.68) is equivalent to

$$
\begin{bmatrix}
H & \tilde{F}_3^T P \\
P\tilde{F}_3 & P
\end{bmatrix} > 0.
\tag{7.69}
$$

By setting $P_{22}B_f = Z_1$ and $P_{22}A_f = Z$, (7.67) and (7.69) lead respectively to the LMIs

$$
\begin{bmatrix}
\Delta_{11} & \Delta_{12} & \sqrt{2}C^T P_{11} & \sqrt{2}C^T P_{22} & -C_1^T Z_1^T & G_1^T P_{11} & G_1^T P_{22} & G_2^T P_{11} & G_2^T P_{22} \\
* & \Delta_{22} & 0 & 0 & 0 & 0 & 0 & 0 & 0 \\
* & 0 & -P_{11} & 0 & 0 & 0 & 0 & 0 & 0 \\
* & 0 & 0 & -P_{22} & 0 & 0 & 0 & 0 & 0 \\
* & 0 & 0 & 0 & -P_{22} & 0 & 0 & 0 & 0 \\
* & 0 & 0 & 0 & 0 & -P_{11} & 0 & 0 & 0 \\
* & 0 & 0 & 0 & 0 & 0 & -P_{22} & 0 & 0 \\
* & 0 & 0 & 0 & 0 & 0 & 0 & -P_{11} & 0 \\
* & 0 & 0 & 0 & 0 & 0 & 0 & 0 & -P_{22}
\end{bmatrix} < 0
\tag{7.70}
$$

and

$$
\begin{bmatrix}
H & B_0^T P_{11} & B_0^T P_{22} - B_1^T Z_1^T \\
P_{11}B_0 & P_{11} & 0 \\
P_{22}B_0 - Z_1 B_1 & 0 & P_{22}
\end{bmatrix} > 0,
\tag{7.71}
$$

where $\Delta_{11} = P_{11}A + A^T P_{11} + P_{11}$, $\Delta_{12} = A^T P_{22} - A_1^T Z_1^T - Z^T$, $\Delta_{22} = -Z - Z^T + D^T D + P_{22}$, and $*$ is derived by matrix symmetry. Therefore, we have the following theorem.

THEOREM 7.5
Under the condition of (7.60), if there exists a solution $(P_{11} > 0, P_{22} > 0, Z, Z_1, \alpha > 0)$ to the following convex optimization problem:

$$\min_{P_{11},P_{22},Z,Z_1,\alpha>0,(7.52),(7.53),\ (7.70)\ and\ (7.71)} \text{Trace}(H),$$

then the corresponding H_2/H_∞ filter is given by (7.54).

REMARK 7.6 In the proofs of Theorems 7.4–7.5, the matrix P is chosen as $diag(P_{11}, P_{22})$ for simplicity, which, however, leads to a conservative result. In order to reduce the conservatism, it is better to choose P of the general form of $\begin{bmatrix} P_{11} & P_{12} \\ P_{12}^T & P_{22} \end{bmatrix}$. However, this will increase the complexity of computation.
∎

7.5 Numerical Example

Example 7.2
Consider a nonlinear stochastic system governed by the following Itô differential equation

$$dx = [Ax + B_0 v + F_0(x)]\, dt + [Cx + F_1(x)]\, d\tilde{B}_0,$$
$$dy = (A_1 + B_1 v)\, dt + C_1 x d\tilde{B}_1, \quad z = Dx, \tag{7.72}$$

where

$$A = \begin{bmatrix} -3 & \frac{1}{2} \\ -1 & -3 \end{bmatrix}, \ B_0 = \begin{bmatrix} 1 \\ 0 \end{bmatrix}, \ C = \begin{bmatrix} 1 & 0 \\ 0 & 0 \end{bmatrix}, \ F_0(x) = 0.3\tanh(x),$$

$$F_1(x) = 0.3\sin x, \quad A_1 = \begin{bmatrix} -1 & 1 \\ 1 & -1 \end{bmatrix},$$

$$B_1 = \begin{bmatrix} 0 \\ 1 \end{bmatrix}, \ C_1 = \begin{bmatrix} 1 & 0 \\ 0 & 1 \end{bmatrix}, \ D = \begin{bmatrix} 0 \\ 1 \end{bmatrix}, \ v(t) = \frac{1}{1+2t}, \ t \ge 0.$$

We adopt the following filter for the estimation of $z(t)$:

$$d\hat{x} = A_f \hat{x}\, dt + B_f dy, \quad \hat{z} = D\hat{x}. \tag{7.73}$$

By setting $\gamma = 0.9$, and using the MATLAB LMI control toolbox, an H_∞ filter is derived from Theorem 7.4 as

$$A_f = \begin{bmatrix} 5.6231 & 3.7259 \\ -0.1617 & 8.2289 \end{bmatrix}, \quad B_f = \begin{bmatrix} 0.1812 & -1.8190 \\ -0.2525 & 0.4635 \end{bmatrix}.$$

From Theorem 7.5, an H_2/H_∞ filter is given by

$$A_f = \begin{bmatrix} 4.1449 & 3.4665 \\ -0.2469 & 6.3382 \end{bmatrix}, \quad B_f = \begin{bmatrix} 0.5270 & -1.2388 \\ -0.3693 & 0.3445 \end{bmatrix}.$$

The initial condition in the simulation is assumed to be

$$\eta_0 = \begin{bmatrix} 0.3 & 0.2 & -0.02 & -0.05 \end{bmatrix}^T.$$

Figures 7.1 and 7.2 show the trajectories of $x_1(t)$ and $\hat{x}_1(t)$, and the trajectories of $x_2(t)$ and $\hat{x}_2(t)$ based on the derived H_∞ and H_2/H_∞ filters, respectively. The estimation errors $\tilde{z}(t)$ for the H_∞ and H_2/H_∞ filters are shown in Figures 7.3 and 7.4, respectively. From Figures 7.3 and 7.4, it is obvious that the performance of the proposed mixed H_2/H_∞ filter is better than that of the H_∞ filter.

□

7.6 Nonlinear H_∞ Filtering: Time-Delay Case

Consider the following nonlinear stochastic time-delay system

$$\begin{cases} dx(t) = [f(x(t), x(t-\tau), t) + k(x(t), x(t-\tau), t)v(t)]\, dt \\ \qquad\quad + [h(x(t), x(t-\tau), t) + l(x(t), x(t-\tau), t)v(t)]\, dB(t), \\ y(t) = n(x(t), x(t-\tau), t) + r(x(t), x(t-\tau), t)v(t), \\ z(t) = m(x(t), x(t-\tau), t), \\ x(t) = \phi(t) \in C^b_{\mathcal{F}_0}([-\tau, 0]; \mathcal{R}^n),\ t \in [0, \infty) \end{cases} \qquad (7.74)$$

where τ is a known constant delay, and $C^b_{\mathcal{F}_0}([-\tau, 0]; \mathcal{R}^n)$ denotes the set of all $\mathcal{F}_0$-measurable bounded $C([-\tau, 0], \mathcal{R}^n)$-valued random variables $\eta(s)$ with $s \in [-\tau, 0]$. We assume $f(0,0,t) = h(0,0,t)$, $n(0,0,t) \equiv 0$ and $m(0,0,t) \equiv 0$. So $x \equiv 0$ is an equilibrium point of (7.74). System (7.74) is a more general state-delayed model, which includes many previously studied models as special cases.

7.6.1 Definitions and lemmas

Since we will deal with the infinite horizon stochastic H_∞ filtering problem, it is necessary to investigate stochastic stability. Hence, we first present the following definition.

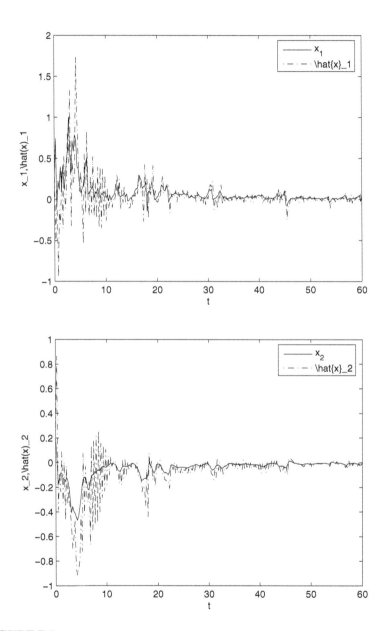

FIGURE 7.1

Trajectories of $x_1(t)$, $\hat{x}_1(t)$, $x_2(t)$, and $\hat{x}_2(t)$ under the proposed H_∞ filter.

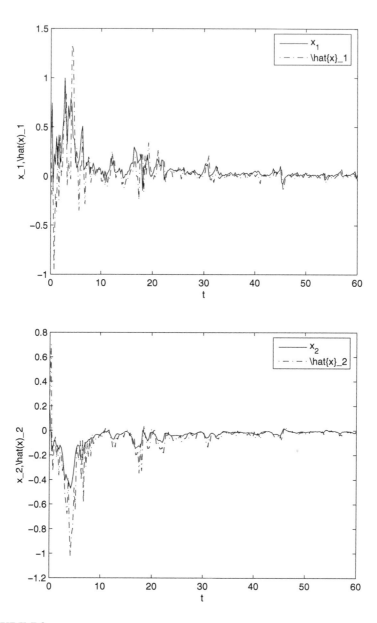

FIGURE 7.2
Trajectories of $x_1(t)$, $\hat{x}_1(t)$, $x_2(t)$, and $\hat{x}_2(t)$ under the proposed H_2/H_∞ filter.

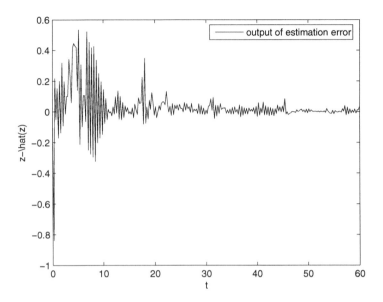

FIGURE 7.3
Trajectory of the estimation error $\tilde{z}(t)$ under the proposed H_∞ filter.

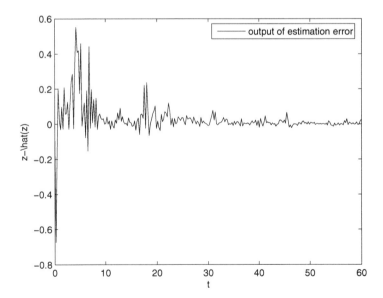

FIGURE 7.4
Trajectory of the estimation error $\tilde{z}(t)$ under the proposed H_2/H_∞ filter.

DEFINITION 7.2   *The nonlinear stochastic time-delay system*

$$\begin{cases} dx(t) = f(x(t), x(t-\tau), t)\, dt + h(x(t), x(t-\tau), t)\, dB(t), \\ x(t) = \phi(t) \in C^b_{\mathcal{F}_0}([-\tau, 0]; \mathcal{R}^n) \end{cases} \tag{7.75}$$

is said to be exponentially mean square stable if there are positive constants ϱ and α such that

$$\mathcal{E}\|x(t)\|^2 \le \varrho\|\phi\|^2 e^{-\alpha t},$$

where $\|\phi\|^2 = \mathcal{E} \max_{-\tau \le t \le 0} \|\phi(t)\|^2$.

Associated with (7.74) and $V : \mathcal{R}^n \times \mathcal{R}^+ \mapsto \mathcal{R}^+$, the infinitesimal generator $\mathcal{L}_1 V : \mathcal{R}^n \times \mathcal{R}^n \times \mathcal{R}^+ \mapsto \mathcal{R}$ is given by

$$\mathcal{L}_1 V(x, y, t) = V_t(x, t) + V_x^T(x, t)[f(x, y, t) + k(x, y, t)v(t)]$$
$$+ \frac{1}{2}[h(x, y, t) + l(x, y, t)v(t)]^T V_{xx}(x, t)[h(x, y, t) + l(x, y, t)v(t)].$$

The following lemma is a generalized version of Lemma 7.1, which may be viewed as a nonlinear SBRL for time-delay systems. Because the proof is very similar to that of Lemma 7.1, it is omitted.

LEMMA 7.6
Consider the following system

$$\begin{cases} dx(t) = [f(x(t), x(t-\tau), t) + k(x(t), x(t-\tau), t)v(t)]\, dt \\ \qquad\quad + [h(x(t), x(t-\tau), t) + l(x(t), x(t-\tau), t)v(t)]\, dB(t), \\ z(t) = m(x(t), x(t-\tau), t), \quad x(t) = \phi(t) \in C^b_{\mathcal{F}_0}([-\tau, 0]; \mathcal{R}^n). \end{cases} \tag{7.76}$$

If there exists a positive definite Lyapunov function $V(x, t) \in C^{2,1}(\mathcal{R}^n \times \mathcal{R}^+; \mathcal{R}^+)$ solving the following HJI

$$\begin{cases} \Gamma(x, y, t) := V_t(x, t) + V_x^T(x, t)f(x, y, t) \\ \qquad\quad + \frac{1}{2}\Upsilon^T[\gamma^2 I - l^T(x, y, t)V_{xx}(x, t)l(x, y, t)]^{-1}\Upsilon \\ \qquad\quad + \frac{1}{2}\|z(t)\|^2 + \frac{1}{2}h^T(x, y, t)V_{xx}(x, t)h(x, y, t) < 0 \\ \gamma^2 I - l^T(x, y, t)V_{xx}(x, t)l(x, y, t) > 0, \forall(x, y, t) \in \mathcal{R}^n \times \mathcal{R}^n \times \mathcal{R}^+ \\ V(0, 0) = 0 \end{cases} \tag{7.77}$$

for some $\gamma > 0$, then the inequality

$$\|z(t)\|^2_{[0,\infty)} \le \gamma^2 \|v(t)\|^2_{[0,\infty)}, \forall v \in \mathcal{L}^2_{\mathcal{F}}(\mathcal{R}^+, \mathcal{R}^{n_v}), v \ne 0 \tag{7.78}$$

holds with initial state $x(s) = 0, a.s., \forall s \in [-\tau, 0]$, where in (7.77),

$$\Upsilon = k^T(x, y, t)V_x(x, t) + l^T(x, y, t)V_{xx}(x, t)h(x, y, t).$$

LEMMA 7.7
Consider the unforced system

$$\begin{cases} dx(t) = f(x(t), x(t-\tau), t)\, dt + h(x(t), x(t-\tau), t)\, dB(t), \\ x(t) = \phi(t) \in C^b_{\mathcal{F}_0}([-\tau, 0]; \mathcal{R}^n). \end{cases} \tag{7.79}$$

If there exists a positive definite Lyapunov function $V(x,t) \in \mathcal{C}^{2,1}(\mathcal{R}^n \times [-\tau, \infty); \mathcal{R}^+)$, $c_1, c_2, c_3, c_4 > 0$ with $c_3 > c_4$ satisfying the following conditions:

(i) $c_1\|x\|^2 \le V(x,t) \le c_2\|x\|^2$, $\forall(x,t) \in \mathcal{R}^n \times [-\tau, \infty)$;

(ii) $\mathcal{L}_1 V(x,y,t)|_{v=0} \le -c_3\|x\|^2 + c_4\|y\|^2$, $\forall t > 0$;

then

$$\mathcal{E}\|x(t)\|^2 \le \frac{c_2 + c_4\tau}{c_1}\|\phi\|^2 e^{-\frac{c_3-c_4}{c_2}t},$$

i.e., (7.79) is exponentially mean square stable.

Proof. Applying Itô's formula to $V(x(t),t)e^{\frac{c_3-c_4}{c_2}t}$ and then taking an expectation computation, we have

$$\mathcal{E}V(x(t),t)e^{\frac{c_3-c_4}{c_2}t} - \mathcal{E}V(x(0),0)$$
$$= \int_0^t \left[\frac{c_3-c_4}{c_2}\mathcal{E}V(x(r),r) + \mathcal{E}\mathcal{L}_1V(x(r),x(r-\tau),r)|_{v=0}\right]e^{\frac{c_3-c_4}{c_2}r}\,dr. \quad (7.80)$$

By conditions (i) and (ii), (7.80) yields

$$\mathcal{E}V(x(t),t)e^{\frac{c_3-c_4}{c_2}t} - \mathcal{E}V(x(0),0) \le \int_0^t [(c_3-c_4)\mathcal{E}\|x(r)\|^2$$
$$+ (-c_3\mathcal{E}\|x(r)\|^2 + c_4\mathcal{E}\|x(t-\tau)\|^2)]e^{\frac{c_3-c_4}{c_2}r}\,dr$$
$$= -c_4\int_0^{t-\tau}\mathcal{E}\|x(r)\|^2 e^{\frac{c_3-c_4}{c_2}r}\,dr - c_4\int_{t-\tau}^t \mathcal{E}\|x(r)\|^2 e^{\frac{c_3-c_4}{c_2}r}\,dr$$
$$+ c_4\int_{-\tau}^0 \mathcal{E}\|x(r)\|^2 e^{\frac{c_3-c_4}{c_2}r}\,dr + c_4\int_0^{t-\tau}\mathcal{E}\|x(r)\|^2 e^{\frac{c_3-c_4}{c_2}r}\,dr$$
$$= -c_4\int_{t-\tau}^t \mathcal{E}\|x(r)\|^2 e^{\frac{c_3-c_4}{c_2}r}\,dr + c_4\int_{-\tau}^0 \mathcal{E}\|x(r)\|^2 e^{\frac{c_3-c_4}{c_2}r}\,dr$$
$$\le c_4\int_{-\tau}^0 \mathcal{E}\|x(r)\|^2 e^{\frac{c_3-c_4}{c_2}r}\,dr \le c_4\|\phi\|^2\tau. \quad (7.81)$$

Because

$$V(x,t) \ge c_1\|x\|^2,$$

we have

$$c_1\mathcal{E}\|x(t)\|^2 \le (c_2\|x(0)\|^2 + c_4\|\phi\|^2\tau)e^{-\frac{c_3-c_4}{c_2}t}$$
$$\le (c_2 + c_4\tau)\|\phi\|^2 e^{-\frac{c_3-c_4}{c_2}t}.$$

So

$$\mathcal{E}\|x(t)\|^2 \le \frac{c_2 + c_4\tau}{c_1}\|\phi\|^2 e^{-\frac{c_3-c_4}{c_2}t}.$$

This lemma is proved. □

In what follows, we construct the following general filtering equation for the estimation of $z(t)$:

$$\begin{cases} d\hat{x}(t) = \hat{f}(\hat{x}(t), \hat{x}(t-\tau), t)\, dt + \hat{G}(\hat{x}(t), \hat{x}(t-\tau), t) y(t)\, dt, \\ \hat{z}(t) = \hat{m}(\hat{x}(t), \hat{x}(t-\tau), t), \; \hat{x}(0) = 0 \end{cases} \tag{7.82}$$

where $\hat{f}$, $\hat{G}$, and $\hat{m}$ are matrices of appropriate dimensions to be determined. (7.82) is more general, and includes the following Luenberger-type filter as a special form.

$$\begin{cases} d\hat{x}(t) = f(\hat{x}(t), \hat{x}(t-\tau), t)\, dt + G(\hat{x}(t), \hat{x}(t-\tau), t)[y(t) - n(\hat{x}(t), \hat{x}(t-\tau), t)]\, dt, \\ \hat{z}(t) = m(\hat{x}(t), \hat{x}(t-\tau), t), \hat{x}(0) = 0. \end{cases}$$

Set $\eta(t) = [x^T(t) \; \hat{x}^T(t)]^T$ and let

$$\tilde{z}(t) = z(t) - \hat{z}(t) = m(x(t), x(t-\tau), t) - \hat{m}(\hat{x}(t), \hat{x}(t-\tau), t)$$

denote the estimation error. Then we get the following augmented system

$$\begin{cases} d\eta(t) = [f_e(\eta(t)) + k_e(\eta(t))v(t)]\, dt + [h_e(\eta(t)) + l_e(\eta(t))v(t)]\, dB(t), \\ \tilde{z}(t) = z(t) - \hat{z}(t) = m(x(t), x(t-\tau), t) - \hat{m}(\hat{x}(t), \hat{x}(t-\tau), t), \\ \eta(t) = \begin{bmatrix} \phi(t) \\ 0 \end{bmatrix}, \phi(t) \in C_{\mathcal{F}_0}^b([-\tau, 0]; \mathcal{R}^n), \forall t \in [-\tau, 0], \end{cases} \tag{7.83}$$

where

$$f_e(\eta(t)) = \begin{bmatrix} f(x(t), x(t-\tau), t) \\ \hat{f}(\hat{x}(t), \hat{x}(t-\tau), t) + \hat{G}(\hat{x}(t), \hat{x}(t-\tau), t)n(x(t), x(t-\tau), t) \end{bmatrix},$$

$$k_e(\eta(t)) = \begin{bmatrix} k(x(t), x(t-\tau), t) \\ \hat{G}(\hat{x}(t), \hat{x}(t-\tau), t)r(x(t), x(t-\tau), t) \end{bmatrix},$$

$$h_e(\eta(t)) = \begin{bmatrix} h(x(t), x(t-\tau), t) \\ 0 \end{bmatrix}, \; l_e(\eta(t)) = \begin{bmatrix} l(x(t), x(t-\tau), t) \\ 0 \end{bmatrix}.$$

According to different notions of internal stability, we define their corresponding H_∞ filters as follows.

DEFINITION 7.3 (Exponential mean square H_∞ filtering) *Find the matrices $\hat{f}$, $\hat{G}$ and $\hat{m}$ in (7.82), such that the following hold:*

(i) *The equilibrium point $\eta \equiv 0$ of the augmented system (7.83) is exponentially mean square stable in the case when $v = 0$.*

(ii) *For a given disturbance attenuation level $\gamma > 0$, the following H_∞ performance holds for $x(t) \equiv 0$ on $t \in [-\tau, 0]$:*

$$\|\tilde{z}\|_{[0,\infty)}^2 \le \gamma^2 \|v\|_{[0,\infty)}^2, \forall v \in \mathcal{L}_{\mathcal{F}}^2(\mathcal{R}^+, \mathcal{R}^{n_v}), v \ne 0. \tag{7.84}$$

DEFINITION 7.4 (Asymptotic mean square H_∞ filtering) *If the equilibrium point $\eta \equiv 0$ of the augmented system (7.83) is ASMS and (7.84) holds, then (7.82) is called an asymptotic mean square H_∞ filter.*

7.6.2 Main results

THEOREM 7.6 (Exponential mean square H_∞ filter)
Suppose there exist a function $V(\eta, t) = V(x, \hat{x}, t) \in \mathcal{C}^{2,1}(\mathcal{R}^{2n} \times [-\tau, \infty); \mathcal{R}^{+})$, and positive constants $c_1, c_2, c_3, c_4 > 0$ with $c_3 > c_4$ such that

$$c_1(\|x\|^2 + \|\hat{x}\|^2) \leq V(\eta, t) \leq c_2(\|x\|^2 + \|\hat{x}\|^2), \forall(\eta, t) \in \mathcal{R}^{2n} \times [-\tau, \infty), \quad (7.85)$$

$$-\frac{1}{2}\|m(x, y, t) - \hat{m}(\hat{x}, \hat{y}, t)\|^2 \leq -c_3(\|x\|^2 + \|\hat{x}\|^2) + c_4(\|y\|^2 + \|\hat{y}\|^2), \forall t > 0. \quad (7.86)$$

For a given disturbance attenuation level $\gamma > 0$, if $V(\eta, t)$ solves the following HJI

$$\begin{cases} \Gamma(x, y, \hat{x}, \hat{y}) := V_t + V_x^T f(x, y, t) + V_{\hat{x}}^T[\hat{f}(\hat{x}, \hat{y}, t) + \hat{G}(\hat{x}, \hat{y}, t)n(x, y, t)] \\ \quad + \frac{1}{2}\Theta^T(x, \hat{x}, y, \hat{y}, t)[\gamma^2 I - l^T(x, y, t)V_{xx}l(x, y, t)]^{-1}\Theta(x, \hat{x}, y, \hat{y}, t) \\ \quad + \frac{1}{2}\|m(x, y, t) - \hat{m}(\hat{x}, \hat{y}, t)\|^2 + \frac{1}{2}h^T(x, y, t)V_{xx}h(x, y, t) < 0, \\ \gamma^2 I - l^T(x, y, t)V_{xx}l(x, y, t) > 0, \forall(x, y, \hat{x}, \hat{y}, t) \in \mathcal{R}^n \times \mathcal{R}^n \times \mathcal{R}^n \times \mathcal{R}^n \times \mathcal{R}^+, \\ V(0, 0) = 0 \end{cases}$$

$$\quad (7.87)$$

for some matrices $\hat{f}, \hat{G}$ and $\hat{m}$ of suitable dimensions, then an exponential mean square H_∞ filter is obtained by (7.82), where

$$\Theta^T(x, \hat{x}, y, \hat{y}, t) = V_x^T k(x, y, t) + V_{\hat{x}}^T \hat{G}(\hat{x}, \hat{y}, t)r(x, y, t) + h^T(x, y, t)V_{xx}l(x, y, t).$$

Proof. In Lemma 7.6, we substitute $V(x, \hat{x}, t)$, $\tilde{z} = m(x, y, t) - \hat{m}(\hat{x}, \hat{y}, t)$,

$$f_e = \begin{bmatrix} f(x, y, t) \\ \hat{f}(\hat{x}, \hat{y}, t) + \hat{G}(\hat{x}, \hat{y}, t)n(x, y, t) \end{bmatrix}, \quad k_e = \begin{bmatrix} g(x, y, t) \\ \hat{G}(\hat{x}, \hat{y}, t)r(x, y, t) \end{bmatrix}$$

$$h_e = \begin{bmatrix} h(x, y, t) \\ 0 \end{bmatrix}, \quad l_e = \begin{bmatrix} l(x, y, t) \\ 0 \end{bmatrix}$$

for $V(x, t)$, z, f, k, h and l, respectively. Then, by some simple computations, (7.84) is obtained.

Next, we show that the augmented system (7.83) is exponentially mean square stable for $v \equiv 0$. Set

$$\mathcal{L}_\eta^{v=0}V(x, \hat{x}, t) := V_t + V_\eta^T f_e + \frac{1}{2}h_e^T V_{\eta\eta}h_e.$$

By (7.87),

$$\mathcal{L}_\eta^{v=0}V(x, \hat{x}, t) < -\frac{1}{2}\|m(x, y, t) - \hat{m}(\hat{x}, \hat{y}, t)\|^2$$

$$-\frac{1}{2}\Theta^T(x, \hat{x}, y, \hat{y}, t)[\gamma^2 I - l^T(x, y, t)V_{xx}l(x, y, t)]^{-1}\Theta(x, \hat{x}, y, \hat{y}, t)$$

$$\leq -\frac{1}{2}\|m(x, y, t) - \hat{m}(\hat{x}, \hat{y}, t)\|^2$$

$$\leq -c_3(\|x\|^2 + \|\hat{x}\|^2) + c_4(\|y\|^2 + \|\hat{y}\|^2). \quad (7.88)$$

Next, in light of Lemma 7.7, we know that (7.83) is internally stable in an exponential mean square sense. The proof is completed. □

The inequality (7.87) is a constrained HJI, which is not easily tested in practice. However, if in (7.74), $l \equiv 0$, i.e., only the state depends on noise, then the constraint condition $\gamma^2 I - l^T(x, y, t)V_{xx}l(x, y, t) > 0$ holds automatically.

THEOREM 7.7 (Asymptotic mean square H_∞ filter)

Assume $V(\eta, t) \in C^{2,1}(\mathcal{R}^{2n} \times \mathcal{R}^+; \mathcal{R}^+)$ has an infinitesimal upper limit, i.e.,

$$\lim_{\|\eta\| \to \infty} \inf_{t > 0} V(\eta, t) = \infty. \tag{7.89}$$

Additionally, we assume that $V(\eta, t) > c\|\eta\|^2$ for some $c > 0$. If $V(\eta, t)$ solves HJI (7.87), then (7.82) is an asymptotic mean square H_∞ filter.

Proof. Obviously, we only need to show that (7.83) is ASMS when $v = 0$. From (7.88), $\mathcal{L}_\eta^{v=0}V(x, \hat{x}, t) < 0$. Hence, (7.83) is globally asymptotically stable in probability one according to Lemma 7.2.

By Itô's formula and the property of stochastic integration, we have for any $0 \leq s \leq t$,

$$\mathcal{E}V(\eta(t), t) = \mathcal{E}V(\eta(s), s) + \int_s^t \mathcal{E}\mathcal{L}_\eta^{v=0}V(\eta(s), s)\, ds$$

$$+ \int_s^t \mathcal{E}h_e^T(\eta(s), s)V_\eta(\eta(s), s)\, dB(s)$$

$$= \mathcal{E}V(\eta(s), s) + \int_s^t \mathcal{E}\mathcal{L}_\eta^{v=0}V(\eta(s), s)\, ds$$

$$\leq \mathcal{E}V(\eta(s), s)$$

$$- \frac{1}{2}\int_s^t \mathcal{E}\|m(x(s), x(s-\tau), s) - \hat{m}(\hat{x}(s), \hat{x}(s-\tau), s)\|^2\, ds$$

$$\leq \mathcal{E}V(\eta(s), s) < \infty. \tag{7.90}$$

By setting $\tilde{\mathcal{F}}_t = \mathcal{F}_t \cup \sigma(y(s), 0 \leq s \leq t)$, it follows from (7.90) that

$$\mathcal{E}[V(\eta(t), t)|\tilde{\mathcal{F}}_s] \leq V(\eta(s), s) \text{ a.s.,}$$

which implies that $\{V(\eta(t), t), \tilde{\mathcal{F}}_t, 0 \leq s \leq t\}$ is a nonnegative supermartingale with respect to $\{\tilde{\mathcal{F}}_t\}_{t \geq 0}$. By Doob's convergence theorem and the fact that $\lim_{t \to \infty} \eta(t) = 0$ a.s., it immediately yields that $V(\eta(\infty), \infty) = \lim_{t \to \infty} V(\eta(t), t) = 0$ a.s. Moreover, $\lim_{t \to \infty} \mathcal{E}V(\eta(t), t) = \mathcal{E}V(\eta(\infty), \infty) = \mathcal{E}V(0, \infty) = 0$. Because $V(\eta, t) \geq c\|\eta\|^2$ for some $c > 0$, it follows that $\lim_{t \to \infty} \mathcal{E}\|\eta(t)\|^2 = 0$ and the theorem is proved. □

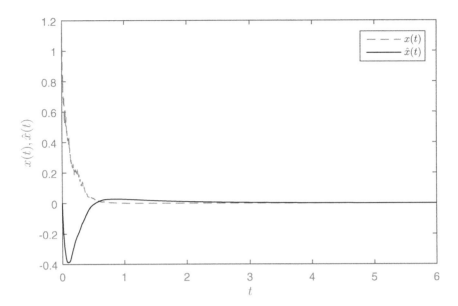

FIGURE 7.5

Trajectories of $x(t)$ and $\hat{x}(t)$.

Example 7.3

Consider the following one-dimensional stochastic time-delay system

$$
\begin{cases}
dx(t) = [(-10x(t) - x(t)x^2(t - \tau)) + x(t - \tau)v(t)]\, dt + x(t)\, dB(t), \\
x(t) = \phi(t) \in C^b_{\mathcal{F}_0}([-\tau, 0]; \mathcal{R}), \\
y(t) = -\frac{25}{2}x(t) - 2x(t)x(t - \tau) + v(t), \\
z(t) = 5x(t).
\end{cases}
\tag{7.91}
$$

Given the disturbance attenuation level $\gamma = 1$, by Theorem 7.6, in order to determine the filtering parameters $\hat{f}, \hat{G}$ and $\hat{m}$, we must solve the HJI (7.87). By setting $V(x, \hat{x}) = x^2 + \hat{x}^2$ and $\hat{m} = -5\hat{x}$, then (7.85) and (7.86) hold obviously. In addition, we can easily test that $\Gamma(x, y, \hat{x}, \hat{y}) = -6.5x^2 - 13.5\hat{x}^2 < 0$ when we take $\hat{f} = -14\hat{x}$, $\hat{G} = 1$, $\hat{m} = -5\hat{x}$. So an exponential mean square H_∞ filter of (7.91) is given by

$$
d\hat{x}(t) = -14\hat{x}(t)\, dt + y(t)\, dt, \quad \hat{z}(t) = -5\hat{x}(t).
$$

Because there may be more than one triple $(\hat{f}, \hat{G}, \hat{m})$ solving HJI (7.87), H_∞ filtering is in general not unique. By setting the external disturbance $v(t) = e^{-t}$, $\tau = 0.2$, and the initial value $\phi(t) = 1.2$ for $t \in [-0.2, 0]$, the simulation result can be seen in Figure 7.5 and Figure 7.6.

□

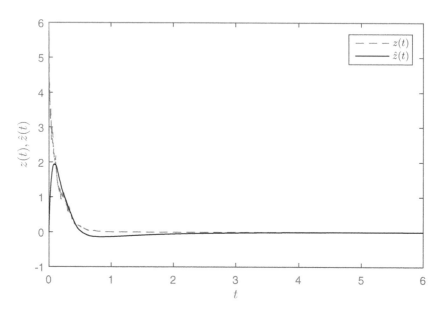

FIGURE 7.6

Trajectories of $z(t)$ and $\hat{z}(t)$.

7.7 Luenberger-Type Linear Time-Delay H_∞ Filtering

We now apply Theorem 7.7 to the design of a linear stochastic time-delay H_∞ filter.
Consider the following linear time-invariant stochastic time-delay system

$$
\begin{cases}
dx(t) = [A_0x(t) + A_1x(t-\tau) + B_1v(t)]\,dt + [C_0x(t) + C_1x(t-\tau) + Dv(t)]\,dB(t), \\
y(t) = l_0x(t) + l_1x(t-\tau) + Kv(t), \\
z(t) = m_0x(t) + m_1x(t-\tau), \\
x(t) = \phi(t) \in C^b_{\mathcal{F}_0}([-\tau,0];\mathcal{R}^n),
\end{cases}
$$

(7.92)

where all coefficient matrices are assumed to be constant. For simplicity of design,
we consider the following Luenberger-type filtering equation

$$
\begin{cases}
d\hat{x}(t) = [A_0\hat{x}(t) + A_1\hat{x}(t-\tau)]\,dt + G[y(t) - l_0\hat{x}(t) - l_1\hat{x}(t-\tau)]\,dt, \\
\hat{z}(t) = m_0\hat{x}(t) + m_1\hat{x}(t-\tau), \quad \hat{x}(0) = 0,
\end{cases}
$$

(7.93)

where only the constant matrix G is to be determined. In this case,

$$
\hat{f}(\hat{x}(t), \hat{x}(t-\tau), t) = A_0\hat{x}(t) + A_1\hat{x}(t-\tau) - G[l_0\hat{x}(t) + l_1\hat{x}(t-\tau)], \quad \hat{G} = G.
$$

Set

$$
V(x, \hat{x}, t) = x^T(t)Px(t) + \int_{t-\tau}^t x^T(\theta)P_1x(\theta)\,d\theta + \hat{x}^T(t)Q\hat{x}(t) + \int_{t-\tau}^t \hat{x}^T(\theta)Q_1\hat{x}(\theta)\,d\theta,
$$

where $P > 0, P_1 > 0, Q > 0, Q_1 > 0$ are to be determined later. Then by some computations, we have from HJI (7.87) that

$$V_t = x^T(t)P_1 x(t) - x^T(t-\tau)P_1 x(t-\tau) + \hat{x}^T(t)Q_1\hat{x}(t) - \hat{x}^T(t-\tau)Q_1\hat{x}(t-\tau),$$

$$V_x^T f(x,y,t) = \begin{bmatrix} x \\ y \\ \hat{x} \\ \hat{y} \end{bmatrix}^T \begin{bmatrix} PA_0 + A_0^T P & \star & 0 & 0 \\ A_1^T P & 0 & 0 & 0 \\ 0 & 0 & 0 & 0 \\ 0 & 0 & 0 & 0 \end{bmatrix} \begin{bmatrix} x \\ y \\ \hat{x} \\ \hat{y} \end{bmatrix},$$

$$V_{\hat{x}}^T Gn(x,y,t) = \begin{bmatrix} x \\ y \\ \hat{x} \\ \hat{y} \end{bmatrix}^T \begin{bmatrix} 0 & 0 & \star & 0 \\ 0 & 0 & \star & 0 \\ QGl_0 & QGl_1 & 0 & 0 \\ 0 & 0 & 0 & 0 \end{bmatrix} \begin{bmatrix} x \\ y \\ \hat{x} \\ \hat{y} \end{bmatrix},$$

$$V_{\hat{x}}^T \hat{f}(\hat{x},\hat{y},t) = \begin{bmatrix} x \\ y \\ \hat{x} \\ \hat{y} \end{bmatrix}^T \begin{bmatrix} 0 & 0 & 0 & 0 \\ 0 & 0 & 0 & 0 \\ 0 & 0 & Q(A_0 - Gl_0) + (A_0^T - l_0^T G^T)Q & \star \\ 0 & 0 & (A_1 - Gl_1)^T Q & 0 \end{bmatrix} \begin{bmatrix} x \\ y \\ \hat{x} \\ \hat{y} \end{bmatrix},$$

$$\frac{1}{2}h^T(x,y,t)V_{xx}^T h(x,y,t) = \begin{bmatrix} x \\ y \\ \hat{x} \\ \hat{y} \end{bmatrix}^T \begin{bmatrix} C_0^T P C_0 & \star & 0 & 0 \\ C_1^T P C_0 & C_1^T P C_1 & 0 & 0 \\ 0 & 0 & 0 & 0 \\ 0 & 0 & 0 & 0 \end{bmatrix} \begin{bmatrix} x \\ y \\ \hat{x} \\ \hat{y} \end{bmatrix},$$

$$\frac{1}{2}\|m(x,y,t) - m(\hat{x},\hat{y},t)\|^2 = \frac{1}{2}\begin{bmatrix} x \\ y \\ \hat{x} \\ \hat{y} \end{bmatrix}^T \begin{bmatrix} m_0^T m_0 & \star & \star & \star \\ m_1^T m_0 & m_1^T m_1 & \star & \star \\ -m_0^T m_0 & -m_0^T m_1 & m_0^T m_0 & \star \\ -m_1^T m_0 & -m_1^T m_1 & m_1^T m_0 & m_1^T m_1 \end{bmatrix} \begin{bmatrix} x \\ y \\ \hat{x} \\ \hat{y} \end{bmatrix},$$

$$\frac{1}{2}\Theta^T(x,\hat{x},y,\hat{y},t)(\gamma^2 I - l^T(x,y,t)V_{xx}l(x,y,t))^{-1}\Theta(x,\hat{x},y,\hat{y},t)$$

$$= \frac{1}{2}\begin{bmatrix} x \\ y \\ \hat{x} \\ \hat{y} \end{bmatrix}^T \begin{bmatrix} C_0^T PD + 2PB_1 \\ C_1^T PD \\ 2QGK \\ 0 \end{bmatrix} \left(\gamma^2 I - 2D^T PD\right)^{-1} \begin{bmatrix} C_0^T PD + 2PB_1 \\ C_1^T PD \\ 2QGK \\ 0 \end{bmatrix}^T \begin{bmatrix} x \\ y \\ \hat{x} \\ \hat{y} \end{bmatrix},$$

where $\star$ is derived by symmetry. Hence, HJI (7.87) is equivalent to

$$\begin{bmatrix} A_{11} & \star & \star & \star \\ A_{21} & A_{22} & \star & \star \\ QGl_0 - \frac{1}{2}m_0^T m_0 & QGl_1 - \frac{1}{2}m_0^T m_1 & A_{33} & \star \\ -\frac{1}{2}m_1^T m_0 & -\frac{1}{2}m_1^T m_1 & (A_1 - Gl_1)^T Q + \frac{1}{2}m_1^T m_0 & \frac{1}{2}m_1^T m_1 - Q_1 \end{bmatrix}$$

$$+ \frac{1}{2}\begin{bmatrix} C_0^T PD + 2PB_1 \\ C_1^T PD \\ 2QGK \\ 0 \end{bmatrix} \left(\gamma^2 I - 2D^T PD\right)^{-1} \begin{bmatrix} C_0^T PD + 2PB_1 \\ C_1^T PD \\ 2QGK \\ 0 \end{bmatrix}^T < 0 \qquad (7.94)$$

and

$$\gamma^2 I - 2D^T PD > 0 \qquad (7.95)$$

with

$$A_{11} = PA_0 + A_0^T P + C_0^T PC_0 + P_1 + \frac{1}{2}m_0^T m_0,$$

$$A_{21} = A_1^T P + C_1^T PC_0 + \frac{1}{2}m_1^T m_0, \quad A_{22} = -P_1 + C_1^T PC_1 + \frac{1}{2}m_1^T m_1,$$

$$A_{33} = Q(A_0 - Gl_0) + (A_0 - Gl_0)^T Q + Q_1 + \frac{1}{2}m_0^T m_0.$$

By Schur's complement, (7.94) and (7.95) are equivalent to

$$\begin{bmatrix} A_{11} & \star & & \star & & \star & \star \\ A_{21} & A_{22} & & \star & & \star & \star \\ G_1 l_0 - \frac{1}{2}m_0^T m_0 & G_1 l_1 - \frac{1}{2}m_0^T m_1 & & A_{33} & & \star & \star \\ -\frac{1}{2}m_1^T m_0 & -\frac{1}{2}m_1^T m_1 & A_1 Q - l_1^T G_1^T + \frac{1}{2}m_1^T m_0 & A_{44} & 0 \\ 2B_1^T P + D^T PC_0 & D^T PC_1 & 2K^T G_1^T & & 0 & A_{55} \end{bmatrix} < 0 \quad (7.96)$$

with $QG = G_1$, $A_{44} = \frac{1}{2}m_1^T m_1 - Q_1$, $A_{55} = -2\gamma^2 I + 4D^T PD$. Obviously, (7.96) is an **LMI** in terms of P, P_1, Q, Q_1 and G_1. By Theorem 7.7, we immediately obtain the following theorem.

THEOREM 7.8
If (7.96) is feasible with solutions $P > 0, P_1 > 0, Q > 0, Q_1 > 0$ and G_1, then (7.93) with the filtering gain $G = Q^{-1}G_1$ is an asymptotic mean square H_∞ filter.

Example 7.4
In (7.92), we take

$$A_0 = \begin{bmatrix} -2.6 & -0.2 \\ 0.4 & -1.8 \end{bmatrix}, A_1 = \begin{bmatrix} -1.8 & 0.2 \\ -0.7 & -0.9 \end{bmatrix}, B_1 = \begin{bmatrix} 0.7 \\ 0.94 \end{bmatrix},$$

$$C_0 = \begin{bmatrix} -0.8 & 0 \\ 0 & -0.9 \end{bmatrix}, C_1 = \begin{bmatrix} -0.3 & 0.4 \\ 0.21 & -1.05 \end{bmatrix}, D = \begin{bmatrix} 0.2 \\ 0.3 \end{bmatrix},$$

$$l_0 = \begin{bmatrix} 1.3 & 0.8 \end{bmatrix}, l_1 = \begin{bmatrix} 1.2 & 3 \end{bmatrix}, K = 0.5,$$

$$m_0 = \begin{bmatrix} -0.11 & 0.3 \end{bmatrix}, m_1 = \begin{bmatrix} 0.28 & 0.63 \end{bmatrix}.$$

By substituting the above data into (7.96) with $\gamma = 2$ and solving the LMI (7.96), we have

$$P = \begin{bmatrix} 1.6095 & -0.0293 \\ -0.0293 & 0.7909 \end{bmatrix} > 0, P_1 = \begin{bmatrix} 3.8622 & -0.5054 \\ -0.5054 & 1.6277 \end{bmatrix} > 0,$$

$$Q = \begin{bmatrix} 1.0009 & 0.0275 \\ 0.0275 & 1.3260 \end{bmatrix} > 0, Q_1 = \begin{bmatrix} 3.6487 & 0.1333 \\ 0.1333 & 3.6199 \end{bmatrix} > 0,$$

$$G_1 = \begin{bmatrix} -0.0772 \\ 0.0235 \end{bmatrix}, \ G = Q^{-1}G_1 = \begin{bmatrix} -0.0777 \\ 0.0194 \end{bmatrix}.$$

Set the external disturbance $v(t) = e^{-t}$ and the time-delay $\tau = 0.2$, and the initial value $\phi(t) = \begin{bmatrix} 0.2 \ 0.5 \end{bmatrix}^T$ for $t \in [-0.2, 0]$. Hence, by Theorem 7.8, the H_∞ filter (7.93) is obtained.

☐

7.8 Notes and References

For affine nonlinear delay-free stochastic systems, using an HJI-based SBRL, the H_∞ and suboptimal mixed H_2/H_∞ filtering problems have been studied. A Nash game approach was applied to the suboptimal H_2/H_∞ filter design where the H_2 performance is minimized in the presence of the worst-case disturbance. For a class of quasi-linear stochastic delay-free systems, the H_∞ filter design was presented in terms of LMIs. Moreover, a new H_2/H_∞ filter, for which the H_2 performance index is taken as the error variance and is minimized when the disturbance is viewed as a white noise, has been investigated via a convex optimization technique. As extensions of the results of delay-free stochastic systems, H_∞ filtering for affine nonlinear and linear stochastic state-delayed systems has been studied, in particular, a useful moment estimate formula (Lemma 7.7) was obtained.

There are many references on stochastic H_∞ and mixed H_2/H_∞ filtering with applications; see, for example, [39, 40, 181, 176, 182, 183, 184]. In our viewpoint, similar to nonlinear stochastic H_∞ control, the main difficulty of the nonlinear stochastic H_∞ filter design still lies in solving the HJIs (7.6) and (7.87). Although a fuzzy linearized approach can be used to deal with such a problem, more rigorous mathematical analysis is needed.

The materials of this chapter mainly come from [220], [222] and [224].

8

Some Further Research Topics in Stochastic H_2/H_∞ Control

In this chapter, we present some research topics on stochastic H_2/H_∞ control. Under each topic, there are some unsolved problems that deserve further research. In our opinion, there are few results in the study of H_2/H_∞ control for stochastic Itô-type systems with random coefficients, affine nonlinear discrete-time stochastic systems with multiplicative noise, continuous- and discrete-time singular systems with multiplicative noise, mean-field stochastic systems and forward-backward stochastic systems.

8.1 Stochastic H_2/H_∞ Control with Random Coefficients

Consider the following stochastic linear control system with random coefficients

$$\begin{cases} dx(t) = [A_1(t,\omega)x(t) + B_1(t,\omega)u(t) + C_1(t,\omega)v(t)]\,dt \\ \qquad + [A_2(t,\omega)x(t) + B_2(t,\omega)u(t) + C_2(t,\omega)v(t)]\,dB(t), \\ z(t) = \begin{bmatrix} C(t,\omega)x(t) \\ D(t,\omega)u(t) \end{bmatrix}, \ x(0) = x_0 \in \mathcal{R}^n, \end{cases} \tag{8.1}$$

or for simplicity (the sample ω is suppressed),

$$\begin{cases} dx(t) = [A_1(t)x(t) + B_1(t)u(t) + C_1(t)v(t)]\,dt \\ \qquad + [A_2(t)x(t) + B_2(t)u(t) + C_2(t)v(t)]\,dB(t), \\ z(t) = \begin{bmatrix} C(t)x(t) \\ D(t)u(t) \end{bmatrix}, \ x(0) = x_0 \in \mathcal{R}^n, \end{cases} \tag{8.2}$$

where in (8.2), we make the following assumptions:

(i) $A_1, A_2 \in \mathcal{L}_{\mathcal{F}}^\infty([0,T], \mathcal{R}^{n \times n})$, $B_1, B_2 \in \mathcal{L}_{\mathcal{F}}^\infty([0,T], \mathcal{R}^{n \times n_u})$, $C_1, C_2 \in \mathcal{L}_{\mathcal{F}}^\infty([0,T], \mathcal{R}^{n \times n_v})$, $C \in \mathcal{L}_{\mathcal{F}}^\infty([0,T], \mathcal{R}^{n_{z_1} \times n})$, $D \in \mathcal{L}_{\mathcal{F}}^\infty([0,T], \mathcal{R}^{n_{z_2} \times n_u})$, $n_{z_1} + n_{z_2} = dim(z)$, $\mathcal{L}_{\mathcal{F}}^\infty([0,T], X)$ is the set of all X-valued $\{\mathcal{F}_t\}_{t \geq 0}$-adapted uniformly bounded processes.

(ii) $D^T(t)D(t) = I$ almost surely and almost everywhere.

Stochastic Itô systems with random coefficients have important applications especially in mathematical finance [109, 140, 196], and the corresponding stochastic LQ control can be found in [22, 23, 86, 148]. It seems that the reference [226] first started to consider the H_∞ control of Itô-type differential equations with random coefficients. In this section, we aim to study the H_2/H_∞ control of (8.2), where the external disturbance is considered in mathematical modeling, which is very realistic in comparison with the sole LQ control. For simplicity of our statement, we adopt the same notations and definitions as in Chapter 2 without repeating them.

8.1.1   SBRL and stochastic LQ lemma

Below, for the following stochastic perturbed system with random coefficients

$$\begin{cases} dx(t) = [A_{11}(t)x(t) + B_{11}(t)v(t)]\, dt + [A_{12}(t)x(t) + B_{12}(t)v(t)]\, dB(t), \\ z_1(t) = C_{11}(t)x(t), \\ x(0) = x_0, \end{cases}$$

$$(8.3)$$

we present the following SBRL:

LEMMA 8.1

For the system (8.3) and a given disturbance attenuation level $\gamma > 0$, if the following backward stochastic Riccati equation (BSRE)

$$\begin{cases} dP = -[A_{11}^T P + PA_{11} + A_{12}^T PA_{12} + A_{12}^T L + LA_{12} - C_{11}^T C_{11} \\ \qquad\quad -(B_{11}^T P + B_{12}^T PA_{12} + B_{12}^T L)^T \\ \qquad\quad \cdot(\gamma^2 I + B_{12}^T PB_{12})^{-1}(B_{11}^T P + B_{12}^T PA_{12} + B_{12}^T L)]\, dt \\ \qquad\quad +L\, dB(t), \\ \gamma^2 I + B_{12}^T PB_{12} > 0, \ a.e. \ a.s.(t,\omega) \in [0,T] \times \Omega \\ P(T) = 0, a.s. \end{cases}$$

$$(8.4)$$

has a pair of $\{\mathcal{F}_t\}_{t \geq 0}$-adapted square integrable solutions $(P(t), L(t)) \in \mathcal{L}_\mathcal{F}^2$ $([0,T], \mathcal{S}_n) \times \mathcal{L}_\mathcal{F}^2([0,T], \mathcal{S}_n)$, $t \in [0,T]$, then we have $\|\tilde{\mathcal{L}}_T\| < \gamma$, where $\tilde{\mathcal{L}}_T$ is defined as in Lemma 2.1.

REMARK 8.1   The BSRE (8.4) is a backward Itô equation with the constraint conditions

$$\gamma^2 I + B_{12}^T PB_{12} > 0, \quad P(T) = 0.$$

When the coefficients of (8.3) are deterministic matrix-valued functions, BSRE (8.4) reduces to GDRE (2.6). In this case, $(P(\cdot), 0)$ is the unique solution to (8.4). ∎

Proof. This lemma can be proved following the line of Theorem 3.1 in [45]. Let $P(t,\omega) \in \mathcal{L}_\mathcal{F}^2([0,T], \mathcal{S}_n)$ be semimartingale satisfying

$$dP(t) = \Gamma(t)\, dt + L(t)\, dB(t).$$

$$(8.5)$$

Applying Itô's formula to $x^T(t)P(t)x(t)$, we have (the variables t and ω are suppressed)

$$
\begin{aligned}
d(x^T P x) &= (dx^T)Px + x^T(dP)x + x^T P(dx) + (dx^T)(dP)x + (dx^T)P(dx) \\
&\quad + x^T(dP)(dx) \\
&= \{x^T(\Gamma + PA_{11} + A_{11}^T P + A_{12}^T PA_{12} + LA_{12} + A_{12}^T L)x \\
&\quad + 2v^T(B_{11}^T P + B_{12}^T PA_{12} + B_{12}^T L)x + v^T B_{12}^T PB_{12}v\} \, dt \\
&\quad + \{\cdots\} \, dB(t).
\end{aligned}
$$

Using the standard completing squares technique and considering the BSRE (8.4), it follows in the case of $x_0 = 0$ that

$$
\begin{aligned}
\mathcal{E} \int_0^T (\gamma^2 \|v\|^2 - \|z_1\|^2) \, dt &= \mathcal{E} \int_0^T [(\gamma^2\|v\|^2 - \|z_1\|^2)\,dt + d(x^T P x)] \\
&= \mathcal{E} \int_0^T (v + K^{-1}\tilde{L}x)^T K(v + K^{-1}\tilde{L}x) \, dt \\
&\quad + \mathcal{E} \int_0^T x^T(\Gamma + PA_{11} + A_{11}^T P + A_{12}^T PA_{12} + LA_{12} \\
&\quad + A_{12}^T L - C_{11}^T C_{11} - \tilde{L}^T K^{-1}\tilde{L})x \, dt, \qquad (8.6)
\end{aligned}
$$

where $K = \gamma^2 I + B_{12}^T PB_{12} > 0$ and $\tilde{L} = B_{11}^T P + B_{12}^T PA_{12} + B_{12}^T L$. If (P, L) is the solution to (8.5), in view of BSRE (8.4), we have:

$$
\Gamma + PA_{11} + A_{11}^T P + A_{12}^T PA_{12} + LA_{12} + A_{12}^T L - C_{11}^T C_{11} - \tilde{L}^T K^{-1}\tilde{L} = 0.
$$

From (8.6), we know that $\|\tilde{\mathcal{L}}_T\| \le \gamma$. As done in Lemma 2.1, it is easy to verify $\|\tilde{\mathcal{L}}_T\| < \gamma$. $\square$

We do not know whether the converse of Lemma 8.1 holds, but for $B_{12} \equiv 0$, we have

LEMMA 8.2 [186]
If $\|\tilde{\mathcal{L}}_T\| < \gamma$, then there is a unique solution $(P(t), L(t)) \in \mathcal{L}_\mathcal{F}^2([0,T], \mathcal{S}_n) \times \mathcal{L}_\mathcal{F}^2([0,T], \mathcal{S}_n)$ to

$$
\begin{cases}
dP = -(A_{11}^T P + PA_{11} + A_{12}^T PA_{12} + A_{12}^T L + LA_{12} - C_{11}^T C_{11} \\
\qquad\quad -\gamma^{-2}PB_{11}B_{11}^T P)\,dt + L\,dB, \\
P(T) = 0, a.s.
\end{cases}
\qquad (8.7)
$$

on $[0, T]$, where $P(t) \le 0$ is uniformly bounded and $L(t)$ is square integrable. Moreover,

$$
\mathcal{E} \int_0^T \|L(t)\|^2 \, dt \le \beta
$$

for some deterministic constant $\beta > 0$, which depends on the uniform lower bound of P and the upper bound of all the coefficients.

The proof of Lemma 8.2 is very complicated, and we refer the reader to Theorem 3.2 of [186]. The following lemma is a special case of Lemma 3.1 of [86].

LEMMA 8.3 [86]
Consider the following stochastic control system with random coefficients

$$\begin{cases} dx(t) = [A_1(t)x(t) + B_1(t)u(t)]\,dt + [A_2(t)x(t) + B_2(t)u(t)]\,dB(t), \\ x(0) = x_0 \in \mathcal{R}^n. \end{cases} \tag{8.8}$$

If the BSRE

$$\begin{cases} dP = -[A_1^T P + PA_1 + A_2^T PA_2 + A_2^T L + LA_2 + C^T C \\ \qquad\quad -(B_1^T P + B_2^T PA_2 + B_2^T L)^T (I + B_2^T PB_2)^{-1} \\ \qquad\quad \cdot(B_1^T P + B_2^T PA_2 + B_2^T L)]\,dt + L\,dB, \\ I + B_2^T PB_2 > 0, \\ P(T) = 0 \end{cases} \tag{8.9}$$

admits a solution $(P, L) \in \mathcal{L}_\mathcal{F}^2([0,T], \mathcal{S}_n) \times \mathcal{L}_\mathcal{F}^2([0,T], \mathcal{S}_n)$, *then we have*

$$\min_{u \in \mathcal{L}_\mathcal{F}^2([0,T], \mathcal{R}^{n_u})} \left\{ J_T(0, x_0; u) := \mathcal{E} \int_0^T (x^T C^T Cx + u^T u)\,dt \right\} = x_0^T P(0)x_0,$$

and the optimal control u^* is given by

$$u^*(t) = -(I + B_2^T PB_2)^{-1}(B_1^T P + B_2^T PA_2 + B_2^T L)x(t).$$

8.1.2   Mixed H_2/H_∞ control

Applying Lemma 8.1 and Lemma 8.3, the following theorem on H_2/H_∞ control can be obtained similarly to Theorem 2.5.

THEOREM 8.1
For system (8.2), if the following four coupled stochastic matrix-valued equations

$$\begin{cases} dP_{1,T} = -\big\{(A_1 + B_1 K_{2,T})^T P_{1,T} + P_{1,T}(A_1 + B_1 K_{2,T}) + (A_2 + B_2 K_{2,T})^T \\ \qquad\quad \cdot P_{1,T}(A_2 + B_2 K_{2,T})^T + (A_2 + B_2 K_{2,T})^T L_1 + L_1(A_2 + B_2 K_{2,T}) \\ \qquad\quad -C^T C - K_{2,T}^T K_{2,T} \\ \qquad\quad -\{P_{1,T} C_1 + [(A_2 + B_2 K_{2,T})^T P_{1,T} + L_1]C_2\}(\gamma^2 I + C_2^T P_{1,T} C_2)^{-1} \\ \qquad\quad \cdot\{C_1^T P_{1,T} + C_2^T[P_{1,T}(A_2 + B_2 K_{2,T}) + L_1]\}\big\}\,dt + L_1\,dB(t), \\ \gamma^2 I + C_2^T P_{1,T} C_2 > 0, \\ P_{1,T}(T) = 0, \end{cases}$$

$$\tag{8.10}$$

$$\begin{cases} dP_{2,T} = -\{(A_1 + C_1K_{1,T})^T P_{2,T} + P_{2,T}(A_1 + C_1K_{1,T}) + (A_2 + C_2K_{1,T})^T \\ \qquad \cdot P_{2,T}(A_2 + C_2K_{1,T}) + (A_2 + C_2K_{1,T})^T L_2 + L_2(A_2 + C_2K_{1,T}) \\ \qquad + C^T C - \{P_{2,T}B_1 + [(A_2 + C_2K_{1,T})^T P_{2,T} + L_2]B_2\} \\ \qquad (I + B_2^T P_{2,T}B_2)^{-1}\{B_1^T P_{2,T} + B_2^T[P_{2,T}(A_2 + C_2K_{1,T}) + L_2]\}\} \ dt \\ \qquad + L_2 \ dB(t), \\ I + B_2^T P_{2,T}B_2 > 0, \\ P_{2,T}(T) = 0, \end{cases}$$

$$(8.11)$$

$$K_{1,T}(t) = -(\gamma^2 I + C_2^T P_{1,T}C_2)^{-1} \\ \qquad \cdot [C_1^T P_{1,T} + C_2^T P_{1,T}(A_2 + B_2K_{2,T}) + C_2^T L_1] \tag{8.12}$$

and

$$K_{2,T}(t) = -(I + B_2^T P_{2,T}B_2)^{-1} \\ \qquad \cdot [B_1^T P_{2,T} + B_2^T P_{2,T}(A_2 + C_2K_{1,T}) + B_2^T L_2] \tag{8.13}$$

have solutions $(P_{1,T}(t), L_1(t), K_{1,T}(t); P_{2,T}(t), L_2(t), K_{2,T}(t))$ *on* $[0, T]$ *with* $P_{i,T}, L_i \in \mathcal{L}_{\mathcal{F}}^2([0,T], \mathcal{S}_n)$, $i = 1, 2$, $K_{1,T} \in \mathcal{L}_{\mathcal{F}}^2([0,T], R^{n_v \times n})$ *and* $K_{2,T} \in \mathcal{L}_{\mathcal{F}}^2([0,T], R^{n_u \times n})$, *then the stochastic* H_2/H_∞ *control is solvable with*

$$u_T^*(t) = K_{2,T}(t)x(t), \ v_T^*(t) = K_{1,T}(t)x(t).$$

The optimal H_2 *cost functional is*

$$\min_{u \in \mathcal{L}_{\mathcal{F}}^2([0,T], \mathcal{R}^{n_u})} J_{2,T}(u, v_T^*) = J_{2,T}(u_T^*, v_T^*) = x_0^T P_{2,T}(0)x_0.$$

The above four coupled matrix-valued equations (8.10)–(8.13) are too complicated. Now we consider the special case, i.e., $B_2 \equiv 0$, $C_2 \equiv 0$. In this case, (8.10)–(8.13) reduce to two coupled BSREs.

COROLLARY 8.1 [186]
For $B_2 \equiv 0$, $C_2 \equiv 0$ *in system (8.2), if the following two coupled BSREs*

$$\begin{cases} dP_{1,T} = -\{A_1^T P_{1,T} + P_{1,T}A_1 + A_2^T P_{1,T}A_2 + A_2^T L_1 + L_1 A_2 - C^T C \\ \qquad - \begin{bmatrix} P_{1,T} & P_{2,T} \end{bmatrix} \begin{bmatrix} \gamma^{-2}C_1C_1^T & B_1B_1^T \\ B_1B_1^T & B_1B_1^T \end{bmatrix} \begin{bmatrix} P_{1,T} \\ P_{2,T} \end{bmatrix}\} \ dt + L_1 \ dB(t), \\ P_{1,T}(T) = 0 \end{cases}$$

$$(8.14)$$

and

$$\begin{cases} dP_{2,T} = -\{A_1^T P_{2,T} + P_{2,T}A_1 + A_2^T P_{2,T}A_2 + A_2^T L_2 + L_2 A_2 + C^T C \\ \qquad - \begin{bmatrix} P_{1,T} & P_{2,T} \end{bmatrix} \begin{bmatrix} 0 & \gamma^{-2}C_1C_1^T \\ \gamma^{-2}C_1C_1^T & B_1B_1^T \end{bmatrix} \begin{bmatrix} P_{1,T} \\ P_{2,T} \end{bmatrix}\} \ dt \\ \qquad + L_2 \ dB(t), \\ P_{2,T}(T) = 0 \end{cases}$$

$$(8.15)$$

have solutions $P_{1,T}(t), L_1(t), P_{2,T}(t), L_2(t) \in \mathcal{L}_{\mathcal{F}}^2([0,T], \mathcal{S}_n)$ on $[0,T]$, then

$$u_T^*(t) = K_{2,T}(t)x(t), \quad v_T^*(t) = K_{1,T}(t)x(t), \tag{8.16}$$

$$J_{2,T}(u_T^*, v_T^*) = x_0^T P_{2,T}(0)x_0,$$

where

$$K_{1,T}(t) = -\gamma^{-2}C_1^T(t)P_{1,T}(t), \quad K_{2,T}(t) = -B_1^T(t)P_{2,T}(t).$$

When A_1, B_1, C_1, C, A_2, B_2 and C_2 are deterministic matrix-valued functions, BSREs (8.10)–(8.13) reduce to GDREs (2.55)–(2.58).

Applying Lemma 8.2, the converse of Corollary 8.1 still holds.

THEOREM 8.2 [186]

For $B_2 \equiv 0$, $C_2 \equiv 0$ in system (8.2), if the finite horizon H_2/H_∞ control admits a pair of solutions $(u_T^*(t), v_T^*(t))$ given by (8.16), where $K_{2,T}(t)$ and $K_{1,T}(t)$ are bounded $\mathcal{F}_t$-adapted processes, then the coupled BSREs (8.14)–(8.15) have a unique solution

$$(P_{1,T}(t) \leq 0, P_{2,T}(t) \geq 0) \in \mathcal{L}_{\mathcal{F}}^2([0,T], \mathcal{S}_n) \times \mathcal{L}_{\mathcal{F}}^2([0,T], \mathcal{S}_n), \quad t \in [0,T].$$

8.1.3 H_∞ control

In what follows, we consider only the H_∞ control problem, i.e., for $\gamma > 0$, we search for $\tilde{u}_T^* \in \mathcal{L}_{\mathcal{F}}^2([0,T], \mathcal{R}^{n_u})$ such that $\|\mathcal{L}_T\| < \gamma$ for $x_0 = 0$ and any non-zero $v \in \mathcal{L}_{\mathcal{F}}^2([0,T], \mathcal{R}^{n_v})$, where $\mathcal{L}_T$ is defined as in Definition 2.1.

LEMMA 8.4

Assume the matrices $(\gamma^2 I + C_2^T P C_2)$ and $(-I + B_2^T P B_2)$ are invertible for a.e. a.s. (t,ω). For linear stochastic system (8.2), if the following BSRE

$$\begin{cases} dP = -[PA_1 + A_1^T P + A_2^T P A_2 + A_2^T L + L A_2 - C^T C \\ \quad -(PC_1 + A_2^T P C_2 + L C_2)(\gamma^2 I + C_2^T P C_2)^{-1} \\ \quad \cdot(C_1^T P + C_2^T P A_2 + C_2^T L) \\ \quad +(PB_1 + A_2^T P B_2 + L B_2)(-I + B_2^T P B_2)^{-1} \\ \quad \cdot(B_1^T P + B_2^T P A_2 + B_2^T L)] \, dt + L \, dB, \\ P(T) \in \mathcal{L}_{\mathcal{F}_T}^2(\Omega, \mathcal{S}_n) \end{cases} \tag{8.17}$$

admits a pair of $\mathcal{F}_t$-adapted solutions $(P, L) \in \mathcal{L}_{\mathcal{F}}^2([0,T], \mathcal{S}_n) \times \mathcal{L}_{\mathcal{F}}^2([0,T], \mathcal{S}_n)$, then we have the following identity:

$$J_{1,T}(u,v) := \mathcal{E} \int_0^T (\gamma^2 \|v\|^2 - \|z\|^2) \, dt = x_0^T P(0)x_0 - \mathcal{E}[x^T(T)P(T)x(T)]$$

$$+ \mathcal{E} \int_0^T [v + \bar{K}_{1,T}(t)x]^T (\gamma^2 I + C_2^T P C_2)[v + \bar{K}_{1,T}(t)x] \, dt$$

$$+\mathcal{E}\int_0^T [u + \bar{K}_{2,T}(t)x]^T (-I + B_2^T P B_2)[u + \bar{K}_{2,T}(t)x]\,dt$$

$$+\mathcal{E}\int_0^T (u^T B_2^T P C_2 v + v^T C_2^T P B_2 u)\,dt, \tag{8.18}$$

where

$$\bar{K}_{1,T} = (\gamma^2 I + C_2^T P C_2)^{-1}(C_1^T P + C_2^T P A_2 + C_2^T L),$$

$$\bar{K}_{2,T} = (-I + B_2^T P B_2)^{-1}(B_1^T P + B_2^T P A_2 + B_2^T L).$$

To prove Lemma 8.4, we need to use identity (6.44).

Proof. By applying Itô's formula to $x^T P(t,\omega)x$, where $P(t,\omega)$ is defined in (8.5) and $x(t)$ is the trajectory of (8.2), we have

$$\mathcal{E}\int_0^T (\gamma^2\|v\|^2 - \|z\|^2)\,dt = \mathcal{E}\int_0^T \{(\gamma^2\|v\|^2 - \|z\|^2)\,dt + d(x^T P x)\}$$

$$+\mathcal{E}[x_0^T P(0)x_0] - \mathcal{E}[x^T(T)P(T)x(T)]$$

$$= \mathcal{E}\int_0^T [\Upsilon_1(v,x) + \Upsilon_2(x) + \Upsilon_3(u,x) + \Upsilon_4(u,v)]\,dt$$

$$+x_0^T P(0)x_0 - \mathcal{E}[x^T(T)P(T)x(T)], \tag{8.19}$$

where

$$\begin{aligned}
\Upsilon_1(v,x) &= \gamma^2\|v\|^2 + v^T C_1^T P x + x^T P C_1 v + x^T A_2^T P C_2 v \\
&\quad +v^T C_2^T P A_2 x + v^T C_2^T L x + x^T L C_2 v + v^T C_2^T P C_2 v \\
&= v^T(\gamma^2 I + C_2^T P C_2)v + v^T(C_1^T P + C_2^T P A_2 + C_2^T L)x \\
&\quad +x^T(P C_1 + A_2^T P C_2 + L C_2)v, \\
\Upsilon_2(x) &= x^T(P A_1 + A_1^T P + A_2^T P A_2 + A_2^T L + L A_2 + \Gamma - C^T C)x, \\
\Upsilon_3(u,x) &= -\|u\|^2 + u^T B_1^T P x + x^T P B_1 u + x^T A_2^T P B_2 u \\
&\quad +u^T B_2^T P A_2 x + u^T B_2^T P B_2 u + u^T B_2^T L x + x^T L B_2 u \\
&= u^T(-I + B_2^T P B_2)u + u^T(B_1^T P + B_2^T P A_2 + B_2^T L)x \\
&\quad +x^T(A_2^T P B_2 + P B_1 + L B_2)u, \\
\Upsilon_4(u,v) &= u^T B_2^T P C_2 v + v^T C_2^T P B_2 u.
\end{aligned}$$

In (8.19), we have used the fact that x_0 is deterministic, and $P(0)$ is $\mathcal{F}_0 = \{\phi, \Omega\}$-measurable and hence is a constant matrix. So $\mathcal{E}[x_0^T P(0)x_0] = x_0^T P(0)x_0$. Applying (6.44) to $\Upsilon_1(v,x)$ and $\Upsilon_3(u,x)$, we obtain, respectively, that

$$\begin{aligned}
\Upsilon_1(v,x) &= (v + \bar{K}_{1,T}x)^T(\gamma^2 I + C_2^T P C_2)(v + \bar{K}_{1,T}x) \\
&\quad -x^T(P C_1 + A_2^T P C_2 + L C_2)(\gamma^2 I + C_2^T P C_2)^{-1} \\
&\quad \cdot(C_1^T P + C_2^T P A_2 + C_2^T L)x,
\end{aligned}$$

$$\tag{8.20}$$

and

$$\begin{aligned}
\Upsilon_3(u, x) &= (u + \bar{K}_{2,T}x)^T(-I + B_2^T PB_2)(u + \bar{K}_{2,T}x) \\
&\quad -x^T(PB_1 + A_2^T PB_2 + LB_2)(-I + B_2^T PB_2)^{-1} \\
&\quad \cdot(B_1^T P + B_2^T PA_2 + B_2^T L)x.
\end{aligned} \tag{8.21}$$

If (8.17) admits a pair of $\mathcal{F}_t$-adapted solutions (P, L), we may set

$$\begin{aligned}
\Gamma = &-[PA_1 + A_1^T P + A_2^T PA_2 + A_2^T L + LA_2 - C^T C \\
&-(PC_1 + A_2^T PC_2 + LC_2)(\gamma^2 I + C_2^T PC_2)^{-1} \\
&\cdot(C_1^T P + C_2^T PA_2 + C_2^T L) \\
&+(PB_1 + A_2^T PB_2 + LB_2)(-I + B_2^T PB_2)^{-1} \\
&\cdot(B_1^T P + B_2^T PA_2 + B_2^T L)].
\end{aligned}$$

By (8.19)–(8.21), we immediately obtain (8.18). □

THEOREM 8.3

Given a scalar $\gamma > 0$, assume $B_2 \equiv 0$ in (8.2). If the following BSRE

$$\begin{cases}
dP = -[PA_1 + A_1^T P + A_2^T PA_2 + A_2^T L + LA_2 - C^T C \\
\quad -(PC_1 + A_2^T PC_2 + LC_2)(\gamma^2 I + C_2^T PC_2)^{-1} \\
\quad \cdot(C_1^T P + C_2^T PA_2 + C_2^T L) - PB_1 B_1^T P] dt + L\, dB, \\
\gamma^2 I + C_2^T PC_2 > 0, \\
P(T) = 0
\end{cases} \tag{8.22}$$

admits a pair of $\mathcal{F}_t$-adapted solutions $(P, L) \in \mathcal{L}_{\mathcal{F}}^2([0, T], \mathcal{S}_n) \times \mathcal{L}_{\mathcal{F}}^2([0, T], \mathcal{S}_n)$, then the H_∞ control problem is solvable and the corresponding H_∞ control law is given by $\tilde{u}_T^* = B_1(t)^T P(t)x(t)$.

Proof. By Lemma 8.4 and identity (8.18), for $x_0 = 0$ and $P(T) = 0$, we have

$$\mathcal{E} \int_0^T (\gamma^2 \|v\|^2 - \|z\|^2)\, dt$$

$$= \mathcal{E} \int_0^T [v + \bar{K}_{1,T}(t)x]^T (\gamma^2 I + C_2^T PC_2)[v + \bar{K}_{1,T}(t)x]\, dt$$

$$-\mathcal{E} \int_0^T [u + \bar{K}_{2,T}(t)x]^T [u + \bar{K}_{2,T}(t)x]\, dt$$

which yields

$$\|z\|_{[0,T]}^2 \le \gamma^2 \|v\|_{[0,T]}^2 + \|u + \bar{K}_{2,T}(x)\|_{[0,T]}^2. \tag{8.23}$$

From (8.23), if we take $u = \tilde{u}_T^* = B_1(t)^T P(t)x(t)$, then $\|\mathcal{L}_T\| \le \gamma$. Following the same line of arguments as in Theorem 2.1, we can further show $\|\mathcal{L}_T\| < \gamma$. □

Similarly, for $C_2 \equiv 0$ in (8.2), we have the following theorem:

THEOREM 8.4
Assume $C_2 \equiv 0$ in (8.2). Given a scalar $\gamma > 0$, if the following BSRE

$$
\begin{cases}
dP = -[PA_1 + A_1^T P + A_2^T PA_2 + A_2^T L + LA_2 - C^T C \\
\quad -\gamma^{-2} PC_1 C_1^T P + (PB_1 + A_2^T PB_2 + LB_2) \\
\quad (-I + B_2^T PB_2)^{-1}(B_1^T P + B_2^T PA_2 + B_2^T L)]\, dt + L\, dB, \quad (8.24) \\
-I + B_2^T PB_2 < 0, \\
P(T) = 0
\end{cases}
$$

admits a pair of $\mathcal{F}_t$-adapted solutions $(P, L) \in \mathcal{L}_{\mathcal{F}}^2([0,T], \mathcal{S}_n) \times \mathcal{L}_{\mathcal{F}}^2([0,T], \mathcal{S}_n)$, then the H_∞ control problem is solvable and the corresponding H_∞ control law is given by $\tilde{u}_T^* = -(-I + B_2^T PB_2)^{-1}(B_1^T P + B_2^T PA_2 + B_2^T L)x(t)$.

COROLLARY 8.2
Consider $C_2 \equiv 0$ and $B_2 \equiv 0$ in (8.2). Given a scalar $\gamma > 0$, if the following BSRE

$$
\begin{cases}
dP = -(PA_1 + A_1^T P + A_2^T PA_2 + A_2^T L + LA_2 - C^T C \\
\quad -\gamma^{-2} PC_1 C_1^T P - PB_1 B_1^T P)\, dt + L\, dB, \quad (8.25) \\
P(T) = 0
\end{cases}
$$

admits a pair of $\mathcal{F}_t$-adapted solutions $(P, L) \in \mathcal{L}_{\mathcal{F}}^2([0,T], \mathcal{S}_n) \times \mathcal{L}_{\mathcal{F}}^2([0,T], \mathcal{S}_n)$, then $\tilde{u}_T^* = B_1^T Px(t)$ is an H_∞ control law, which makes $\|\mathcal{L}_T\| < \gamma$.

REMARK 8.2 BSRE (8.25) can be written as

$$
\begin{cases}
dP = -\{PA_1 + A_1^T P + A_2^T PA_2 + A_2^T L + LA_2 - C^T C \\
\quad -P\bar{B}_1 R^{-1} \bar{B}_1^T P\}\, dt + L\, dB, \quad (8.26) \\
P(T) = 0
\end{cases}
$$

with $\bar{B}_1 = \begin{bmatrix} C_1 & B_1 \end{bmatrix}$, $R = \begin{bmatrix} \gamma^2 I & 0 \\ 0 & I \end{bmatrix}$, which is a BSRE coming from the indefinite stochastic LQ control with the state weighting matrix $Q = -C^T C < 0$ and the control weighting matrix $R > 0$; see [86]. ∎

8.1.4 Some unsolved problems

From the above discussions, we can see that the stochastic H_∞ and H_2/H_∞ controller designs of (8.2) depend on the solvability of some coupled BSREs. There exist several problems that remain to be studied.

- We should point out that, up to now, we even have no efficient method to solve the coupled BSREs (8.14)–(8.15), let alone (8.10)–(8.13), which is a key obstacle in designing an H_2/H_∞ controller for systems with random co-efficients.

- Compared with Lemma 8.2, it is not clear whether the converse of Lemma 8.1 holds, which merits further study. Under $B_{12} \equiv 0$ in (8.3), Lemma 8.1 and Lemma 8.2 reveal the equivalent relation between the $\mathcal{L}_2$-gain less than $\gamma > 0$ and the solvability of BSRE (8.7), which contributes to the BSDE theory.

- Infinite horizon stochastic H_2/H_∞ control with random coefficients is a more complicated research topic than the finite horizon one, because (i) we know little about the solvability of coupled infinite time horizon SBREs, which is a very challenging problem in the field of BSDEs and (ii) it is not an easy task to study the stability of BSDEs.

8.2 Nonlinear Discrete-Time Stochastic H_2/H_∞ Control

Although the nonlinear continuous-time stochastic H_∞ control, mixed H_2/H_∞ control and filtering have been basically solved in [15, 213, 214, 220], there is an essential difficulty in the study of the H_2/H_∞ control and filtering of nonlinear discrete-time stochastic systems with multiplicative noise. Even for the H_∞ control, there is a big gap between the nonlinear discrete stochastic H_∞ control [15] and deterministic nonlinear discrete H_∞ control [128]. In this section, we lay a foundation for this study and analyze the challenge in the infinite horizon H_2/H_∞ design of an affine discrete stochastic system

$$\begin{cases} x_{k+1} = f(x_k) + g(x_k)u_k + h(x_k)v_k \\ \qquad + [f_1(x_k) + g_1(x_k)u_k + h_1(x_k)v_k]w_k, \\ z_k = \begin{bmatrix} m(x_k) \\ u_k \end{bmatrix}, \; x_0 \in \mathcal{R}^n, \; k \in \mathcal{N} \end{cases} \tag{8.27}$$

where $\{w_k\}_{k \geq 0}$ is an independent random sequence, $\mathcal{E}w_k = 0$, $\mathcal{E}w_i w_j = 0$ for $i \neq j$ and $\mathcal{E}w_i^2 = 1$ for $i \geq 0$. f, g, h, f_1, g_1 and h_1 are measurable functions of suitable dimensions with $f(0) = f_1(0) = 0$. The difference equation (8.27) is defined on a complete filtered space $(\Omega, \mathcal{F}, \{\mathcal{F}_k\}_{k \in \mathcal{N}}, \mathcal{P})$ with $\mathcal{F}_k = \sigma(w_0, w_1, \cdots, w_{k-1})$. Note that in this chapter, we define $\mathcal{F}_k = \sigma(w_0, w_1, \cdots, w_{k-1})$ instead of $\mathcal{F}_k = \sigma(w_0, w_1, \cdots, w_k)$ as in Chapters 3–5 only for convenience of subsequent discussions. System (8.27) is a stochastic version of the deterministic discrete-time affine system

$$\begin{cases} x_{k+1} = f(x_k) + g(x_k)u_k + h(x_k)v_k, \\ z_k = \begin{bmatrix} m(x_k) \\ u_k \end{bmatrix}, x_0 \in \mathcal{R}^n, \; k \in \mathcal{N}, \end{cases} \tag{8.28}$$

for which its H_∞ control was perfectly solved in [128].

8.2.1 Dissipation, l_2-gain and SBRL

As in [213], to investigate the H_∞ or H_2/H_∞ control, we need to introduce some definitions and preliminaries. Consider the following system

$$\begin{cases} x_{k+1} = F(x_k, u_k, w_k), \ x_0 \in \mathcal{R}^n, \\ z_k = G(x_k, u_k, w_k), \ x_0 \in \mathcal{R}^n, \ k \in \mathcal{N}. \end{cases} \tag{8.29}$$

DEFINITION 8.1 *System (8.29) is said to be dissipative with the supply rate $w(u_k, z_k)$ which satisfies*

$$\sum_{i=j}^{k} \mathcal{E}|w(u_i, z_i)| < \infty, \quad \forall k \geq j \geq 0, \ u \in l_w^2(\mathcal{N}, \mathcal{R}^{n_u}),$$

if there is a nonnegative function $V(x) : \mathcal{R}^n \mapsto \mathcal{R}^+, V(0) = 0$, called the storage function, such that for all $k \geq j \geq 0$, we have

$$\mathcal{E}V(x_{k+1}) - \mathcal{E}V(x_j) \leq \sum_{i=j}^{k} \mathcal{E}w(u_i, z_i). \tag{8.30}$$

The inequality (8.30) is called a dissipation inequality.

From the following remark, we can see that in Definition 8.1, it is more convenient to use (8.30) than using the following special form

$$\mathcal{E}V(x_{k+1}) - V(x_0) \leq \sum_{i=0}^{k} \mathcal{E}w(u_i, z_i), \ x_0 \in \mathcal{R}^n, \ k \geq 0 \tag{8.31}$$

as in deterministic systems; see (2.2) of [29].

REMARK 8.3 In deterministic nonlinear systems, the dissipation inequality

$$V(x_{k+1}) - V(x_0) \leq \sum_{i=0}^{k} w(u_i, z_i)$$

is equivalent to the following [29]

$$V(x_{k+1}) - V(x_k) \leq w(u_k, z_k), \quad k \geq 0. \tag{8.32}$$

However, in stochastic case, (8.31) is not equivalent to

$$\mathcal{E}V(x_{k+1}) - \mathcal{E}V(x_k) \leq \mathcal{E}w(u_k, z_k), \quad k \geq 0, \tag{8.33}$$

because we cannot set $x_0 = x_k$, $z_0 = z_k$ and $u_0 = u_k$ for $k > 0$ due to different adaptiveness requirements. It is easy to see that the dissipation inequality (8.30) is equivalent to (8.33). ∎

Corresponding to Definition 6.2, the following is introduced.

DEFINITION 8.2   *An available storage function with supply rate $w(\cdot,\cdot)$ on $[j,\infty)$, $j \geq 0$, is defined by*

$$
V_{a,j}(x) = - \inf_{u \in l_w^2([j,k],\mathcal{R}^{n_u}),k \geq j \geq 0, x(j)=x \in \mathcal{R}^n} \sum_{i=j}^{k} \mathcal{E}w(u_i,z_i)
$$

$$
= \sup_{u \in l_w^2([j,k],\mathcal{R}^{n_u}),k \geq j \geq 0, x(j)=x \in \mathcal{R}^n} -\sum_{i=j}^{k} \mathcal{E}w(u_i,z_i). \quad (8.34)
$$

PROPOSITION 8.1
If system (8.29) with supply rate w is dissipative on $[j,\infty)$, $j \geq 0$, then the available storage function $V_{a,j}(x)$ is finite for each $x \in \mathcal{R}^n$. Moreover, for any possible storage function V_j,

$$
0 \leq V_{a,j}(x) \leq V_j(x), \quad \forall x \in \mathcal{R}^n.
$$

$V_{a,j}$ is itself a possible storage function. Conversely, if $V_{a,j}$ is finite for each $x \in \mathcal{R}^n$, then system (8.29) is dissipative on $[j,\infty)$.

Proof. The proof is the same with that of Proposition 6.2; see also Theorem 1 of [16]. □
From now on, the supply rate is taken as

$$
w(u,z) = \gamma^2 \|u\|^2 - \|z\|^2,
$$

which is closely related to finite gain systems. If in (8.30), $w(u_k,z_k) = z_k^T u_k$, then system (8.29) is said to be passive; If $w(u_k,z_k) = z_k^T u_k - S(x_k)$ in (8.30) for some positive function $S(\cdot)$, then system (8.29) is called strictly passive. When

$$
\mathcal{E}V(x_{k+1}) - V(x_j) = \sum_{i=j}^{k} \mathcal{E}(z_i^T u_i),
$$

system (8.29) is said to be lossless. A detailed discussion of losslessness, feedback equivalence and global stabilization of deterministic discrete nonlinear systems can be found in [29].

DEFINITION 8.3   *For a given $\gamma > 0$, if in (8.29), $\{u_k\}_{k \geq 0}$ is taken as an external disturbance sequence and $x_0 = 0$, then system (8.29) is said to have an l_2-gain less than or equal to $\gamma > 0$ if*

$$
\sum_{i=0}^{k} \mathcal{E}\|z_i\|^2 \leq \gamma^2 \sum_{i=0}^{k} \mathcal{E}\|u_i\|^2, \quad \forall u = \{u_k\}_{k \geq 0} \in l_w^2(\mathcal{N},\mathcal{R}^{n_u}), \ k \in \mathcal{N} \quad (8.35)
$$

or equivalently

$$\sum_{i=0}^{k} \mathcal{E}w(u_i, z_i) \geq 0, \ \forall u = \{u_k\}_{k \geq 0} \in l_w^2(\mathcal{N}, \mathcal{R}^{n_u}), \ k \in \mathcal{N}.$$

For $x_0 \neq 0$, (8.35) should be replaced by

$$\sum_{i=0}^{k} \mathcal{E}\|z_i\|^2 \leq \beta(x_0) + \gamma^2 \sum_{i=0}^{k} \mathcal{E}\|u_i\|^2, \ \forall u = \{u_k\}_{k \geq 0} \in l_w^2(\mathcal{N}, \mathcal{R}^{n_u}), \ k \in \mathcal{N}$$
$$(8.36)$$

for some nonnegative function $\beta(\cdot) \geq 0$ with $\beta(0) = 0$; see [93, 94]. The following is a nonlinear discrete SBRL.

LEMMA 8.5
System (8.29) with supply rate $w(u, z) = \gamma^2\|u\|^2 - \|z\|^2$ is dissipative iff system (8.29) has its l_2-gain less than or equal to $\gamma > 0$.

Proof. If (8.29) is dissipative, by setting $j = 0$ and $x_0 = 0$ in (8.30), then

$$0 \leq \mathcal{E}V(x_{k+1}) \leq \sum_{i=0}^{k} \mathcal{E}w(u_i, z_i), \ \forall k > 0,$$

which implies (8.35), i.e., system (8.29) has its l_2-gain no larger than $\gamma > 0$. Conversely, if system (8.29) has an l_2-gain $\leq \gamma$, then $\sum_{i=0}^{k} \mathcal{E}\|z_i\|^2$ and $\sum_{i=0}^{k} \mathcal{E}\|u_i\|^2$ are all well defined. By Proposition 8.1, $V_{a,\cdot}$ is an available storage function.

$$\mathcal{E}V_{a,k+1}(x_{k+1}) - V_{a,j}(x)$$

$$= \inf_{u \in l_w^2([j,N],\mathcal{R}^{n_u}), N \geq j \geq 0, x(j)=x \in \mathcal{R}^n} \sum_{i=j}^{N} \mathcal{E}w(u_i, z_i)$$

$$- \inf_{u \in l_w^2([k+1,N],\mathcal{R}^{n_u}), N \geq k+1 > j, x_{k+1} \in l_w^2([k+1,N],\mathcal{R}^n)} \sum_{i=k+1}^{N} \mathcal{E}w(u_i, z_i)$$

$$\leq \inf_{u \in l_w^2([j,N],\mathcal{R}^{n_u}), N \geq j \geq 0, x(j)=x \in \mathcal{R}^n} \sum_{i=j}^{k} \mathcal{E}w(u_i, z_i)$$

$$\leq \sum_{i=j}^{k} \mathcal{E}w(u_i, z_i), \qquad (8.37)$$

where we have used $\inf_n a_n - \inf_n b_n \leq \inf_n(a_n - b_n)$. This implies that system (8.29) is dissipative. $\square$

PROPOSITION 8.2
Consider system (8.29). The inequality (8.33) is equivalent to that for any

fixed $x \in \mathcal{R}^n$ and $u \in \mathcal{R}^{n_u}$, the following discrete HJI holds.

$$V(x) \geq \sup_{u \in \mathcal{R}^{n_u}} \left\{ \mathcal{E}V(F(x, u, w_k)) - \gamma^2 \|u\|^2 + \mathcal{E}G(x, u, w_k) \right\}, \ \forall k \geq 0. \quad (8.38)$$

Proof. If (8.38) holds, then

$$V(x) \geq \mathcal{E}V(F(x, u, w_k)) - \gamma^2 \|u\|^2 + \mathcal{E}G(x, u, w_k), \forall x \in \mathcal{R}^n, \ \forall u \in \mathcal{R}^{n_u}, \ \forall k \geq 0. \quad (8.39)$$

In (8.39), let $x = x_k$, $u = u_k$, $k \geq 0$; the following holds almost surely.

$$V(x_k) \geq [\mathcal{E}V(F(x, u, w_k))]|_{x=x_k, u=u_k} - \gamma^2 \|u_k\|^2 + [\mathcal{E}G(x, u, w_k)]|_{x=x_k, u=u_k}. \quad (8.40)$$

By Markovian property,

$$\mathcal{E}[V(F(x_k, u_k, w_k))|\mathcal{F}_{k-1}] = [\mathcal{E}V(F(x, u, w_k))]|_{x=x_k, u=u_k},$$

$$\mathcal{E}[G(x_k, u_k, w_k)|\mathcal{F}_{k-1}] = [\mathcal{E}G(x, u, w_k)]|_{x=x_k, u=u_k}.$$

Hence, (8.40) becomes

$$V(x_k) \geq \mathcal{E}[V(F(x_k, u_k, w_k))|\mathcal{F}_{k-1}] - \gamma^2 \|u_k\|^2 + \mathcal{E}[G(x, u, w_k)|\mathcal{F}_{k-1}]. \quad (8.41)$$

Taking the mathematical expectation operator on both sides of (8.41), we have

$$\mathcal{E}V(x_k) \geq \mathcal{E}[V(F(x_k, u_k, w_k))] - \gamma^2 \mathcal{E}\|u_k\|^2 + \mathcal{E}[G(x_k, u_k, w_k)]$$
$$= \mathcal{E}[V(F(x_{k+1}))] - \gamma^2 \mathcal{E}\|u_k\|^2 + \mathcal{E}[G(x_k, u_k, w_k)]$$

which yields (8.33).

Conversely, for any $x \in \mathcal{R}^n$, $u \in \mathcal{R}^{n_u}$, by taking $x_k = x$, $u_k = u$ in (8.33), it follows that

$$\mathcal{E}[V(F(x, u, w_k))] - V(x) \leq \mathcal{E}w(u, G(x, u, w_k)) = \gamma^2 \|u\|^2 - \mathcal{E}G(x, u, w_k),$$

which derives (8.38) due to arbitrariness of u. $\square$

Combining Remark 8.3 with Proposition 8.2, the following SBRL is obtained.

THEOREM 8.5
System (8.29) has an l_2-gain $\leq \gamma$ iff HJI (8.38) holds.

Since HJI (8.38) is a necessary and sufficient condition for system (8.29) to have an l_2-gain $\leq \gamma$, it does not have any conservatism. Moreover, HJI (8.38) does not contain $\mathcal{E}x_k$ and $\mathcal{E}u_k$, which can be checked for any given numerical example.

8.2.2   Observability and detectability

We first extend observability and detectability of linear discrete stochastic systems introduced in Chapter 3 to the following nonlinear state-measurement system:

$$\begin{cases} x_{k+1} = F(x_k, w_k), \ x_0 \in \mathcal{R}^n, \\ y_k = G(x_k, w_k), \end{cases} \quad (8.42)$$

where in (8.42), $F(0, w_k) \equiv 0$ for $k \geq 0$, i.e., $x = 0$ is the equilibrium point of

$$x_{k+1} = F(x_k, w_k), \; x_0 \in \mathcal{R}^n. \tag{8.43}$$

DEFINITION 8.4 *System (8.43) is said to be locally asymptotically stable if there exists a neighborhood $U_0 \subset \mathcal{R}^n$, such that for any $x_0 \in U_0$, we have*

$$\lim_{k \to \infty} x_k(\omega) = 0, \; a.s..$$

Furthermore, if $U_0 = \mathcal{R}^n$, then the globally asymptotic stability is defined.

DEFINITION 8.5 *System (8.42) is said to be locally (respectively, globally) zero-state observable if there exists a neighborhood $U_0 \subset \mathcal{R}^n$ (respectively, $\mathcal{R}^n$), such that*

$$y_k = 0, \; a.s., \; k \geq 0 \Rightarrow x_0 = 0.$$

DEFINITION 8.6 *System (8.42) is said to be locally (respectively, globally) zero-state detectable if there exists a neighborhood $U_0 \subset \mathcal{R}^n$ (respectively, $\mathcal{R}^n$), such that*

$$y_k = 0, \; a.s., \; k \geq 0 \Rightarrow \lim_{k \to \infty} x_k(\omega) = 0, \; a.s..$$

8.2.3 Review of martingale theory

Now, we first review some results on martingale theory. The following lemma is the convergence theorem for discrete submartingale; see Theorem 2.2 in [155].

LEMMA 8.6
If $\{X_k, \mathcal{F}_k\}_{k \in \mathcal{N}}$ is a super-martingale such that

$$\sup_k \mathcal{E}\|X_k\| < \infty,$$

then $\{X_k\}_{k \in \mathcal{N}}$ converges almost surely to a limit X_∞ with $\mathcal{E}\|X_\infty\| < \infty$.

The following is the well known Doob's decomposition theorem; see Lemma 7.10 in [110].

LEMMA 8.7
Suppose $\{Y_k, \mathcal{F}_k\}_{k \geq 0}$ is a super-martingale. Then, there exist an increasing predictable sequence $\{A_k\}_{k \geq 0}$ with $A_0 = 0$ and a martingale $\{M_k\}_{k \geq 0}$ such that

$$Y_k = M_k - A_k \quad a.s.. \tag{8.44}$$

Moreover, if $\{Y_k, \mathcal{F}_k\}_{k \geq 0}$ is a nonnegative super-martingale, then A_k converges to A_∞ as $k \to \infty$ and A_∞ is integrable.

We first give the following lemma for the convergence property of the super-martingale.

LEMMA 8.8
Suppose $\{Y_k\}_{k \geq 0}$ is a nonnegative super-martingale. Then,

$$\lim_{k \to \infty} [Y_k - \mathcal{E}(Y_{k+1}|\mathcal{F}_k)] = 0, \quad a.s.. \tag{8.45}$$

Proof. By Doob's decomposition Lemma 8.7, we know that Y_k can be written as

$$Y_k = M_k - A_k, \quad k \geq 0,$$

where $\{M_k, \mathcal{F}_k\}_{k \geq 0}$ is a martingale and $\{A_k\}_{k \geq 0}$ is increasing with $A_0 = 0$. So

$$\begin{aligned} 0 \leq Y_k - \mathcal{E}(Y_{k+1}|\mathcal{F}_k) &= \mathcal{E}(A_{k+1}|\mathcal{F}_k) - A_k \\ &\leq \mathcal{E}(A_\infty|\mathcal{F}_k) - A_k. \end{aligned} \tag{8.46}$$

Since

$$\lim_{k \to \infty} \mathcal{E}(A_\infty|\mathcal{F}_k) = \mathcal{E}(A_\infty|\mathcal{F}_\infty) = A_\infty \quad a.s.,$$

letting $k \to \infty$ on both sides of (8.46), we can obtain (8.45). $\square$

8.2.4 LaSalle-type theorems

The classical LaSalle theorem describes the limit behavior of dynamic systems [129], which is a seminal work in stability theory. Various extended LaSalle theorems to Itô-type SDEs can be found in [104, 132]. In this section, we aim to give a LaSalle-type theorem studied in [124, 221] for the nonhomogeneous system

$$x_{k+1} = F_k(x_k, w_k), \quad x_0 \in \mathcal{R}^n, \quad k \in \mathcal{N}, \tag{8.47}$$

which, we believe, is surely useful in ultimately solving the H_2/H_∞ control of the system (8.27). In (8.47), $\{w_k\}_{k \geq 0}$ is an independent random variable sequence as in (8.27). To this end, we first introduce a class of important Lyapunov functions that will be used in stability study.

DEFINITION 8.7   *A sequence of positive measurable functions $\{V_k\}_{k \geq 0}$: $\mathcal{R}^n \mapsto \mathcal{R}^+$ is called the strong Lyapunov function sequence if there exist a deterministic real-valued sequence $\{\gamma_k \geq 0, k \in \mathcal{N}\}$ and $W : \mathcal{R}^n \mapsto \mathcal{R}^+$ such that*

$$\Delta V_k(x) := \mathcal{E}V_{k+1}[F_k(x, w_k)] - V_k(x) \leq \gamma_k - W(x), \quad \forall x \in \mathcal{R}^n, k \in \mathcal{N}. \tag{8.48}$$

Now, we first review some results of conditional expectation and martingale theory. The following lemma is a special case of Theorem 6.4 in [110].

LEMMA 8.9
If $\mathcal{R}^d$-valued random variable η is independent of the σ-field $\mathcal{G} \subset \mathcal{F}$, and $\mathcal{R}^n$-valued random variable ξ is $\mathcal{G}$-measurable, then, for every bounded function $f : \mathcal{R}^n \times \mathcal{R}^d \mapsto \mathcal{R}$, there exists

$$\mathcal{E}[f(\xi, \eta)|\mathcal{G}] = \mathcal{E}[f(x, \eta)]_{x=\xi} \quad a.s..$$

The following lemma shows the convergence of $V_k(x_k)$ and $W(x_k)$.

LEMMA 8.10
For the strong Lyapunov function sequence $\{V_k\}_{k\in\mathcal{N}}$, if

$$\sum_{k=0}^{\infty} \gamma_k < \infty,$$

then

$$\lim_{k\to\infty} \mathcal{E}V_k(x_k) \quad \text{exists and is finite}$$

and

$$\lim_{k\to\infty} \mathcal{E}W(x_k) = 0,$$

where $\{x_k\}_{k\in\mathcal{N}}$ is the state trajectory of (8.47).

Proof. Denote $V_k^{(N)}(x) = V_k(x)1_{\{|x|\leq N\}}$, then $V_k^{(N)}$ is bounded. For each $x_0 \in \mathcal{R}^n$ and $k \geq 0$, we have

$$\mathcal{E}[V_{k+1}^{(N)}(x_{k+1})|\mathcal{F}_k] = \mathcal{E}[V_{k+1}^{(N)}(F_k(x_k, w_k))|\mathcal{F}_k].$$

Since x_k is $\mathcal{F}_k$-measurable and w_k is independent of $\mathcal{F}_k$, by Lemma 8.9, we have

$$\mathcal{E}[V_{k+1}^{(N)}(x_{k+1})|\mathcal{F}_k] = \mathcal{E}[V_{k+1}^{(N)}(F_k(x, w_k))]_{x=x_k}.$$

By the dominated convergence theorem and letting $N \to \infty$, we have

$$\mathcal{E}[V_{k+1}(x_{k+1})|\mathcal{F}_k] = \mathcal{E}[V_{k+1}(F_k(x, w_k))]_{x=x_k}.$$

By (8.48), we have

$$\mathcal{E}[V_{k+1}(F_k(x, w_k))]_{x=x_k} \leq [V_k(x) + \gamma_k - W(x)]_{x=x_k}$$
$$\leq V_k(x_k) + \gamma_k - W(x_k).$$

We obtain

$$\mathcal{E}[V_{k+1}(x_{k+1})|\mathcal{F}_k] \leq V_k(x_k) + \gamma_k - W(x_k). \tag{8.49}$$

By taking expectation on both sides of (8.49), we obtain

$$\mathcal{E}[V_{k+1}(x_{k+1})] \leq \mathcal{E}[V_k(x_k)] + \gamma_k - \mathcal{E}W(x_k). \tag{8.50}$$

With $W(x) \geq 0$ and accordingly $\mathcal{E}W(x_k) \geq 0$ in mind, we have

$$\mathcal{E}[V_{k+1}(x_{k+1})] \leq \mathcal{E}[V_k(x_k)] + \gamma_k. \tag{8.51}$$

By iterations, we have

$$\mathcal{E}[V_k(x_k)] \leq V_0(x_0) + \sum_{j=0}^{k-1} \gamma_j. \tag{8.52}$$

Since $\sum_{i=0}^{\infty} \gamma_k < \infty$, we have

$$\sup_k \mathcal{E}V_k(x_k) < \infty.$$

Denote

$$\alpha_k = \mathcal{E}V_k(x_k) + \sum_{i=k}^{\infty} \gamma_i.$$

Since $\sum_{i=1}^{\infty} \gamma_i$ is convergent, by (8.52), we can obtain that

$$\alpha_{k+1} \leq \alpha_k,$$

i.e., $\{\alpha_k\}_{k\geq 0}$ is a positive and decreasing sequence. Hence, $\lim_{k\to\infty} \alpha_k$ exists. Note that

$$\mathcal{E}V_k(x_k) = \alpha_k - \sum_{i=k}^{\infty} \gamma_i,$$

so $\mathcal{E}V_k(x_k)$ is convergent. By iterating the inequality (8.51), it follows that

$$\mathcal{E}[V_{k+1}(x_{k+1})] + \sum_{i=0}^{k} \mathcal{E}W(x_i) \leq \mathcal{E}[V_0(x_0)] + \sum_{i=0}^{k} \gamma_i.$$

From the above discussion, we see that $\sum_{i=0}^{\infty} \mathcal{E}W(x_i)$ is convergent, which implies

$$\lim_{k\to\infty} \mathcal{E}W(x_k) = 0,$$

and the proof is hence completed. $\square$

The following theorem is called the LaSalle-type theorem for the discrete-time stochastic system (8.47).

THEOREM 8.6

Suppose $\{V_k\}_{k\in\mathcal{N}}$ is a strong Lyapunov function sequence, and

(i) $\sum_{k=0}^{\infty} \gamma_k < \infty;$

(ii)

$$\liminf_{\|x\| \to \infty} \inf_{k \in \mathcal{N}} V_k(x) = \infty. \tag{8.53}$$

Then

$$\lim_{k \to \infty} V_k(x_k) \quad \text{exists and is finite almost surely,}$$

and

$$\lim_{k \to \infty} W(x_k) = 0 \quad a.s..$$

Proof. If we define

$$Y_k := V_k(x_k) + \sum_{i=k}^{\infty} \gamma_i,$$

then

$$\mathcal{E}(Y_{k+1}|\mathcal{F}_k) \le Y_k - W(x_k).$$

From the above inequality, we know that $\{Y_k, \mathcal{F}_k\}_{k \ge 0}$ is a nonnegative super-martingale. Moreover, by Lemma 8.10, we know that

$$\sup_{k \ge 0} \mathcal{E}\|Y_k\| < \infty.$$

By Lemma 8.6, Y_k converges to Y_∞ almost surely as $k \to \infty$, and $\mathcal{E}\|Y_\infty\| < \infty$. Since $\sum_{i=1}^{\infty} \gamma_k$ is convergent, $V_k(x_k)$ is also convergent and the limit is finite almost surely.

As for $W(x_k) \to 0$ a.s., it can be shown by the following inequality

$$0 \le W(x_k) \le Y_k - \mathcal{E}(Y_{k+1}|\mathcal{F}_k)$$

and Lemma 8.8. This ends the proof. $\square$

If $\{w_k\}_{k \ge 0}$ is an i.i.d. random variable sequence, for the autonomous stochastic system

$$x_{k+1} = F(x_k, w_k), \quad x_0 \in \mathcal{R}^n, \tag{8.54}$$

we set

$$\Delta V_k(x) = \Delta V(x) := \mathcal{E}[V(F(x, w_0))] - V(x). \tag{8.55}$$

The following corollary shows that our result includes that of [129].

COROLLARY 8.3
For system (8.54), if

$$\Delta V(x) \le 0, \tag{8.56}$$

then $\lim_{k\to\infty} V(x_k)$ *exists and is finite almost surely. Moreover, if* $\Delta V(x)$ *is a continuous function and random variable* ξ *is the limit point of* $\{x_k\}_{k\geq 0}$, *then*

$$P\{\xi \in G\} = 1,$$

where

$$G = \{x : \Delta V(x) = 0\}.$$

Proof. The convergence of $\{V(x_k)\}$ can be obtained directly by Theorem 8.6 with $\gamma_k = 0$ and $W(x) = -\Delta V(x) \geq 0$. Moreover, we also have

$$\lim_{k\to\infty} \Delta V(x_k) = -\lim_{k\to\infty} W(x_k) = 0 \quad a.s..$$

By the continuity of $\Delta V(x)$, we have

$$\Delta V(\xi) = 0 \quad a.s.,$$

which implies $\xi \in G$. This ends the proof. □

REMARK 8.4   Corollary 8.3 contains the classical LaSalle's theorem as a special case. This is because if F is a deterministic function (see [129]), then $\Delta V(x)$ given by (8.55) is equivalent to

$$\Delta V(x) = V(F(x)) - V(x).$$

∎

Example 8.1
Consider the following one-dimensional second-order linear difference equation

$$\begin{cases} x_{k+1} = ax_k + bx_{k-1} + \frac{1}{k}w_k, \\ x_0, x_1 \in \mathcal{R}, \ k = 1, 2, \cdots, \end{cases} \tag{8.57}$$

where $\{w_k\}_{k\geq 0}$ is as in (8.27). By introducing another variable $y_{k+1} = x_k$, (8.57) can be transformed into a two-dimensional system as follows:

$$\begin{cases} x_{k+1} = ax_k + by_k + \frac{1}{k}w_k, \\ y_{k+1} = x_k, \\ x_0, x_1 \in \mathcal{R}, \ y_1 = x_0, k = 0, 1, 2, \cdots. \end{cases} \tag{8.58}$$

Set $\eta_k = \begin{bmatrix} x_k \\ y_k \end{bmatrix}$, then (8.58) can be written as a standard second-order difference equation

$$\eta_{k+1} = F_k(\eta_k, w_k) = \begin{bmatrix} a & b \\ 1 & 0 \end{bmatrix} \eta_k + \begin{bmatrix} \frac{1}{k} \\ 0 \end{bmatrix} w_k. \tag{8.59}$$

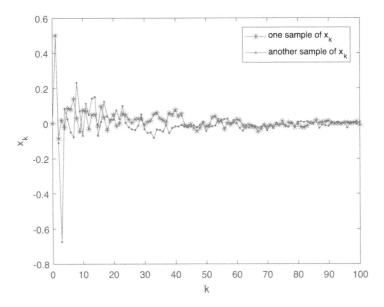

FIGURE 8.1

Trajectory of x_k for (8.57) with $a = b = 0.3$.

If $|a| + |b| < 1$, we take the positive number c in $(\underline{d}^2, \bar{d}^2)$ with

$$\underline{d}^2 = \max\{b^2, [1 + b^2 - a^2 - \sqrt{(1 + b^2 - a^2)^2 - 4b^2}]/2\},$$

$$\bar{d}^2 = \min\{1 - a^2, [1 + b^2 - a^2 + \sqrt{(1 + b^2 - a^2)^2 - 4b^2}]/2\},$$

and let

$$V(\eta) = V(x, y) = \eta^T \begin{bmatrix} 1 & 0 \\ 0 & c \end{bmatrix} \eta = x^2 + cy^2.$$

Then

$$\mathcal{E}V(F_k(\eta, w_k)) - V(\eta) = \mathcal{E}[V(ax + by + \frac{1}{k}w_k, x)] - V(x, y)$$

$$= \frac{1}{k^2} - W(x, y),$$

where $W(x, y) = (1 - c - a^2)x^2 - 2abxy + (c - b^2)y^2 \geq 0$, and $G = \{(x, y) : W(x, y) = 0\}, = \{(0, 0)\}$, i.e., $W(x, y)$ is a positive definite function. Since $\sum_{k=0}^{\infty} \frac{1}{k^2} < \infty$, by Theorem 8.6, we know that $x_k \to 0$ almost surely as $k \to \infty$; see Figure 8.1 for the simulation. If $|a| + |b| \geq 1$, the solutions of (8.57) are not necessarily convergent; see Figure 8.2 for $a = b = 0.612$ and Figure 8.3 for $a = b = 0.5$.

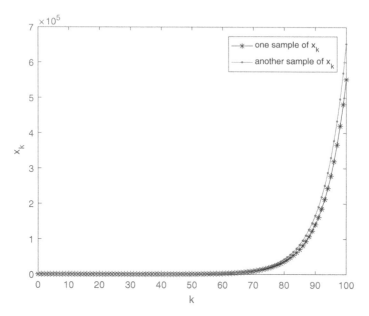

FIGURE 8.2

Trajectory of x_k for system (8.57) with $|a| + |b| > 1$.

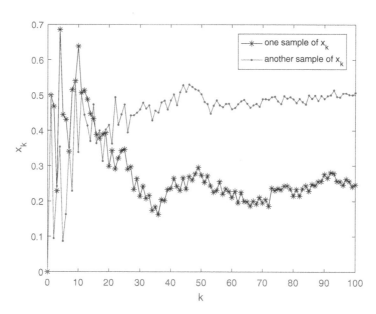

FIGURE 8.3

Trajectory of x_k for system (8.57) with $|a| + |b| = 1$.

REMARK 8.5 The following is the derivation of the constrained conditions $|a| + |b| < 1$ and $c \in (\underline{d}^2, \bar{d}^2)$ in Example 8.1. The quadratic form of $W(x, y)$ can be written as

$$W(x, y) = (1 - a^2 - c)x^2 - 2abxy + (c - b^2)y^2$$
$$= (x, y) \begin{pmatrix} 1 - a^2 - c & -ab \\ -ab & c - b^2 \end{pmatrix} \begin{pmatrix} x \\ y \end{pmatrix}.$$

It is well-known that $W(x, y)$ is a positive definite quadratic form iff

$$1 - a^2 - c > 0, \tag{8.60}$$

$$c - b^2 > 0, \tag{8.61}$$

$$\begin{vmatrix} 1 - a^2 - c & -ab \\ -ab & c - b^2 \end{vmatrix} > 0. \tag{8.62}$$

The inequality (8.62) yields

$$c^2 - (b^2 - a^2 + 1)c + b^2 < 0. \tag{8.63}$$

A necessary and sufficient condition for the solvability of (8.63) is

$$\Delta := [-(b^2 - a^2 + 1)]^2 - 4b^2 > 0.$$

By (8.60) and (8.61), $1 - a^2 > c > 0$, which implies $b^2 - a^2 + 1 > 0$. Therefore, from $\Delta > 0$, we have

$$b^2 - a^2 + 1 > 2|b|,$$

i.e.,

$$(|b| - 1)^2 > |a|^2. \tag{8.64}$$

Again, by (8.60) and (8.61), it is easy to see that $|b| < 1$. So, from (8.64), we have

$$1 - |b| > |a|,$$

i.e., $|a| + |b| < 1$ is a necessary condition for $W(x, y) > 0$.
   Solving the inequality (8.63), we have

$$[1 + b^2 - a^2 - \sqrt{(1 + b^2 - a^2)^2 - 4b^2}]/2 < c$$
$$< [1 + b^2 - a^2 + \sqrt{(1 + b^2 - a^2)^2 - 4b^2}]/2. \tag{8.65}$$

By (8.60) and (8.61), we obtain

$$b^2 < c < 1 - a^2. \tag{8.66}$$

Combining (8.65) and (8.66), c must satisfy

$$c \in (\underline{d}^2, \bar{d}^2),$$

where $\underline{d}^2 = \max\{b^2, [1 + b^2 - a^2 - \sqrt{(1 + b^2 - a^2)^2 - 4b^2}]/2\}$, and $\bar{d}^2 = \min\{1 - a^2, [1 + b^2 - a^2 + \sqrt{(1 + b^2 - a^2)^2 - 4b^2}]/2\}$. ∎

8.2.5   Difficulties in affine nonlinear discrete H_2/H_∞ control

Generally speaking, most research issues concerned with discrete-time systems are easier to settle than their continuous-time counterparts. However, there are some exceptions among which the affine nonlinear discrete H_∞ and H_2/H_∞ control problems are such examples. We note that reference [16] considered a special case of (8.27) with only state-dependent noise:

$$\begin{cases} x_{k+1} = f(x_k) + g(x_k)u_k + h(x_k)v_k + f_1(x_k)w_k, \\ z_k = \begin{bmatrix} m(x_k) \\ u_k \end{bmatrix}, \; x_0 \in \mathcal{R}^n, \; k \in \mathcal{N}, \end{cases} \tag{8.67}$$

where

$$f(x) = A + HF(x)E_1, \; f_1 = (A_1 + HF(x)E_3)x, \tag{8.68}$$

$$g(x) = B_2 + HF(x)E_2, \; h(x) = B_1, \; F^T(x)F(x) \leq I. \tag{8.69}$$

In (8.68) and (8.69), A, A_1, H, H_1 E_1, E_2, E_3, B_1 and B_2 are constant matrices of appropriate dimensions. For the system (8.67), reference [16] obtained an LMI-based sufficient condition for the existence of an H_∞ controller assuming a constant state feedback $u_k = Kx_k$; see Theorem 3 of [16]. There does not seem to have any other studies on the general H_∞ control of the system (8.27).

Although we have presented many preliminaries in the preceding sections, we still have no methods to efficiently solve the H_∞ control as well as the mixed H_2/H_∞ control problems for system (8.27). We summarize the main difficulties as follows:

- Firstly, it is easy to find that the completing squares technique is no longer applicable except for special Lyapunov functions such as quadratic Lyapunov functions. This is because nonlinear discrete systems cannot be iterated. In addition, different from Itô systems where an infinitesimal generator $\mathcal{L}V(x)$ can be used, how to give practical criteria which are not dependent on the mathematical expectation of the trajectory is challenging.

- The methodology used in [128] seems to be invalid for stochastic systems. One perhaps should first develop new techniques such as a discrete stochastic dynamic programming principle to first solve affine quadratic optimal control. Then, perhaps the mixed H_2/H_∞ control can be solved.

8.3   Singular Stochastic H_2/H_∞ Control

Singular systems which are able to describe a larger class of practical systems than normal linear systems have received considerable attention in recent years. Indeed, in the past years, the control of singular systems has been extensively studied and a lot of concepts and results for normal linear systems such as stability [48, 102,

168, 192] and H_∞ control [192, 193, 229] have been extended to singular systems. Furthermore, the study of singular systems has undergone a similar development and a class of stochastic singular systems called the Markov jumping singular systems have been investigated; see [26], [185], [192], [194], [195] and the references therein. However, there are few reports on the H_2/H_∞ control of stochastic singular systems with state-dependent noise, due to many challenging essential issues.

8.3.1 Lemma and definition

Consider the following n-dimensional linear time-invariant stochastic Itô singular system

$$\begin{cases} Edx(t) = Ax(t)\,dt + A_1 x(t)\,dB(t), \\ Ex(0) = x_0 \in \mathcal{R}^n, \end{cases} \tag{8.70}$$

where $Ex(0) \in \mathcal{R}^n$ is the initial condition which is deterministic, E, A and A_1 are constant $n \times n$ matrices and $rank(E) = r \leq n$. In order to guarantee the existence and uniqueness of the solution to system (8.70), we give the following lemma.

LEMMA 8.11
If there are a pair of nonsingular matrices $M \in \mathcal{R}^{n \times n}$ and $N \in \mathcal{R}^{n \times n}$ for the triplet (E, A, A_1) such that one of the following conditions is satisfied, then (8.70) has a unique solution.

(i)

$$MEN = \begin{bmatrix} I_{n_1} & 0 \\ 0 & \tilde{N} \end{bmatrix}, \ MAN = \begin{bmatrix} \tilde{A}_1 & 0 \\ 0 & I_{n_2} \end{bmatrix}, \ MA_1N = \begin{bmatrix} F_1 & F_2 \\ 0 & 0 \end{bmatrix}, \tag{8.71}$$

where $\tilde{N} \in \mathcal{R}^{n_2 \times n_2}$ is a nilpotent matrix, i.e., $\tilde{N}^i \neq 0$ for $i = 1, 2, \cdots, h-1$, while $\tilde{N}^h = 0$. $F_1 \in \mathcal{R}^{n_1 \times n_1}, F_2 \in \mathcal{R}^{n_1 \times n_2}, n_1 + n_2 = n$.

(ii)

$$MEN = \begin{bmatrix} I_r & 0 \\ 0 & 0 \end{bmatrix}, \ MAN = \begin{bmatrix} \tilde{A}_1 & 0 \\ 0 & I_{n-r} \end{bmatrix}, \ MA_1N = \begin{bmatrix} F_1 & F_2 \\ 0 & F_3 \end{bmatrix}, \tag{8.72}$$

where $\tilde{A}_1, F_1 \in \mathcal{R}^{r \times r}, \ F_2 \in \mathcal{R}^{r \times (n-r)}$ and $F_3 \in \mathcal{R}^{(n-r) \times (n-r)}$.

Proof. The proof of item (ii) of Lemma 8.11 can be found in Lemma 2.2 of [96]. As for (i), if we let $\xi(t) = N^{-1}x(t) = [\xi_1(t)^T \ \ \xi_2(t)^T]^T, \xi_1(t) \in \mathcal{R}^{n_1}, \xi_2(t) \in \mathcal{R}^{n_2}$, then under the condition (8.71), (8.70) is equivalent to

$$d\xi_1(t) = \tilde{A}_1\xi_1(t)\,dt + [F_1\xi_1(t) + F_2\xi_2(t)]\,dB(t) \tag{8.73}$$

and

$$\tilde{N}d\xi_2(t) = \xi_2(t)\,dt. \tag{8.74}$$

Taking the Laplace transform on both sides of (8.74), we have

$$(s\tilde{N} - I)\xi_2(s) = \tilde{N}\xi_2(0). \tag{8.75}$$

From (8.75), we obtain

$$\xi_2(s) = (s\tilde{N} - I)^{-1}\tilde{N}\xi_2(0). \tag{8.76}$$

The inverse Laplace transform of $\xi_2(s)$ yields

$$\xi_2(t) = -\sum_{i=1}^{h-1} \delta^{i-1}(t)\tilde{N}^i\xi_2(0), \tag{8.77}$$

where the Dirac function $\delta(t)$ has the Laplace transformation of $L[\delta^i(t)] = s^i$.

On the other hand, by substituting (8.77) into (8.73), we obtain an ordinary SDE with respect to $\xi_1(t)$, which has a unique solution. □

The response of system (8.70) may contain an impulse term, which is to be eliminated to guarantee that the solution of (8.70) is well posed.

DEFINITION 8.8

(i) *The stochastic Itô singular system (8.70) is said to be impulse-free if the conditions in Lemma 8.11-(i) together with* $\deg \det(sE - A) = \operatorname{rank}(E)$ *or the conditions in Lemma 8.11-(ii) hold.*

(ii) *The stochastic Itô singular system (8.70) is said to be asymptotically mean square admissible if it has a unique solution that is impulse-free and ASMS.*

Definition 8.8-(i) considers the effect of the diffusion matrix A_1, so it is more reasonable than most previous definitions on an impulse-free diffusion system.

8.3.2   Asymptotical mean square admissibility

THEOREM 8.7
If system (8.70) has a unique impulse-free solution, then it is asymptotically mean square admissible if the following deterministic singular system is mean square admissible.

$$\bar{E}\dot{\tilde{X}} = \bar{A}\tilde{X}, \tag{8.78}$$

where

$$\bar{E} = H_n^T(E \otimes E)H_n, \quad \tilde{X} = (H_n^T H_n)^{-1}H_n^T vec(X),$$
$$\bar{A} = H_n^T(A \otimes E + E \otimes A + A_1 \otimes A_1)H_n,$$
$$X = \mathcal{E}[x(t)x(t)^T], \quad vec(X) = H_n\tilde{X}$$

with $x(t)$ the trajectory of (8.70) and H_n the $\mathcal{H}$-representation matrix as defined in (1.32).

Proof. For (8.70), by applying Itô's formula and letting $X = \mathcal{E}[x(t)x(t)^T]$, we have

$$E\dot{X}E^T = AXE^T + EXA^T + A_1XA_1^T. \tag{8.79}$$

By Lemma 1.2, (8.79) is equivalent to

$$(E \otimes E)vec(\dot{X}) = (A \otimes E + E \otimes A + A_1 \otimes A_1)vec(X). \tag{8.80}$$

From Lemma 1.1, we have

$$(E \otimes E)H_n\dot{\widetilde{X}} = (A \otimes E + E \otimes A + A_1 \otimes A_1)H_n\widetilde{X}, \tag{8.81}$$

where $\widetilde{X}$ is an $\frac{n(n+1)}{2}$-dimensional vector. Pre-multiplying H_n^T on both sides of (8.81), we derive

$$H_n^T(E \otimes E)H_n\dot{\widetilde{X}} = H_n^T(A \otimes E + E \otimes A + A_1 \otimes A_1)H_n\widetilde{X}. \tag{8.82}$$

Let

$$\bar{E} = H_n^T(E \otimes E)H_n, \quad \bar{A} = H_n^T(A \otimes E + E \otimes A + A_1 \otimes A_1)H_n,$$

then (8.82) is changed into (8.78). Because

$$\lim_{t \to \infty} \widetilde{X} = 0 \Leftrightarrow \lim_{t \to \infty} vec(X) = 0 \Leftrightarrow \lim_{t \to \infty} X = 0 \Leftrightarrow \lim_{t \to \infty} \mathcal{E}\|x(t)\|^2 = 0,$$

if (8.78) is admissible, by Definition 8.8, (8.70) is asymptotically mean square admissible. This completes the proof. $\square$

COROLLARY 8.4

If system (8.70) has a unique impulse-free solution, then it is asymptotically mean square admissible if there exist matrices $\bar{P} > 0$, and $\bar{Q}$ such that the inequality

$$(\bar{P}\bar{E} + \bar{S}\bar{Q})^T\bar{A} + \bar{A}^T(\bar{P}\bar{E} + \bar{S}\bar{Q}) < 0 \tag{8.83}$$

holds, where $\bar{S}$ is any matrix with full column rank and satisfies $\bar{E}^T\bar{S} = 0$.

Proof. If there exist $\bar{P} > 0$ and $\bar{Q}$ such that (8.83) holds, then system (8.78) is admissible; see Theorem 2.2 in [192]. By Theorem 8.7, (8.70) is admissible. $\square$

REMARK 8.6 Compared with [96], we directly use Itô's formula to transform (8.70) into (8.78) without relying on any aided matrices, which makes the computation process simple and clear. (8.83) is a strict LMI, which can be easily solved via the LMI toolbox. ∎

REMARK 8.7 In general, we can directly apply the existing results on deterministic linear systems to study the problems of linear stochastic systems once a linear stochastic system is transformed into a deterministic one. However, this is not true for singular systems. To see this, consider system (8.70) with the following data:

$$E = \begin{bmatrix} 1 & 1 \\ 2 & 2 \end{bmatrix}, \quad A = \begin{bmatrix} -1 & 0 \\ 1 & 1 \end{bmatrix}, \quad A_1 = \begin{bmatrix} 0.5 & 0.2 \\ 1 & 0.4 \end{bmatrix}.$$

In this case, there exist

$$M = \begin{bmatrix} 0 & 0.5 \\ 1 & -0.5 \end{bmatrix}, \quad N = \begin{bmatrix} -0.5 & -1 \\ 1.5 & 1 \end{bmatrix},$$

such that

$$MEN = \begin{bmatrix} 1 & 0 \\ 0 & 0 \end{bmatrix}, \quad MAN = \begin{bmatrix} 0.5 & 0 \\ 0 & 1 \end{bmatrix}, \quad MA_1N = \begin{bmatrix} 0.5 & -0.3 \\ 0 & 0 \end{bmatrix}.$$

By Lemma 8.11-(i), system (8.70) has a solution, but it can be verified that system (8.78) has no solution since $\det(s\bar{E} - \bar{A}) \equiv 0$. Therefore, (8.70) having a solution cannot ensure that (8.78) has a solution and vice versa. The above phenomena reveal the essential difference between singular and normal stochastic systems. ∎

The following is another sufficient condition for asymptotic mean square admissibility without assuming that system (8.70) has an impulse-free solution.

THEOREM 8.8
System (8.70) is asymptotically mean square admissible if there exist matrices $P > 0$ and Q such that

$$A^T PE + E^T PA + A_1^T PA_1 + A^T SQ + Q^T S^T A < 0 \qquad (8.84)$$

and one of the following conditions holds.
(i) $rank(E, A_1) = rank(E)$.
(ii) The condition in Lemma 8.11-(ii) holds.
In (8.84), S is any matrix with full column rank and satisfies $E^T S = 0$.

Proof. We only need to prove this theorem under condition (i), because the proof for the result under condition (ii) is very similar. We first show that system (8.70) has an impulse-free solution. Under the condition (i), there exist nonsingular matrices M, N such that

$$MEN = \begin{bmatrix} I_r & 0 \\ 0 & 0 \end{bmatrix}, \quad MAN = \begin{bmatrix} \bar{A}_{11} & \bar{A}_{12} \\ \bar{A}_{21} & \bar{A}_{22} \end{bmatrix}, \quad MA_1N = \begin{bmatrix} \bar{A}_1 & \bar{A}_2 \\ 0 & 0 \end{bmatrix}. \qquad (8.85)$$

In view of $P > 0$, $A_1^T P A_1 \geq 0$, (8.84) results in that

$$A^T PE + E^T PA + A^T SQ + Q^T S^T A < 0. \tag{8.86}$$

By letting $\bar{P} = PE + SQ$, (8.86) becomes

$$A^T \bar{P} + \bar{P}^T A < 0, \tag{8.87}$$

$$E^T \bar{P} = \bar{P}^T E \geq 0. \tag{8.88}$$

Therefore, by Theorem 2.1 and Lemma 2.3 in [192], (E, A) is impulse-free and $\bar{A}_{22}$ is invertible. By letting

$$M_1 = \begin{bmatrix} I & -\bar{A}_{12}\bar{A}_{22}^{-1} \\ 0 & \bar{A}_{22}^{-1} \end{bmatrix} M, \ N_1 = N \begin{bmatrix} I & 0 \\ -\bar{A}_{22}^{-1}\bar{A}_{21} & I \end{bmatrix},$$

$$\tag{8.89}$$

we have

$$M_1 E N_1 = \begin{bmatrix} I_r & 0 \\ 0 & 0 \end{bmatrix}, \ M_1 A N_1 = \begin{bmatrix} \bar{A}_{11} - \bar{A}_{12}\bar{A}_{22}^{-1}\bar{A}_{21} & 0 \\ 0 & I \end{bmatrix},$$

$$M_1 A_1 N_1 = \begin{bmatrix} \bar{A}_1 - \bar{A}_2 \bar{A}_{22}^{-1} \bar{A}_{21} & \bar{A}_2 \\ 0 & 0 \end{bmatrix}. \tag{8.90}$$

It is easy to see that (8.90) satisfies condition (ii) in Lemma 8.11, so (8.70) has an impulse-free solution. Let

$$\xi(t) = N_1^{-1} x(t) = [\xi_1(t)^T \quad \xi_2(t)^T]^T, \tag{8.91}$$

where $\xi_1(t) \in \mathcal{R}^r, \xi_2(t) \in \mathcal{R}^{n-r}$, then system (8.70) is equivalent to

$$\begin{cases} d\xi_1(t) = (\bar{A}_{11} - \bar{A}_{12}\bar{A}_{22}^{-1}\bar{A}_{21})\xi_1(t)\, dt + (\bar{A}_1 - \bar{A}_2\bar{A}_{22}^{-1}\bar{A}_{21})\xi_1(t)\, dB(t), \\ \xi_2(t) = 0. \end{cases}$$

By (8.90), S can be expressed as

$$S = M_1^T \begin{bmatrix} 0 \\ I \end{bmatrix} H, \tag{8.92}$$

where H is any nonsingular matrix. Write

$$M_1^{-T} P M_1^{-1} = \begin{bmatrix} \bar{P}_{11} & \bar{P}_{12} \\ \bar{P}_{12}^T & \bar{P}_{22} \end{bmatrix}, \ HQN_1 = \begin{bmatrix} \bar{Q}_1 & \bar{Q}_2 \end{bmatrix}, \tag{8.93}$$

where the partition is compatible with that of A in (8.90). Using (8.84), (8.90) and (8.93), it is easy to verify that

$$A^T PE + E^T PA + A_1^T PA_1 + A^T SQ + Q^T S^T A = N_1^{-T} \begin{bmatrix} U & * \\ * & * \end{bmatrix} N_1^{-1}$$

$$< 0, \tag{8.94}$$

where $'*'$ represents an element which is not essential in the following discussion, and

$$U = (\bar{A}_{11} - \bar{A}_{12}\bar{A}_{22}^{-1}\bar{A}_{21})^T \bar{P}_{11} + \bar{P}_{11}(\bar{A}_{11} - \bar{A}_{12}\bar{A}_{22}^{-1}\bar{A}_{21})$$
$$+ (\bar{A}_1 - \bar{A}_2\bar{A}_{22}^{-1}\bar{A}_{21})^T \bar{P}_{11}(\bar{A}_1 - \bar{A}_2\bar{A}_{22}^{-1}\bar{A}_{21}).$$

$$(8.95)$$

(8.94) implies $U < 0$. Using $U < 0$ and $\bar{P}_{11} > 0$ together with Lyapunov-type Theorem 1.5, we derive

$$\lim_{t\to\infty} \mathcal{E}\|\xi_1(t)\|^2 = 0. \tag{8.96}$$

By (8.91) and (8.96), we have

$$\lim_{t\to\infty} \mathcal{E}\|x(t)\|^2 = 0. \tag{8.97}$$

From Definition 8.8, system (8.70) is asymptotically mean square admissible. □

REMARK 8.8   When $A_1 = 0$, system (8.70) degenerates to a deterministic singular system [192]. Theorem 8.8 can be regarded as an extension of the corresponding result in [192]. When $E = I$, system (8.70) reduces to a normal stochastic system. In this case, Theorem 8.8 is consistent with Theorem 1.5 via taking $E = I$ and $Q = 0$ in (8.84).   ∎

REMARK 8.9   For deterministic continuous-time singular systems, the usual Lyapunov function candidate is chosen as

$$V(x(t)) = x(t)^T E^T P x(t), \tag{8.98}$$

where $E^T P = P^T E \geq 0$. It is natural to consider (8.98) as a Lyapunov function candidate for stochastic Itô singular systems as done in Theorem 3.1 of [27]. However, we would like to point out that such a choice is not applicable for system (8.70), because we do not know how to compute $dx(t)$ in the following

$$dV(x(t)) = d(x^T(t)E^T)Px(t) + x^T(t)P^T d(Ex(t))$$
$$+ d(x^T(t)E^T)Pd(x(t)).$$

In general, for stochastic singular systems, we take $V(x(t)) = x(t)^T E^T P E x(t)$ with P a symmetric positive definite matrix. In this case, the infinitesimal generator of the system considered in [27] can be given as

$$\mathcal{L}V(x(t), i) = x^T(t)[A(i)^T P(i)E + E^T P(i)A(i)$$
$$+ F^T(i)P(i)F(i) + \sum_{j=1}^{N} \lambda_{ij} E^T P(j)E]x(t)$$
$$+ f^T P(i)Ex(t) + x^T(t)E^T P(i)f.$$

∎

By means of the Moore–Penrose inverse E^+, the following sufficient condition was given in [228].

THEOREM 8.9
Under the assumption of $rank(E, A_1) = rank(E)$, system (8.70) is asymptotically mean square admissible if there exists a matrix P such that

$$E^T P = P^T E \geq 0, \tag{8.99}$$

$$A^T P + P^T A + A_1^T (E^+)^T E^T P(E^+) A_1 < 0. \tag{8.100}$$

Theorem 8.9 improves Theorem 1 of [82] as it does not require regular assumption $\det(sE - A) \not\equiv 0$.

COROLLARY 8.5
Under the conditions of Theorem 8.9, system (8.70) is asymptotically mean square admissible if there exist a matrix $P \in \mathcal{R}^{n \times n} > 0$ and a nonsingular matrix $Q \in \mathcal{R}^{(n-r) \times (n-r)}$, such that the following LMI holds:

$$A^T (PE + U^T QV^T) + (PE + U^T QV^T)^T A + A_1^T (E^+)^T E^T PEE^+ A_1 < 0,$$

where $U \in \mathcal{R}^{(n-r) \times n}$ and $V \in \mathcal{R}^{n \times (n-r)}$ are of full row rank and column rank that are bases of the left and right null space of E, respectively.

Proof. By setting $P = \bar{P} := PE + U^T QV^T$ in (8.99) and (8.100), it is easily tested that (8.99) and (8.100) hold. Hence, the result follows from Theorem 8.9 immediately. $\square$

We now investigate the discrete counterpart of system (8.70).

For the discrete-time stochastic singular system with state-dependent noise described by

$$\begin{cases} Ex_{k+1} = Ax_k + A_1 x_k w_k, \\ Ex_0 \in \mathcal{R}^n, \end{cases} \tag{8.101}$$

where $\{w_k\}_{k \in \mathcal{N}}$ is a one-dimensional independent white noise process, $\mathcal{E}(w_k) = 0$ and $\mathcal{E}(w_k w_s) = \delta_{ks}$ with δ_{ks} being a Kronecker delta. Similar to Corollary 8.4 and Theorem 8.8, the following theorems can be readily obtained.

THEOREM 8.10
System (8.101) is asymptotically mean square admissible if there exist a matrix $\bar{P} > 0$ and a symmetric nonsingular matrix $\bar{Q}$ such that

$$\bar{A}^T (\bar{P} - \bar{S}\bar{Q}\bar{S}^T)\bar{A} - \bar{E}^T \bar{P}\bar{E} < 0, \tag{8.102}$$

where $\bar{S}$ is any matrix with full column rank and satisfies $\bar{E}^T \bar{S} = 0$, $\bar{A} = H_n^T (A \otimes A + A_1 \otimes A_1) H_n$, and $\bar{E}$ is the same as in Theorem 8.7.

THEOREM 8.11

System (8.101) is asymptotically mean square admissible if there exist matrices $P > 0$, Q and a full column rank matrix S satisfying $E^T S = 0$, such that

$$A^T PA - E^T PE + A_1^T PA_1 + A^T SQ + Q^T S^T A < 0, \tag{8.103}$$

and one of the following conditions holds.
(i) $rank[E \ A_1] = rank(E)$.
(ii) the condition in Lemma 8.11-(ii) holds.

8.3.3   An illustrative example

We consider the oil catalytic cracking model [48] given by

$$\dot{x}_1(t) = R_{11}x_1(t) + R_{12}x_2(t) + B_1u(t) + C_1\eta,$$
$$0 = R_{21}x_1(t) + R_{22}x_2(t) + B_2u(t) + C_2\eta, \tag{8.104}$$

where $x_1(t)$ is a vector to be regulated, which includes regeneration temperature, valve position, blower capacity, etc., $x_2(t)$ is the vector reflecting business benefits, administration, policy, etc., $u(t)$ is the regulation value, and η represents external disturbances. For convenience, we consider the case of $u(t) = 0, \eta = 0$, then (8.104) can be expressed as

$$E\dot{x}(t) = Rx(t), \tag{8.105}$$

where $x(t) = [x_1(t)^T \quad x_2(t)^T]^T$ is the state vector,

$$E = \begin{bmatrix} 1 & 0 \\ 0 & 0 \end{bmatrix}, \ R = \begin{bmatrix} R_{11} & R_{12} \\ R_{21} & R_{22} \end{bmatrix}.$$

It is obvious that (8.105) is a deterministic singular system. However, it might happen that R is subject to some random environmental effects such as $R = A + A_1$ "*noise.*" In this case, (8.105) becomes

$$\frac{Edx(t)}{dt} = Ax(t) + A_1x(t) \text{ "}noise.\text{"} \tag{8.106}$$

It turns out that a reasonable mathematical interpretation for the "noise" term is the so-called white noise $\dot{w}(t)$. By (8.106), we have

$$Edx(t) = Ax(t)\,dt + A_1x(t)\,dB(t), \tag{8.107}$$

which is in the form of system (8.70). In (8.107), A is called the drift matrix reflecting the effect on the system state, while A_1 is called the diffusion matrix reflecting the noise intensity. In what follows, we will verify the effectiveness of Theorem 8.8. The following data are taken for (8.107):

$$E = \begin{bmatrix} 1 & 0 \\ 0 & 0 \end{bmatrix}, \ A = \begin{bmatrix} -1 & 0 \\ 0 & 1 \end{bmatrix}, \ A_1 = \begin{bmatrix} 0.2 & 0.2 \\ 0 & 0.8 \end{bmatrix}. \tag{8.108}$$

(8.108) satisfies the condition of Lemma 8.11-(ii), so system (8.107) has an impulse-free solution. By solving (8.84) with $S = \begin{bmatrix} 0 & 1 \end{bmatrix}^T$, we obtain

$$P = \begin{bmatrix} 76.0663 & 0 \\ 0 & 124.4199 \end{bmatrix}, \quad Q = \begin{bmatrix} -3.0427 & -103.5456 \end{bmatrix}.$$

Therefore, system (8.107) is asymptotically mean square admissible.

8.3.4 Problems in H_2/H_∞ control

We consider the following singular system with only the state-dependent noise:

$$\begin{cases} Edx(t) = [A_1 x(t) + B_1 u(t) + C_1 v(t)] \, dt + A_2 x(t) \, dB(t), \\ Ex(0) = x_0, \\ z(t) = \begin{bmatrix} Cx(t) \\ Du(t) \end{bmatrix}, \quad D^T D = I, \, t \in [0, \infty), \end{cases} \tag{8.109}$$

where all coefficient matrices are constant and the square matrix E satisfies $rank(E) = r \leq n$.

DEFINITION 8.9 *The infinite horizon H_2/H_∞ control of (8.109) is to search for $(u_\infty^*(t), v_\infty^*(t)) \in \mathcal{L}_\mathcal{F}^2(\mathcal{R}^+, \mathcal{R}^{n_u}) \times \mathcal{L}_\mathcal{F}^2(\mathcal{R}^+, \mathcal{R}^{n_v})$ such that:*

(i) *When $u_\infty^*(t)$ is applied to (8.109), system (8.109) is asymptotically mean square admissible in the absence of v.*

(ii) *For a given disturbance attenuation $\gamma > 0$, under the constraint of*

$$\begin{cases} Edx(t) = [A_1 x(t) + B_1 u_\infty^*(t) + C_1 v(t)] \, dt + A_2 x(t) dB(t), \\ z(t) = \begin{bmatrix} Cx(t) \\ Du_\infty^*(t) \end{bmatrix}, \quad D^T D = I, \end{cases} \tag{8.110}$$

we have

$$\|\mathcal{L}_\infty\| = \sup_{v \in \mathcal{L}_\mathcal{F}^2(\mathcal{R}^+, \mathcal{R}^{n_v}), v \neq 0, u = u_\infty^*, x_0 = 0} \frac{\|z\|_{[0,\infty)}}{\|v\|_{[0,\infty)}} < \gamma.$$

(iii) *When the worst-case $v_\infty^*(t) \in \mathcal{L}_\mathcal{F}^2(\mathcal{R}^+, \mathcal{R}^{n_v})$ is applied to (8.109), $u_\infty^*(t)$ minimizes the output energy*

$$\|z(t)\|_{[0,\infty)}^2 = \mathcal{E} \int_0^\infty \|z(t)\|^2 \, dt.$$

If the above (u_∞^*, v_∞^*) exists, then we say that the infinite horizon H_2/H_∞ control has a solution pair (u_∞^*, v_∞^*).

Up to now, little research has been done concerning the stochastic H_∞, H_2 and mixed H_2/H_∞ control of system (8.109). Of course, one may give some sufficient conditions for the H_∞ control, however, there have been no results corresponding to Theorem 2.15, Theorem 2.16 and Theorem 2.17. We believe that before solving the mixed H_2/H_∞ control, one must introduce some new concepts such as exact observability and exact detectability, and solve the singular LQ control problem.

8.4   Mean-Field Stochastic H_2/H_∞ Control

In recent years, mean-field theory has attracted a great deal of attention, which is to study collective behaviors resulting from individuals' mutual interactions in various physical and sociological dynamical systems. The new feature for mean-field models is that the system dynamic involves not only the state $x(t)$, control input $u(t)$ and external disturbance $v(t)$, but also their expectations $\mathcal{E}x(t)$, $\mathcal{E}u(t)$ and $\mathcal{E}v(t)$. For example, the following dynamic system

$$
\begin{cases}
dx(t) = [A_1(t)x(t) + \bar{A}_1(t)\mathcal{E}x(t) + B_1(t)u(t) + C_1(t)v(t) + \bar{C}_1(t)\mathcal{E}v(t)]\,dt \\
\qquad + [A_2(t)x(t) + \bar{A}_2(t)\mathcal{E}x(t) + C_2(t)v(t) + \bar{C}_2(t)\mathcal{E}v(t)]\,dB(t), \\
x(0) = x_0 \in \mathcal{R}^n, \\
z(t) = \begin{bmatrix} C(t)x(t) \\ D(t)u(t) \end{bmatrix}, \quad D^T(t)D(t) = I,\ t \in [0,T]
\end{cases}
$$

(8.111)

is referred to as a continuous-time SDE of McKean–Vlasov type or mean-field SDE. According to mean-field theory, the mean-field term is used to model the interactions among agents, which approaches the expected value when the number of agents goes to infinity. The mean-field approach has been widely applied to various fields such as engineering, finance, economics and game theory in the past few years. The investigation of mean-field SDE can be traced back to McKean–Vlasov SDE in the 1960s [133]. Since then, many authors have made their contributions to McKean–Vlasov type SDEs and their applications [9, 100, 119]. Recently, finite and infinite horizon continuous-time mean-field LQ control problems were discussed in [197] and [99], respectively. Reference [66] presented four methods to solve the discrete-time mean-field LQ control problem and the corresponding infinite horizon case was investigated in [144]. However, up to now, there is no ideal result (necessary and sufficient condition) about the stochastic mean-field H_2/H_∞ control. Below, we study the finite horizon H_2/H_∞ control of the system (8.111).

8.4.1   Definition for H_2/H_∞ control

DEFINITION 8.10   *Consider system (8.111). For a given disturbance attenuation level $\gamma > 0$, $0 < T < \infty$, find a feedback control $u_T^*(t) = U(t)x(t) + \tilde{U}(t)\mathcal{E}x(t) = U(t)[x(t) - \mathcal{E}x(t)] + [U(t) + \tilde{U}(t)]\mathcal{E}x(t) \in \mathcal{L}_{\mathcal{F}}^2([0,T], \mathcal{R}^{n_u})$ such that*

1)

$$\|\mathcal{L}_T\| = \sup_{v \in \mathcal{L}_{\mathcal{F}}^2([0,T],\mathcal{R}^{n_v}), v \neq 0, u = u_T^*, x_0 = 0} \frac{\|z\|_{[0,T]}}{\|v\|_{[0,T]}} < \gamma, \quad (8.112)$$

where $\mathcal{L}_T$ is an operator associated with system

$$\begin{cases} dx(t) = [(A_1 + B_1 U)(t)x(t) + (\bar{A}_1 + B_1 \tilde{U})(t)\mathcal{E}x(t) + (C_1 v + \bar{C}_1 \mathcal{E}v)(t)] \, dt \\ \qquad + (A_2 x + \bar{A}_2 \mathcal{E}x + C_2 v + \bar{C}_2 \mathcal{E}v)(t) \, dB(t), \\ x(0) = x_0 \in \mathcal{R}^n, \\ z(t) = \begin{bmatrix} C(t)x(t) \\ D(t)u(t) \end{bmatrix}, \quad D^T(t)D(t) = I, \ t \in [0,T]. \end{cases}$$

$$(8.113)$$

2) When the worst-case disturbance $v_T^*(t) = V(t)x(t) + \tilde{V}(t)\mathcal{E}x(t)$ is applied to (8.111), $u_T^* = U(t)x(t) + \tilde{U}(t)\mathcal{E}x(t)$ minimizes the output energy

$$J_{2,T}(u, v_T^*) := \|z(t)\|_{[0,T]}^2 = \mathcal{E} \int_0^T \|z(t)\|^2 \, dt.$$

It should be noticed that, different from Definition 2.1, here, we select u_T^* and v_T^* from the sets of $u_T^* = U(t)x(t) + \tilde{U}(t)\mathcal{E}x(t)$ and $v_T^*(t) = V(t)x(t) + \tilde{V}(t)\mathcal{E}x(t)$, respectively, which will make our discussions convenient.

8.4.2 Finite horizon SBRL

Consider the following system

$$\begin{cases} dx(t) = (A_{11}x + \bar{A}_{11}\mathcal{E}x + B_{11}v + \bar{B}_{11}\mathcal{E}v)(t) \, dt \\ \qquad + (A_{12}x + \bar{A}_{12}\mathcal{E}x + B_{12}v + \bar{B}_{12}\mathcal{E}v)(t) \, dB(t), \\ x(0) = x_0, \\ z_1(t) = C_{11}(t)x(t), \ t \in [0,T] \end{cases} \quad (8.114)$$

with its disturbed operator $\tilde{\mathcal{L}}_T$ as defined in Chapter 2. Taking the mathematical expectation in (8.114), we have

$$\begin{cases} d\mathcal{E}x(t) = [(A_{11} + \bar{A}_{11})(t)\mathcal{E}x(t) + (B_{11} + \bar{B}_{11})(t)\mathcal{E}v(t)] \, dt, \\ \mathcal{E}x(0) = x_0, \ t \in [0,T] \end{cases} \quad (8.115)$$

and

$$\begin{cases} d[x(t) - \mathcal{E}x(t)] = \{A_{11}(t)[x(t) - \mathcal{E}x(t)] + B_{11}(t)[v(t) - \mathcal{E}v(t)]\} \, dt \\ \qquad + \{A_{12}(t)[x(t) - \mathcal{E}x(t)] + (A_{12} + \bar{A}_{12})(t)\mathcal{E}x(t) \\ \qquad + B_{12}(t)[v(t) - \mathcal{E}v(t)] + (B_{12} + \bar{B}_{12})(t)\mathcal{E}v(t)\} \, dB(t), \\ x_0 - \mathcal{E}x_0 = 0. \end{cases}$$

$$(8.116)$$

LEMMA 8.12

Consider system (8.114). Assume that $P(t), Q(t) : [0,T] \mapsto S_n$ are arbitrary differentiable matrix-valued functions of t. Then, for any $x_0 \in \mathcal{R}^n$, we have

$$J_1^T(x(t), v(t), x_0, 0) := \gamma^2 \|v(t)\|_{[0,T]}^2 \, |_{x(0)=x_0} - \|z_1(t)\|_{[0,T]}^2 \, |_{x(0)=x_0}$$

$$= \mathcal{E} \int_0^T \begin{bmatrix} x(t) - \mathcal{E}x(t) \\ v(t) - \mathcal{E}v(t) \end{bmatrix}^T G_\gamma(P(t)) \begin{bmatrix} x(t) - \mathcal{E}x(t) \\ v(t) - \mathcal{E}v(t) \end{bmatrix} dt$$

$$+ \int_0^T \begin{bmatrix} \mathcal{E}x(t) \\ \mathcal{E}v(t) \end{bmatrix}^T M_\gamma(P(t), Q(t)) \begin{bmatrix} \mathcal{E}x(t) \\ \mathcal{E}v(t) \end{bmatrix} dt$$

$$- \mathcal{E}\{[x(T) - \mathcal{E}x(T)]^T P(T)[x(T) - \mathcal{E}x(T)]\}$$

$$+ x_0^T Q(0) x_0 - \mathcal{E}x^T(T) Q(T) \mathcal{E}x(T), \qquad (8.117)$$

where

$$G_\gamma(P(t)) = \begin{bmatrix} \dot{P} + A_{11}^T P + P A_{11} + A_{12}^T P A_{12} - C_{11}^T C_{11} & P B_{11} + A_{12}^T P B_{12} \\ B_{11}^T P + B_{12}^T P A_{12} & \gamma^2 I_{n_v} + B_{12}^T P B_{12} \end{bmatrix}(t),$$

$$M_\gamma(P(t), Q(t)) = \begin{bmatrix} \dot{Q} + \bar{A}^T Q + Q\bar{A} + \bar{C}^T P \bar{C} - C_{11}^T C_{11} & Q\bar{B} + \bar{C}^T P \bar{D} \\ \bar{B}^T Q + \bar{D}^T P \bar{C} & \gamma^2 I_{n_v} + \bar{D}^T P \bar{D} \end{bmatrix}(t),$$

and

$$\bar{A}(t) = (A_{11} + \bar{A}_{11})(t), \quad \bar{B}(t) = (B_{11} + \bar{B}_{11})(t),$$

$$\bar{C}(t) = (A_{12} + \bar{A}_{12})(t), \quad \bar{D}(t) = (B_{12} + \bar{B}_{12})(t).$$

Proof. Using Itô's formula and considering (8.115) and (8.116), we have

$$\mathcal{E} \int_0^T d\{[x(t) - \mathcal{E}x(t)]^T P(t)[x(t) - \mathcal{E}x(t)]\}$$

$$= \mathcal{E} \int_0^T \begin{bmatrix} x - \mathcal{E}x \\ v - \mathcal{E}v \end{bmatrix}^T (t) \begin{bmatrix} \dot{P} + A_{11}^T P + P A_{11} + A_{12}^T P A_{12} & P B_{11} + A_{12}^T P B_{12} \\ B_{11}^T P + B_{12}^T P A_{12} & B_{12}^T P B_{12} \end{bmatrix}(t)$$

$$\cdot \begin{bmatrix} x - \mathcal{E}x \\ v - \mathcal{E}v \end{bmatrix}(t) dt + \int_0^T \begin{bmatrix} \mathcal{E}x \\ \mathcal{E}v \end{bmatrix}^T (t) \begin{bmatrix} \bar{C}^T P \bar{C} & \bar{C}^T P \bar{D} \\ \bar{D}^T P \bar{C} & \bar{D}^T P \bar{D} \end{bmatrix}(t) \begin{bmatrix} \mathcal{E}x \\ \mathcal{E}v \end{bmatrix}(t) dt$$

$$= \mathcal{E}\{[x(T) - \mathcal{E}x(T)]^T P(T)[x(T) - \mathcal{E}x(T)]\}$$

and

$$\int_0^T d[\mathcal{E}x^T(t) Q(t) \mathcal{E}x(t)] = \mathcal{E}x^T(T) Q(T) \mathcal{E}x(T) - x_0^T Q(0) x_0$$

$$= \int_0^T \begin{bmatrix} \mathcal{E}x(t) \\ \mathcal{E}v(t) \end{bmatrix}^T \begin{bmatrix} \dot{Q}(t) + \bar{A}(t)^T Q(t) + Q(t)\bar{A}(t) & Q(t)\bar{B}(t) \\ \bar{B}(t)^T Q(t) & 0 \end{bmatrix} \begin{bmatrix} \mathcal{E}x(t) \\ \mathcal{E}v(t) \end{bmatrix} dt.$$

Therefore,

$$\mathcal{E} \int_0^T d\{[x(t) - \mathcal{E}x(t)]^T P(t)[x(t) - \mathcal{E}x(t)]\} + \int_0^T d(\mathcal{E}x^T(t) Q(t) \mathcal{E}x(t))$$

$$= \mathcal{E} \int_0^T \begin{bmatrix} x - \mathcal{E}x \\ v - \mathcal{E}v \end{bmatrix}^T (t) \begin{bmatrix} \dot{P} + A_{11}^T P + P A_{11} + A_{12}^T P A_{12} & P B_{11} + A_{12}^T P B_{12} \\ B_{11}^T P + B_{12}^T P A_{12} & B_{12}^T P B_{12} \end{bmatrix} (t)$$

$$\cdot \begin{bmatrix} x - \mathcal{E}x \\ v - \mathcal{E}v \end{bmatrix} (t)\, dt$$

$$+ \int_0^T \begin{bmatrix} \mathcal{E}x \\ \mathcal{E}v \end{bmatrix}^T (t) \begin{bmatrix} \dot{Q} + \bar{A}^T Q + Q\bar{A} + \bar{C}^T P \bar{C} & Q\bar{B} + \bar{C}^T P \bar{D} \\ \bar{B}^T Q + \bar{D}^T P \bar{C} & \bar{D}^T P \bar{D} \end{bmatrix} (t) \begin{bmatrix} \mathcal{E}x \\ \mathcal{E}v \end{bmatrix} (t)\, dt$$

$$= \mathcal{E}\{ [x(T) - \mathcal{E}x(T)]^T P(T)[x(T) - \mathcal{E}x(T)] \} + \mathcal{E}x^T(T) Q(T) \mathcal{E}x(T) - x_0^T Q(0) x_0.$$

Rewriting $v(t)$ and $x(t)$ as $v(t) = v(t) - \mathcal{E}v(t) + \mathcal{E}v(t)$ and $x(t) = x(t) - \mathcal{E}x(t) + \mathcal{E}x(t)$, respectively, we have

$$\gamma^2 \|v(t)\|_{[0,T]}^2 = \int_0^T \begin{bmatrix} \mathcal{E}x(t) \\ \mathcal{E}v(t) \end{bmatrix}^T \begin{bmatrix} 0 & 0 \\ 0 & \gamma^2 I_{n_v} \end{bmatrix} \begin{bmatrix} \mathcal{E}x(t) \\ \mathcal{E}v(t) \end{bmatrix} dt$$

$$+ \mathcal{E} \int_0^T \begin{bmatrix} x(t) - \mathcal{E}x(t) \\ v(t) - \mathcal{E}v(t) \end{bmatrix}^T \begin{bmatrix} 0 & 0 \\ 0 & \gamma^2 I_{n_v} \end{bmatrix} \begin{bmatrix} x(t) - \mathcal{E}x(t) \\ v(t) - \mathcal{E}v(t) \end{bmatrix} dt$$

and

$$\|z_1(t)\|_{[0,T]}^2 = \int_0^T \begin{bmatrix} \mathcal{E}x(t) \\ \mathcal{E}v(t) \end{bmatrix}^T \begin{bmatrix} C_{11}(t)^T C_{11}(t) & 0 \\ 0 & 0 \end{bmatrix} \begin{bmatrix} \mathcal{E}x(t) \\ \mathcal{E}v(t) \end{bmatrix} dt$$

$$+ \mathcal{E} \int_0^T \begin{bmatrix} x(t) - \mathcal{E}x(t) \\ v(t) - \mathcal{E}v(t) \end{bmatrix}^T \begin{bmatrix} C_{11}(t)^T C_{11}(t) & 0 \\ 0 & 0 \end{bmatrix} \begin{bmatrix} x(t) - \mathcal{E}x(t) \\ v(t) - \mathcal{E}v(t) \end{bmatrix} dt.$$

Therefore,

$$J_1^T(x(t), v(t), x_0, 0) = \gamma^2 \|v(t)\|_{[0,T]}^2 \, |_{x(0) = x_0} - \|z_1(t)\|_{[0,T]}^2 \, |_{x(0) = x_0}$$

$$= \mathcal{E} \int_0^T (\gamma^2 v^T v - z_1^T z_1)(t)\, dt + \mathcal{E} \int_0^T d[(x - \mathcal{E}x)^T(t) P(t)(x - \mathcal{E}x)(t)]$$

$$+ \int_0^T d[\mathcal{E}x^T(t) Q(t) \mathcal{E}x(t)] - \mathcal{E}[(x - \mathcal{E}x)^T(T) P(T)(x - \mathcal{E}x)(T)]$$

$$+ x_0^T Q(0) x_0 - \mathcal{E}x^T(T) Q(T) \mathcal{E}x(T)$$

$$= \mathcal{E} \int_0^T \begin{bmatrix} x(t) - \mathcal{E}x(t) \\ v(t) - \mathcal{E}v(t) \end{bmatrix}^T G_\gamma(P(t)) \begin{bmatrix} x(t) - \mathcal{E}x(t) \\ v(t) - \mathcal{E}v(t) \end{bmatrix} dt$$

$$+ \int_0^T \begin{bmatrix} \mathcal{E}x(t) \\ \mathcal{E}v(t) \end{bmatrix}^T M_\gamma(P(t), Q(t)) \begin{bmatrix} \mathcal{E}x(t) \\ \mathcal{E}v(t) \end{bmatrix} dt$$

$$- \mathcal{E}\{ [x(T) - \mathcal{E}x(T)]^T P(T)[x(T) - \mathcal{E}x(T)] \} + x_0^T Q(0) x_0$$

$$- \mathcal{E}x^T(T) Q(T) \mathcal{E}x(T),$$

and the proof is complete. $\square$

Here and in what follows, we use the following notations for convenience:

$$\mathcal{L}(P(t)) = \dot{P}(t) + A_{11}^T(t) P(t) + P(t) A_{11}(t) + A_{12}^T(t) P(t) A_{12}(t) - C_{11}^T(t) C_{11}(t),$$

$$\mathcal{M}(P(t)) = P(t) B_{11}(t) + A_{12}^T(t) P(t) B_{12}(t),$$

$$\mathcal{H}(P(t)) = \gamma^2 I_{n_v} + B_{12}^T(t)P(t)B_{12}(t),$$
$$\tilde{\mathcal{L}}(P(t), Q(t)) = \dot{Q}(t) + \bar{A}(t)^T Q(t) + Q(t)\bar{A}(t) + \bar{C}(t)^T P(t)\bar{C}(t) - C_{11}^T(t)C_{11}(t),$$
$$\tilde{\mathcal{M}}(P(t), Q(t)) = Q(t)\bar{B}(t) + \bar{C}^T(t)P(t)\bar{D}(t),$$
$$\tilde{\mathcal{H}}(P(t), Q(t)) = \gamma^2 I_{n_v} + \bar{D}^T(t)P(t)\bar{D}(t).$$

Then, G_γ and M_γ in (8.117) can be expressed as

$$G_\gamma(P(t)) = \begin{bmatrix} \mathcal{L}(P(t)) & \mathcal{M}(P(t)) \\ \mathcal{M}(P(t))^T & \mathcal{H}(P(t)) \end{bmatrix}$$

and

$$M_\gamma(P(t), Q(t)) = \begin{bmatrix} \tilde{\mathcal{L}}(P(t), Q(t)) & \tilde{\mathcal{M}}(P(t), Q(t)) \\ \tilde{\mathcal{M}}(P(t), Q(t))^T & \tilde{\mathcal{H}}(P(t), Q(t)) \end{bmatrix}.$$

THEOREM 8.12 (SBRL)

For a stochastic system of mean-field type (8.114), we have $\|\tilde{\mathcal{L}}_T\| < \gamma$ for some $\gamma > 0$ if the GDRE

$$\begin{cases} \mathcal{L}(P(t)) - \mathcal{M}(P(t))\mathcal{H}(P(t))^{-1}\mathcal{M}(P(t))^T = 0, \\ \tilde{\mathcal{L}}(P(t), Q(t)) - \tilde{\mathcal{M}}(P(t), Q(t))\tilde{\mathcal{H}}(P(t), Q(t))^{-1}\tilde{\mathcal{M}}(P(t), Q(t))^T = 0, \\ P(T) = Q(T) = 0, \\ \mathcal{H}(P(t)) > 0, \\ \tilde{\mathcal{H}}(P(t), Q(t)) > 0 \end{cases}$$

(8.118)

has a unique global solution $(P^1(t), Q^1(t) \leq 0)$ on $[0, T]$.

Proof. Suppose $(P^1(t), Q^1(t))$ is the solution of (8.118). From Lemma 8.12 and $P^1(T) = Q^1(T) = 0$, for $x_0 = 0$, we have

$$J_1^T(x(t), v(t), 0, 0) := \gamma^2 \|v(t)\|_{[0,T]}^2 |_{x(0)=0} - \|z_1(t)\|_{[0,T]}^2 |_{x(0)=0}$$

$$= \mathcal{E} \int_0^T \begin{bmatrix} x(t) - \mathcal{E}x(t) \\ v(t) - \mathcal{E}v(t) \end{bmatrix}^T G_\gamma(P^1(t)) \begin{bmatrix} x(t) - \mathcal{E}x(t) \\ v(t) - \mathcal{E}v(t) \end{bmatrix} dt$$

$$+ \int_0^T \begin{bmatrix} \mathcal{E}x(t) \\ \mathcal{E}v(t) \end{bmatrix}^T M_\gamma(P^1(t), Q^1(t)) \begin{bmatrix} \mathcal{E}x(t) \\ \mathcal{E}v(t) \end{bmatrix} dt. \qquad (8.119)$$

In what follows, for notational simplicity, the time variable t is suppressed. By completing squares, for any $v(t) \in \mathcal{L}_{\mathcal{F}}^2([0, T], \mathcal{R}^{n_v})$ with $v(t) \neq 0$, we have

$$\mathcal{E} \int_0^T \begin{bmatrix} x - \mathcal{E}x \\ v - \mathcal{E}v \end{bmatrix}^T G_\gamma(P^1) \begin{bmatrix} x - \mathcal{E}x \\ v - \mathcal{E}v \end{bmatrix} dt$$

$$= \mathcal{E} \int_0^T (x - \mathcal{E}x)^T [\mathcal{L}(P^1) - \mathcal{M}(P^1)\mathcal{H}(P^1)^{-1}\mathcal{M}(P^1)^T](x - \mathcal{E}x) dt$$

$$+ \mathcal{E} \int_0^T [(v - \mathcal{E}v) + \mathcal{H}(P^1)^{-1}\mathcal{M}(P^1)^T(x - \mathcal{E}x)]^T$$

$$\cdot \mathcal{H}(P^1)[(v - \mathcal{E}v) + \mathcal{H}(P^1)^{-1}\mathcal{M}(P^1)^T(x - \mathcal{E}x)]\, dt$$

$$= \mathcal{E}\int_0^T [(v - \mathcal{E}v) - (v^* - \mathcal{E}v^*)]^T \mathcal{H}(P^1)[(v - \mathcal{E}v) - (v^* - \mathcal{E}v^*)]\, dt, \quad (8.120)$$

where

$$v^* - \mathcal{E}v^* = -\mathcal{H}(P^1)^{-1}\mathcal{M}(P^1)^T(x - \mathcal{E}x)$$

and

$$\int_0^T \begin{bmatrix} \mathcal{E}x \\ \mathcal{E}v \end{bmatrix}^T M_\gamma(P^1, Q^1) \begin{bmatrix} \mathcal{E}x \\ \mathcal{E}v \end{bmatrix} dt$$

$$= \int_0^T \mathcal{E}x^T[\tilde{\mathcal{L}}(P^1, Q^1) - \tilde{\mathcal{M}}(P^1, Q^1)\tilde{\mathcal{H}}(P^1, Q^1)^{-1}\tilde{\mathcal{M}}(P^1, Q^1)^T]\mathcal{E}x\, dt$$

$$+ \int_0^T [\mathcal{E}v + \tilde{\mathcal{H}}(P^1, Q^1)^{-1}\tilde{\mathcal{M}}(P^1, Q^1)^T\mathcal{E}x]^T \tilde{\mathcal{H}}(P^1, Q^1)[\cdots]\, dt$$

$$= \int_0^T (\mathcal{E}v - \mathcal{E}v^*)^T \tilde{\mathcal{H}}(P^1, Q^1)(\mathcal{E}v - \mathcal{E}v^*)\, dt, \quad (8.121)$$

where

$$\mathcal{E}v^* = -\tilde{\mathcal{H}}(P^1, Q^1)^{-1}\tilde{\mathcal{M}}(P^1, Q^1)^T\mathcal{E}x.$$

Combining (8.119) and (8.120) with (8.121) yields

$$J_1^T(x, v, 0, 0) = \gamma^2\|v\|_{[0,T]}^2 - \|z\|_{[0,T]}^2 \geq 0,$$

which leads to $\|\tilde{\mathcal{L}}_T\| \leq \gamma$.

Next, to prove $\|\tilde{\mathcal{L}}_T\| < \gamma$, we define the following operators

$$\tilde{\mathbf{L}}_1 : \mathcal{L}_{\mathcal{F}}^2([0,T], \mathcal{R}^{n_v}) \mapsto \mathcal{L}_{\mathcal{F}}^2([0,T], \mathcal{R}^{n_v}), \quad \tilde{\mathbf{L}}_1(\mathcal{E}v) = \mathcal{E}v - \mathcal{E}v^*,$$
$$\mathbf{L}_1 : \mathcal{L}_{\mathcal{F}}^2([0,T], \mathcal{R}^{n_v}) \mapsto \mathcal{L}_{\mathcal{F}}^2([0,T], \mathcal{R}^{n_v}), \quad \mathbf{L}_1(v - \mathcal{E}v) = (v - \mathcal{E}v) - (v^* - \mathcal{E}v^*)$$

with their realizations

$$\begin{cases} d\mathcal{E}x = [(A_{11} + \bar{A}_{11})\mathcal{E}x + (B_{11} + \bar{B}_{11})\mathcal{E}v]\, dt, \\ \mathcal{E}x_0 = 0, \end{cases}$$

$$\mathcal{E}v - \mathcal{E}v^* = \mathcal{E}v + \tilde{\mathcal{H}}(P^1, Q^1)^{-1}\tilde{\mathcal{M}}(P^1, Q^1)^T\mathcal{E}x,$$

$$\begin{cases} d(x - \mathcal{E}x) = [A_{11}(x - \mathcal{E}x) + B_{11}(v - \mathcal{E}v)]\, dt \\ \qquad\qquad + [A_{12}(x - \mathcal{E}x) + (A_{12} + \bar{A}_{12})\mathcal{E}x \\ \qquad\qquad + B_{12}(v - \mathcal{E}v) + (B_{12} + \bar{B}_{12})\mathcal{E}v]\, dB, \\ x_0 - \mathcal{E}x_0 = 0, \end{cases}$$

$$(v - \mathcal{E}v) - (v^* - \mathcal{E}v^*) = (v - \mathcal{E}v) + \mathcal{H}(P^1)^{-1}\mathcal{M}(P^1)^T(x - \mathcal{E}x).$$

Then $\tilde{\mathbf{L}}_1^{-1}$ and $\mathbf{L}_1^{-1}$ exist, which are respectively determined by

$$\begin{cases} d\mathcal{E}x = \{[(A_{11} + \bar{A}_{11}) - (B_{11} + \bar{B}_{11})\tilde{\mathcal{H}}(P^1, Q^1)^{-1}\tilde{\mathcal{M}}(P^1, Q^1)^T]\mathcal{E}x \\ \qquad\qquad + (B_{11} + \bar{B}_{11})(\mathcal{E}v - \mathcal{E}v^*)\}\, dt, \\ \mathcal{E}x_0 = 0 \end{cases}$$

with

$$\mathcal{E}v = -\tilde{\mathcal{H}}(P^1, Q^1)^{-1}\tilde{\mathcal{M}}(P^1, Q^1)^T \mathcal{E}x + (\mathcal{E}v - \mathcal{E}v^*),$$

and

$$\begin{cases} d(x - \mathcal{E}x) = \{[A_{11} - B_{11}\mathcal{H}(P^1)^{-1}\mathcal{M}(P^1)^T](x - \mathcal{E}x) \\ \qquad + B_{11}[(v - \mathcal{E}v) - (v^* - \mathcal{E}v^*)]\} \, dt \\ \qquad + \{[A_{12} - B_{12}\mathcal{H}(P^1)^{-1}\mathcal{M}(P^1)^T](x - \mathcal{E}x) \\ \qquad + B_{12}[(v - \mathcal{E}v) - (v^* - \mathcal{E}v^*)] \\ \qquad + [\bar{C} - \bar{D}\tilde{\mathcal{H}}(P^1, Q^1)^{-1}\tilde{\mathcal{M}}(P^1, Q^1)^T]\mathcal{E}x + \bar{D}[\mathcal{E}v - \mathcal{E}v^*]\} \, dB, \\ x_0 - \mathcal{E}x_0 = 0 \end{cases}$$

with

$$v - \mathcal{E}v = -\mathcal{H}(P^1)^{-1}\mathcal{M}(P^1)^T(x - \mathcal{E}x) + [(v - \mathcal{E}v) - (v^* - \mathcal{E}v^*)].$$

Suppose $\mathcal{H}(P^1) \geq cI$ and $\tilde{\mathcal{H}}(P^1, Q^1) \geq cI$ for some $c > 0$. Because $\tilde{\mathbf{L}}_1^{-1}$ and $\mathbf{L}_1^{-1}$ exist, there exists a constant $\epsilon > 0$ such that

$$\begin{aligned} J_1^T(x, v, 0, 0) = \mathcal{E} &\int_0^T [(v - \mathcal{E}v) - (v^* - \mathcal{E}v^*)]^T \mathcal{H}(P^1)[\cdots] \, dt \\ &+ \int_0^T [\mathcal{E}v - \mathcal{E}v^*]^T \tilde{\mathcal{H}}(P^1, Q^1)[\mathcal{E}v - \mathcal{E}v^*] \, dt \\ &\geq c[\|\mathbf{L}_1(v - \mathcal{E}v)\|_{[0,T]}^2 + \|\tilde{\mathbf{L}}_1(\mathcal{E}v)\|_{[0,T]}^2] \\ &\geq \epsilon[\|v - \mathcal{E}v\|_{[0,T]}^2 + \|\mathcal{E}v\|_{[0,T]}^2] \\ &= \epsilon\|v\|_{[0,T]}^2, \end{aligned}$$

which leads to $\|\tilde{\mathcal{L}}_T\| < \gamma$.

According to the above discussion and Lemma 8.12, we know that

$$\begin{aligned} \min_{v \in \mathcal{L}_{\mathcal{F}}^2([0,T], \mathcal{R}^{n_v})} J_1^T(x(t), v(t), x(t_0), t_0) &= J_1^T(x^*(t), v^*(t), x(t_0), t_0) \\ &= \mathcal{E}x(t_0)^T Q^1(t_0)\mathcal{E}x(t_0) \\ &\leq J_1^T(x(t), 0, x(t_0), t_0) = -\|z_1\|_{[0,T]}^2 \leq 0 \qquad (8.122) \end{aligned}$$

for an arbitrary $x(t_0) \in \mathcal{R}^n$ and $t_0 \geq 0$, where x^* is a trajectory corresponding to v^*. Hence, $Q^1 \leq 0$ from (8.122). The proof is completed. $\square$

REMARK 8.10

At least at the present stage, we are not able to show that $\|\tilde{\mathcal{L}}_T\| < \gamma$ does necessarily imply the invertibility of $\mathcal{H}(P^1)$ and $\tilde{\mathcal{H}}(P^1, Q^1)$ simultaneously, so Theorem 8.12 is only a sufficient but not a necessary condition for $\|\tilde{\mathcal{L}}_T\| < \gamma$. However, for the equation (8.114), if $B_{12} = \bar{B}_{12} = 0$, then

$$\begin{cases} dx = (A_{11}x + \bar{A}_{11}\mathcal{E}x + B_{11}v + \bar{B}_{11}\mathcal{E}v) \, dt + (A_{12}x + \bar{A}_{12}\mathcal{E}x) \, dB, \\ z_1 = C_{11}x, x_0 \in \mathcal{R}^n, t \in [0, T]. \end{cases} \qquad (8.123)$$

A necessary and sufficient condition can be given in the next theorem. ∎

THEOREM 8.13 (SBRL)
For stochastic system of mean-field type (8.123), $\|\tilde{\mathcal{L}}_T\| < \gamma$ for some $\gamma > 0$ iff the following GDRE (the time variable t is suppressed)

$$\begin{cases} -\dot{P} = A_{11}^T P + P A_{11} + A_{12}^T P A_{12} - C_{11}^T C_{11} - \gamma^{-2} P B_{11} B_{11}^T P, \\ -\dot{Q} = \bar{A}^T Q + Q \bar{A} + \bar{C}^T P \bar{C} - C_{11}^T C_{11} - \gamma^{-2} Q \bar{B} \bar{B}^T Q, \\ P(T) = Q(T) = 0 \end{cases} \quad (8.124)$$

has a unique global solution (P^1, Q^1) with $Q^1 \le 0$ on $[0, T]$.

Proof. The sufficiency is derived from Theorem 8.12. Next, we proceed to prove the necessity, i.e., $\|\tilde{\mathcal{L}}_T\| < \gamma$ implies that (8.124) has a unique solution (P^1, Q^1) on $[0, T]$. Otherwise, (8.124) has a finite escape time on $[0, T]$. By the standard theory of differential equations, there exists a unique solution Q on a maximal interval $(t_0, T]$ with $t_0 \ge 0$, and Q is unbounded when $t \to t_0$. Similarly, there exists a unique solution P on a maximal interval $(\tilde{t}_0, T]$ with $\tilde{t}_0 \ge 0$. Note that (8.124) is not coupled but decoupled ODEs, for which the first equation on $P(t)$ should be solved in advance. Hence, generally speaking, $(t_0, T] \subset (\tilde{t}_0, T]$. That is, we only need to show the global existence of $Q(t)$ on $[0, T]$.

Let $0 < \epsilon < T - t_0$,

$$J_1^T(x, v, x_{t_0+\epsilon}, t_0 + \epsilon) = \gamma^2 \|v\|_{[t_0+\epsilon, T]}^2 - \|z_1\|_{[t_0+\epsilon, T]}^2$$

$$= \gamma^2 \mathcal{E} \int_{t_0+\epsilon}^T [(v - \mathcal{E}v) - (v^* - \mathcal{E}v^*)]^T [(v - \mathcal{E}v) - (v^* - \mathcal{E}v^*)] \, dt$$

$$+ \gamma^2 \int_{t_0+\epsilon}^T (\mathcal{E}v - \mathcal{E}v^*)^T (\mathcal{E}v - \mathcal{E}v^*) \, dt + \mathcal{E}x(t_0 + \epsilon)^T Q(t_0 + \epsilon) \mathcal{E}x(t_0 + \epsilon),$$

where $v^* = -\gamma^{-2} B_{11}^T P x + (\gamma^{-2} B_{11}^T P - \gamma^{-2} \bar{B}^T Q) \mathcal{E}x$. So

$$\min_{v \in \mathcal{L}_{\mathcal{F}}^2([t_0+\epsilon, T], \mathcal{R}^{n_v})} J_1^T(x(t), v(t), x(t_0 + \epsilon), t_0 + \epsilon) = J_1^T(x^*(t), v^*(t), x(t_0 + \epsilon), t_0 + \epsilon)$$

$$= \mathcal{E}x(t_0 + \epsilon)^T Q(t_0 + \epsilon) \mathcal{E}x(t_0 + \epsilon) \le J_1^T(x, 0, x(t_0 + \epsilon), t_0 + \epsilon)$$

$$= \mathcal{E} \int_{t_0+\epsilon}^T -\|z\|^2 \, dt \le 0. \quad (8.125)$$

It follows that $Q(t_0 + \epsilon) \le 0$ for any $0 < \epsilon < T - t_0$.

Next, we shall prove that there exists $\delta > 0$ such that $J_1^T(x, v, x(t_0 + \epsilon), t_0 + \epsilon) \ge -\delta \|x(t_0 + \epsilon)\|^2$ for any $v \in \mathcal{L}_{\mathcal{F}}^2([t_0 + \epsilon, T], \mathcal{R}^{n_v})$. Let $x(t, v, x_{t_0+\epsilon}, t_0 + \epsilon)$ denote the solution of (8.123) with the initial state $x_{t_0+\epsilon} := x(t_0 + \epsilon)$, and $(X(t), Y(t))$ the solution of

$$\begin{cases} \dot{X} + A_{11}^T X + X A_{11} + A_{12}^T X A_{12} - C_{11}^T C_{11} = 0, \\ \dot{Y} + \bar{A}^T Y + Y \bar{A} + \bar{C}^T X \bar{C} - C_{11}^T C_{11} = 0, \\ X(T) = 0, \ Y(T) = 0, \ t \in [0, T]. \end{cases}$$

By linearity

$$x(t, v, x_{t_0+\epsilon}, t_0 + \epsilon) = x(t, 0, x_{t_0+\epsilon}, t_0 + \epsilon) + x(t, v, 0, t_0 + \epsilon),$$

we know that

$$
\begin{aligned}
&J_1^T (x, v, x_{t_0+\epsilon}, t_0 + \epsilon) - J_1^T (x, v, 0, t_0 + \epsilon) \\
&= (x_{t_0+\epsilon} - \mathcal{E}x_{t_0+\epsilon})^T X(t_0 + \epsilon)(x_{t_0+\epsilon} - \mathcal{E}x_{t_0+\epsilon}) \\
&\quad + \mathcal{E} \int_{t_0+\epsilon}^T \{ [x(t, 0, x_{t_0+\epsilon}, t_0 + \epsilon) - \mathcal{E}x(t, 0, x_{t_0+\epsilon}, t_0 + \epsilon)]^T X B_{11}(v - \mathcal{E}v) \\
&\quad + (v - \mathcal{E}v)^T B_{11}^T X [x(t, 0, x_{t_0+\epsilon}, t_0 + \epsilon) - \mathcal{E}x(t, 0, x_{t_0+\epsilon}, t_0 + \epsilon)] \} \, dt \\
&\quad + \int_{t_0+\epsilon}^T [\mathcal{E}x(t, 0, x_{t_0+\epsilon}, t_0 + \epsilon)^T Y \bar{B} \mathcal{E}v + \mathcal{E}v^T \bar{B}^T Y \mathcal{E}x(t, 0, x_{t_0+\epsilon}, t_0 + \epsilon)] \, dt \\
&\quad + \mathcal{E}x_{t_0+\epsilon}^T Y(t_0 + \epsilon)\mathcal{E}x_{t_0+\epsilon}.
\end{aligned}
\tag{8.126}
$$

Take $0 < c^2 < \gamma^2 - \|\tilde{\mathcal{L}}_T\|^2$, then

$$
\begin{aligned}
J_1^T (x, v, 0, t_0 + \epsilon) &\geq \gamma^2 \|\bar{v}\|_{[0,T]}^2 - \|z_1\|_{[0,T]}^2 \geq c^2 \|\bar{v}\|_{[0,T]}^2 = c^2 \|v\|_{[t_0+\epsilon,T]}^2 \\
&= c^2 (\|v - \mathcal{E}v\|_{[t_0+\epsilon,T]}^2 + \|\mathcal{E}v\|_{[t_0+\epsilon,T]}^2),
\end{aligned}
\tag{8.127}
$$

where

$$
\bar{v} = \begin{cases} v, & t \in [t_0 + \epsilon, T], \\ 0, & t \in [0, t_0 + \epsilon). \end{cases}
$$

Therefore, by completing squares, and in view of (8.126) and (8.127), it follows that

$$
\begin{aligned}
J_1^T (x, v, x_{t_0+\epsilon}, t_0 + \epsilon) &\geq (x_{t_0+\epsilon} - \mathcal{E}x_{t_0+\epsilon})^T X(t_0 + \epsilon)(x_{t_0+\epsilon} - \mathcal{E}x_{t_0+\epsilon}) \\
&\quad + \mathcal{E}x_{t_0+\epsilon}^T Y(t_0 + \epsilon)\mathcal{E}x_{t_0+\epsilon} \\
&\quad - \mathcal{E} \int_{t_0+\epsilon}^T \|c^{-1} B_{11}^T X [x(t, 0, x_{t_0+\epsilon}, t_0 + \epsilon) - \mathcal{E}x(t, 0, x_{t_0+\epsilon}, t_0 + \epsilon)]\|^2 \, dt \\
&\quad - \int_{t_0+\epsilon}^T \|c^{-1} \bar{B}^T Y \mathcal{E}x(t, 0, x_{t_0+\epsilon}, t_0 + \epsilon)\|^2 \, dt.
\end{aligned}
\tag{8.128}
$$

It is well known that there exists $\alpha_1 > 0$ such that

$$
\mathcal{E} \int_{t_0+\epsilon}^T \|x(t, 0, x_{t_0+\epsilon}, t_0 + \epsilon) - \mathcal{E}x(t, 0, x_{t_0+\epsilon}, t_0 + \epsilon)\|^2 \leq \alpha_1 \|x_{t_0+\epsilon} - \mathcal{E}x_{t_0+\epsilon}\|^2
$$

and

$$
\int_{t_0+\epsilon}^T \|\mathcal{E}x(t, 0, x_{t_0+\epsilon}, t_0 + \epsilon)\|^2 \leq \alpha_1 \|\mathcal{E}x_{t_0+\epsilon}\|^2.
$$

So, there exists $\beta_2 > 0$ such that

$$
\begin{aligned}
&\mathcal{E} \int_{t_0+\epsilon}^T \|c^{-1} B_{11}^T X [x(t, 0, x_{t_0+\epsilon}, t_0 + \epsilon) - \mathcal{E}x(t, 0, x_{t_0+\epsilon}, t_0 + \epsilon)]\|^2 \, dt \\
&\leq \beta_2 \|x_{t_0+\epsilon} - \mathcal{E}x_{t_0+\epsilon}\|^2,
\end{aligned}
\tag{8.129}
$$

and

$$\int_{t_0+\epsilon}^{T} \|c^{-1}\bar{B}^T Y \mathcal{E}x(t,0,x_{t_0+\epsilon},t_0+\epsilon)\|^2 \, dt \le \beta_2 \|\mathcal{E}x_{t_0+\epsilon}\|^2. \tag{8.130}$$

Similar to the discussion of Lemma 8.12, there exists $\beta_1 > 0$ satisfying

$$(x_{t_0+\epsilon} - \mathcal{E}x_{t_0+\epsilon})^T X(t_0+\epsilon)(x_{t_0+\epsilon} - \mathcal{E}x_{t_0+\epsilon}) + \mathcal{E}x_{t_0+\epsilon}^T Y(t_0+\epsilon)\mathcal{E}x_{t_0+\epsilon}$$

$$= -\mathcal{E}\int_{t_0+\epsilon}^{T} [x(t,0,x_{t_0+\epsilon},t_0+\epsilon) - \mathcal{E}x(t,0,x_{t_0+\epsilon},t_0+\epsilon)]^T C_{11}^T C_{11}[\cdots] \, dt$$

$$- \int_{t_0+\epsilon}^{T} [\mathcal{E}x(t,0,x_{t_0+\epsilon},t_0+\epsilon)]^T C_{11}^T C_{11}[\mathcal{E}x(t,0,x_{t_0+\epsilon},t_0+\epsilon)] \, dt$$

$$= -\mathcal{E}\int_{t_0+\epsilon}^{T} x(t,0,x_{t_0+\epsilon},t_0+\epsilon)^T C_{11}^T C_{11}x(t,0,x_{t_0+\epsilon},t_0+\epsilon) \, dt$$

$$\ge -\beta_1\|x_{t_0+\epsilon}\|^2. \tag{8.131}$$

So, from (8.128), (8.129), (8.130) and (8.131), it yields

$$J_1^T(x,v,x_{t_0+\epsilon},t_0+\epsilon) \ge -(\beta_1+\beta_2)\|x_{t_0+\epsilon}\|^2 = -\delta\|x_{t_0+\epsilon}\|^2.$$

Therefore, $-\delta I_{n \times n} \le Q(t_0+\epsilon) \le 0$ for $0 < \epsilon < T - t_0$. So $Q(t_0+\epsilon)$ cannot tend to ∞ as $\epsilon \to \infty$, which contradicts the unboundedness of Q. The above discussion shows that (8.124) has a unique solution (P^1, Q^1) on $[0,T]$. The proof is completed. $\square$

REMARK 8.11 It seems almost sure that $P(t) \le 0$ in (8.118), unfortunately, it is not an easy thing to prove this fact. ∎

When $\bar{A}_{11} = 0$, $\bar{B}_{11} = 0$, $\bar{A}_{12} = 0$, system (8.123) reduces to

$$\begin{cases} dx = (A_{11}x + B_{11}v) \, dt + A_{12}x \, dB, \\ z_1 = C_{11}x, x_0 \in \mathcal{R}^n, t \in [0,T], \end{cases} \tag{8.132}$$

and GDRE (8.124) becomes

$$\begin{cases} -\dot{P} = A_{11}^T P + PA_{11} + A_{12}^T PA_{12} - C_{11}^T C_{11} - \gamma^{-2}PB_{11}B_{11}^T P, \\ -\dot{Q} = A_{11}^T Q + QA_{11} + A_{12}^T PA_{12} - C_{11}^T C_{11} - \gamma^{-2}QB_{11}B_{11}^T Q, \\ P(T) = Q(T) = 0, \ t \in [0,T]. \end{cases} \tag{8.133}$$

COROLLARY 8.6
The equation (8.133) has a unique solution $(P^1(t), Q^1(t))$ with $Q^1(t) \le 0$ iff

$$\begin{cases} -\dot{P} = A_{11}^T P + PA_{11} + A_{12}^T PA_{12} - C_{11}^T C_{11} - \gamma^{-2}PB_{11}B_{11}^T P, \\ P(T) = 0, \ t \in [0,T] \end{cases} \tag{8.134}$$

has a unique solution $P(t) \leq 0$ on $[0, T]$.

Proof. The result can be derived based on Lemma 2.1 and Theorem 8.13. However, a direct proof without applying Lemma 2.1 and Theorem 8.13 will be an interesting exercise. $\square$

8.4.3 Mean-field stochastic LQ control

Consider the following stochastic control system

$$\begin{cases} dx(t) = (Ax + \tilde{A}\mathcal{E}x + Fu)(t)\, dt + (Bx + \tilde{B}\mathcal{E}x)(t)\, dB(t), \\ z(t) = \begin{bmatrix} Cx \\ Du \end{bmatrix}(t),\ D^T(t)D(t) = I,\ t \in [0, T] \end{cases} \tag{8.135}$$

as well as the associated cost functional

$$\begin{aligned} J_2(x_0, u) = \|z\|_{[0,T]}^2 &= \mathcal{E} \int_0^T z^T(t)z(t)\, dt \\ &= \mathcal{E} \int_0^T \begin{bmatrix} x - \mathcal{E}x \\ u - \mathcal{E}u \end{bmatrix}^T \begin{bmatrix} C^T C & 0 \\ 0 & I_{n_u} \end{bmatrix} \begin{bmatrix} x - \mathcal{E}x \\ u - \mathcal{E}u \end{bmatrix} dt \\ &\quad + \int_0^T \begin{bmatrix} \mathcal{E}x \\ \mathcal{E}u \end{bmatrix}^T \begin{bmatrix} C^T C & 0 \\ 0 & I_{n_u} \end{bmatrix} \begin{bmatrix} \mathcal{E}x \\ \mathcal{E}u \end{bmatrix} dt. \end{aligned} \tag{8.136}$$

Similar to the proof of Theorem 8.13, we easily have the following theorem on the stochastic LQ control of (8.135)–(8.136).

THEOREM 8.14

For the stochastic LQ control of (8.135)–(8.136), there exists $u^* \in \mathcal{L}_\mathcal{F}^2([0, T], \mathcal{R}^{n_u})$ such that $\min\limits_{u \in \mathcal{L}_\mathcal{F}^2([0,T], \mathcal{R}^{n_u})} J_2(x_0, u) = J_2(x_0, u^*) = x_0^T \hat{Q}^1(0)x_0 \geq 0$ iff the following GDRE

$$\begin{cases} -\dot{\hat{P}} = A^T \hat{P} + \hat{P}A + B^T \hat{P}B + C^T C - \hat{P}FF^T \hat{P}, \\ -\dot{\hat{Q}} = \bar{A}_1^T \hat{Q} + \hat{Q}\bar{A}_1 + \bar{B}_1^T \hat{P}\bar{B}_1 + C^T C - \hat{Q}FF^T \hat{Q}, \\ \hat{P}(T) = \hat{Q}(T) = 0,\ t \in [0, T] \end{cases} \tag{8.137}$$

has a unique solution $(\hat{P}^1, \hat{Q}^1 \geq 0)$, where

$$\bar{A}_1 = A + \tilde{A}, \qquad \bar{B}_1 = B + \tilde{B}$$

and

$$u^* = -F^T \hat{P}^1 x + (F^T \hat{P}^1 - F^T \hat{Q}^1)\mathcal{E}x.$$

8.4.4 H_2/H_∞ control with (x, v)-dependent noise

Before presenting the main results, we introduce the coupled Riccati equations as follows:

$$
\begin{cases}
-\dot{P} = (A_1 + B_1 U)^T P + P(A_1 + B_1 U) + A_2^T P A_2 \\
\qquad -C^T C - U^T U - \mathcal{M}_u(P)\mathcal{H}_u(P)^{-1}\mathcal{M}_u(P)^T, \\
-\dot{Q} = [\mathcal{A} + B_1(U + \tilde{U})]^T Q + Q[\mathcal{A} + B_1(U + \tilde{U})] + \mathcal{C}^T P \mathcal{C} \\
\qquad -C^T C - (U + \tilde{U})^T (U + \tilde{U}) - \tilde{\mathcal{M}}_u(P,Q)\tilde{\mathcal{H}}_u(P,Q)^{-1}\tilde{\mathcal{M}}_u(P,Q)^T, \\
\mathcal{H}_u(P) > 0, \\
\tilde{\mathcal{H}}_u(P,Q) > 0, \\
P(T) = Q(T) = 0,
\end{cases}
\tag{8.138}
$$

$$
\begin{cases}
V = -\mathcal{H}_u(P)^{-1}\mathcal{M}_u(P)^T, \\
\tilde{V} = \mathcal{H}_u(P)^{-1}\mathcal{M}_u(P)^T - \tilde{\mathcal{H}}_u(P,Q)^{-1}\tilde{\mathcal{M}}_u(P,Q)^T,
\end{cases}
\tag{8.139}
$$

$$
\begin{cases}
-\dot{\hat{P}} = (A_1 + C_1 V)^T \hat{P} + \hat{P}(A_1 + C_1 V) + A_2^T \hat{P} A_2 + C^T C - \hat{P} B_1 B_1^T \hat{P}, \\
-\dot{\hat{Q}} = [\mathcal{A} + \mathcal{B}(V + \tilde{V})]^T \hat{Q} + \hat{Q}[\mathcal{A} + \mathcal{B}(V + \tilde{V})] \\
\qquad + \mathcal{C}^T \hat{P} \mathcal{C} + C^T C - \hat{Q} B_1 B_1^T \hat{Q}, \\
\hat{P}(T) = \hat{Q}(T) = 0,
\end{cases}
\tag{8.140}
$$

$$
\begin{cases}
U = -B_1^T \hat{P}, \\
\tilde{U} = B_1^T \hat{P} - B_1^T \hat{Q},
\end{cases}
\tag{8.141}
$$

where

$$
\mathcal{M}_u(P) = PC_1 + A_2^T PC_2, \quad \mathcal{H}_u(P) = \gamma^2 I_{n_v} + C_2^T PC_2,
$$

$$
\tilde{\mathcal{M}}_u(P,Q) = Q\mathcal{B} + \mathcal{C}^T P \mathcal{D}, \quad \tilde{\mathcal{H}}_u(P,Q) = \gamma^2 I_{n_v} + \mathcal{D}^T P \mathcal{D}.
$$

In the above,

$$
\mathcal{A} = A_1 + \bar{A}_1, \quad \mathcal{B} = C_1 + \bar{C}_1, \quad \mathcal{C} = A_2 + \bar{A}_2, \quad \mathcal{D} = C_2 + \bar{C}_2.
$$

THEOREM 8.15

For system (8.111), the finite horizon H_2/H_∞ control has a solution (u_T^*, v_T^*) as

$$
u_T^* = Ux + \tilde{U}\mathcal{E}x, \qquad v_T^* = Vx + \tilde{V}\mathcal{E}x
$$

with $U, \tilde{U} \in \mathcal{R}^{n_u \times n}$ and $V, \tilde{V} \in \mathcal{R}^{n_v \times n}$ being matrix-valued functions and $Q^1 \leq 0, \hat{Q}^1 \geq 0$, if the coupled Riccati equations (8.138)–(8.141) have the solution $(P^1, Q^1; \hat{P}^1, \hat{Q}^1; U, \tilde{U}; V, \tilde{V})$ on $[0, T]$.

Proof. For the solution $(P^1, Q^1; \hat{P}^1, \hat{Q}^1; U, \tilde{U}; V, \tilde{V})$, we can obtain system (8.113) by substituting $u_T^* = Ux + \tilde{U}\mathcal{E}x$ for u of system (8.111). By Theorem 8.12 and

(8.138), it yields that $\|\mathcal{L}_T\| < \gamma$. Keeping (8.138) in mind, by the technique of completing squares and Lemma 8.12, we immediately get $Q^1 \leq 0$ and

$$
\begin{aligned}
J_{1,T}(u_T^*, v) &:= \gamma^2 \|v\|_{[0,T]}^2 - \|z\|_{[0,T]}^2 \\
&= \mathcal{E} \int_0^T \begin{bmatrix} x - \mathcal{E}x \\ v - \mathcal{E}v \end{bmatrix}^T G_0(P^1) \begin{bmatrix} x - \mathcal{E}x \\ v - \mathcal{E}v \end{bmatrix} dt \\
&\quad + \int_0^T \begin{bmatrix} \mathcal{E}x \\ \mathcal{E}v \end{bmatrix}^T M_0(P^1, Q^1) \begin{bmatrix} \mathcal{E}x \\ \mathcal{E}v \end{bmatrix} dt + x_0^T Q^1(0) x_0 \\
&= x_0^T Q^1(0) x_0 + \gamma^2 \mathcal{E} \int_0^T [(v - \mathcal{E}v) - (v_T^* - \mathcal{E}v_T^*)]^T \\
&\quad \cdot [(v - \mathcal{E}v) - (v_T^* - \mathcal{E}v_T^*)] \, dt \\
&\quad + \gamma^2 \int_0^T (\mathcal{E}v - \mathcal{E}v_T^*)^T (\mathcal{E}v - \mathcal{E}v_T^*) \\
&\geq J_{1,T}(u_T^*, v_T^*) = x_0^T Q^1(0) x_0,
\end{aligned}
$$

where

$$
G_0(P^1) = \begin{bmatrix} \mathcal{L}_u(P^1) & \mathcal{M}_u(P^1) \\ \mathcal{M}_u(P^1)^T & \mathcal{H}_u(P^1) \end{bmatrix},
$$

$$
M_0(P^1, Q^1) = \begin{bmatrix} \tilde{\mathcal{L}}_u(P^1, Q^1) & \tilde{\mathcal{M}}_u(P^1, Q^1) \\ \tilde{\mathcal{M}}_u(P^1, Q^1)^T & \tilde{\mathcal{H}}_u(P^1, Q^1) \end{bmatrix}
$$

with

$$
\begin{aligned}
\mathcal{L}_u(P) &= \dot{P} + (A_1 + B_1 U)^T P + P(A_1 + B_1 U) + A_2^T P A_2 - C^T C - U^T U, \\
\tilde{\mathcal{L}}_u(P, Q) &= \dot{Q} + [\mathcal{A} + B_1(U + \tilde{U})]^T Q + Q[\mathcal{A} + B_1(U + \tilde{U})] \\
&\quad + \mathcal{C}^T P \mathcal{C} - C^T C - (U + \tilde{U})^T (U + \tilde{U}).
\end{aligned}
$$

So, $v_T^* = Vx + \tilde{V}\mathcal{E}x$ with $(V, \tilde{V})$ given by (8.139) is the worst-case disturbance. Furthermore, using the technique of completing squares and considering Theorem 8.14 and (8.140) yield $\hat{Q}^1 \geq 0$ and

$$
\begin{aligned}
J_{2,T}(u, v_T^*) &= \|z\|^2 = x_0^T \hat{Q}^1(0) x_0 + \int_0^T (\mathcal{E}u - \mathcal{E}u_T^*)^T (\mathcal{E}u - \mathcal{E}u_T^*) \, dt \\
&\quad + \mathcal{E} \int_0^T [(u - \mathcal{E}u) - (u_T^* - \mathcal{E}u_T^*)]^T [(u - \mathcal{E}u) - (u_T^* - \mathcal{E}u_T^*)] \, dt \\
&\geq J_{2,T}(u_T^*, v_T^*) = x_0^T \hat{Q}^1(0) x_0.
\end{aligned}
$$

Therefore, (u_T^*, v_T^*) solves the mean-field H_2/H_∞ control problem of system (8.111), and the proof is complete. $\square$

For system (8.111), if $C_2 = 0$, $\bar{C}_2 = 0$, then

$$
\begin{cases}
dx = (A_1 x + \bar{A}_1 \mathcal{E}x + C_1 v + \bar{C}_1 \mathcal{E}v + B_1 u) \, dt \\
\qquad + (A_2 x + \bar{A}_2 \mathcal{E}x) \, dB, \\
z = \begin{bmatrix} Cx \\ Du \end{bmatrix}, D^T D = I, x_0 \in \mathcal{R}^n, t \in [0, T].
\end{cases}
\tag{8.142}
$$

The following theorem is an extension of Theorem 2.1-(i)⇔ (iii) to mean-field stochastic systems.

THEOREM 8.16

For system (8.142), the finite horizon H_2/H_∞ control has a solution (u_T^*, v_T^*) as

$$u_T^* = Ux + \tilde{U}\mathcal{E}x, \qquad v_T^* = Vx + \tilde{V}\mathcal{E}x$$

with $U, \tilde{U} \in \mathcal{R}^{n_u \times n}$ and $V, \tilde{V} \in \mathcal{R}^{n_v \times n}$ being matrix-valued functions, iff the coupled GDREs

$$
\begin{cases}
-\dot{P} = A_1^T P + P A_1 + A_2^T P A_2 \\
\quad -C^T C - [P\ \hat{P}] \begin{bmatrix} \gamma^{-2}C_1 C_1^T & B_1 B_1^T \\ B_1 B_1^T & B_1 B_1^T \end{bmatrix} \begin{bmatrix} P \\ \hat{P} \end{bmatrix}, \\
-\dot{Q} = \mathcal{A}^T Q + Q\mathcal{A} + \mathcal{C}^T P \mathcal{C} \\
\quad -C^T C - [Q\ \hat{Q}] \begin{bmatrix} \gamma^{-2}\mathcal{B}\mathcal{B}^T & B_1 B_1^T \\ B_1 B_1^T & B_1 B_1^T \end{bmatrix} \begin{bmatrix} Q \\ \hat{Q} \end{bmatrix}, \\
P(T) = Q(T) = 0
\end{cases}
\tag{8.143}
$$

and

$$
\begin{cases}
-\dot{\hat{P}} = A_1^T \hat{P} + \hat{P} A_1 + A_2^T \hat{P} A_2 \\
\quad +C^T C - [P\ \hat{P}] \begin{bmatrix} 0 & \gamma^{-2}C_1 C_1^T \\ \gamma^{-2}C_1 C_1^T & B_1 B_1^T \end{bmatrix} \begin{bmatrix} P \\ \hat{P} \end{bmatrix}, \\
-\dot{\hat{Q}} = \mathcal{A}^T \hat{Q} + \hat{Q}\mathcal{A} + \mathcal{C}^T \hat{P} \mathcal{C} \\
\quad +C^T C - [Q\ \hat{Q}] \begin{bmatrix} 0 & \gamma^{-2}\mathcal{B}\mathcal{B}^T \\ \gamma^{-2}\mathcal{B}\mathcal{B}^T & B_1 B_1^T \end{bmatrix} \begin{bmatrix} Q \\ \hat{Q} \end{bmatrix}, \\
\hat{P}(T) = \hat{Q}(T) = 0
\end{cases}
\tag{8.144}
$$

have a solution $(P^1, Q^1; \hat{P}^1, \hat{Q}^1)$ with $Q^1 \leq 0, \hat{Q}^1 \geq 0$ on $[0, T]$. In this case,

$$
\begin{cases}
U = -B_1^T \hat{P}^1, \\
\tilde{U} = B_1^T \hat{P}^1 - B_1^T \hat{Q}^1, \\
V = -\gamma^{-2}C_1^T P^1, \\
\tilde{V} = \gamma^{-2}C_1^T P^1 - \gamma^{-2}\mathcal{B}^T Q^1,
\end{cases}
\tag{8.145}
$$

and

$$J_{2,T}(u_T^*, v_T^*) = x_0^T \hat{Q}^1(0)x_0.$$

Proof. Sufficiency is a corollary of Theorem 8.15 by setting $C_2 = 0, \bar{C}_2 = 0$. Next, we prove necessity: Implement $u_T^* = Ux + \tilde{U}\mathcal{E}x$ in (8.142), then

$$
\begin{cases}
dx = [(A_1 + B_1 U)x + (\bar{A}_1 + B_1\tilde{U})\mathcal{E}x + C_1 v + \bar{C}_1 \mathcal{E}v]\, dt \\
\quad + (A_2 x + \bar{A}_2 \mathcal{E}x)\, dB, \\
z = \begin{bmatrix} Cx \\ D(Ux + \tilde{U}\mathcal{E}x) \end{bmatrix}, D^T D = I, t \in [0, T], x_0 = 0.
\end{cases}
\tag{8.146}
$$

By Definition 8.10, $\|\mathcal{L}_T\| < \gamma$ for system (8.146). In view of Theorem 8.13, the following GDRE

$$\begin{cases} -\dot{P} = (A_1 + B_1 U)^T P + P(A_1 + B_1 U) + A_2^T P A_2 \\ \qquad -C^T C - U^T U - \gamma^{-2} P C_1 C_1^T P, \\ -\dot{Q} = [A^T + B_1(U + \tilde{U})]^T Q + Q[A^T + B_1(U + \tilde{U})] + C^T P \mathcal{C} \qquad (8.147) \\ \qquad -C^T C - (U + \tilde{U})^T (U + \tilde{U}) - \gamma^{-2} Q \mathcal{B} \mathcal{B}^T Q, \\ P(T) = Q(T) = 0 \end{cases}$$

has a unique solution (P^1, Q^1) with $Q^1 \leq 0$ on $[0, T]$. From the proof of Theorem 8.13, we can see that $v_T^* = -\gamma^{-2} C_1^T P^1 (x - \mathcal{E}x) - \gamma^{-2} \mathcal{B}^T Q^1 \mathcal{E}x$. Substituting $v = v_T^* = Vx + \tilde{V} \mathcal{E}x = -\gamma^{-2} C_1^T P^1 (x - \mathcal{E}x) - \gamma^{-2} \mathcal{B}^T Q^1 \mathcal{E}x$ into (8.142) yields

$$\begin{cases} dx = \{(A_1 + C_1 V)x + [\bar{A}_1 + C_1 \tilde{V} + \bar{C}_1 (V + \tilde{V})] \mathcal{E}x + B_1 u\}\, dt \\ \qquad + (A_2 x + \bar{A}_2 \mathcal{E}x)\, dB, \\ z = \begin{bmatrix} Cx \\ Du \end{bmatrix}, D^T D = I, x_0 \in \mathcal{R}^n. \end{cases} \qquad (8.148)$$

By Theorem 8.14, the following GDRE

$$\begin{cases} -\dot{\hat{P}} = (A_1 - \gamma^{-2} C_1 C_1^T P^1)^T \hat{P} + \hat{P}(A_1 - \gamma^{-2} C_1 C_1^T P^1) + A_2^T \hat{P} A_2 \\ \qquad + C^T C - \hat{P} B_1 B_1^T \hat{P}, \\ -\dot{\hat{Q}} = (\mathcal{A} - \gamma^{-2} \mathcal{B} \mathcal{B}^T Q^1)^T \hat{Q} + \hat{Q}(\mathcal{A} - \gamma^{-2} \mathcal{B} \mathcal{B}^T Q^1) + C^T \hat{P} \mathcal{C} \\ \qquad + C^T C - \hat{Q} B_1 B_1^T \hat{Q}, \\ \hat{P}(T) = \hat{Q}(T) = 0 \end{cases}$$

$$\qquad (8.149)$$

has a unique solution $(\hat{P}^1, \hat{Q}^1)$ with $\hat{Q}^1 \leq 0$.

Additionally, by a series of computations, it yields

$$J_{2,T}(u, v_T^*) = \|z\|_{[0,T]}^2$$

$$= x_0^T \hat{Q}^1(0) x_0 + \int_0^T (\mathcal{E}u - \mathcal{E}u_T^*)^T (\mathcal{E}u - \mathcal{E}u_T^*)\, dt$$

$$+ \mathcal{E} \int_0^T [(u - \mathcal{E}u) - (u_T^* - \mathcal{E}u_T^*)]^T [(u - \mathcal{E}u) - (u_T^* - \mathcal{E}u_T^*)]\, dt.$$

Therefore,

$$J_{2,T}(u_T^*, v_T^*) = \min_{u \in \mathcal{L}_{\mathcal{F}}^2([0,T], \mathcal{R}^{n_u})} J_{2,T}(u, v_T^*) = x_0^T \hat{Q}^1(0) x_0,$$

where $u_T^* = U(x - \mathcal{E}x) + (U + \tilde{U}) \mathcal{E}x = -B_1^T \hat{P}^1(x - \mathcal{E}x) - B_1^T \hat{Q}^1 \mathcal{E}x$. Substituting $U = -B_1^T \hat{P}^1$ and $U + \tilde{U} = -B_1^T \hat{Q}^1$ into (8.147), (8.143) is obtained. The proof is complete. $\square$

8.4.5 Further research problems

The materials for the discussion in this section come from [142]. We also note that the output feedback H_∞ control for discrete-time mean-field stochastic systems can be found in [141]. From our above discussions, we can see that it is complicated to deal with the finite horizon H_2/H_∞ control of mean-field system (8.111). In our viewpoint, there are at least the following problems deserving extensive investigation.

- Because the infinite horizon H_2/H_∞ control is inevitably related to stability, observability and detectability, the infinite horizon H_2/H_∞ control for the following time-invariant linear mean-field system

$$\begin{cases} dx(t) = [A_1 x(t) + \bar{A}_1 \mathcal{E}x(t) + B_1 u(t) + C_1 v(t) + \bar{C}_1 \mathcal{E}v(t)]\, dt \\ \qquad + [A_2 x(t) + \bar{A}_2 \mathcal{E}x(t) + C_2 v(t) + \bar{C}_2 \mathcal{E}v(t)]\, dB(t), \\ x(0) = x_0 \in \mathcal{R}^n, \\ z(t) = \begin{bmatrix} Cx(t) \\ Du(t) \end{bmatrix}, \quad D^T D = I, \ t \in [0, \infty) \end{cases}$$

  is difficult and remains unsolved.

- Even for the finite horizon H_2/H_∞ control, Theorem 8.12 and Theorem 8.15 provide only sufficient conditions for the SBRL and H_2/H_∞ control, respectively. Obviously, there is a great gap by comparing Theorem 8.12 with Lemma 2.1, and Theorem 8.15 with Theorem 2.3. Necessary and sufficient conditions are to be obtained for the H_2/H_∞ control.

- In [68] and [223], finite horizon LQ control and H_2/H_∞ control of linear discrete-time stochastic mean-field systems were respectively discussed, however, the infinite horizon H_2/H_∞ control remains unsolved. In particular, the relationship between the H_2/H_∞ control and the Nash game merits further clarification.

8.5 Notes and References

This chapter lists some unsolved H_2/H_∞ problems for stochastic systems with random coefficients, discrete-time affine systems with multiplicative noise, continuous time singular stochastic systems as well as mean-field stochastic systems. All these problems are important but difficult and deserve further study in the future. In addition, it is also noted that some researchers have recently extended the Nash game-based H_2/H_∞ approach to BSDEs [230, 231], which is without doubt valuable, but how to give a practical design method is a very challenging problem.

References

[1] S. Aberkane, J. C. Ponsart, M. Rodrigues, D. Sauter. Output feedback control of a class of stochastic hybrid systems. *Automatica*, 44: 1325–1332, 2008.

[2] M. Ait Rami and X. Y. Zhou. Linear matrix inequalities, Riccati equations, and indefinite stochastic linear quadratic control. *IEEE Trans. Automat. Contr.*, 45: 1131–1142, 2000.

[3] A. Albert. Conditions for positive and nonnegative definiteness in terms of pseudo–inverse. *SIAM J. Appl. Math.*, 17: 434–440, 1969.

[4] M. Ait Rami, J. B. Moore, and X. Y. Zhou. Indefinite stochastic linear quadratic control and generalized differential Riccati equation. *SIAM J. Contr. Optim.*, 40(4): 1296–1311, 2001.

[5] M. Ait Rami, X. Chen, J. B. Moore, and X. Y. Zhou. Solvability and asymptotic behavior of generalized Riccati equations arising in indefinite stochastic LQ controls. *IEEE Trans. Automat. Contr.*, 46: 428–440, 2001.

[6] M. Ait Rami, X. Y. Zhou, and J. B. Moore. Well-posedness and attainability of indefinite stochastic linear quadratic control in infinite time horizon. *Systems and Control Letters*, 41: 123–133, 2000.

[7] M. Ait Rami, X. Chen, and X. Y. Zhou. Discrete-time indefinite LQ control with state and control dependent noises. *J. Global Optimization*, 23: 245–265, 2002.

[8] F. Amato, G. Carannante, G. De Tommasi, and A. Pironti. Input-output finite-time stability of linear systems: Necessary and sufficient conditions. *IEEE Trans. Automat. Contr.*, 57(12): 3051–3063, 2012.

[9] N. U. Ahmed and X. Ding. Controlled McKean-Vlasov equations, *Commun. Appl. Anal.*, 5(2): 183–206, 2001.

[10] B. D. O. Anderson and J. B. Moore. Detectability and stabilizability of time-varying discrete-time linear systems. *SIAM J. Contr. Optim.*, 19: 20–32, 1981.

[11] B. D. O. Anderson and J. B. Moore. *Optimal Control-Linear Quadratic Methods*. Prentice-Hall, New York, 1989.

[12] G. Ackley. *Macroeconomic Theory*. New York: Macmillan, 1969.

[13] M. D. S. Aliyu and E. K. Boukas. Mixed nonlinear H_2/H_∞ filtering. *Int. J. Robust Nonlinear Control*, 19: 394–417, 2009.

[14] X. An and W. Zhang. A note on detectability of stochastic systems with applications. *Proceedings of the 25th Chinese Control Conference*, Heilongjiang, China, 32–35, 7–11 August, 2006.

[15] N. Berman and U. Shaked. H_∞-like control for nonlinear stochastic control. *Systems and Control Letters*, 55: 247–257, 2006.

[16] N. Berman and U. Shaked. H_∞ control for discrete-time nonlinear stochastic systems. *IEEE Trans. Automat. Contr.*, 51(6): 1041–1046, 2006.

[17] N. Berman and U. Shaked. H_∞ nonlinear filtering. *Int. J. Robust Nonlinear Control*, 6: 281–296, 1996.

[18] A. Beghi and D. D'alessandro. Discrete-time optimal control with control-dependent noise and generalized Riccati difference equations. *Automatica*, 34(8): 1031–1034, 1998.

[19] J. B. Burl. *Linear Optimal Control:H_2 and H_∞ Methods*. Addison Wesley Longman, Inc., 1999.

[20] W. L. Brogan, *Modern Control Theory*. 3rd edition, Englewood Cliffs, New Jersey: Prentice-Hall, Inc., 1991.

[21] D. S. Bernstein and W. M. Haddad. LQG control with an H_∞ performance bound: A Riccati equation approach. *IEEE Trans. Automat. Contr.*, 34(3): 293–305, 1989.

[22] J. M. Bismut. Linear quadratic optimal stochastic control with random coefficients. *SIAM J. Contr. Optim.*, 14, 419–444, 1976.

[23] A. Bensoussan. Lecture on stochastic control: Part I. *Lect. Notes. Math.*, 972: 1–39, 1983.

[24] S. Boyd, L. El Ghaoui, E. Feron, and V. Balakrishnan. *Linear Matrix Inequalities in System and Control Theory*. Philadelphia, PA: SIAM, 1994.

[25] E. K. Boukas, Q. Zhang, and G. Yin. Robust production and maintenance planning in stochastic manufacturing systems. *IEEE Trans. Automat. Contr.*, 40(6): 1098–1102, 1995.

[26] E. K. Boukas, S. Xu, and J. Lam. On stability and stabilizability of singular stochastic systems with delays. *J. Optimization Theory and Applications*, 127: 249–262, 2005.

[27] E. K. Boukas. Stabilization of stochastic singular nonlinear hybrid systems. *Nonlinear Analysis*, 64: 217–228, 2006.

[28] V. Borkar and S.Mitter. A note on stochastic dissipativeness. *Lect. Notes. Contr. Inf.*, 286: 41–49, 2003.

[29] C. I. Byrnes and W. Lin. Losslessness, feedback equivalence, and the global stabilization of discrete-time nonlinear systems. *IEEE Trans. Automat. Contr.,* 39: 83–98, 1994.

[30] C. I. Byrnes, A. Isidori, and J. C. Willems. Passivity, feedback equivalence, and the global stabilization of minimum phase nonlinear systems. *IEEE Trans. Automat. Contr.*, 36: 1228–1240, 1991.

[31] T. Basar and P. Bernhard. H^∞-*Optimal Control and Related Minimax Design Problems: A Dynamic Game Approach.* Birkhäuser Boston, Boston, MA, 1995.

[32] S. Bittanti and P. Colaneri. *Periodic Systems-Filtering and Control.* London: Springer-Verlag, 2009.

[33] M. Chilali and P. Gahinet. H_∞ design with pole placement constraints: An LMI approach. *IEEE Trans. Automat. Contr.,* 41: 358–367, 1996.

[34] T. Coleman, M. Branch, and A. Grace. *Optimization Toolbox for Use with Matlab.* The MathWorks Inc. Natick, Mass, 1999.

[35] F. Carravetta and G. Navelli. Suboptimal stochastic linear feedback control of linear systems with state- and control-dependent noise: The incomplete information case. *Automatica,* 43: 751–757, 2007.

[36] B. S. Chen and W. Zhang. Stochastic H_2/H_∞ control with state-dependent noise. *IEEE Trans. Automat. Contr.*, 49(1): 45–57, 2004.

[37] B. S. Chen, C. S. Tseng, and H. J. Uang. Mixed H_2/H_∞ fuzzy output feedback control design for nonlinear dynamic systems: An LMI approach. *IEEE Trans. Fuzzy Systems,* 8: 249–265, 2000.

[38] B. S. Chen, Y. T. Chang, and Y. C. Wang. Robust H_∞-stabilization design in gene networks under stochastic molecular noises: Fuzzy interpolation approach. *IEEE Trans. Fuzzy Systems,* 38(1): 25–42, 2008.

[39] B. S. Chen and W. S. Wu. Robust filtering circuit design for stochastic gene networks under intrinsic and extrinsic molecular noises. *Mathematical Biosciences,* 211: 342–355, 2008.

[40] B. S. Chen and C. H. Wu. Robust optimal reference-tracking design method for stochastic synthetic biology systems: T-S Fuzzy approach. *IEEE Trans. Fuzzy Systems,* 18: 1144–1158, 2010.

[41] E. F. Costa and J. B. R. do Val. On the observability and detectability of continuous-time Markov jump linear systems. *SIAM J. Contr. Optim.,* 41: 1295–1314, 2002.

[42] E. F. Costa and J. B. R. do Val. On the detectability and observability of discrete-time Markov jump linear systems. *Systems and Control Letters*, 44: 135–145, 2001.

[43] O. L. V. Costa and de P. L. Wanderlei. Indefinite quadratic with linear costs optimal control of Markov jump with multiplicative noise systems. *Automatica*, 43: 587–597, 2007.

[44] O. L. V. Costa and R. P. Marques. Mixed H_2/H_∞-control of discrete-time Markovian jump linear systems. *IEEE Trans. Automat. Contr.*, 43: 95–100, 1998.

[45] S. Chen, X. Li, and X. Y. Zhou. Stochastic linear quadratic regulators with indefinite control weight costs, *SIAM J. Contr. Optim.*, 36: 1685–1702, 1998.

[46] M. C. De Oliviera, J. C. Geromel, and J. Bernussou. An LMI optimization approach to multiobjective controller design for discrete-time systems. *Proc. 38th IEEE Conf. Decision and Control,* Phoenix, AZ, 3611–3616, 1999.

[47] C. E. de Souza and L. Xie. On the discrete-time bounded real lemma with application in the characterization of static state feedback H_∞ controllers. *Systems and Control Letters*, 18: 61–71, 1992.

[48] L. Dai. *Singular Control Systems*. Lect. Notes. Contr. Inf., 118, New York: Springer, 1989.

[49] V. V. Dombrovskii and E. A. Lyashenko. A linear quadratic control for discrete systems with random parameters and multiplicative noise and its application to investment portfolio optimization. *Automat. Remote Control*, 64: 1558–1570, 2003.

[50] V. Dragan and T. Morozan. Observability and detectability of a class of discrete-time stochastic linear systems. *IMA J. Math. Control Inf.*, 23: 371–394, 2006.

[51] V. Dragan and T. Morozan. Stochastic observability and applications. *IMA J. Math. Control Inf.*, 21: 323–344, 2004.

[52] V. Dragan, T. Morozan, and A. M. Stoica. *Mathematical Methods in Robust Control of Discrete-Time Linear Stochastic Systems*. New York: Springer, 2010.

[53] V. Dragan, T. Morozan, and A. M. Stoica. *Mathematical Methods in Robust Control of Linear Stochastic Systems*. New York: Springer, 2006.

[54] V. Dragan, A. Halanay, and A.Stoica. A small gain theorem for linear stochastic systems. *Systems and Control Letters*, 30: 243–251, 1997.

[55] C. Du and L. Xie. H_∞ *Control and Filtering of Two-dimensional Systems*. Springer-Verlag, 2002.

[56] T. Damm. Rational matrix equations in stochastic control. *Lecture Notes in Control and Information Sciences 297*, Berlin-Heidelberg: Springer, 2004.

[57] T. Damm. On detectability of stochastic systems. *Automatica*, 43: 928–933, 2007.

[58] T. Damm. State-feedback H^∞ control of linear systems with time-varying parameter uncertainty. *Linear Algebra and Its Applications*, 351–352: 185–210, 2002.

[59] C. Du, L. Xie, J. N. Teoh, and G. Guo. An improved mixed H_2/H_∞ control design for hard disk drives. *IEEE Trans. Control Systems Technology*, 13(5): 832–839, 2005.

[60] G. Da Prato and J. Zabczyk. *Stochastic Equations in Infinite Dimensions, Encyclopedia of Mathematics and Its Applications*. Cambridge, MA: Cambridge University Press, 1992.

[61] S. M. Djouadi, C. D. Charalambous, and D. W. Repperger. A convex programming approach to the multiobjective H_2/H_∞ problem. *Proceedings of 2002 American Control Conference*, Anchorage, Alaska, USA, 4315–4320, 8–10 May, 2002.

[62] J. C. Doyle, K. Glover, P. P. Khargonekar, and B. Francis. State-space solutions to standard H_2 and H_∞ problems. *IEEE Trans. Automat. Contr.*, 34: 831–847, 1989.

[63] W. L. De Koning. Infinite horizon optimal control of linear discrete-time systems with stochastic parameters. *Automatica*, 18(4): 443–453, 1982.

[64] W. L. De Koning. Detectability of linear discrete-time systems with stochastic parameters. *Int. J. Control*, 38(5): 1035–1046, 1983.

[65] A. El Bouhtouri, D. Hinrichsen, and A. J. Pritchard. H_∞-type control for discrete-time stochastic systems. *Int. J. Robust Nonlinear Control*, 9: 923–948, 1999.

[66] R. Elliott, X. Li, and Y. Ni. Discrete time mean-field stochastic linear-quadratic optimal control problems. *Automatica*, 49(11): 3222–3233, 2013.

[67] L. El Ghaoui and M. Ait Rami. Robust state-feedback stabilization of jump linear systems via LMIs. *Int. J. Robust Nonlinear Control*, 6: 1015–1022, 1996.

[68] R. Elliott, X. Li, and Y. Ni. Discrete time mean-field stochastic linear-quadratic optimal control problems. *Automatica*, 49: 3222–3233, 2013.

[69] B. A. Francis and J. C. Doyle. Linear control theory with an H_∞ optimality criterion. *SIAM J. Contr. Optim.*, 25: 815–844, 1987.

[70] P. Florchinger. A passive system approach to feedback stabilization of nonlinear control stochastic systems. *SIAM J. Contr. Optim.*, 37: 1848–1864, 1999.

[71] M. D. Fragoso, O. L. V. Costa, and C. E. de Souza. A new approach to linearly perturbed Riccati equations arising in stochastic control. *Appl. Math. Optim.*, 37: 99–126, 1998.

[72] M. D. Fragoso and O. L. V. Costa. A unified approach for stochastic and mean square stability of continuous-time linear systems with Markovian jumping parameters and additive disturbance. *SIAM J. Contr. Optim.*, 44(4): 1165–1191, 2005.

[73] Y. Feng and B. D. O. Anderson. An iterative algorithm to solve state-perturbed stochastic algebraic Riccati equations in LQ zero-sum games. *Systems and Control Letters*, 59: 50–56, 2010.

[74] X. Feng, A. A. Loparo, Y. Ji and H. J. Chizeck. Stochastic stability properties of jump linear systems. *IEEE Trans. Automat. Contr.*, 37: 38–53, 1992.

[75] M. Fu and L. Xie. The sector bounded approach to quantized feedback control. *IEEE Trans. Automat. Contr.*, 50(11): 1698–1711, 2005.

[76] H. Gao, J. Lam, L. Xie, and C. Wang. New approach to mixed H_2/H_∞ filtering for polytopic discrete-time systems. *IEEE Trans. Signal Processing*, 53(8): 3183–3192, 2005.

[77] L. Guo. *Time-Varying Stochastic Systems-Stability, Estimation and Control*. Jilin: Jilin Science and Technology Press, 1993.

[78] E. Gershon, U. Shaked, and U. Yaesh. *Control and Estimation of State-Multiplicative Linear Systems*. London: Springer-Verlag, 2005.

[79] E. Gershon and U. Shaked. Static H_2 and H_∞ output-feedback of discrete-time LTI systems with state multiplicative noise. *Systems and Control Letters*, 55: 232–239, 2006.

[80] E. Gershon, D. J. N. Limebeer, U. Shaked, and I. Yaesh. Robust H_∞ filtering of stationary continuous-time linear systems with stochastic uncertainties. *IEEE Trans. Automat. Contr.*, 46: 1788–1793, 2001.

[81] E. Gershon and U. Shaked. Advanced Topics in Control and Estimation of State-Multiplicative Noisy Systems. *Lecture Notes in Con-*

trol and Information Sciences, Vol. 439, Berlin Heidelberg: Springer-Verlag, 2013

[82] Z. Gao and X. Shi. Observer-based controller design for stochastic descriptor systems with Brownian motions. *Automatica,* 49: 2229–2235, 2013.

[83] P. Gahinet, A. Nemirovski, A. J. Laub, and M. Chilali. *LMI Control Toolbox.* Natick, MA: Math Works, 1995.

[84] D. Hinrichsen and A. J. Pritchard. Stochastic H_∞. *SIAM J. Contr. Optim.*, 36: 1504–1538, 1998.

[85] R. A. Horn and C. R. Johnson. *Matrix Analysis.* Cambridge: Cambridge University Press, 1985.

[86] Y. Hu and X. Zhou. Indefinite stochastic Riccati equations. *SIAM J. Contr. Optim.*, 42: 123–127, 2003.

[87] R. Z. Has'minskii. *Stochastic Stability of Differential Equations.* Alphen: Sijtjoff and Noordhoff, 1980.

[88] T. Hou, W. Zhang, and H. Ma. Finite horizon H_2/H_∞ control for discrete-time stochastic systems with Markovian jumps and multiplicative noise. *IEEE Trans. Automat. Contr.*, 55(5): 1185–1191, 2010.

[89] T. Hou, W. Zhang, and H. Ma. Infinite horizon H_2/H_∞ optimal control for discrete-time Markov jump systems with (x,u,v)-dependent noise. *J. Global Optimization*, 57(4): 1245–1262, 2013.

[90] T. Hou, W. Zhang, and H. Ma. A game-based control design for discrete-time Markov jump systems with multiplicative noise. *IET Control Theory and Application*, 7(5): 773–783, 2013.

[91] T. Hou, H. Ma, and W. Zhang. Spectral tests for observability and detectability of periodic Markov jump systems with nonhomogeneous Markov chain. *Automatica*, 63(1): 175–181, 2016.

[92] J. W. Helton and M. R. James. *Extending H^∞ Control to Nonlinear Systems.* SIAM: Philadelphia, 1999.

[93] D. J. Hill and P. J. Moylan. Stability results for nonlinear feedback systems. *Automatica,* 13: 377–382, 1977.

[94] D. J. Hill and P. J. Moylan. The stability of nonlinear dissipative systems. *IEEE Trans. Automat. Contr.*, 21: 708–711, 1976.

[95] J. Hamilton. A new approach to the economic analysis of nonstationary time series and the business cycle. *Econometrica,* 57(2): 357–384, 1989.

[96] L. Huang and X. Mao. Stability of singular stochastic systems with Markovian switching. *IEEE Trans. Automat. Contr.*, 56: 424–429, 2011.

[97] Y. Huang, W. Zhang, and H. Zhang. Infinite horizon LQ optimal control for discrete-time stochastic systems. *Asian J. Control*, 10(5): 608–615, 2008.

[98] Y. Huang, W. Zhang, and G. Feng. Infinite horizon H_2/H_∞ control for stochastic systems with Markovian jumps. *Automatica*, 44(3): 857–863, 2008.

[99] J. Huang, X. Li, and J. Yong. A linear-quadratic optimal control problem for mean-field stochastic differential equations in infinite horizon. *Math. Control Relat. Fields*, 5(1): 97–139, 2015.

[100] M. Huang, P. E. Caines, and R. P. Malhamé. Large-population cost-coupled LQG problems with nonuniform agents: Individual-mass behavior and decentralized ε-nash equilibria. *IEEE Trans. Automat. Contr.*, 52(9): 1560–1571, 2007.

[101] A. Isidori and A. Astolfi. Disturbance attenuation and H_∞-control via measurement feedback in nonlinear systems. *IEEE Trans. Automat. Contr.*, 37: 1283–1293, 1992.

[102] J. Y. Ishihara and M. H. Terra. On the Lyapunov theorem for singular systems. *IEEE Trans. Automat. Contr.*, 47: 1926–1930, 2002.

[103] Y. C. Ji and W. B. Gao. Nonlinear H_∞ control and estimation of optimal H_∞ gain. *Systems and Control Letters*, 24: 321–332, 1995.

[104] H. J. Kushner. Stochastic stability. *Lecture Notes in Math.* 294, Springer, Berlin, 97–124, 1972.

[105] D. L. Kleinman. On the stability of linear stochastic systems. *IEEE Trans. Automat. Contr.*, 14: 429–430, 1969.

[106] N. V. Krylov. *Introduction to the Theory of Diffusion Processes*. Translations of Mathematical Monographs 142, AMS, Providence, RI, 1995.

[107] R. E. Kalman. Contributions to the theory of optimal control. *Bol. Soc. Mexicana*, 5: 102–119, 1960.

[108] P. P. Khargonekar and M. A. Rotea. Mixed H_2/H_∞ control: A convex optimization approach. *IEEE Trans. Automat. Contr.*, 36: 824–837, 1991.

[109] M. Kohlmann and S. Tang. Minimization of risk and linear quadratic optimal control theory. *SIAM J. Contr. Optim.*, 42: 1118–1142, 2003.

[110] O. Kallenberg. *Foundations of Modern Probability*. New York: Springer-Verlag, 2002.

[111] P. P. Khargonekar, K. M. Nagpal, and K. R. Poolla. H_∞ control with transients. *SIAM J. Contr. Optim.*, 29: 1373–1393, 1991.

[112] P. P. Khargonekar, M. A. Rotea and E. Baeyens. Mixed H_2/H_∞ filtering. *Int. J. Robust Nonlinear Control,* 6: 313–330, 1996.

[113] D. J. N. Limebeer, B. D. O. Anderson, and B. Hendel. A Nash game approach to mixed H_2/H_∞ control. *IEEE Trans. Automat. Contr.,* 39: 69–82, 1994.

[114] D. J. N. Limebeer, B. D. O. Anderson, P. P. Khargonekar, and M. Green. A game theoretic approach to H^∞ control for time-varying systems. *SIAM J. Contr. Optim.*, 30: 262–283, 1992.

[115] R. S. Liptser and A. N. Shiryayev. *Statistics of Random Processes*. New York: Springer-Verlag, 1977.

[116] W. M. Lu. H_∞-control of nonlinear time-varying systems with finite time horizon. *Int. J. Control*, 64: 241–262, 1996.

[117] Y. Liu, *Backward Stochastic Differential Equation and Stochastic Control System*. Ph.D. thesis, Jinan: Shandong University, 1999.

[118] Z. Y. Li, Y. Wang, B. Zhou, and G. R. Duan. Detectability and observability of discrete-time stochastic systems and their applications. *Automatica*, 45: 1340–1346, 2009.

[119] J. M. Lasry and P. L. Lions. Mean-field games. *Japan. J. Math.,* 2(1): 229–260, 2007.

[120] Z. Y. Li, Y. Wang, B. Zhou, and G. R. Duan. On unified concepts of detectability and observability for continuous-time stochastic systems. *Applied Mathematics and Computation,* 217: 521–536, 2010.

[121] Z. Y. Li, B. Zhou, Y. Wang, and G. R. Duan. On eigenvalue sets and convergence rate of Itô stochastic systems with Markovian switching. *IEEE Trans. Automat. Contr.,* 56(5): 1118–1124, 2011.

[122] X. Li and X. Y. Zhou. Indefinite stochastic LQ controls with Markovian jumps in a finite horizon. *Communications in Information and Systems*, 2(3): 265–282, 2002.

[123] X. Lin and W. Zhang. A maximum principle for optimal control of discrete-time stochastic systems with multiplicative noise. *IEEE Trans. Automat. Contr.,* 60(4): 1121–1126, 2015.

[124] X. Lin and W. Zhang. The LaSalle theorem for the stochastic difference equations. *Proceedings of the 33rd Chinese Control Conference*, Nanjing, China, 5300–5305, 28–30 July, 2014.

[125] X. Li and H. H. T. Liu. Characterization of $\mathcal{H}_-$ index for linear time-varying systems. *Automatica*, 49: 1449–1457, 2013.

[126] Z. Lin, J. Liu, W. Zhang, and Y. Niu. Stabilization of interconnected nonlinear stochastic Markovian jump systems via dissipativity approach. *Automatica*, 47(12): 2796–2800, 2011.

[127] W. Lin. Mixed H_2/H_∞ control via state feedback for nonlinear systems. *Int. J. Control*, 64(5): 899–922, 1996.

[128] W. Lin and C. I. Byrnes. H_∞ control of discrete-time nonlinear systems. *IEEE Trans. Automat. Contr.*, 41(4): 494–510, 1996.

[129] J. P. LaSalle. *The Stability and Control of Discrete Processes*. New York: Springer-Verlag, 1986.

[130] X. Mao. *Stochastic Differential Equations and Applications*. 2nd Edition, Horwood, 2007.

[131] X. Mao and C. Yuan. *Stochastic Differential Equations with Markovian Switching*. London, U.K.: Imperial College Press, 2006.

[132] X. Mao. Stochastic versions of the LaSalle theorem. *J. Differential Equations*, 153(1): 175–195, 1999.

[133] H. P. McKean, A class of Markov processes associated with nonlinear parabolic equations, Pro. National Acad. Sci. USA , 56(6): 1907–1911, 1966.

[134] L. Ma, Z. Wang, B. Shen, and Z. Guo. A game theory approach to mixed H_2/H_∞ control for a class of stochastic time-varying systems with randomly occurring nonlinearities. *Systems and Control Letters*, 60(12): 1009–1015, 2011.

[135] H. Mukaidani. Robust guaranteed cost control for uncertain stochastic systems with multiple decision makers. *Automatica*, 45: 1758–1764, 2009.

[136] H. Mukaidani. Soft-constrained stochastic Nash games for weakly coupled large-scale systems. *Automatica*, 45: 1272–1279, 2009.

[137] L. Ma, Z. Wang, B. Shen, and Z. Guo. Finite-horizon H_2/H_∞ control for a class of nonlinear Markovian jump systems with probabilistic sensor failures. *Int. J. Control*, 84(11): 1847–1857, 2011.

[138] R. Muradore and G. Picci. Mixed H_2/H_∞ control: the discrete-time case. *Systems and Control Letters*, 54: 1–13, 2005.

[139] H. Ma, W. Zhang, and T. Hou. Infinite horizon H_2/H_∞ control for discrete-time time-varying Markov jump systems with multiplicative noise. *Automatica*, 48: 1447–1454, 2012.

[140] J. Ma and J. Yong. *Forward-Backward Stochastic Differential Equations and Their Applications*. New York: Springer, 1999.

[141] L. Ma and W. Zhang. Output feedback H_∞ control for discrete-time mean-field stochastic systems. *Asian J. Control*, 17(6): 2241–2251, 2015.

[142] L. Ma, T. Zhang, W. Zhang, and B. S. Chen. Finite horizon mean-field stochastic H_2/H_∞ control for continuous-time systems with (x,v)-dependent noise. *J. Franklin Institute*, 352: 5393–5414, 2015.

[143] Y. H. Ni, W. Zhang, and H. T. Fang. On the observability and detectability of linear stochastic systems with Markov jumps and multiplicative noise. *J. Systems Science and Complexity*, 23(1): 100–113, 2010.

[144] Y. Ni, R. Elliott, and X. Li. Discrete time mean-field stochastic linear-quadratic optimal control problems, II: infinite horizon case. *Automatica*, 57(11): 65–77, 2015.

[145] B. Øksendal. *Stochastic Differential Equations: An Introduction with Applications*. 6th Edition, Springer-Verlag, 2003.

[146] I. R. Petersen, V. A. Ugrinovskii, and A. V. Savkin. *Robust Control Design Using H_∞ Methods*. New York: Springer-Verlag, 2000.

[147] E. Pardoux and S. Peng. Adapted solution of a backward stochastic differential equation. *Systems and Control Letters*, 14: 55–61, 1990.

[148] S. Peng. Stochastic Hamilton-Jaco-Bellman equations. *SIAM J. Contr. Optim.*, 30: 284–304, 1992.

[149] R. Penrose. A generalized inverse of matrices. *Proc. Cambridge Philos. Soc.*, 51: 406–413, 1955.

[150] M. A. Peters and A. A. Stoorvogel. Mixed H_2/H_∞ control in a stochastic framework. *Linear Algebra and its Applications*, 205: 971–996, 1992.

[151] P. V. Pakshin and V. A. Ugrinovskii. Stochastic problems of absolute stability. *Automation and Remote Control*, 67(11): 1811–1846, 2006.

[152] M. A. Peters and P. A. Iglesias. A spectral test for observability and reachability of time-varying systems. *SIAM J. Contr. Optim.*, 37(5): 1330–1345, 1999.

[153] L. Qian and Z. Gajic. Variance minimization stochastic power control in CDMA systems. *IEEE Trans. Wireless Communications*, 5(1): 193–202, 2006.

[154] W. J. Rugh. *Linear System Theory*. New Jersey: Prentice-Hall, Inc, 1993.

[155] D. Revus and M. Yor. *Continuous Martingales and Brownian Motion.* New-York: Springer, 1999.

[156] J. P. Richard. Time-delay systems: an overview of some recent advances and open problems. *Automatica*, 39: 1667–1694, 2003.

[157] A. Rantzer. On the Kalman-Yakubovich-Popov lemma. *Systems and Control Letters*, 28: 7–10, 1996.

[158] R. Ravi, K. M. Nagpal, and P. P. Khargonekar. H_∞ control of linear time-varying systems: A state-space approach. *SIAM J. Contr. Optim.*, 29: 1394–1413, 1991.

[159] M. Sznaier and H. Rotsein. An exact solution to general 4-blocks discrete-time mixed H_2/H_∞ problems via convex optimization. *Proc. American Control*, 2251–2256, 1994.

[160] T. Shimomura and T. Fujii. An iterative method for mixed H_2/H_∞ control design with uncommon LMI solutions. *Proceedings of the 1999 American Control Conference*, San Diego, CA, USA, 3292–3296, 2–4 June, 1999.

[161] L. Shen, J. Sun, and Q. Wu. Observability and detectability of discrete-time stochastic systems with Markovian jump. *Systems and Control Letters*, 62: 37–42, 2013.

[162] B. Shen, Z. Wang, and H. Shu. *Nonlinear Stochastic Systems with Incomplete Information: Filtering and Control*, Springer, London, 2013.

[163] H. S. Shu, and G. L. Wei. H_∞ analysis of nonlinear stochastic time-delay systems. *Chaos, Solitons and Fractals*, 26: 637–647, 2005.

[164] H. Schneider. Positive operator and an inertia theorem. *Numerische Mathematik*, 7(1): 11–17, 1965.

[165] L. Sheng, W. Zhang, and M. Gao. Relationship between Nash equilibrium strategies and H_2/H_∞ control of stochastic Markov jump systems with multiplicative noise. *IEEE Trans. Automat. Contr.*, 59(9): 2592–2597, 2014.

[166] C. Song and Q. Xu. *Modern Western Economics: Macro-economics (in Chinese)*. Shanghai: Fudan University Press, 2004.

[167] U. H. Thygesen. On dissipation in stochastic systems. *Proc. American Control Conference*, San Diego, California, 1430–1434, 1999.

[168] K. Takaba, N. Morihara, and T. Katayama. A generalized Lyapunov theorem for descriptor systems. *Systems and Control Letters*, 24: 49–51, 1995.

[169] V. A. Ugrinovskii. Robust H^∞ control in the presence of stochastic uncertainty. *Int. J. Control*, 71: 219–237, 1998.

[170] V. M. Ungureanu. Stability, stabilization and detectability for Markov jump discrete-time linear systems with multiplicative noise in Hilbert spaces. *Optimization*, 63(11): 1689–1712, 2014.

[171] V. M. Ungureanu and V. Dragan. Stability of discrete-time positive evolution operators on ordered Banach spaces and applications. *J. Difference Equations and Applications*, 19(6): 952–980, 2013.

[172] V. M. Ungureanu. Stochastic uniform observability of linear differential equations with multiplicative noise. *J. Math. Anal. Appl.*, 343: 446–463, 2008.

[173] A. J. van der Schaft. L_2-gain analysis of nonlinear systems and nonlinear state feedback H_∞ control. *IEEE Trans. Automat. Contr.*, 37: 770–784, 1992.

[174] L. G. Van Willigenburg and W. L. De Koning. Optimal reduced-order compensators for time-varying discrete-time systems with deterministic and white parameters. *Automatica*, 35: 129–138, 1999.

[175] L. G. Van Willigenburg and W. L. De Koning. Minimal representation of matrix valued white stochastic processes and U-D factorization of algorithms for optimal control. *Int. J. Control*, 86(2): 309–321, 2013.

[176] G. L. Wei and H. S. Shu. H_∞ filtering on nonlinear stochastic systems with delay. *Chaos, Solitons and Fractals*, 33: 663–670, 2007.

[177] J. L. Willems and J. C. Willems. Feedback stabilizability for stochastic systems with state and control dependent noise. *Automatica*, 12: 277–283, 1976.

[178] J. C. Willems. Dissipative dynamic systems Part I: General Theory. *Arch. Rational Mech.*, 45: 321–393, 1972.

[179] Z. J. Wu, M. Y. Cui, X. J. Xie, and P. Shi. Theory of stochastic dissipative systems. *IEEE Trans. Automat. Contr.*, 56(7): 1650–1655, 2011.

[180] W. M. Wonham. On a matrix Riccati equation of stochastic control. *SIAM J. Contr.*, 6: 681–697, 1968.

[181] C. H. Wu, W. Zhang, and B. S. Chen. Multiobjective H_2/H_∞ synthetic gene network design based on promoter libraries. *Mathematical Biosciences*, 233(2): 111–125, 2011.

[182] Z. Wang, Y. R. Liu, and X. H. Liu, H_∞ filtering for uncertain stochastic time-delay systems with sector bounded nonlinearities, *Automatica*, 44: 1268–1277, 2008.

[183] Z. Wang and B. Huang. Robust H_2/H_∞ filtering for linear systems with error variance constraints. *IEEE Trans. Automat. Contr.,*, 48(8): 2463–2467, 2000.

[184] G. Wei, Z. Wang, J. Lam, K. Fraser, G. P. Rao, and X. Liu. Robust filtering for stochastic genetic regulatory networks with time-varying delay. *Mathematical Biosciences*, 220: 73–80, 2009.

[185] L. Wu, P. Shi, and H. Gao. State estimation and sliding-mode control of Markovian jump singular systems. *IEEE Trans. Automat. Contr.*, 55: 1213–1219, 2010.

[186] M. Wang. Stochastic H_2/H_∞ control with random coefficients. *Chin. Ann. Math.*, 34B(5): 733–752, 2013.

[187] N. Xiao, L. Xie, and L. Qiu. Feedback stabilization of discrete-time networked systems over fading channels. *IEEE Trans. Automat. Contr.*, 57: 2176–2189, 2012.

[188] L. Xie. Output feedback H_∞ control of systems with parameter uncertainty. *Int. J. Control*, 63(4): 741–750, 1996.

[189] L. Xie, C.E. de Souza, and Y. Wang. Robust control of discrete time uncertain dynamical systems. *Automatica*, 29(4): 1133–1137, 1993.

[190] X. J. Xie and N. Duan. Output tracking of high-order stochastic nonlinear systems with application to benchmark mechanical system. *IEEE Trans. Automat. Contr.*, 55: 1197–1202, 2010.

[191] S. Xu and T. Chen. Reduced-order H_∞ filtering for stochastic systems. *IEEE Trans. Signal Processing*, 50(12): 2998–3007, 2002.

[192] S. Xu and J. Lam. *Robust Control and Filtering of Singular Systems*. Berlin: Springer, 2006.

[193] S. Xu and Y. Zou. H_∞ filtering for singular systems. *IEEE Trans. Automat. Contr.*, 48: 2217–2222, 2003.

[194] Y. Xia, J. Zhang, and E. K. Boukas. Control for discrete singular hybrid systems. *Automatica*, 44: 2635–2641, 2008.

[195] Y. Xia, E. K. Boukas, P. Shi, and J. Zhang. Stability and stabilization of continuous-time singular hybrid systems. *Automatica*, 45: 1504–1509, 2009.

[196] J. Yong and X. Y. Zhou. *Stochastic Control: Hamiltonian Systems and HJB Equations*. New York: Springer, 1999.

[197] J. Yong. A linear-quadratic optimal control problem for mean-field stochastic differential equation. *SIAM J. Contr. Optim.*, 51(4): 2809–2838, 2013.

[198] G. Yin and X. Y. Zhou. Markowitz's mean-variance portfolio selection with regime switching: From discrete-time models to their continuous-time limits. *IEEE Trans. Automat. Contr.*, 49(3): 349–360, 2004.

[199] K. You and L. Xie. Minimum data rate for mean square stabilization of discrete LTI systems over lossy channels. *IEEE Trans. Automat. Contr.*, 55(10): 2373–2378, 2010.

[200] E. Yaz. Linear state estimation for non-linear stochastic systems with noisy non-linear observations. *Int. J. Control*, 48: 2465–2475, 1988.

[201] H. Zhang, X. Song, and L. Shi. Convergence and mean square stability of suboptimal estimator for systems with measurement packet dropping. *IEEE Trans. Automat. Contr.*, 57: 1248–1253, 2012.

[202] H. Zhang and L. Xie. *Control and Estimation of Systems with Input-Output Delays. Lecture Notes in Control and Information Sciences*, Vol. 355, Berlin Heidelberg: Springer-Verlag, 2007.

[203] X. Zhao and F. Deng. Moment stability of nonlinear discrete stochastic systems with time-delays based on $\mathcal{H}$-representation technique. *Automatica*, 50(2): 530–536, 2014.

[204] X. Zhao and F. Deng. Moment stability of nonlinear stochastic systems with time-delays based on $\mathcal{H}$-representation technique. *IEEE Trans. Automat. Contr.*, 59(3): 814–819, 2014.

[205] W. Zhang and B. S. Chen. On stabilization and exact observability of stochastic systems with their applications. *Automatica*, 40(1): 87–94, 2004.

[206] W. Zhang, H. Zhang, and B. S. Chen. Generalized Lyapunov equation approach to state-dependent stochastic stabilization/detectability criterion. *IEEE Trans. Automat. Contr.*, 53(7): 1630–1642, 2008.

[207] W. Zhang, Y. Huang, and L. Xie. Infinite horizon stochastic H_2/H_∞ control for discrete-time systems with state and disturbance dependent noise. *Automatica*, 44(9): 2306–2316, 2008.

[208] W. Zhang, Y. Huang, and H. Zhang. Stochastic H_2/H_∞ control for discrete-time systems with state and disturbance dependent noise. *Automatica*, 43: 513–521, 2007.

[209] W. Zhang, H. Zhang, and B. S. Chen. Stochastic H_2/H_∞ control with (x,u,v)-dependent noise: Finite horizon case. *Automatica*, 42(11): 1891–1898, 2006.

[210] W. Zhang and X. An. Finite-time control of linear stochastic systems. *Int. J. Innovative Computing, Information and Control*, 4(3): 689–696, 2008.

[211] W. Zhang and L. Xie. Interval stability and stabilization of linear stochastic systems. *IEEE Trans. Automat. Contr.*, 54(4): 810–815, 2009.

[212] W. Zhang and B. S. Chen. $\mathcal{H}$-representation and applications to generalized Lyapunov equations and linear stochastic systems. *IEEE Trans. Automat. Contr.*, 57(12): 3009–3022, 2012.

[213] W. Zhang and B. S. Chen. State feedback H_∞ control for a class of nonlinear stochastic systems. *SIAM J. Contr. Optim.*, 44: 1973–1991, 2006.

[214] W. Zhang and G. Feng. Nonlinear stochastic H_2/H_∞ control with (x, u, v)-dependent noise: Infinite horizon case. *IEEE Trans. Automat. Contr.*, 53: 1323–1328, 2008.

[215] W. Zhang, Y. Zhao, and L. Sheng. Some remarks on stability of stochastic singular systems with state-dependent noise. *Automatica*, 51(1): 273–277, 2015.

[216] W. Zhang. *Study on Algebraic Riccati Equation Arising from Infinite Horizon Stochastic LQ Optimal Control.* Ph.D. Thesis, Hangzhou: Zhejiang University, 1998.

[217] W. Zhang and C. Tan. On detectability and observability of discrete-time stochastic Markov jump systems with state-dependent noise. *Asian J. Control*, 15(5): 1366–1375, 2013.

[218] W. Zhang, W. Zheng, and B. S. Chen. Detectability, observability and Lyapunov-type theorems of linear discrete time-varying stochastic systems with multiplicative noise. *Int. J. Control*, http://dx.doi.org/10.1080/00207179.2016.1257152.

[219] W. Zhang, B.S. Chen, H. Tang, L. Sheng, and M. Gao. Some Remarks on general nonlinear stochastic H_∞ control with state, control and disturbance-dependent noise. *IEEE Trans. Automat. Contr.*, 59(1): 237–242, 2014.

[220] W. Zhang, B. S. Chen, and C. S. Tseng. Robust H_∞ filtering for nonlinear stochastic systems. *IEEE Trans. Signal Processing*, 53(2): 289–298, 2005.

[221] W. Zhang, X. Lin, and B. S. Chen. LaSalle-type theorem and its applications to infinite horizon optimal control of discrete-time nonlinear stochastic systems. *IEEE Trans. Automat. Contr.*, 62(1): 250–261, 2017.

[222] W. Zhang, B. S. Chen, L. Sheng, and M. Gao. Robust H_2/H_∞ filter design for a class of nonlinear stochastic systems with state-dependent noise. *Mathematical Problems in Engineering*, 2012:1-16, doi:10.1155/2012/750841, 2012.

[223] W. Zhang, L. Ma, and T. Zhang. Discrete-time mean-field stochastic H_2/H_∞ control. *J. Systems Science and Complexity*, accepted for publication.

[224] W. Zhang, G. Feng, and Q. Li. Robust H_∞ filtering for general nonlinear stochastic state-delayed systems. *Mathematical Problems in Engineering*, 2012: 1-15, doi:10.1155/2012/231352, 2012.

[225] W. Zhang, Y. Li, and X. Liu. Infinite horizon indefinite stochastic linear quadratic control for discrete-time systems. *J. Control Theory and Technology*, 13(3): 230–237, 2015.

[226] W. Zhang, B. S. Chen, and H. Tang. Some remarks on stochastic H_∞ control of linear and nonlinear Itô-type differential systems. *Proceedings of the 30th Chinese Control Conference*, Yantai, China, 5958–5963, 22–24 July, 2011.

[227] K. Zhou and J. C. Doyle. *Essentials of Robust Control*. Prentice Hall Inc., 1998.

[228] Y. Zhao and W. Zhang. New results on stability of singular stochastic Markov jump systems with state-dependent noise. *Int. J. Robust Nonlinear Control*, 26: 2169–2186, 2016.

[229] L. Zhang, B. Huang, and J. Lam. LMI synthesis of H_2 and mixed H_2/H_∞ controllers for singular systems. *IEEE Trans. Circuits and Systems II*, 50: 615–626, 2003.

[230] Q. Zhang. H_2/H_∞ control problems of backward stochastic systems. *J. Systems Science and Complexity*, 27(5): 99–910, 2014.

[231] Q. Zhang. Backward stochastic H_2/H_∞ control with random jumps. *Asian J. Control*, 16(4): 1238–1244, 2014.

[232] G. Zames. Feedback and optimal sensitivity: Model reference transformation, multiplicative seminorms and approximate inverses. *IEEE Trans. Automat. Contr.*, 26: 301–320, 1981.

Index

Milton Keynes UK
Ingram Content Group UK Ltd.
UKHW020319111024
449327UK00040B/1381